DNA
SYNTHESIS
AND
ITS
REGULATION

Other volumes in this series include:

Volume One: Mechanisms of Virus Disease
Edited by William S. Robinson and C. Fred Fox © 1974

Volume Two: Developmental Biology: Pattern Formation, Gene Regulation
Edited by Daniel McMahon and C. Fred Fox © 1975

VOLUME
THREE

ICN ICN-UCLA SYMPOSIA
ON
MOLECULAR & CELLULAR BIOLOGY

DNA
SYNTHESIS
AND
ITS
REGULATION

EDITED BY
MEHRAN GOULIAN
University of California, San Diego
PHILIP HANAWALT
Stanford University

C. FRED FOX
University of California, Los Angeles
Series Editor

W. A. Benjamin, Inc.
Menlo Park, California. Reading, Massachusetts.
London. Don Mills, Ontario. Sydney

Dedicated to Reiji Okazaki
1930 — 1975

ISBN 0-8053-3350-9
ABCDEFGHIJKL—HA—79876

W.A. Benjamin, Inc.
2727 Sand Hill Road
Menlo Park, California 94025

PREFACE

This volume documents the proceedings of a conference on *DNA Synthesis and Its Regulation*, held at Squaw Valley, California in March 1975 with nearly 300 participants. This meeting was one of the 1975 Winter Symposia on topics in molecular and cellular biology, sponsored by ICN Pharmaceuticals, Inc., and organized through the Molecular Biology Institute of the University of California, Los Angeles.

In the past few years research in the field of DNA replication has begun to accelerate in several important directions. The development of a number of *in vitro* systems has facilitated the purification and study by complementation analysis of the specific components required for the process, as well as their organization into multicomponent complexes. Improvement in the isolation of functional complexes of replicons has permitted the visualization by electron microscopy of structures which may closely resemble the *in vivo* situation. Much more has been learned about the polymerases and unwinding proteins; the role of the latter fascinating class in DNA transactions is on the verge of being understood. The replication cycles and control of various specialized classes of genomes such as mictochondrial DNA, small bacteriophage, and certain plasmids are now known in considerable detail. This conference brought together investigators and their students—all with diverse interests in DNA replication—and provided an opportunity for the mutual exchange of ideas and research results in this exciting and health-relevant field.

These proceedings include the presentations of essentially all the invited symposium speakers. We have also included a selected group of short papers derived from poster sessions, and we regret that we were unable to include all of the submitted manuscripts in this volume. We have taken some liberties in the topical reorganization of manuscripts into coherent parts.

We express our thanks to the session conveners for guiding discussion, keeping it on topic, and generally holding to the alloted time—to permit a few afternoons in the blizzard! We appreciate the financial sponsorship by ICN Pharmaceuticals Inc., of these symposia, and we acknowledge the American Cancer Society and the Energy Research and Development Administration for assisting with travel support for the principal invited speakers in this meeting.

We thank the students who handled the projection and audio equipment, Jim Loehr, Debbie Miyajima, Mike McKeown, Tom Simon, and Wayne Yee. Finally, we are indebted to Fran Stusser for her heroic organizational efforts before, during, and after the meeting, and to Pat Seawell for administrative assistance during the meeting.

M. Goulian

P. Hanawalt

TABLE OF CONTENTS

Part I.   DNA Polymerases, Binding, and Unwinding Proteins   1

The Role of the *Escherichia coli* DNA Binding Protein in DNA
Synthesis   2
*Malcolm L. Gefter, Ian J. Molineux, Andrew Pauli, and Linda
Sherman*

DNA Polymerase III from *Bacillus subtilis* and the Mechanism of
Arylhydrazinopyrimidine Inhibition   14
*Nicholas R. Cozzarelli, Robert L. Low, Stephan A. Rashbaum,
and Craig L. Peebles*

In vitro DNA Synthesis on Primed Covalently Closed Double-
stranded Templates. I. Studies with *Escherichia coli* DNA
Polymerase I.   58
Leroy F. Liu and James C. Wang

HeLa Cell DNA Polymerases: Properties and Possible Functions   .   64
Arthur Weissbach, Silvio Spadari, and Karl-Werner Knopf

Assay and Partial Purification of the DNA Untwisting Activity
from Rat Liver   83
James J. Champoux and Joyce M. Durnford

Isolation and Properties of a Nicking-closing Protein from
Mammalian Nuclei   94
Hans-Peter Vosberg and Jerome Vinograd

Part II. Replicon Complexes and Their Stabilization   121

RNA Molecules Attached to DNA in Isolated Bacterial Nucleoids:
Their Possible Role in Stabilizing the Condensed Chromosome   .   122
Ralph M. Hecht and David Pettijohn

Viscometric Analysis of Conformational Transitions in the
Escherichia coli Chromosome   138
Karl Drlica and A. Worcel

Properties of Membrane-associated Folded Chromosomes of
Escherichia coli During the DNA Replication Cycle 159
Oliver A. Ryder, Ruth Kavenoff, and Douglas W. Smith

Membrane-DNA Complex in *Bacillus subtilis* 187
Sumi Imada, Lynn E. Carroll, and Noboru Sueoka

The Unit Chromosomal Fiber: Evidence for its Universal Nature   201
Jack D. Griffith

Part III.   Systems for Replication of DNA *in vitro* **209**

Multienzyme Systems in the Replication of φX174 and G4 DNAs   210
*Jean-Pierre Bouché, Randy Schekman, Joel Weiner, Kasper
Zechel, and Arthur Kornberg*

In vitro Synthesis of DNA   227
Sue Wickner and Jerard Hurwitz

In vitro Synthesis of Bacteriophage T7 DNA 239
*W. E. Masker, D. C. Hinkle, D. Mark, P. A. Modrich, and
Charles Richardson*

Reconstruction of the T4 Bacteriophage DNA Replication
Apparatus from Purified Components 241
*Bruce Alberts, C. Fred Morris, David Mace, Navin Sinha, Michael
Bittner, and Larry Moran*

Stimulation of DNA Replication *in vitro* by *Escherichia coli*
Periplasmic Factors 270
Douglas W. Smith, Catherine Kemper, and J. Weaver Zyskind

Studies on the *in vitro* Replication of the *Escherichia coli*
Chromosome 296
Thomas Kornberg

Part IV.   Intermediates in DNA Replication **303**

See also Part VIII-1; paper by R. Okazaki *et al.*, added as the
volume went to press.

Mutants of *Escherichia coli* Defective in the Joining of Nascent
DNA Fragments 304
E. Bruce Konrad and I. R. Lehman

Evidence That Both Growing DNA Chains at a Replication Fork
are Synthesized Discontinuously 309
Rolf Sternglanz, Helen F. Wang, and James J. Donegan

CONTENTS

DNA Synthesis in Human Lymphocytes 322
*Ben Y. Tseng, Wolfgang Oertel, Richard M. Fox, and
Mehran Goulian*

Evidence for RNA Linked to Nascent DNA in Eukaryotic
Organisms 334
*M. Anwar Waqar, Robert Minkoff, Alice Tsai, and Joel A.
Huberman*

Initiator RNA in Discontinuous Synthesis of Polyoma DNA . . 357
Peter Reichard

Discontinuous SV40 DNA Synthesis and the Detection of Gap
Circle Intermediates 361
Philip J. Laipis, Arup Sen, Arnold J. Levine, and Carel Mulder

Part V. **Replication of Small Bacterial Viruses** 369

Structure and Replication of Replicative Forms of the ϕX-related
Bacteriophage G4 370
Dan S. Ray and Jeanene Dueber

A Physical Map of G4 and the Origins of G4 Double and Single
Stranded DNA Replication 386
G. Nigel Godson

Replication of ϕX174 in *Escherichia coli*: Structure of the
Replicating Intermediate and the Effect of Mutations in the
Host *lig* and *rep* Genes 398
*David T. Denhardt, Shlomo Eisenberg, Barbara Harbers,
H. E. David Lane, and Grant McFadden*

Bacteriophage M13 Replication in *Escherichia coli dnaB* Strains . 423
J. Michael Bowes

Replication of Mycoplasmaviruses 445
Jack Maniloff and Jyotirmoy Das

Part VI. **Phage λ Replication** 451

Structural Analysis of Intracellular λ DNA 452
Manuel S. Valenzuela and Ross B. Inman

Molecular Mechanisms in the Control of λ DNA Replication:
Interaction Between Phage and Host Functions 460
A. Skalka, M. Greenstein, and R. Reuben

CONTENTS

Role of *oop* RNA Primer in Initiation of Coliphage Lambda
DNA Replication 486
Sidney Hayes and Waclaw Szybalski

Part VII. Plasmid and Plastid Replication 513

Modes of Plasmid DNA Replication in *Escherichia coli*. . . . 514
D. R. Helinski, M. A. Lovett, P. H. Williams, L. Katz,
J. Collins, Y. Kupersztoch-Portnoy, S. Sato, R. W. Leavitt,
R. Sparks, V. Hershfield, D. G. Guiney, and D. G. Blair

Control of RTF and r-determinants Replication in Composite
R Plasmids 537
Robert H. Rownd, Daniel Perlman, Nobuichi Goto, and
Edward R. Appelbaum

Mechanisms of Mitochondrial DNA Replication in Mouse L-cells. 560
Arnold J. Berk and David A. Clayton

Replication of Mitochondrial DNA and its Initiation Site. . . . 578
Katsuro Koike, Midori Kobayashi, and Shigeaki Tanaka

The Replication of mtDNA in *Tetrahymena* 586
Piet Borst, Rob W. Goldbach, Annika C. Arnberg, and
Ernst F. J. Van Bruggen

Part VIII. Initiation and Control of the Replication Cycle in *Escherichia coli* 601

Regulation of the Initiation of DNA Replication in *Escherichia*
coli. Isolation of I-RNA and the Control of I-RNA Synthesis . 602
Walter Messer, Ludolf Dankwarth, Ruth Tippe-Schindler,
John E. Womack, and Gabriele Zahn

Specific Labelling of Initiation RNA in *Escherichia coli* 618
John E. Womack and Walter Messer

Cryolethal Suppressors of Thermosensitive *dnaA* Mutations . . 624
James A. Wechsler and Malgorzata Zdzienicka

Triton X-100 Dependent *in vitro* DNA Synthesis in an
Escherichia coli dnaC Mutant 640
Michael H. Hanna, L. Stephanie Soucek, and Philip L. Carl

Part IX. Topology and Replication Control in Bacteria 649

Bidirectional Replication in *Bacillus subtilis* 650
R. G. Wake

CONTENTS

The Role of Semiconservative DNA Replication in Bacterial
Cell Development 677
Terrance Leighton, George Khachatourians, and Neal Brown

Effect of an Arylazopyrimidine on DNA Synthesis in Toluene-
treated *Salmonella typhimurium* 688
Charles P. Van Beveren and Andrew Wright

Replication Fork Velocity in *Escherichia coli* at Various Cell
Growth Rates 695
M. Chandler, M. Funderburgh, and L. Caro

Part X. **Eucaryote DNA Replication Systems** 701

Temperature Sensitive DNA — Mutants of a Mammalian Tissue
Culture Cell Line 702
Donald J. Roufa and Michael A. Haralson

The Inhibition of DNA Strand Elongation by Cycloheximide
in *Physarum polycephalum* 713
Helen H. Evans, Thomas E. Evans, and Eugene N. Brewer

Studies of DNA Synthesis in Permeabilized Mouse L Cells . . . 719
Nathan A. Berger and Elizabeth S. Johnson

Part XI. **Viral Nucleic Acid Replication in Eucaryotes** 729

Role of Reverse Transcriptase in the Life Cycle of RNA Tumor
Viruses 730
Inder M. Verma and Wade Gibson

A Thermolabile Reverse Transcriptase from a Temperature-
sensitive Mutant of Murine Leukemia Virus. 753
*Steven R. Tronick, John R. Stephenson, Inder M. Verma, and
Stuart A. Aaronson*

Correlation of the Binding Properties of α and $\alpha\beta$ DNA
Polymerase of Avian Myeloblastosis Virus with Their Different
Mode of Ribonuclease H Activity 760
Duane Gradgenett

Rolling Circular DNA Associated with a Human Hepatitis B
Candidate Virus (Dane Particles) 766
P. P. Hung, J. C. Mao, C. M. Ling, L. R. Overby, and T. Kakefuda

CONTENTS

Part XII.   DNA Repair Synthesis   773

The Repair Mode of DNA Replication in *Escherichia coli*. .   .   .   774
Philip Hanawalt, Ann Burrell, Priscilla Cooper, and Warren Masker

Excision Repair of DNA   791
Lawrence Grossman

DNA Polymerase I Involvement in Repair in Toluene-treated
Escherichia coli   815
John W. Dorson and Robb E. Moses

DNA Synthesis in Phleomycin-treated Lysates of Go Lymphocyte
Nuclei   822
Harvard Reiter and Pei-Ling Hsu

Part XIII. Replication Intermediates   831

Discontinuous Replication in Prokaryotic Systems   832
Reiji Okazaki, Tuneko Okazaki, Susumu Hirose, Akio Sugino,
Tohru Ogawa, Yoshikazu Kurosawa, Kazuo Shinozaki,
Fuyuhiko Tamanoi, Tetsunori Seki, Yasunori Machida,
Asao Fujiyama, and Yuji Kohara

List of Contributors   836

Subject Index   871

Part I
DNA Polymerases, Binding, and Unwinding Proteins

THE ROLE OF THE ESCHERICHIA COLI DNA BINDING PROTEIN IN DNA SYNTHESIS

Malcolm L. Gefter, Ian J. Molineux,
Andrew Pauli, and Linda Sherman

Department of Biology
Massachusetts Institute of Technology
Cambridge, Massachusetts 02139

ABSTRACT

Escherichia coli DNA binding protein specifically
stimulates the polymerase and the $3' \rightarrow 5'$ exonuclease
activities of DNA polymerase II of E. coli. The specificity for polymerase resides in the ability of the
DNA binding protein to interact specifically, in the
presence and absence of DNA, with DNA polymerase II.
The enhancement of enzymatic activity is due to the fact
that in the presence of DNA binding protein, DNA polymerase does not dissociate from its substrate during
its action. DNA binding protein influences the enzymatic activity of DNA polymerases I and III, exonuclease I
and RNA polymerase. The central role of DNA binding
protein in DNA metabolism is discussed.

INTRODUCTION

While examining the properties of E. coli DNA polymerases II and III, it was observed that in addition to
the requirement for the presence of template and primer
in order for the reaction to proceed at all, the nature
of the template had a profound influence on the rate of
reaction. As the template strand was made longer and
longer, the rate of the reaction for copying that template decreased to almost non-detectable levels. The
rate of copying could be restored, in the case of DNA polymerase II, back to its original level, by the addition to

the DNA of DNA binding protein[1] (1). Furthermore, it
appeared that this observed release of inhibition of
propagation of DNA chains was not restricted to this one
example in that it had already been observed with T4
DNA polymerase and "gene 32" protein (2) and was later
shown to occur also with T7 DNA polymerase and the T7-
induced DNA unwinding protein (3). One remarkable
feature of these reactions was that they were specific
in that the polymerase and unwinding protein had to be
from the same source in order for the effect to be observed. Subsequently, we have shown that the E. coli
DNA binding protein exists as a tetramer, having a mass
of 90,000 daltons and it can interact with DNA polymerase II to form a complex containing one binding
protein monomer (22,000 daltons) and DNA polymerase II
(4).

We will describe here the details of the interaction
of binding protein with DNA and with DNA polymerase. We
will demonstrate that the enzymatic properties of the
polymerase are dramatically altered by interaction with
the binding protein and suggest a general model for the
specific role of DNA binding protein in DNA replication.

RESULTS

The interaction of DNA binding protein with DNA was
studied by monitoring the fluorescence of the protein at
350 nm in the presence and absence of DNA. As seen in
Figure 1, the fluorescence of the protein decreases as
the concentration of DNA is increased until about 70%
of the original fluorescence is abolished. Using this
observation, a stoichiometry of binding was calculated
to be 1 protein monomer to 8 nucleotides. The apparent
dissociation constant for the interaction of DNA binding

---

[1]The original name given to this protein by Alberts
and colleagues is DNA unwinding protein. We prefer
"binding protein" since, to date, we have not observed
unwinding of DNA by this protein.

3

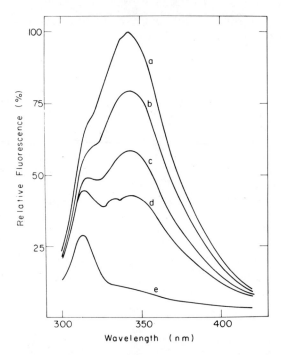

Figure 1. Fluorescence emission spectra of binding
protein in the presence of single-stranded DNA.

(a) Binding protein 0.11 µM; (b) +0.18 µM DNA;
(c) +0.51 µM DNA; (d) +0.66 µM DNA; (e) No protein +
0.66 µM DNA.

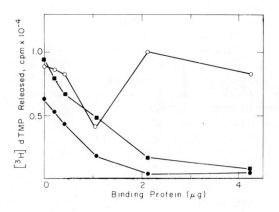

Figure 2. Degradation of [^{3}H] poly(dT) by DNA polymerase
in the presence of the binding protein.

●, DNA polymerase I; O, DNA polymerase II; ■ ,
DNA polymerase III.

protein and single-stranded DNA at 5°C in 0.02 M Tris-HC1, pH 7.4, was also calculated. The apparent K_{diss} is equal to 1.0 ($\pm$0.8) x 10^{-9} M. Similar dissociation constants were obtained when homoploymers replaced single-stranded DNA and no significant influence of sequence or composition on binding was observed. Measurements of the dissociation of protein from DNA using sedimentation analysis as was done for the T7 DNA binding protein (3), a K_{off} of 4 x 10^{-5} sec^{-1} was determined, and thus, the forward rate constant of 2 x 10^4 liter $mole^{-1}$ sec^{-1} is calculated. The half life of the complex under those conditions is 300 min. Alteraction of the composition of the reaction mixture by the addition of magnesium ions reduces the half life of the complex. Under the conditions employed in the subsequent experiminets, we can safely assume that the DNA:DNA binding protein complex forms readily and is a stable entity.

Given the specificity of the binding protein in terms of stimulation of DNA polymerases, we examined also the effect on the 3' → 5' polymerase-associated exonuclease activities. In this case as well, the phenomenon of specificity was observed; however, the effect was more complicated. As seen in Figure 2, addition of small amounts (insufficient to saturate all of the input DNA) of binding protein inhibited the activity of all three polymerase-associated nucleases tested. Increasing the concentration of protein (to greater than saturation) restored the activity of the DNA polymerase II activity but abolished the activity of polymerases I and III. The fact that in the presence of excess, or free protein, activity was restored, suggested an interaction between the polymerase and the binding protein.

It can be observed that DNA binding protein and E. coli DNA polymerase II indeed form a complex in the absence of DNA. The properties of this complex are in keeping with a 1:1 complex of protein monomer and DNA polymerase II monomer (see Figure 3). It has been suggested that this binary complex is responsible for the digestion of a DNA binding protein: DNA complex (4).

5

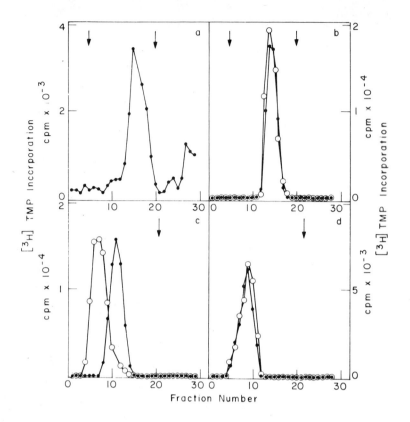

Figure 3. Detection of DNA polymerase complexed to
binding protein by sedimentation through
glycerol gradients.

Centrifugation for for 20 hours. Binding protein
was assayed by its ability to stimulate DNA polymerase II
activity at 12° with poly(dA)·(dT)$_{10}$ as template (4).
DNA polymerase was assayed on "gapped" calf thymus DNA
at 30°.   (a) Sedimentation of binding protein (37 µg).
(b) Sedimentation of DNA polymerase I (60 units) in the
presence (O) or absence (●) of binding protein (37 µg).
(c) Sedimentation of DNA polymerase II (5 units) in
the presence (O) or absence (●) of binding protein
(37 µg).   (d) Sedimentation of DNA polymerase III (5
units) in the presence (O) or absence (●) of binding
protein. Internal and external markers of the RNA-
directed DNA polymerase from murine leukemia virus
(70,000 daltons) and avian myeloblastosis virus (150,000
daltons) were included in the gradients and assayed by
their ability to catalyze the synthesis of poly(dG)
using poly(rC)·(dG)$_{12-16}$ as template. Only the 70,000-
dalton enzyme was included internally in panels c and d.

Other enzymes, including E. coli exonuclease I, have been shown also to form complexes with binding protein and similarly their activities are enhanced when assayed on DNA complexed with binding protein (5). As suggested from the kinetic experiments and confirmed by physical studies, the formation of binary complex is specific for DNA polymerase II in that DNA polymerases I and III as well as the T4-induced DNA polymerase do not associate with the E. coli DNA binding protein.

We have further shown that the ability of binding protein to bind to both DNA and enzyme are independent. A ternary complex consisting of DNA, DNA binding protein, and DNA polymerase II has been observed. As seen in Figure 4, DNA polymerase II can be bound to circular single-stranded DNA in the presence, but not in the absence of binding protein. This result suggests that, in part, the restoration of the rate of digestion of DNA by polymerase II observed in Figure 2 may be due to both the formation of binary complex as well as the possibility of ternary complex formation; i. e., the presence of binding protein on the DNA substrate prevents dissociation from the DNA once digestion has commenced. This has indeed been shown to be the case.

In the absence of binding protein, DNA polymerase II will digest single-stranded DNA. However, the mechanism of digestion is such that the enzyme dissociates from its substrate following its action. In the presence of binding protein, each molecule of DNA is digested completely before dissociation takes place. This has been demonstrated by initiating digestion on DNA labelled with one isotope and then adding the same DNA labelled with another isotope. These results are summarized in Figure 5. The conclusion is that the mechanism of enzyme action is altered by the DNA binding protein. It is reasonable to suggest that in the absence of binding proteins the rate limiting step in the reaction is re-association of enzyme to DNA. Ths binding protein abolishes this step.

Having observed the formation of complexes and alteration of enzyme activity by binding protein, we re-investigated the details of the synthetic reaction.

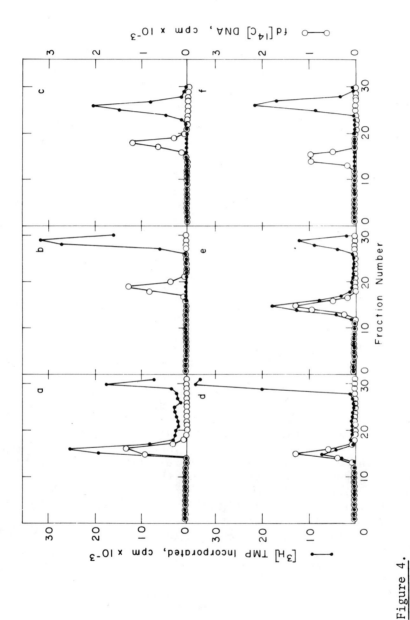

<u>Figure 4.</u>

Glycerol gradient analysis of DNA polymerase sedimenting in the presence of 0.4 μg of fd [^{14}C] DNA (panels a, b, and c) and in the presence of both 0.4 g of fd [^{14}C] DNA and 2.5 μg of binding protein (panels d, e, and f). Centrifugation was for 2 hours, and sedimentation is plotted from right to left. DNA polymerase was assayed on "gapped" calf thymus DNA (4) at 30°. O, fd [^{14}C] DNA; ●, DNA polymerase; Panels a and d, DNA polymerase I; panels b and e, DNA polymerase II; panels c and f, DNA polymerase III.

8

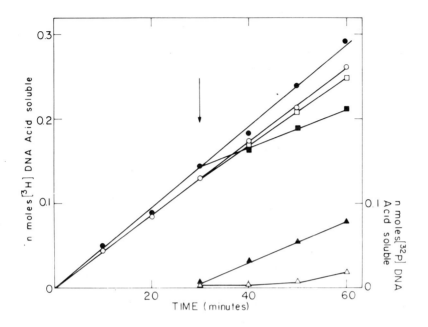

Figure 5.

   Nucleolytic digestion of phage P22 denatured DNA by
E. coli DNA polymerase II in the presence (open symbols)
and absence (closed symbols) of binding protein. Reactions (2.0 ml) contained, in addition to tris, Mg^{++} and
2-mercaptoethanol, 3.0 µg denatured [3H] DNA, 1 unit
DNA polymerase II and, where applicable, 0.26 µg
binding protein. After 30 minutes at 30°C, the reactions
were divided into two. Half the reaction mixtures were
incubated at 30°C for a further 30 minutes (O———O;
●———●); to the other halves, 1.2 µg [32P] DNA was
added and incubation continued for 30 minutes at 30°C.
Production of acid-soluble radioactivity in both 3H-
(□, ■) and 32P- (∆, ▲) nucleotides was analyzed
simultaneously. Reaction mixtures containing binding
protein are represented by open symbols throughout;
reactions performed in the absence of binding protein
are represented by closed symbols throughout. The
arrow marks the time at which the [32P] DNA was added.

catalyzed by DNA polymerase II in the presence and absence of binding protein. In particular, the questions of chain elongation rates and dissociation of enzymes during synthesis were addressed.

As a model system, we have used the single-stranded circular DNA of bacteriophage fd to which a specific complementary restriction enzyme-generated fragment is annealed. The rate of elongation of such a primer is enhanced by greater than 100-fold by the addition of binding protein . Furthermore, we have shown that in the absence of DNA binding protein, DNA polymerase II dissociates from growing DNA molecules. Reactions run in the presence of excess substrate over enzyme molecules have shown that all substrate molecules (^{32}P-labelled primers) appear to grow at a uniform rate. This is not the case in the presence of DNA-binding protein. As seen in Figure 6, a portion of the primers have been elongated while the remainder have not. This implies that the enzyme molecules once bound to a template-primer in the presence of binding protein do not dissociate during synthesis. Under such conditions, the rate of chain growth per enzyme molecule can be shown to be about 1,000 nucleotides/minute at 30°C. The lack of enzyme dissociation during synthesis is sufficient to explain the enhancement of rates observed when DNA binding protein is added to a DNA polymerase II reaction. Furthermore, since the polymerase binds directly to the binding protein both during synthesis and degradation of DNA, we may also explain the specificity observed in these reactions for polymerase and binding protein. We conclude that through its binding properties the DNA binding protein alters the mechanism of the DNA polymerase II catalyzed reactions.

DISCUSSION

We have shown that DNA binding protein isolated from E. coli contains two binding sites, one for DNA and one for DNA polymerase II. These two binding sites can function independently of each other. We have presented evidence that when DNA polymerase II is copying

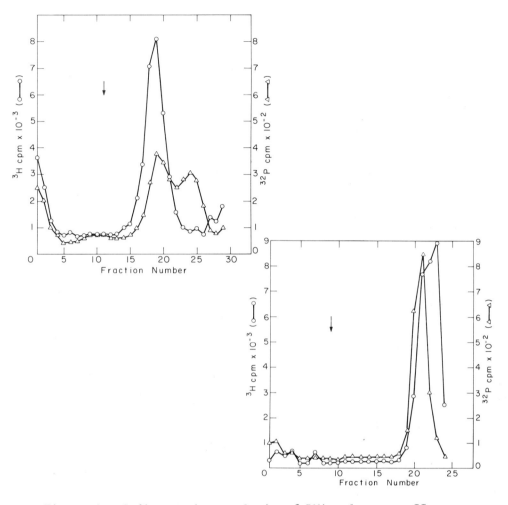

Figure 6.   Sedimentation analysis of DNA polymerase II
product made in the presence and absence of
DNA binding protein.

Specific primer molecules (150 nucleotides in length),
labelled with ^{32}P ($\triangle$———$\triangle$———$\triangle$) were annealed to fd
DNA.   DNA polymerase II was then used to elongate the
primers in the presence of ^{3}H dTTP (0———0———0).
Reactions were set up such that primer molecules
exceeded the number of enzyme molecules used.   The
reaction product was layered on a sucrose gradient at
pH 13 and sedimented in the presence of circular fd DNA
marker (arrow).   Sedimentation was from right to left.
The top figure represents the product made in the
presence of protein and the bottom in its absence.
Reactions were run for equivalent times; however, the
specific activity of the TTP employed was 100-fold
higher in the bottom figure.

a DNA template which itself is complexed with DNA
binding protein, the DNA polymerase does not dissociate
from the template during synthesis. The effect of
this interaction between the two proteins is to enhance
the rate of polymerization by more than 100-fold.
By virtue of the specific interaction between the two
proteins, the DNA binding protein is able to restrict
the activities of both E. coli polymerases I and III.
In addition, we have shown that other proteins such as
RNA polymerase and endonuclease I are prohibited from
interacting with DNA when they are complexed with
binding protein. Previous results of others (6)
have shown that DNA binding protein promotes denaturation and renaturation of DNA molecules. Thus, to this
point, we can conclude that the protein is likely
to be involved in the processes of DNA repair and in
recombination.

Does the protein have a role in DNA replication?

It has been shown that DNA polymerase II carries
out the major elongation step in DNA replication;
however, it does not copy a DNA complexed with DNA
binding protein. Since DNA polymerase III displays
the same properties as DNA polymerase II in that it
fails to copy long DNA templates, we might expect that
a similar interaction as is shown with DNA polymerase II
and DNA binding protein also does take place with DNA
polymerase III. It is obvious that DNA polymerase III
as such does not function in this reaction. However,
it has been shown that in the presence of other protein
factors isolated in the laboratories of Hurwitz (7)
and Kornberg (8), DNA polymerase III does copy a
DNA molecule complexed with DNA binding protein. Thus,
it appears that DNA polymerase III in the process of
replication complexes with binding protein through an
intermediary factor such that replication is achieved
on a DNA binding protein DNA complex. It is reasonable
to conclude that the DNA binding protein is also
involved in the process of DNA replication.

In considering all of the roles of this protein in
DNA metabolism, we must ascribe central importance to it.
It promotes denaturation and renaturation of DNA, it

12

prevents transcription of single-stranded DNA and pro-
hibits digestion endonucleolytically of single-stranded
DNA. It enhances the binding capacity of DNA polymerase to DNA and, more importantly, this interaction
results in high rates of replication in the absence of
dissociation of polymerase from DNA templates. All of
these properties of the DNA binding proteins clearly
demonstrate its importance in DNA metabolism of cells.

ACKNOWLEDGEMENTS

This work was supported in part by grant no.
B36649 from the National Science Foundation and in
part by grant no. 5-R01-GM20363-02 from the National
Institutes of Health.

REFERENCES

1. Sigal, N., Delius, H., Kornberg, T., Gefter, M. L.,
      and Alberts, B. (1972). Proc. Nat. Acad.
      Sci. USA 69, 3537-3541.

2. Huberman, J. A., Kornberg, A., and Alberts, B. M.
      (1971). J. Mol. Biol. 62, 39-52.

3. Reuben, R. C. and Gefter, M. L. (1973). Proc.
      Nat. Acad. Sci. USA 70, 1846-1850.

4. Molineux, I. J. and Gefter, M. L. (1974). Proc.
      Nat. Acad. Sci. USA 71, 3858-3862.

5. Molineux, I. J. and Gefter, M. L. (1975). J.
      Mol. Biol., in press.

6. Alberts, B. M. and Frey, L. (1970). Nature 227,
      1313-1318.

7. Hurwitz, J. and Wickner, S. (1974). Proc. Nat.
      Acad. Sci. USA 71, 6-10.

8. Wickner, W., Schekman, R., Geider, K., and Kornberg,
      A. (1973). Proc. Nat. Acad. Sci. USA 70,
      1764-1769.

DNA POLYMERASE III FROM BACILLUS SUBTILIS AND THE MECHANISM OF ARYLHYDRAZINOPYRIMIDINE INHIBITION

Nicholas R. Cozzarelli, Robert L. Low,
Stephan A. Rashbaum, and Craig L. Peebles
Departments of Biochemistry and
Biophysics and Theoretical Biology
The University of Chicago
Chicago, Illinois 60637

ABSTRACT. Bacillus subtilis DNA polymerase III has been purified about 4,500 fold. When subjected to disc gel electrophoresis, the most purified fraction exhibits a single major protein band which migrates with the same mobility as polymerase activity. Polyacrylamide gel electrophoresis of the polymerase under denaturing conditions yields a single, 165,000 dalton band. The hydrodynamic properties of the enzyme are ionic strength dependent. The average values from determinations in high and low salt are 7.6 S for the sedimentation coefficient and 52 Å for the Stokes radius. These two parameters indicate a molecular weight for the native enzyme of 160,000 daltons. Therefore, the enzyme appears to contain a single, long, polypeptide chain. DNA polymerase III also catalyzes pyrophosphate exchange into deoxyribonucleoside triphosphates, pyrophosphorolysis of DNA, hydrolysis of DNA, and hydrolysis of RNA. The intrinsic exonuclease of the polymerase attacks single-stranded DNA from the 3'-terminus yielding 5'-mononucleotides. All five enzymatic activities of DNA polymerase III are inhibited by arylhydrazinopyrimidines. These drugs inhibit by forming a long-lived ternary complex with the enzyme and the DNA substrate. The drugs bind to the template in a base-specific manner such that a uracil containing drug pairs with cytosine and an isocytosine analogue pairs with thymine. Whereas the half-life of the ternary complex containing wild-type polymerase is 15 min at 0°, the dissociation rate for a drug-resistant mutant form of polymerase III was at least an order of magnitude faster.

INTRODUCTION

Three distinct DNA polymerases have been isolated from
Bacillus subtilis, DNA polymerases I, II, and III (1-3).
They all catalyze the same basic reaction of condensation
of deoxyribonucleoside triphosphates onto the 3'-hydroxyl
primer terminus under the direction of a template, but
differ in physical, enzymatic, and functional properties
(1). The three enzymes are somewhat similar to the correspondingly numbered Escherichia coli polymerases but there
are marked differences; for example, B. subtilis polymerase III is sensitive to the antibacterial agents, 6-(p-
hydroxyphenylhydrazino)-uracil [OHPh(NH)$_2$Ura] and 6-(p-
hydroxyphenylhydrazino)-isocytosine [OHPh(NH)$_2$Iso],
whereas the E. coli enzyme is resistant (4). The structural gene for B. subtilis polymerase I has been mapped by
3-factor transformation crosses (1). Analysis of mutants
has shown that polymerase I functions in the repair of
chromosomal damage and the joining of Okazaki fragments (4,
and R. Okazaki, personal communication). This paper
focuses on B. subtilis DNA polymerase III, an enzyme
required for DNA replication (1,5). We report here the
purification of the enzyme to near homogeneity and some of
its functional, physical, and enzymatic properties.

METHODS

The methodology and bacterial strains used are indicated in the figure and table legends or have been
described (1,6). By definition, a unit of enzyme catalyzes
the incorporation into DNA of 10 nmoles of nucleotide in
30 min under standard assay conditions (1).

RESULTS

Polymerase III mutants: A DNA polymerase III or polC
mutant was first isolated on the basis of resistance to
OHPh(NH)$_2$Ura and OHPh(NH)$_2$Iso (5). Since the enzyme purified from this mutant, F22, was resistant to the drugs in
vitro, polC is a structural gene for polymerase III. Some
drug-resistant polC mutants fail to grow at high temperatures and cause a marked increase in the mutation rate of
outside markers (1). Bazill and Gross have reported that
a temperature-sensitive mutant called mut-1 had impaired

15

polymerase III activity and was also a mutator strain (7).
We have found that the polymerase III purified from this
strain is about five times more thermolabile than the wild-
type enzyme (data not shown). When a polC25 mutant (a
drug-resistant, temperature-sensitive, mutator strain) was
incubated at 51°, a non-permissive temperature, the rate
of DNA synthesis was markedly reduced compared to the isogenic control, whereas the rate of protein synthesis was
the same in both strains (Fig. 1).

Brown (personal communication) has reported that dnaF
(ref. 8) cotransduces with ura and cysC and that a dnaF
strain has lowered polymerase III activity. We have confirmed this linkage in crosses mediated by the transducing
phage, AR9. As shown in Table 1, five independent polC
mutations, including mut-1 and polC25, cotransduce with ura
and cysC1 at about a 45% frequency. In the crosses with
drug-resistant, temperature-sensitive mutants, the two
phenotypes segregated together. PolC is also linked to
cysC1 by transformation; with DNA carefully prepared to
avoid shearing the cotransformation frequency was about
10%.

Purification and physical properties of DNA polymerase III:
DNA polymerase III was purified from polA⁻ cells [strain
BC26(F)] by the procedure outlined in Table 2. The purified polymerase has similar properties to the enzyme preparations obtained by different procedures (1,3). The
Fraction VII enzyme had about 4,500 times the specific
activity of the starting material. The enzyme is nearly
homogeneous. When a concentrate of Fraction VII was analyzed by polyacrylamide gel electrophoresis at pH 8.0, one
major protein band was observed after staining with Coomassie blue (Fig. 2A) which had the same mobility as DNA
polymerase activity (Fig. 2B). The protein band was eluted
from a third parallel gel and run on a polyacrylamide gel
containing sodium dodecyl sulfate. A single sharp band
was observed (Fig. 2C) with a molecular weight of about
165,000 daltons (in two experiments the molecular weights
were 165,000 and 167,000). There is no evidence of a low
molecular weight subunit.

The molecular weight of native B. subtilis polymerase
III was determined from the sedimentation coefficient and
the Stokes radius by the method of Siegel and Monty (9).
The determinations were done in a low salt buffer (0.05 M

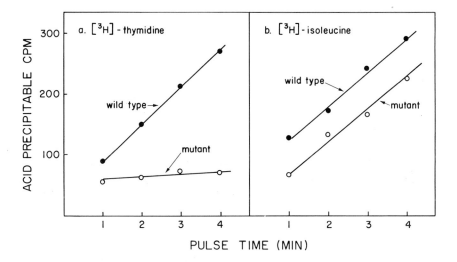

Fig. 1.   DNA and protein synthesis in a polymerase
III mutant at a restrictive temperature. The wild-type
strain (BD170) is polA59 and the mutant strain (BD293)
is polA59, polC25. The cells were grown out at 37° and
10 min after a shift to 51° were labeled with [³H]thymi-
dine or [³H]isoleucine and the acid-precipitable counts
measured 1-4 min thereafter.

TABLE 1

MAPPING OF polC MUTANTS BY TRANSDUCTION

Donor strain	Recipient strain	Selected marker	Number analyzed	Linkage of polC
				%
	GSY1035	ura	161	48
F25	BC34	ura	61	39
	JH158	cysCl	39	46
F26	BC34	ura	211	45
	JH158	cysCl	40	60
	GSY1035	ura	26	46
F27	BC34	ura	56	52
	JH158	cysCl	29	38
2355	BC34	ura	60	47
	JH158	cysCl	40	48
BD54(F)	BC34	ura	39	59
	JH158	cysCl	36	36

The transduction was mediated by phage AR9. The mapping procedure and strains F25, F26, and F27 have been described (1). The polymerase III mutational sites in strains F25, F26, F27, 2355, and BD54(F) are polC25, polC26, polC27, mut-1 (ref. 7), and dnaF (ref. 8). Strains BC34, GSY1035, and 2355 were obtained from J. Copeland, T. Matsushita, and G. Bazill, respectively, and strains JH158 and BD54(F) from N. Brown. AR9 was obtained from D. Dubnau.

TABLE   2

PURIFICATION OF DNA POLYMERASE III

Fraction and step	Activity	Protein	Specific activity
	units	mg	units/mg
I. High speed supernatant[a]	12,500	11,000	1.1
II. Streptomycin sulfate	9,700	2,400	4.0
III. DEAE-cellulose I	7,600	810	9.4
IV. Sephadex G-200	4,400	220	20
V. DEAE-cellulose II	3,900	59	66
VI. Phosphocellulose	1,500	3.7	410
VII. DNA-cellulose	320	0.06[b]	5,000[b]

[a] From 200 g of strain BC26(F).

[b] To conserve purified enzyme the protein concentration in
this fraction was measured by a microfluorometric assay
and by a scan of a stained polyacrylamide gel rather than
our usual procedure (1). The two determinations differed
by 40% and the mean value is shown. The DNA-cellulose
step can be replaced by hydroxylapatite chromatography
(see Fig. 5); the enzyme is obtained in higher yield and
is free of contaminating nuclease but has a lower specific
activity.

POLYACRYLAMIDE GEL ELECTROPHORESIS OF POLYMERASE III

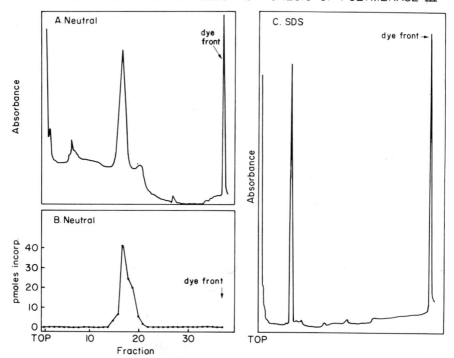

Fig. 2. Polyacrylamide gel electrophoresis of DNA polymerase III. Fraction VII was concentrated by ultra-filtration and applied to three 4%, pH 8.0, (0.6 x 8.5) cylindrical polyacrylamide gels. One gel containing 4 μg of protein was stained with Coomassie blue (A) while a parallel gel was cut into 37 slices which were assayed for polymerase activity (B). The R_f of the polymerase was 0.46. Thirteen μg of protein were applied to the third gel and the section corresponding to an R_f of 0.43 - 0.49 was cut out. The protein eluted from this section was run on a 5%, pH 7.4 polyacrylamide gel containing 0.2% sodium dodecyl sulfate (C) along with a series of external protein standards.

potassium phosphate, pH 7.0) and a high salt buffer (0.2 M
potassium phosphate, pH 7.0). In Fig. 3 the sedimentation
pattern of polymerase III in a high salt glycerol gradient
is displayed. The polymerase activity sedimented as a
single band with the same mobility as lactic dehydrogenase.
There is no evidence of aggregation since the profiles of
the polymerase and dehydrogenase were the same. The calculated sedimentation coefficient, $S_{20,w}$, for the polymerase
is 7.4 S. In low salt, the coefficient is 7.8. The
results of filtration through Sephadex G-200 in both low
and high salt are shown in Fig. 4. The distribution
coefficient, or K_d, of the polymerase is clearly salt
sensitive with the calculated Stokes radii being 56 Å and
48 Å in high and low salt, respectively. The enzyme eluted
ahead of catalase in high salt but behind it in low salt.
The average value of the relative frictional coefficient,
f/f_o, from the two sets of data is 1.45. This result is
consistent with (but is not proof of) a markedly asymmetric
molecule (10). The calculated molecular weights (rounded
to two significant figures) of 170,000 and 150,000 daltons
in high and low salt average to 160,000 daltons. Since
this value is in good agreement with the value of 165,000
daltons obtained from denaturing gels, polymerase III
probably consists of a single polypeptide of very high
molecular weight.

Additional reactions catalyzed by DNA polymerase III: The
polymerase not only catalyzes synthesis but also exchange
of pyrophosphate into deoxyribonucleoside triphosphates
and pyrophosphorolysis of DNA (Table 3). Exchange was
measured in the standard polymerase assay mixture supplemented with 1.0 mM [^{32}P]PP$_i$; the deoxyribonucleoside triphosphates were omitted in the pyrophosphorolysis experiments. The activity of the exchange reaction is about 40%
of the synthetic activity with the full deoxyribonucleoside
triphosphate complement but 410% of the synthetic activity
with only dTTP present; the rate of pyrophosphorolysis
is about 20% of the synthetic rate. Both partial reactions
require DNA and are inhibited by OHPh(NH)$_2$Ura. There is,
therefore, very likely an activated nucleotidyl intermediate in the polymerase III reaction mechanism as found for
most but not all polymerases (4,11).
    Because of the utility of ribonucleotide incorporation
by DNA polymerases in sequencing DNA, we tested the

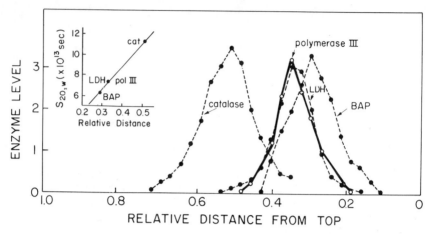

Fig. 3.   Sedimentation of polymerase III through a
high-salt, 35-50% glycerol gradient. The buffer contained
0.2 M potassium phosphate, pH 7.0, 5 mM dithiothreitol,
and 0.5 mM EDTA and the sedimentation was at 2° for
34 hours at 54,000 rpm in a Spinco SW56 rotor. The
ordinate units are arbitrary. BAP, LDH, cat, and pol III
are abbreviations for bacterial alkaline phosphatase,
lactic dehydrogenase, catalase, and polymerase III,
respectively.

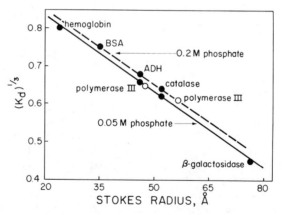

Fig. 4.   Determination of the Stokes radius of DNA
polymerase III in low and high salt. A 0.95 x 57 cm
Sephadex G-200 column was eluted with a buffer containing
14 mM 2-mercaptoethanol, 10% glycerol, 0.5 mM EDTA, and
0.05 M or 0.2 M potassium phosphate, pH 7.0. N4
[3H]phage was used to measure the void volume. ADH and
BSA are abbreviations for alcohol dehydrogenase and bovine
serum albumin. Polymerase III (O), ADH, and catalase
were chromatographed on both columns.

TABLE 3

PYROPHOSPHATE EXCHANGE AND PYROPHOSPHOROLYSIS
CATALYZED BY DNA POLYMERASE III

| Conditions | Incorporation | |
	Acid-precipitable ^{3}H	Norit-adsorbable ^{32}P
	pmoles	
Complete mixture[a]	280	120
+ OHPh(NH)$_2$Ura	61	16
− DNA	<1	<2
− dATP, dCTP, dGTP	49	200
− dNTP[b]	−	54
− dNTP, + OHPh(NH)$_2$Ura	−	8
− dNTP, − DNA	−	<2
− PP$_i$	280	−

[a] Contains 5 µM each of dATP, dCTP, dGTP, and [^{3}H]dTTP, 1.0 mM [^{32}P]PP$_i$, and gapped DNA in the standard polymerase reaction mixture (1).

[b] Abbreviation for all four deoxyribonucleoside triphosphates.

Pyrophosphorolysis was measured by the production of Norit-adsorbable ^{32}P in the absence of any deoxyribonucleoside triphosphates. Exchange was estimated from the Norit-adsorbable ^{32}P in the complete reaction mixture. No ^{32}P was incorporated into the acid-precipitable product.

incorporation of rGTP, rCTP, rUTP, and rATP by B. subtilis
polymerase III. Only rGTP was utilized appreciably; rGTP
was incorporated 1/7 as well as dGTP in the presence of
0.5 mM $MnCl_2$ and negligibly in the presence of 6 mM $MgCl_2$.

We have shown previously that DNA polymerase III
contains a Mg^{++} requiring, intrinsic exonuclease (6).
There is no endonuclease activity as determined by the
absence of cleavage of single-stranded and double-stranded
circular DNA. The nuclease must be intrinsic since it comigrates with the polymerase after chromatography on
hydroxylapatite (Fig. 5) and DNA-cellulose and sedimentation through a glycerol gradient. The two activities are
similarly inhibited by N-ethylmaleimide, high ionic
strength, and arylhydrazinopyrimidines; the drug induced
ternary complex isolated by agarose chromatography contains
both activities (6).

The activity of the nuclease on several substrates is
shown in Table 4. The nuclease is much more active on
single-stranded than double-stranded DNA and prefers small
single-stranded DNA, such as sonicated DNA, to long DNA,
even at saturating levels of both. Introduction of 3'-
phosphoryl groups markedly reduces hydrolysis and, in fact,
3'-phosphoryl terminated DNA inhibits the nuclease. The
nuclease is quite an active enzyme with the appropriate
substrate. With poly(dT), even at a non-saturating concentration of 0.15 mM, hydrolysis is three times faster than
synthesis under optimal conditions. Polymerase III also
has some RNase activity as indicated by the low level of
hydrolysis of poly(rU). Poly(rU) inhibits the hydrolysis
of single-stranded DNA and poly(rU) hydrolysis is inhibited
by $OHPh(NH)_2Ura$; the RNase is thus intrinsic to polymerase
III.

There are three lines of evidence that hydrolysis is
initiated predominantly from the 3'-end. First, 3'-
phosphoryl terminated DNA is not an effective substrate.
Second, exonuclease III-treated DNA and λ DNA which contain
protruding 5'-single-stranded regions are not effective
inhibitors of single-stranded DNA hydrolysis (6). Third,
^3H is released orders of magnitude faster than ^{32}P from
T7 DNA labeled uniformly with ^3H but with ^{32}P only at the
5'-terminus. E. coli polymerase III contains a 5'→3'
exonuclease which has approximately 1/15 of the 3'→5'
exonuclease activity (12). The 5'→3' exonuclease activity
in B. subtilis polymerase III would have to be even more

24

TABLE   4

SUBSTRATE SPECIFICITY OF THE POLYMERASE III NUCLEASE

Substrate	Relative nuclease activity
1. Denatured E. coli DNA	1.0
2. Native E. coli DNA	0.01
3. Sonicated, denatured E. coli DNA	2.9
4. 3'-Phosphoryl-terminated, short, denatured E. coli DNA	0.02
5. Phosphatase treated substrate 4	6.9
6. Equal mixture of 4 and 5	1.8
7. Poly(dT)	67
8. Poly(rU)	0.09

The nuclease reaction conditions and the preparation of the E. coli DNA derived substrates have been described (6). The reactions contained 25 μM E. coli DNA, 150 μM poly(dT), or 70 μM poly(rU). The [^{3}H]poly(dT) was the product of a terminal deoxynucleotidyl transferase reaction (chain length about 400) and the [^{3}H]poly(rU) was synthesized with polynucleotide phosphorylase. The activity with substrate 1 was set equal to unity and on this scale the polymerase activity assayed under optimal conditions equals 22.

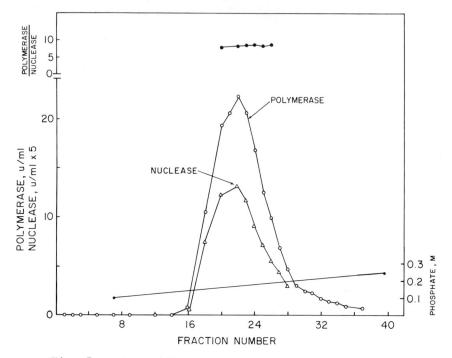

Fig. 5. Copurification of the polymerase and
nuclease activities of DNA polymerase III on hydroxylapatite. Thirty-eight units of Fraction VI polymerase
were applied to 0.8 x 2.4 cm hydroxylapatite column and
the column was eluted with a 12-ml linear gradient of
potassium phosphate, pH 6.8, containing 5 mM
glutathione, and 10% glycerol. The nuclease substrate
was sonicated, heat-denatured, [^{3}H]E. coli DNA.

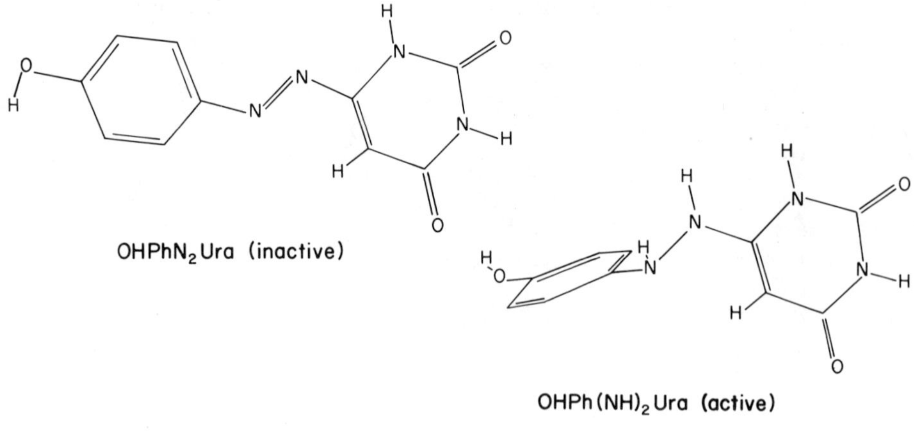

OHPhN$_2$Ura (inactive)

OHPh(NH)$_2$Ura (active)

Fig. 6. Structure of OHPhN$_2$Ura and OHPh(NH)$_2$Ura as
determined by x-ray crystallography (15,16).

feeble, but it may be present. In the digestion of the
double-labeled T7 DNA the ratio of the two isotopes
released remained constant with time and the products were
[^{3}H]5'-mononucleotides but [^{32}P]oligonucleotides, probably
trinucleotides. The oligonucleotides could be the product
of hydrolysis from the 5'-end, a remnant of such a product,
or a limit digest from 3'→5' hydrolysis. The very low rate
of release of the 5'-terminus has thus far frustrated
attempts at resolution of these alternatives.

An attractive hypothesis for the regular occurrence of
intrinsic nucleases in DNA polymerases postulates that the
nuclease edits or removes mismatched bases incurred during
replication (13). The nuclease associated with B. subtilis
DNA polymerase III has properties similar to the editing
nucleases of E. coli and T4 DNA polymerases. Namely, it
degrades single-stranded DNA exonucleolytically from the
3'-end releasing 5'-mononucleotides. In light of the
evidence that mutator T4 DNA polymerases may result from
inefficient editing (14), such lax proof-reading provides
a working hypothesis for the greatly increased mutation
rate in certain B. subtilis DNA polymerase III mutants
(1,7).

The mechanism of arylhydrazinopyrimidine inhibition of DNA
polymerase III: The arylhydrazinopyrimidine inhibitors
of B. subtilis DNA polymerase III have been invaluable in
the study of this enzyme (4). They have made the isolation
of polC mutants easy, have helped identify polymerase III
as the replicative DNA polymerase, have provided a diag-
nostic test for enzyme activities which are intrinsic to
DNA polymerase III, and have contributed to the understanding of the polymerase III reaction mechanism. The
drugs are interesting in their own right as the most
specific inhibitors of DNA synthesis known and their
mechanism of action has been established.

The structures of 6-(p-hydroxyphenylazo)-uracil
(OHPhN$_2$Ura) and OHPh(NH)$_2$Ura as determined by x-ray crystallography are shown in Fig. 6 (15,16). The azo form of
the drugs used to inhibit bacterial DNA synthesis are
reduced intracellularly to the active hydrazino derivatives (17). The reduction of the azo link has several
consequences of importance to the activity of the drugs
(16). OHPh(NH)$_2$Ura will be used illustratively. First,
the reduction breaks up the conjugated double bond system

thereby permitting the phenyl ring to move out of the plane
of the uracil ring; the phenyl ring has been postulated
to bind to the enzyme. Second, the hydrazino hydrogen
proximal to the uracil ring is very likely bonded to cytosine in the drug-DNA interaction. Third, the planarity
required for the binding of the drug moiety to cytosine is
stabilized by electron delocalization from the hydrazino
nitrogen adjacent to uracil.

In 1973 we proposed a model for arylhydrazinopyrimidine inhibition called ternary scavenging which is depicted
schematically in Fig. 7 for $OHPh(NH)_2Ura$ (18). The drug
forms a ternary complex with DNA polymerase III and
template-primer DNA. The drug is bound to the enzyme in a
position overlapping the deoxyribonucleoside triphosphate
binding site; the binding may be stabilized by a hydrophobic interaction with the phenyl ring of the drug. The
drug is also bound via hydrogen bonds to the template
cytosine residue adjacent to the complement of the primer
terminus base. The hydrogen-bonded primer strand is also
required for the formation of the complex as is Mg^{++} or
polyamines. $OHPh(NH)_2Iso$ forms a similar ternary complex
but hydrogen bonds selectively to thymine template residues. A key postulate of the model is that the inhibitory
complex dissociates slowly thereby trapping the enzyme in
an inactive state. Consequently, the formation of the
ternary complex efficiently scavenges free enzyme, and thus
the name of the model.

The rather compelling evidence for the model and the
structure of the drug-base pairs have been detailed in
recent reports (16-18,3) and only a brief and selected
summary will be given here. The inhibition by $OHPh(NH)_2Ura$
is selectively and competitively attenuated by dGTP and the
inhibition by $OHPh(NH)_2Iso$ is equally selectively overcome
by dATP. The base specificity of complex formation was
also shown by studies with defined templates. One clear
demonstration of the scavenging aspect of inhibition is
the properties of inhibition of the polymerase III nuclease. The hydrolysis of single-stranded [3H]DNA is barely
diminished by even high concentrations of $OHPh(NH)_2Ura$.
However, the addition with the drug of small amounts of
unlabeled DNA containing 3'-hydroxyl terminated gaps, such
as phage λDNA, to scavenge the enzyme caused full inhibition of the nuclease (Fig. 8). Nicked DNA was much less
effective. The lack of scavenging by purely single-

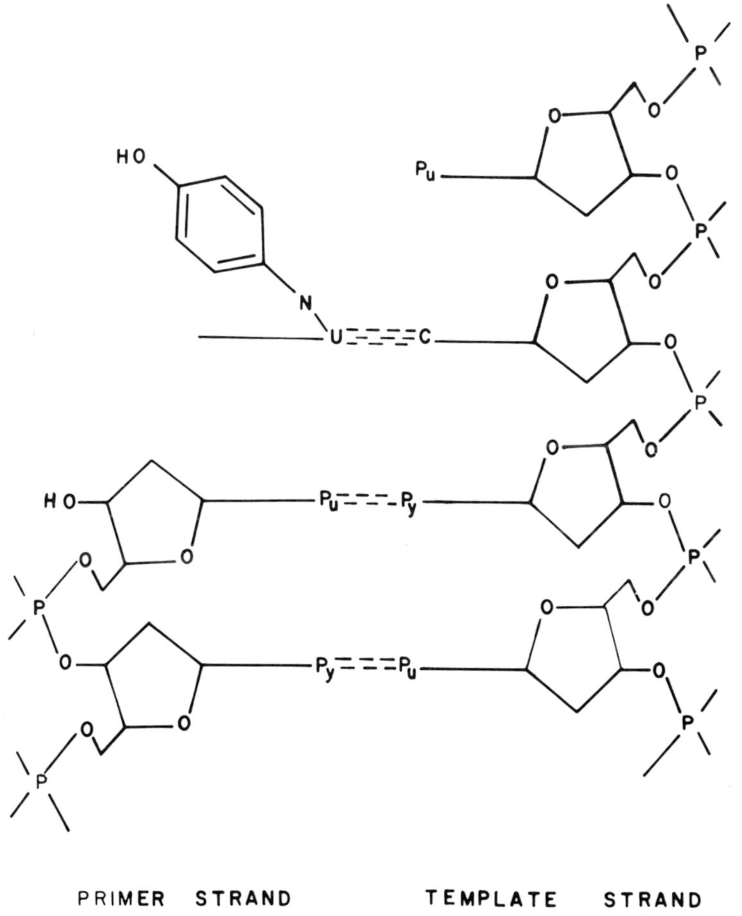

PRIMER STRAND   TEMPLATE STRAND

Fig. 7.   Schematic representation of $OHPh(NH)_2Ura$,
DNA, DNA polymerase III ternary complex. The drug and
DNA are bound to the enzyme which is not shown for
clarity. The orientation of the drug phenyl ring, away
from the primer terminus, is the conformation of crystalline $OHPh(NH)_2Ura$ (16) and is opposite to the position of
the deoxyribose ring of dGTP, the competitor of the drug.
Pu = purine and Py = pyrimidine base.

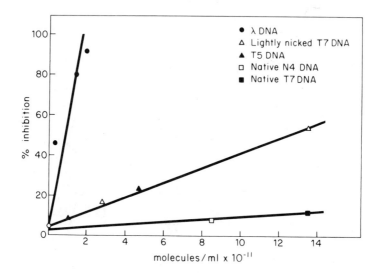

Fig. 8.   Sequestering DNA requirement for
OHPh(NH)$_2$Ura inhibition of the polymerase III nuclease.
The nicked T7 DNA contained 4 single-strand breaks per
molecule introduced by DNase I treatment.   The reaction
contained the indicated unlabeled DNA and 7 μM heat-
denatured, E. coli [^{3}H]DNA as has been described (6).

stranded and double-stranded DNA emphasizes the importance of the primer for complex formation (Fig. 8). The results with OHPh(NH)$_2$Iso are qualitatively similar but the requirement for the primer terminus may be less strict since OHPh(NH)$_2$Iso causes about a 20% inhibition of poly(dT) hydrolysis in the absence of added sequestering DNA. Poly(dT) can not form "hairpin" structures and in the control experiment with OHPh(NH)$_2$Ura the inhibition was negligible. Direct evidence for the ternary complex was provided by its isolation using agarose gel filtration of the polymerase reaction mixture. In the presence of drug the enzyme elutes in the void volume with the DNA but in its absence it is included into the gel (Fig. 9). The exclusion of the enzyme is prevented by dGTP while rGTP, dATP, dCTP, and dTTP have no effect. The data with dGTP and dTTP are also shown in Fig. 9.

We wish to make three additional points about the mechanism of action of the drugs. First, the base specificity of OHPh(NH)$_2$Ura pairing to cytosine and OHPh(NH)$_2$Iso to thymine is dictated at the binding step of the drug to the enzyme and template-primer, and not at some subsequent step in the reaction sequence. With a template-primer of poly(d[A-T]), OHPh(NH)$_2$Iso but not OHPh(NH)$_2$Ura promoted exclusion of the enzyme by agarose (Fig. 10), as expected from the presence of a template T but not C residue. In the converse experiment with poly(dC):oligo(dG) (Fig. 11), OHPh(NH)$_2$Ura but not OHPh(NH)$_2$Iso promoted complex formation. With poly(dC) alone, and no primer, the complex was not formed (data not shown).

Second, using two methods we showed that the dissociation rate of the ternary complex was very slow, thereby confirming a key prediction of the ternary scavenging model (6). To obtain an accurate estimate of the breakdown rate, additional measurements were made at low temperature. In Fig. 12, the percentage of polymerase remaining in the ternary complex on agarose gels is plotted as a function of dissociation time; the ordinate is a logarithmic scale. The decay is first-order at both 4° and 0°; the half-lives of decomposition are 9 and 15 min, respectively. Using the Arrhenius equation, the activation energy calculated from these and other data at higher temperatures is approximately 18 ± 5 kcal/mole. The dissociation rate of the inhibitory ternary complex is of the same order as the dissociation rate of the lac operator-repressor complex

31

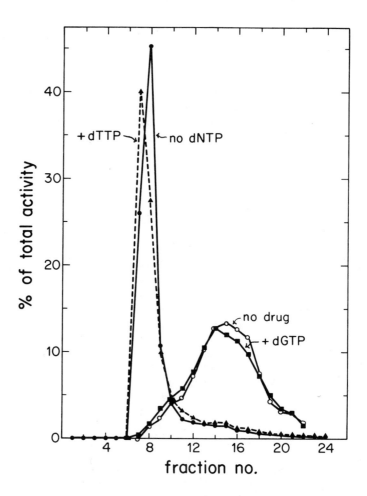

Fig. 9.   The isolation of the ternary complex by agarose chromatography and its specific antagonism by dGTP.   The procedures used in the experiments reported in Fig. 9–13 have been detailed (6).   The preincubation mixture (0.04 mM gapped DNA and 0.2 unit of enzyme) and 0.8 x 10 cm Sepharose-4B column buffers contained 10 μM OHPh(NH)$_2$Ura (closed symbols) or no drug (O) and 0.5 mM dGTP (■), 0.5 mM dTTP (▲), or no nucleotide (●,O).   The relative polymerase activity of the column fractions is plotted.

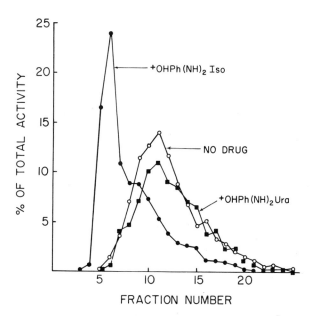

Fig. 10. The drug specificity of ternary complex formation with poly(d[A-T]). The preincubation mixture [0.24 mM poly(d[A-T])] and the 0.5 x 5 cm Sepharose-4B column buffer contained 0.1 mM OHPh(NH)$_2$Ura (■), 0.1 mM OHPh(NH)$_2$Iso (●), or no drug (O).

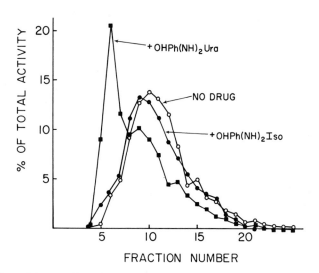

Fig. 11. The drug specificity of ternary complex formation with poly(dC):oligo(dG). The conditions and symbols are the same as in Fig. 10 except 0.24 mM poly(dC):(dG)$_{10}$ replaced the poly(d[A-T]).

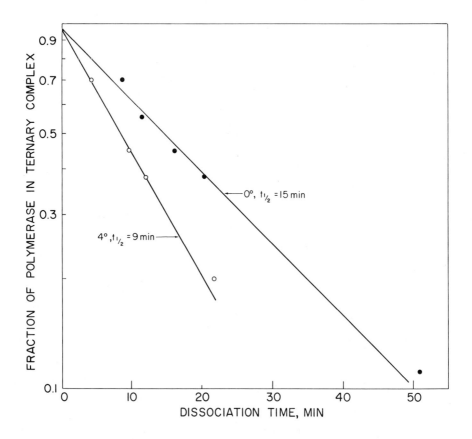

Fig. 12. The dissociation rate of the ternary complex. The preincubation mixture contained 0.04 mM gapped DNA, polymerase III, and 30 μM OHPh(NH)₂Ura. The agarose column buffers contained no drug but 0.5 mM dGTP to minimize reassociation of the enzyme. A series of 0.8 x 10 cm columns were run at various flow rates at 0° and 4° and the percent of the enzyme in the void volume is plotted vs. the time for void volume elution. The 4° curve is a plot of published data (6).

measured under similar conditions (19).

Third, the DNA polymerase III purified from a drug-resistant mutant, F22, has a K_i for $OHPh(NH)_2Ura$ of 20 μM which is 40 times higher than that of the wild-type enzyme (5). The increase in K_i could result from a decreased rate of formation of the complex or an augmented dissociation rate. Attempts to measure the dissociation rate using the agarose column procedure at 0° were unsuccessful, because less than 10% of the enzyme was recovered in the ternary complex even with 1 mM $OHPh(NH)_2Ura$ in the column buffer. Since this high concentration of drug greatly inhibits the mutant enzyme, the mutant enzyme ternary complex seems markedly unstable. This conclusion was confirmed by the kinetic experiment depicted in Fig. 13. In the wild-type enzyme control, the reversal by dGTP followed a lag of about 50 sec at 10° which is the result of the slow break-down of the ternary complex (18). In contrast, dGTP immediately reverses the inhibition of the mutant enzyme. In five separate experiments, the lag averaged minus 1 sec. Therefore, the ternary complex with the mutant enzyme has a much higher dissociation rate than the one with the wild-type enzyme. While inhibition of the mutant enzyme requires much higher $OHPh(NH)_2Ura$ levels, the drug still binds selectively to template cytosine since inhibition is alleviated specifically by dGTP (5). Also, the deoxyribo-nucleoside triphosphate binding site appears intact since the K_m for dGTP is the same for mutant and wild-type enzymes (5). Therefore, we have speculated that the mutant enzyme has lost an interaction with the drug which is not shared with dGTP, namely, a hydrophobic interaction with the phenyl ring of the drug (5,16). The data in Fig. 13 indicate that the consequence of this loss is a marked increase in the dissociation rate of the ternary complex. This single mutational change makes B. subtilis DNA poly-merase III more like other polymerases which are resistant to the drugs.

ACKNOWLEDGMENTS

This work was supported by grants from the National Institutes of Health (GM 21397 and CA 14599). We thank James Copeland and David Dubnau for valuable advice on genetic mapping.

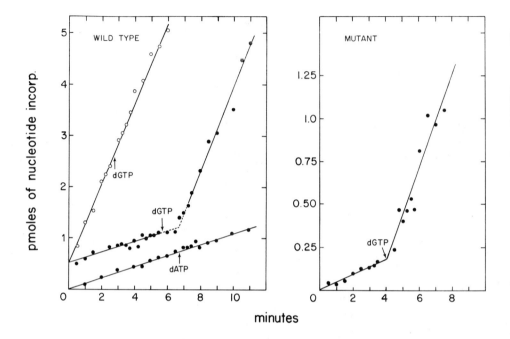

Fig. 13. The time lag in reversal of OHPh(NH)$_2$Ura inhibition by dGTP at 10° for mutant (strain F22) and wild-type polymerase III. The standard reaction mixture contained 0.3 mM gapped DNA and either 5 μM dGTP, 30 μM OHPh(NH)$_2$Ura, and the wild-type enzyme, or 1 μM dGTP, 200 μM OHPh(NH)$_2$Ura, and F22 enzyme. At the arrows the indicated nucleotides were added to 230 μM for the wild-type enzyme and to 1.25 mM for the mutant enzyme. The best straight lines were determined by a linear regression analysis of the data. Further details on the experiment with the wild-type enzyme have been presented (6).

REFERENCES

1. Gass, K.B. and Cozzarelli, N.R., J. Biol. Chem. 248, 7688 (1973).
2. Ganesan, A.T., Yehle, C.O. and Yu, C.C., Biochem. Biophys. Res. Commun. 50, 155 (1973).
3. Clements, J.E., D'Ambrosio, J. and Brown, N.C., J. Biol. Chem. 250, 522 (1975).
4. Kornberg, A., DNA Synthesis, W.H. Freeman and Co., San Francisco (1974).
5. Cozzarelli, N.R. and Low, R.L., Biochem. Biophys. Res. Commun. 51, 151 (1973).
6. Low, R.L., Rashbaum, S.A. and Cozzarelli, N.R., Proc. Nat. Acad. Sci. (USA) 71, 2973 (1974).
7. Bazill, G.W. and Gross, J.D., Nature New Biol. 243, 241 (1973).
8. Karamata, D. and Gross, J.D., Molec. Gen. Genetics 108, 277 (1970).
9. Siegel, L.M. and Monty, K.J., Biochim. Biophys. Acta 112, 346 (1966).
10. Scheraga, H.A., Protein Structure, Academic Press, New York and London (1961).
11. Chang, L.M.S. and Bollum, F.J., J. Biol. Chem. 248, 3398 (1973).
12. Livingston, D.M. and Richardson, C.C., J. Biol. Chem. 250, 470 (1975).
13. Brutlag, D. and Kornberg, A., J. Biol. Chem. 247, 241 (1972).
14. Muzyczka, N., Poland, R.L. and Bessman, M.J., J. Biol. Chem. 247, 7116 (1972).
15. Coulter, C.L. and Cozzarelli, N.R., Acta Crystallogr. B30, 2176 (1974).
16. Coulter, C.L. and Cozzarelli, N.R., J. Mol. Biol. 91, 329 (1975).
17. MacKenzie, J.M., Neville, M.M., Wright, G.E. and Brown, N.C., Proc. Nat. Acad. Sci. (USA) 70, 512 (1973).
18. Gass, K.B., Low, R.L. and Cozzarelli, N.R., Proc. Nat. Acad. Sci. (USA) 70, 103 (1973).
19. Riggs, A.D., Bourgeois, S. and Cohn, M., J. Mol. Biol. 53, 401 (1970).

IN VITRO DNA SYNTHESIS ON PRIMED COVALENTLY CLOSED DOUBLE-STRANDED TEMPLATES. I. STUDIES WITH ESCHERICHIA COLI DNA POLYMERASE I.

Leroy F. Liu and James C. Wang
Department of Chemistry
University of California
Berkeley, California 94720

ABSTRACT. In order to study DNA synthesis on intact double-stranded template molecules, methods have been developed to prepare negative superhelical DNA samples with short DNA or RNA fragments paired to the strands of the covalently closed DNA. With Escherichia coli DNA polymerase I, provided that the primer fragments possess 3' hydroxyl termini, extension of the primer fragments can occur, yielding covalently closed circular molecules with enlarged loops or "eyes."

Several other competing processes have also been observed. Template switching by the enzyme may occur; extension of the primer fragment and copying of the strand the primer is annealed to may be followed by the copying of the other strand. It is also found that endonucleolytic cleavage of the strands of the covalently closed DNA may occur. Several experiments indicate that the endonucleolytic activity is an activity of the polymerase itself. Efficient endonucleolytic scission is observed only when the primer fragments have 3'-hydroxyl end groups and that DNA synthesis is occurring. Antibodies specific for polymerase I also block the endonucleolytic activity. Furthermore, the requirements for the endonucleolytic activity are similar to the requirements for polymerizing activity.

It appears that loop enlargement is facilitated by the negative superhelical turns. This is interpreted as due to the free energy of

of superhelix formation, which favors the removal
of superhelical turns. Since loop enlargement
requires concomitant unwinding of the circular
strands, it is driven by the favorable free energy accompanying the simultaneous reduction in
the number of superhelical turns. This driving
force diminishes with increasing loop size however. Thus after synthesis has proceeded for a
short time, processes other than loop enlargement
predominate, yielding nicked circular and linear
molecules, some with arms. Synthesis in the
late phase is primarily nick translation on
the nicked templates.

INTRODUCTION

In recent years, extensive studies on the
synthesis of DNA in vitro have been carried out.
The multifaucet approaches and the rapid progress made in this field have been summarized in
a number of recently published books and reviews
(1-5). There have been relatively few studies
however, on in vitro DNA synthesis with intact
double-stranded DNA as the template. While synthesis by purified DNA polymerases can occur on
double-stranded templates containing single-
chain scissions (nicks) or single-stranded regions (gaps), none of the known DNA polymerases
can initiate synthesis on intact double-stranded
templates.

It should be possible to circumvent the difficulty of initiation on an intact duplex by
the introduction of a "primer" into the duplex,
i.e., by the pairing of a short polynucleotide
fragment to one of the strands of the double
helix. If the primer has a 3'-hydroxyl end group,
one would expect that extension of this primer
fragment could be carried out by known polymerases in the presence of deoxynucleotide triphosphates.

Recently Tomizawa et al. have shown that
the covalently closed DNA of colicin El can be
replicated in vitro with colicin El infected
cell lysate (6-8). In this system, initiation

of DNA synthesis is preceeded by the synthesis of an RNA primer hydrogen-bonded to one of the two DNA strands. The replication of intact T7 DNA has also been achieved (9), though the process of initiation is not known. Champoux and McConaughy have reported the synthesis of DNA on covalently closed simian virus 40 DNA templates primed by successive treatment of the DNA with RNA polymerase and pancreatic RNase (10). We will discuss their results in a later section.

In this report, we describe our studies on DNA synthesis by Escherichion coli DNA polymerase I on primed double-stranded templates. Methods have been developed for the preparation of covalently closed DNA duplex molecules containing either DNA or RNA primer fragments with 3'-hydroxyl termini. As evidenced by electron microscopy and sedimentation analysis, substantial net synthesis can occur on such primed templates. Copying of approximately one-quarter of the total length of the parental DNA can be achieved at the present time. We have observed that DNA synthesis by E. coli polymerase I on such templates also leads to endonucleolytic cleavage of the originally intact circular strands. A variety of control experiments, including the use of anti-polymerase I antibodies, all indicate that this endonucleolytic activity is an intrinsic activity of E. coli polymerase I.

MATERIALS AND METHODS

Materials: Unlabelled nucleoside triphosphates were purchased from P-L Biochemicals. ^{14}C-labelled ATP and ^{3}H-labelled dATP were purchased from Schwarz. E. coli DNA polymerase I, with a specific activity of 18,000 units/mg, and purified E. coli polymerase I specific antibodies, were generous gifts of Prof. A. Kornberg. E. coli RNA polymerase was prepared according to the Burgess procedure (11). E. coli alkaline phosphatase, pancreatic DNase I and RNase A were purchased from Worthington. Hpa II restriction enzyme was a gift of Dr. J. Mertz. Single-

stranded phage fd DNA was a gift of Dr. V. McKay.
Phage PM2 DNA samples of varying degrees of
superhelicity were prepared according to previously published procedures (12). Phage PM2 DNA
fragments were obtained by controlled digestion
of the DNA with pancreatic DNase I. The average
single-stranded molecular weight of the fragments was $\sim 10^5$.

Preparation of DNA-primed superhelical PM2 DNA:
The annealing mixture contained ~ 6 M CsCl, 0.05
M potassium phosphate, 80 µg/ml of a superhelical
DNA, and 80 µg/ml of PM2 DNA fragments, which had
been heat-denatured immediately before use.
The pH of the annealing medium, as read from a
Beckman pH meter with an amber glass combination
electrode, was between 11.2 and 11.3. This pH
range was selected based on previous data (13).
After 10 hrs. at 20°C, the reaction was stopped
by neutralizing with tris·HCl. To remove unbound
fragments, the DNA was dialyzed against 0.1 M
tris, pH8, 0.01 M Na$_3$ EDTA, and then layered on
top of linear 1-4 M NaCl gradients (containing
0.01 M Na$_3$ EDTA) in nitrocellulose tubes for the
SW 25.1 rotor (Spinco). After centrifuging for
16 hrs. at 22 Krpm and 0°C, fractions were collected from the gradients. Fractions containing
superhelical PM2 DNA were pooled, concentrated
by dialysis against dry sucrose, and then dia--
lyzed against an appropriate medium.

Preparation of RNA-primed superhelical PM2 DNA:
The reaction mixture (2.2 ml) contained 35 mM
tris, pH 7.8, 50 mM KCl, 10 mM MgCl$_2$, 0.1 mM
EDTA, 10 mM 2-mercaptoethanol, 50 µg/ml of bovine
plasma albumin, 95 µg/ml of PM2 DNA, 35 µg/ml
of E. coli RNA polymerase, 170 µM each of GTP
and UTP, 3 µM of CTP, and 60 µM of ^{14}C-labelled
ATP (10 mCi/mmole). The CTP concentration was
low in order to limit the extent of RNA synthesis
to several percent of the total DNA nucleotides.
After 10 min. at 37°C, the reaction was stopped
by adding sodium dedecyl sulfate to a final concentration of 0.5 %, and the mixture was

41

extracted with buffer saturated, distilled (under
a nitrogen atmosphere) phenol.

Preparation of superhelical DNA with bound RNA
fragments with 3'-phosphoryl termini (the
"Hayashi Complex"): This complex was prepared
by extensive RNA synthesis on a superhelical PM2
DNA template, followed by digestion of the product with RNase A after the removal of the RNA
polymerase by sodium dodecyl sulfate treatment
and phenol extraction (14). Detailed conditions
were as previously described (15).

DNA synthesis on primed superhelical DNAs: The
reaction mixture contained 0.045 $\underline{M}$ tris, pH7.8,
0.09 $\underline{M}$ KCl, 6.5 mM MgCl2, 5 mM potassium phosphate, 0.1 mM Na₃EDTA, 25 µM each of the four
deoxyribonucleoside triphosphates, 50 µg/ml of
bovine plasma albumin, and 2.5% glycerol. The
amounts of polymerase I and DNA, and the incubation temperature and time are given in the appropriate legends.

Other methods: Electron microscopy was done according to the procedures of Davis et al. (16).
Fluorescence measurements were done as described
by Morgan and Pulleyblank (17).

RESULTS

Preparation of DNA primed covalently closed
double-stranded DNA: It is well-established that
when a negative superhelical DNA is titrated
with alkali, disruption of base-pairing occurs
at a pH at which a double-stranded DNA devoid of
superhelical turns is still thermodynamically
stable (18,13). Therefore if a negative superhelical DNA and single-stranded fragments of the
same DNA are annealed at this pH, pairing between
the fragments and the strands of the closed circular DNA can occur. After neutralization the
primed covalently closed DNA can be separated
from excess fragments by zone sedimentation.
Detailed procedures have been described in the

42

in the Materials and Methods section.

In the present work, the fragments were obtained from pancreatic DNase digested phage PM2 DNA. Thus the fragments are terminated with 3'hydroxyl groups (19). Since fragments with complementary sequences were present at equal concentrations during annealing with the superhelical DNA, the product contained loops or "eyes", with the primer fragments bound to both sides of the loops. This is shown diagrammatically in Figure 1a.

Preparation of RNA primed covalently closed double-stranded DNA: In an aqueous medium, RNA-DNA duplex is generally less stable than DNA-DNA duplex (20). Thus an RNA fragment bound to one of the two strands of an ordinary DNA is expected to be unstable. For a negative superhelical DNA however, binding of an RNA (or DNA) fragment to the DNA causes a concomitant unwinding of the strands of the covalently closed circular DNA, which is favored by the free energy of superhelix formation (21,22). It is this favorable free energy term which stabilizes the RNA-DNA pairing for the binding of an RNA fragment to a negative superhelical DNA (15).

It was first observed by Hayashi (14), and later confirmed by others (15,10), that RNA synthesis on negative superhelical DNAs results in RNA associated with the DNA templates. The RNA associates with the DNA in a pancreatic RNase A resistant form, in agreement with the expectation that the RNA is paired with one of the DNA strands.

To prepare RNA primed covalently closed DNA duplex then, one can simply carry out a limited amount of RNA synthesis on a negative superhelical DNA with RNA polymerase, which is followed by extraction with sodium dodecyl sulfate and phenol to remove the enzyme. The substrate so obtained contains RNA primer fragments with the desired 3'-hydroxyl termini. Exposure of the substrate to pancreatic RNase A should be avoided, since the RNase generated 3'-phosphoryl rather

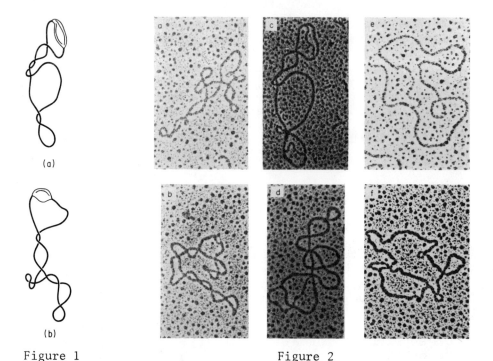

(a)

(b)

Figure 1   Figure 2

Figure 1. Diagrammetic drawings illustrating
superhelical DNA molecules primed by DNA fragments (a) and RNA fragments (b).

Figure 2. Selected electron micrographs of
RNA primed superhelical PM2 DNA molecules after
incubation with DNA polymerase at 37°C for 0 min.
(a and b), 2 min. (c), 3 min. (d) and 4 min. (e
and f). Prior to incubation with DNA polymerase,
a large fraction of the primed superhelical DNA
molecules have loops too small to be visible, as
shown in (a). That primer fragments are associated with such molecules can best be shown after
incubation with the DNA polymerase. The molecule shown in (c) has 2 loops. The molecule shown
in (d) has a loop with one single-stranded side
and one double-stranded side. The formamide layering procedure was used in obtaining all the
micrographs except (f), for which the aqueous
layering procedure was used. DNA synthesis was
carried out as described in Materials and Methods,
with 25 µg/ml of RNA primed PM2 DNA (v_c^o = 0.12)
and 3 µg/ml of DNA polymerase.

than hydroxyl end groups (23). This point will be discussed later.

Unlike the case with DNA-primed substrate, the RNA-primed substrate is expected to contain loops ("eyes") with the RNA fragment bound to only one side of each loop, as illustrated diagrammatically in Figure 1b.

DNA synthesis on primed templates: That net DNA synthesis by E. coli polymerase I can occur on either DNA or RNA fragment primed covalently closed DNA templates is best illustrated by the electron micrographs in Figure 2.

The covalently closed PM2 DNA sample used in this series of measurements originally had a v_C^o value of 0.12,* and was primed by RNA synthesis as described above in the Materials and Methods section.

The fraction of superhelical DNA molecules which had been primed, i.e., have at least one primer fragment associated per DNA molecule, can be best determined by measuring the fraction of molecules converted to the nicked forms after incubation with DNA polymerase I under DNA synthesis conditions. This unique endonucleolytic activity of polymerase I, to be described in detail later, is active only on primed molecules. Typically, our DNA or RNA primed samples contain over 60% primed superhelical molecules.

---

*The quantity v_C^o is defined as the number of bound ethidium per nucleotide needed to remove all the negative superhelical turns under "standard" conditions, commonly taken as neutral 3 M CsCl at 20°C. It is a measure of the degree of superhelicity of the DNA. If ϕ_E is the unwinding angle of the DNA helix per bound ethidium, the superhelical density σ_o of the DNA is related to v_C^o by $\sigma_o = -\phi_E v_C^o/18$. Taking the recent result that $\phi_E = 26°$ (13), a v_C^o value of 0.12 corresponds to a σ_o of -0.17 or 0.17 negative superhelical turns per ten base pairs.

Prior to treatment with DNA polymerase, the primed DNA molecules have a twisted appearance, and rather small "eyes," usually one or two per DNA molecule, can be seen in some of the molecules (Figure 2b). After incubation with DNA polymerase I and the deoxynucleotide triphosphates, enlargement of the eyes is evident, especially in the early phase of the incubation (Figure 2, c-f). The molecules with large loops have invariably a less twisted appearance, indicating that loop enlargement is coupled to the removal of negative superhelical turns.

Neglecting the effect of mass increment due to DNA synthesis, the decrease in the degree of superhelicity accompanying loop enlargement is expected to decrease the sedimentation coefficient s of a DNA of this degree of superhelicity (15). Sedimentation measurements of samples taken out at different incubation times indicate that this is the case. In neutral 3 $\underline{M}$ CsCl at 20°C, the primed PM2 DNA sample sediments with an s of 23 S (Figure 3a) After a brief incubation (1 min.) with DNA polymerase and the triphosphates a slow sedimenting shoulder appears (Figure 3b). The sedimentation coefficient of this slower sedimenting species decreases as the incubation time increases (Figure 3c and 3d). After incubating for 4 min., this species sediments at 18 S. Therafter no further decrease in its sedimentation coefficient occurs.

Measurement of the average loop size by microscopy clearly indicate that loop enlargement stops after a short period of incubation (less than 7 min. for the samples described above). It appears that the limit in loop enlargement is closely related to the degree of superhelicity of the DNA. Measurements were done for two PM2 DNA samples, with ν_c^o values of 0.075 and 0.149 respectively prior to priming. The corresponding values of the negative superhelical density of these samples, taking the unwinding angle of eithidium as 26°, are 0.108 and 0.215 respectively. After priming (with DNA fragments) and DNA synthesis (30 min. at 20°C) it was observed

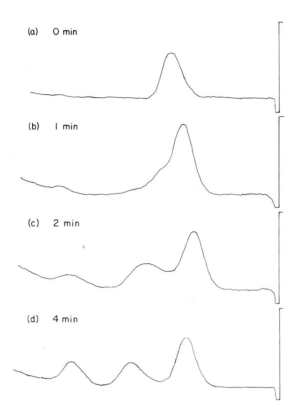

Figure 3. Band sedimentation patterns of RNA primed PM2 DNA samples after incubation with DNA polymerase I at 37°C. Incubation time is indicated in each tracing. Incubation conditions were identical to those described in the legend to Figure 2. Sedimentation was done in neutral 3 $\underline{M}$ CsCl at 20°C and 31, 410 rpm. The vertical line on the right-hand side of each tracing corresponds to the bottom of the centrifuge cell.

that the average loop sizes for the samples corresponded to the unpairing of $9 \pm 2\%$ and $20 \pm 2\%$
respectively of the strands of the covalently
closed DNA samples. Samples after incubating at
37°C for 30 min. gave essentially identical
results, though at this higher incubation temperature only a small percentage of the molecules
had loops after 30 min. The average sizes of
the loops for the two samples indicate that
cessation of loop enlargement for each sample
occurs not far from the point at which the number of superhelical turns of the DNA is zero.

Template switching and secondary initiation:
For the DNA primed substrate, extension of the
DNA fragments in a loop in the 5' to 3' direction by the polymerase is expected to give an
enlarged loop with single-stranded regions located diagonally on each side of the loop. For
the RNA primed substrate, extension of a primer
fragment is expected to give a loop with one double-stranded side and one single-stranded side.

An RNA primed PM2 DNA, after DNA synthesis
under standard conditions, was prepared for
visualization in the electron microscope by both
the formamide layering procedure and the aqueous
layering procedure. Micrographs obtained for
the sample prepared by the formamide procedure
show that while loops with one double-stranded
side and one single-stranded side can be seen,
loops with both sides double-stranded are definitely present. The existence of loops with both
sides double-stranded is further supported by
micrographs obtained for the sample prepared by
the aqueous procedure. Since collapsing of
single-stranded regions is expected when the
aqueous procedure is used, loops with both sides
double-stranded can easily be identified by this
procedure.

Similar attempts were also made to examine
a sample primed by DNA fragments. It is intrin-
sically more difficult however, to determine by
microscopy whether a part of a short DNA segment
is single-stranded. We are therefore uncertain

whether a significant fraction of the looped molecules contain partially single-stranded loops. It can nevertheless be concluded, by comparing the sizes of loops of samples prepared by the foramide procedure and the aqueous procedure, that the loops of some of the molecules are essentially all double-stranded.

There are two plausible interpretations for these results. Firstly, short oligonucleotides, either present due to incomplete removal of the excess fragments used in the preparation of the primed substrate, or formed by the DNA polymerase during incubation of the primed substrate with the enzyme, might cause secondary initiation of DNA synthesis on the single-stranded regions, converting these regions to the double-stranded form. Secondly, template switching might occur at the forks, yielding completely double-stranded loops for both the DNA and RNA fragments primed substrates. (For a review on these reactions, cf. Ref. 1.)

An experiment was done to test whether template switching was occurring on the RNA primed substrate used in the sedimentation (Figure 3) and microscopy (Figure 2) studies. DNA synthesis (in the presence of ^{3}H labelled dATP) was stopped at 4 min. and 7 min. respectively for two samples. After phenol extraction, the samples were dialyzed first against 2 $\underline{M}$ NaCl, 0.01 $\underline{M}$ Na_3EDTA, and then against 0.03 $\underline{M}$ Na Acetate, pH 4.5, 0.05 $\underline{M}$ KCl. The samples were first denatured by heating at 100°C for 5 min., followed by quick cooling in an ice bath.

Stock solutions of $ZnCl_2$ and a single-strand specific endonuclease S1 from Aspergillus orizae were then added rapidly in succession, to give a final Zn(II) concentration of 1.65 m$\underline{M}$, and an appropriate amount of the nuclease in each sample. The samples were then incubated at 37°C and assayed for acid precipitable counts at different time intervals. For DNA synthesized in the first four minutes on the RNA primed substrate, the percent of acid precipitable counts after 5, 10 and 20 min. incubation with the S1 nuclease are 12%, 10% and 8% respectively. The corresponding values for DNA synthesized in the first 7 min. are 20%, 17% and 15% respectively.

For DNA synthesized on nicked PM2 DNA template (7 min. at 37%C), the percent of acid precipitable counts after treatment with S1 nuclease in an identical manner are 1.2%, 0.8% and 0% respectively after 5, 10 and 20 min. incubation. Similar values were obtained with ^{3}H labelled phage λ DNA.

These results show that a significant fraction of the DNA synthesized on RNA primed superhelical PM2 DNA is rapidly renaturable, indicating that template switching is occurring on such a substrate. A quantitative estimation of the fraction of growing chains involving template switching is different however, since even in the early phase of DNA synthesis nicking of the covalently closed strands is occurring (to be discussed later), and therefore some of the DNA synthesized in 4 as well as 7 min. is probably by nick translation, which gives little rapidly renaturable material under our synthesis conditions.

Endonucleolytic clevage of the template strands:
So far we have focused our discussion on loop enlargement. As mentioned earlier, loop enlargement stops after a short period of incubation of the substrate with the enzyme. DNA synthesis however, continues after the cessation of loop enlargement. This is shown in Figure 4. We shall now discuss the nature of the products of the later phase of synthesis.

Figure 5 depeicts the sedimentation patterns of samples from the same series of experiments depicted in Figure 3. With increasing incubation time, the amount of the 18 S species decreases. This is accompanied by an increase in the amounts of a number of slower sedimenting species, with sedimentation coefficients around 13 S, 12 S, and 10 S respectively. Note that the 10 S species appears to form only during the late phase of incubation.

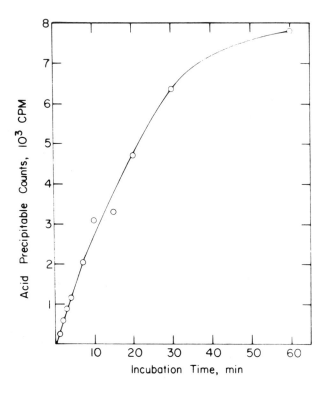

Figure 4. Total amount of DNA synthesized on RNA primed PM2 DNA as a function of incubation time. Incubation conditions were identical to those described in the legend to Figure 2. ^{3}H-dATP (200 mCi/mmole) was used, and each assay was done with aliquot containing 0.4 µg of PM2 DNA.

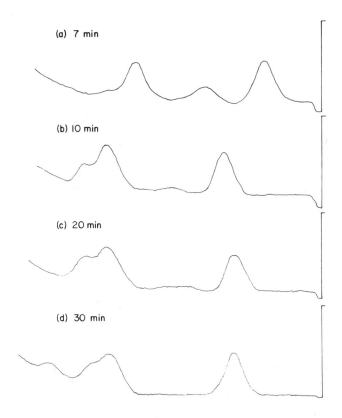

(a) 7 min

(b) 10 min

(c) 20 min

(d) 30 min

Figure 5. Band sedimentations of RNA primed PM2 DNA samples after longer incubation with DNA polymerase I at 37°C. See legend to Figure 3.

Figure 6 depicts the dependence of the relative amounts of the various species on incubation time. The amount of the 23 S species, which consists of superhelical PM2 DNA molecules without primer fragments and PM2 DNA molecules with very small loops, decreases from 100% to a limiting value of 30% in about 10 min. (Curve A). This limiting value is interpreted as resulting from the presence of 30% of unprimed DNA molecules in the sample used in this series of measurements. The amount of covalently closed PM2 molecules with enlarged loops, which sediment as a band with sedimentation coefficients greater than that of the nicked circular DNA but smaller than that of the original superhelical DNA, increases rapidly in the beginning, and reaches a maximum of ∿37% in ∿2 min. Thereafter it starts to decrease (Curve B). The time dependence of the total amount of the slower sedimenting materials with sedimentation coefficients in the range of 10 to 13 S is also plotted in Figure 6 (Curve C). The shape of Curve C is sigmoidal: In the region Curve B is showing a decrease, a steep rise in Curve C is noted.

Since nicked PM2 DNA has an s of 13 S in neutral $3\underline{M}$ CsCl at 20°C, the slower sedimenting species described above are most likely various forms of nicked PM2 DNA. This is in agreement with results obtained by electron microscopy. Micrographs of samples taken at longer incubation times clearly show that endonucleolytic cleavage of the template strands had occurred. Few molecules with loops can be seen. Aside from highly twisted molecules which are presumably unprimed covalently closed circular DNA molecules as discussed above, the predominating species are nicked circles, nicked circles with short arms, linear molecules of approximately full PM2 DNA length, linear molecules of approximately half PM2 DNA length, and shorter linear fragments. Some of the linear molecules also have short arms. Figure 7 depicts several examples of the various species.

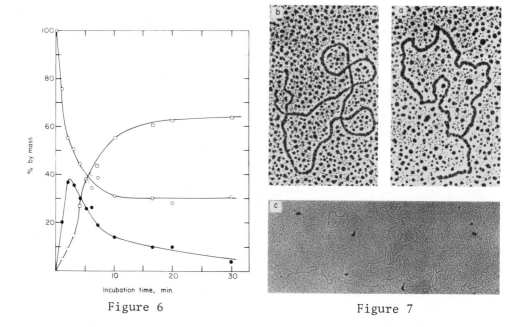

Figure 6   Figure 7

Figure 6. Dependence of the relative amounts of the various species resolvable by band sedimentation (see Figures 3 and 5) as a function of incubation time. Curve A, open circles: the species sedimenting at 23 S. Curve B, filled circles: the species sedimenting between 13 S and 23 S. Curve C, open squares: sum of all species sedimenting with sedimentation coefficients equal or lower than 13 S. Quantitation was done by integrating the area under each band. No correction was made to account for radial dilution.

Figure 7. Electron micrographs showing some of the species after incubation of an RNA-primed superhelical PM2 DNA (which had a v_C^o of 0.12 prior to priming) with DNA polymerase I at 37°C for 7 min. Incubation conditions were identical to those described in the legend to Figure 2. In micrograph c three linear molecules can be seen, two of which show short arms.

Comparing the sedimentation patterns with the microscopy results, we believe that the 13 S species is nicked circles, the 12 S species is linear molecules of full PM2 length, and the 10 S species is linear molecules of approximately half PM2 length. For either the nicked circular or the linear form, molecules with or without short arms are not expected to resolve in the sedimentation measurements.

The endonucleolytic scission is due to E. coli polymerase I itself: In order to decide whether the observed endonucleolytic scission is due to the polymerase itself or is due to a contaminating activity, we have performed a series of experiments. The results indicate strongly that the endonucleolytic scissions are introduced by the polymerase itself.

First, efficient nicking was observed only with primed DNA in the presence of all four triphosphates. No nicking was detectable when unprimed PM2 DNA was incubated with the enzyme and all four triphosphates. With single stranded circular fd DNA as the substrate, no nicking was observed under our standard DNA synthesis conditions. If the polymerase concentration was increased 18 fold however, after incubating for 30 min. at 37°C ∿50% of the fd DNA was converted to the linear form. The nicking of fd DNA was not effected by the addition of triphosphates. In the presence of all four triphosphates though, the single-stranded fd DNA was converted to a mixture of double-stranded linear and circular molecules, presumably due to the presence of some DNA fragments to allow chain initiation. Since the endonucleolytic activity on fd DNA is two orders of magnitude lower than the endonucleolytic activity observed with primed double-stranded PM2 DNA under DNA synthesis conditions, we have not pursued the fd studies further.

Secondly, endonucleolytic cleavage of primed PM2 DNA under DNA synthesis conditions is blocked by the addition of antibodies specific for E. coli polymerase I.

Thirdly, efficient endonucleolytic scission of primed closed circles, as well as DNA synthesis on such templates, requires 3'-hydroxyl groups at the termini of the primer fragments. An RNA primed PM2 DNA sample was prepared by extensive RNA synthesis off the DNA by RNA polymerase, followed by RNase treatment. The RNase treatment is expected to give DNA-bound DNA fragments with 3'-phosphoryl termini. When this sample was used as the substrate for E. coli polymerase I, under standard incubation conditions less than 10% of the closed DNA circles were nicked, as shown by the sedimentation patterns depicted in Figure 8 (a and b). The same sample, when first treated with E. coli alkaline phosphatase to remove the terminal phosphate groups, and then incubated with polymerase I under standard conditions, gave 70% nicked species (Figure 8, c and d).

Finally, the endonucleolytic activty has requirements parallel to the requirements of the polymerizing activity. In addition to the requirements of 3'-hydroxyl groups on primer fragments mentioned above, efficient endonucleolytic scission requires all four deoxyribonucleotide triphosphates and Mg(II) ions, with an optimal Mg(II) concentration of 8.5 mM in 45 mM tris buffer, pH 7.8, 90 mM KCl. Similar to the polymerizing activity of polymerase I, the endonucleolytic activity is not inhibited by 0.1 M KCl or 15 mM N-ethyl-maleimide, but is inhibited by 10 mM pyrophosphate. These results are tabulated in Table 1.

While endonucleolytic scission of the primed template molecules is much stimulated by DNA synthesis, DNA synthesis is not absolutely required for the endonucleolytic activity. When the polymerase concentration was increased by 18 fold even in the absence of the triphosphates 60% of a DNA primed superhelical DNA sample was nicked after 30 min. at 37°C.

Removal of the primer fragments by polymerase I: To test the removal of the primer fragments by

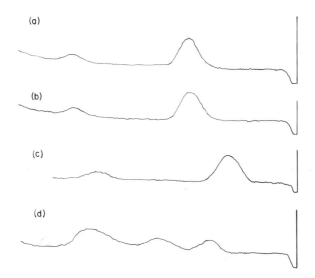

(a)

(b)

(c)

(d)

Figure 8. Experiments showing the requirement
of the 3'-hydroxyl termini of the primer fragments for the endonucleolytic activity of E. coli
polymerase I. The PM2 DNA sample (ν_C^0=0.12 prior
to priming), had been treated successively
with RNA polymerase and RNase A to give RNA
primer fragments with 3'-phosphoryl termini.
Samples of this DNA (40 µg/ml), in 35 mM tris,
pH 7.8, 10 mM MgCl$_2$, were treated as follows:
(a) Control, incubated first at 37°C for 30 min.,
then incubated for 30 min. at 20°C after the addition of 4 deoxyribonucleoside triphosphates
(25 µM each). (b) Sample first incubated at
37°C for 30 min., then incubated for 30 min. at
20°C after the addition of E. coli polymerase I
(3 µg/ml) and 4 triphosphates. (c) Sample first
incubated at 37°C for 30 min. with E. coli
alkaline phosphatase (which had been heated at
90°C for 20 min. to inactivate contaminating
nucleases, final concentration of the enzyme was
∿0.5 units/ml), then incubated at 20°C for 30
min. after the addition of 4 triphosphates. (d)
same as (c) except that the incubation at 20°C
was done in the presence of E. coli polymerase I
(3 µg/ml) and the 4 triphosphates. The band sedimentation patterns of samples (a)-(d) are shown
above.

TABLE 1

RELATIVE AMOUNT OF NICKED DNA AFTER
INCUBATION WITH E. COLI POLYMERASE I

	Relative Amount of Nicked DNA [a]
Complete system[b]	100
- polymerase I	0
- all 4 triphosphates	6
- dATP - dCTP - dTTP	3
- dATP - dGTP - dTTP	5
- dGTP - dCTP - dTTP[c]	9
- dATP - dGTP - dCTP[c]	16
- dATP	11
- Mg^{++}	-4
- KCl	81
+ 15 mM N-ethyl maleimide	98
+ antibodies against polymerase I	1
+ 10 mM PP$_i$[c]	11

[a]Measured by the ethidium fluorescence method as described in Ref. 17. The error in these measurements is $\sim\pm5$.

[b]The medium is described in the Materials and Methods section, with 3 μg/ml of the polymerase. Incubation was carried out at 37°C for 3 min. The actual percent of conversion of the primed superhelical DNA to the nicked forms was $\sim70\%$. Unless noted otherwise, the DNA used was RNA primed PM2 DNA ($v_c^o = 0.075$) at a final concentration of 70 μg/ml.

[c]The DNA used in these measurements was RNA primed PM2 DNA ($v_c^o = 0.12$ prior to priming) at a final concentration of 25 μg/ml.

the known exonucleolytic activities of E. coli
polymerase I, and PM2 DNA sample ($\nu_c^o = 0.075$)
primed with ^{14}C-labelled RNA was incubated with
the enzyme at 37°C in the presence and in the
absence of all four deoxyribonucleotide triphosphates. The DNA and polymerase concentrations in the incubation mixtures were 70 µg/ml
and 3 µg/ml respectively. After 30 min., the
solutions were chilled, and measured amounts of
a CsCl stock solution and ^{3}H-labelled λ DNA (as
a marker) were added to each solution. The
solutions were banded in a Model L ultracentrifuge at 35 Krpm and 10°C for 48 hrs., and drops
were then collected from each gradient and counted in a scintillation counter. Comparing with
the control which was not treated with DNA polymerase, 70% of the RNA counts banding at the
DNA position had been removed either in the presence or in the absence of the triphosphates.
It is not suprising that the removal of the fragments is little affected by the triphosphates,
since it is likely that the 5' to 3' exonuclease
is reaponsible for the removal of the primer
fragments. While this nuclease is known to be
stimulated by DNA synthesis (cf. Ref. 1), for
our substrate there is probably little synthesis
behind the 5' termini of the primer fragments.
    We have no quantitative data on the removal
of DNA primer fragments by polymerase I. Qualitatively, for a PM2 DNA ($\nu_c^o = 0.075$) primed with
DNA fragments ∿800 nucleotides long, after incubation with polymerase I at 37°C for 30 min. in
the absence of the triphosphates, electron microscopy showed that the loops disappeared, most
likely due to the removal of the primer fragments,
and the DNA resumed a highly twisted appearence.
This earlier experiment was carried out in a
medium containing 30 mM tris, pH 7.8, 8 mM $MgCl_2$,
5 mM potassium phosphate rather than the medium
given in the Materials and Methods section.
Thus the two experiments on the removal of RNA
and DNA fragments are not directly comparable.

DISCUSSION

Our results show that with either DNA or RNA primed negative superhelical DNA molecules, provided that the primer fragments have 3'-hydroxyl termini, elongations of the primer fragments by E. coli polymerase I can occur, resulting in loop enlargement.

Loop enlargement however, ceases after a short time. It appears that the cessation of loop enlargement occurs near the point at which the DNA no longer contains superhelical turns. We therefore view the effect of the negative superhelical turns as providing a driving force favoring the unwinding of the parental strands (18,15, 21,22). Since elongation of the primer chain requires concomitant unwinding of the template strands, advancing of the fork is facilitated by any factor which favors the unwinding of the strands.

In the absence of a driving force favoring the unwinding of the template strands, for E. coli polymerase I two other processes become important: template switching and endonucleolytic nicking. Since template switching has been discussed by others (cf. Ref. 1), we shall only discuss endonucleolytic nicking.

It is well-established that E. coli polymerase I has two exonucleolytic activities. There has been few reports however, on endonucleolytic activity of E. coli polymerase I. During the progress of this work, Champoux and McConaughy (10) have reported their studies on DNA synthesis by E. coli polymerase I on RNA primed superhelical simian virus 40 DNA. By banding the products in CsCl-propidium gradient, they found that the amount of newly synthesized DNA associated with nicked SV40 DNA increased steadily with increasing incubation time, while with unprimed control SV40 DNA the amount leveled off in about 10 min. They inferred from this that nicks were generated by polymerase I. The RNA primed SV40 DNA they used however, had been extensively digested with RNase A. We have

observed little synthesis or endonucleolytic
nicking on such a template, because of the lack
of 3'-hydroxyl groups.

We are uncertain at this time on the mechanism of the endonucleolytic scission. Micrographs of the products suggest that endonucleolytic scissions occur near the forks on the molecules, since a high percentage of molecules with
arms were observed, and that the average length
of the arms was observed to be approximately the
same as the average length of one side of the
loops. It is tempting to associate the endonucleolytic activity observed to the known 5'
to 3' exonucleolytic activity. The latter is
known to give oligonucleotides as products and
is also stimulated by DNA synthesis. If the two
activities are indeed related, then one would
expect that endonucleolytic scission occurs at
the strand not paired to the primer fragment,
i.e., the strand cut is the one being displaced
by synthesis. By a combination of such endonucleolytic cutting, template switching, branch
migration and nick translation, we could interpret the formation of all the major products
(curcular and linear molecules with and without
arms, linear half molecules etc.). Since the
combination of these processes gives a large number of possibilities, firm conclusions are difficult, and we will omit such discussions here.

While it is easy to envision the usefulness
of an endonucleolytic activity in repair type
function, such an activity seems to be highly
undesirable for replication. At least for
colicin El, it appears that E. coli polymerase I
is used to carry out its replication (24). It
is plausible that the endonucleolytic activity
is detectable only in the absence of a factor,
normally present in vivo, which facilitates
strand separation. Alternatively, the endonucleolytic activity might be attenuated by complex formation with other factors. Further experimentation is needed to elucidate these points.

ACKNOWLEDGEMENT

We are indebted to Prof. A. Kornberg and his coworkers for providing us with highly purified E. coli polymerase I and antibodies specific for this enzyme. This work was supported by a grant (GM 14621) from the U.S. Public Health Services, and a fellowship (to LFL) from the Li Foundation.

REFERENCES

1. Kornberg, A., DNA Synthesis. Freeman, San Francisco, (1974).
2. Wells, R.D. and Inman, R.B., eds. DNA Synthesis in Vitro, University Park Press, Baltimore, (1973).
3. Goulian, M., Ann. Rev. Biochem. 40, 855 (1971); Prog. Nucl. Acid Res. and Molec. Biol. 12, 29 (1972).
4. Klein, A. and Bonhoeffer, F., Ann. Rev. Biochem. 41, 301 (1972).
5. Becker, A. and Hurwitz, J., Prog. Nucl. Acid Res. and Molec. Biol. 11, 423 (1971).
6. Sakakibara, Y. and Tomizawa, J.-I., Proc. Nat. Acad. Sci. U.S. 71, 802 (1974).
7. Sakakibara, Y. and Tomizawa, J.-I., Proc. Nat. Acad. Sci. U.S. 71, 1403 (1974).
8. Tomizawa, J.-I., Sakakibara, Y. and Kakefuda, T., Proc. Nat. Acad. Sci. U.S. 71, 2260 (1974).
9. Hinkle, D.C. and Richardson, C.C., J. Biol. Chem. 249, 2974 (1974).
10. Champoux, J. and McConaughy, B.L., Biochem. 14 307 (1975).
11. Burgess, R.R., J. Biol. Chem. 244, 6160 (1969).
12. Wang, J.C., in Procedures in Nucleic Acid Research (Cantoni, G.L. and Davies, D.R., eds.), Vol. 2., p. 407. Harper and Row, New York (1971).
13. Wang, J.C., J. Mol. Biol. 89, 783 (1974).
14. Hayashi, M., Proc. Nat. Acad. Sci. U.S. 54, 1736 (1965).
15. Wang, J.C., J. Mol. Biol. 87, 797 (1974).

16. Davis, R., Simon, M.N. and Davidson, N. in Methods in Enzymology (Grossman, L. and Moldave, K., eds.), vol. 21, part D, p. 413, Academic Press, N.Y. (1971).

17. Morgan, A.R. and Pulleyblank, D.E., Biochem. Biophys. Res. Comm. 61, 346 (1974).

18. Vinograd, J., Lebowitz, J. and Watson, R., J. Mol. Biol. 33, 173 (1968).

19. Matsuda, M., and Ogoshi, H., J. Biochem. 59, 230 (1966).

20. Chamberlin, M., Fed. Proc. 24, 1446 (1965).

21. Bauer, W. and Vinograd, J., J. Mol. Biol. 47, 419 (1970).

22. Hsieh, T.-S. and Wang, J. C., Biochemistry 14, 527 (1975).

23. Sheraga, H.A. and Rupley, J.A., Adv. Enzymol. 24, 161 (1962).

24. Goebel, W., Nature New Biol. 237 67 (1972).

HELA CELL DNA POLYMERASES:
PROPERTIES AND POSSIBLE FUNCTIONS

Arthur Weissbach, Silvio Spadari* and Karl-Werner Knopf
Roche Institute of Molecular Biology
Nutley, New Jersey 07110

ABSTRACT. Four distinct DNA polymerases have been identified in vertebrate and mammalian cells. These enzymes,
named polymerase α, β, γ, or mitochondrial, have been partially purified from HeLa cells and examined for properties
related to their possible role in DNA replication. The
HeLa polymerases are found not to be related to one another
and to show unique responses to natural RNA primers, histones, and certain DNA binding proteins. These results,
combined with in vivo studies of synchronized cells indicate that DNA polymerase α is associated with DNA replication. The functions of DNA polymerases β and γ remain
unknown.

INTRODUCTION

The study of the DNA polymerases of the eukaryotic
cell, particularly of vertebrates and mammals, has accel-
erated in the past few years since the discovery of the
"reverse transcriptases" of the RNA tumor viruses. Found
in every cell, the DNA polymerases seemed to represent a
cataloging problem of encyclopedic proportions. Until
recently, the types, properties and intracellular distribution of these enzymes were unclear and it was difficult
to relate the findings of one laboratory to another. Fortunately enough information has become available to permit
placement of most of the known DNA polymerases into a
framework and the establishment of a nomenclature scheme.

---

*Present address: Laboratorio Di Genetica Biochimica
Ed Evoluzionistica, Pavia, Italy

This nomenclature[1], which is shown in Table 1, assigns Greek letters to three of the major eukaryotic DNA polymerases. The enzymes are referred to as "DNA polymerase α" or "α-polymerase" (but not "α-DNA polymerase") in an effort to avoid confusion with pre-existing nomenclatures. The DNA polymerase found in mitochondria is named because of its intracellular location and the viral induced DNA polymerases are identified by the name of the associated virus. Since this paper will present data concerning DNA polymerases α, β, and γ it will be useful to briefly summarize some of the known properties of these enzymes.

DNA polymerase α, an enzyme which has been highly purified and studied by many workers (1-6) is the predominant activity in growing cells. Found in the cytoplasm and nucleus as a high molecular weight form (110-220,000, 6-8 S) it can also exist as aggregates in low ionic strength (7, 8). The purified enzyme requires Mg^{++} and sulfhydryl groups, copies gapped duplex DNA efficiently and is free of nuclease activities.

DNA polymerase β is a small molecular weight enzyme (40-50,000, 3-4 S) first clearly identified in the nucleus of HeLa cells in 1971 (6) and found shortly thereafter in calf thymus and rat liver cells (9-11). It is localized primarily in the nucleus and can exist as aggregates which show higher sedimentation values (12, 13). The enzyme is insensitive to sulfhydryl group inhibitors and utilizes Mg^{++} or Mn^{++}. Though it prefers gapped duplex DNA as a template it will copy the polynucleotide strand of oligomer-homopolymers such as $(A)_n \cdot dT_{15}$ at a significant rate (12). The purified enzyme has no detectable nuclease activities associated with it (14).

DNA polymerase γ is the newest of the DNA polymerases and has been found in most cells examined (15-19). This enzyme represents only 1% or so of the total cellular DNA polymerase activity and is present in the nucleus and cytoplasm. Purified preparations of the enzyme copy synthetic ribohomopolymers at a much higher rate than DNA templates but the enzyme has not been demonstrated to copy natural RNA (20). The enzyme has a molecular weight of

[1]The nomenclature was suggested at a meeting held at the Massachusetts Institute of Technology with D. Baltimore, F. J. Bollum, R. C. Gallo and A. Weissbach.

TABLE 1

NOMENCLATURE OF VERTEBRATE DNA POLYMERASES

DNA Polymerase	Characteristic
α	Large molecular weight (110–220,000) – cytoplasm and nucleus
β	Small molecular weight (40–50,000) – nucleus
γ	Minor activity in nucleus and cytoplasm which copies ribohomopolymers efficiently – (110,000 daltons)
mt	Mitochondrial-copies circular duplex DNA – (100,000 daltons)
Virus-induced	Large DNA Viruses – RNA dependent DNA polymerases of the RNA tumor viruses

110,000 in HeLa cells and requires Mn^{++} or Mg^{++} and sulf-hydryl groups for maximal activity.

The identification of these three classes of DNA polymerases has naturally led to the question of the role of the α, β and γ polymerases in DNA replication. This paper will present studies of HeLa cell DNA polymerases which are pertinent to this question and which deal with the relationship between the various DNA polymerases.

METHODS

Materials. The preparation of the purified HeLa cell DNA polymerases and conditions for their assay have been described (6, 8, 15, 26). The calf thymus binding protein (35) was the kind gift of Dr. Arnold Levine, Princeton, University. Dr. Howard Green, Massachusetts Institute of Technology, supplied the murine cell P-8 binding protein (32). Calf thymus histones were obtained from the Worthington Biochemical Corporation. The purified histone fractions were gifts of Dr. M. J. Modak (Sloan-Kettering Institute) and Dr. Milas Boublik (Roche Institute of Molecular Biology). The histones, after being dissolved in 0.4 M HCl and neutralized to pH 6.8 with potassium phosphate, were mixed with activated DNA and dialyzed against 0.5 M NaCl. Aliquots of this mixture were used for the studies shown in Table 3. Purified E. coli RNA polymerase was the gift of Dr. H. Kung (Roche Institute of Molecular Biology).

Preparation of Templates. Activated DNA was prepared by the action of pancreatic DNase I on salmon sperm DNA (cf 26).

RNA:DNA primer-templates were prepared by the action of E. coli RNA polymerase on either single- or double-stranded HeLa DNA. The partial DNA:RNA hybrid was prepared in 1 ml of a solution containing 0.05 M Tris-HCl (pH 7.9), 10 mM $MgCl_2$, 0.5 mM dithiothreitol, 80 µM EDTA, 200 µg of bovine albumin, 2 mM KPO_4 (pH 7.5), 1×10^{-5} M rabbit liver soluble RNA, 0.1 mM ribonucleoside triphos-phates, 120 µg of HeLa denatured or native DNA and 3.4 units of DNA-dependent RNA polymerase (E. coli). After incubation at 37°C for 20 min approximately 1 to 2% of the DNA was transcribed as judged by [α-^{32}P] ATP incor-poration in parallel experiments. The reaction mixture

was then extracted with phenol-chloroform (1:1) saturated
with 0.3 M Tris-HCl (pH 7.5). The phenol-chloroform was
subsequently removed by extraction with ether. The DNA:
RNA hybrid so isolated was then used in the DNA polymerase
reactions shown in Table 2. In control reactions, either
ribonucleoside triphosphates or RNA polymerase was left
out of the incubation mixture but isolation of the DNA
template was carried out as described above.

The DNA polymerase reaction mixtures (50 μl) contained 0.05 M Tris-HCl (pH 7.9), 10 mM MgCl$_2$, 0.5 mM
dithiothreitol, 80 μM EDTA, 10 μg of bovine serum albumin,
40 μM deoxynucleoside triphosphates (specific activity
as stated) and the designated DNA polymerase and DNA
template.

RESULTS

Dissimilarity of DNA Polymerases. There has been continuing speculation of a possible relationship between the
various DNA polymerases within a given cell. This is an
important issue to resolve since such proposed kinships
would affect the interpretation of data concerning polymerase activities. Chang and Bollum (21) first reported an
antigenic relationship between DNA polymerases α and β
obtained from a variety of cells. The postulate that the
α and β polymerases shared common peptide sequences or
subunits was supported by other workers, notably Hecht (22)
on the basis of sedimentation studies which purported to
show salt mediated conversions of DNA polymerase α to β.

We have studied this problem using partially purified HeLa cell DNA polymerases (8) and find no immunological relationship between DNA polymerases α, β and γ. Using
a rabbit antiserum prepared against a purified α-polymerase, we measured the effect of this antiserum upon other
DNA polymerases. The results are shown in Figure 1. It
can be seen that DNA polymerase β (nuclear D-DNA polymerase I) and γ (R-DNA polymerase) are not inhibited by antiserum which inhibits the α polymerase (D-DNA polymerase II)
by 70%. A partially purified DNA polymerase α from
Chinese hamster cells, however, is partially inhibited by
antiserum directed against the HeLa α polymerase. We conclude from these studies that DNA polymerases α, β and γ
are not related but that the α-polymerase in HeLa cells
and Chinese hamster cells do share common peptide

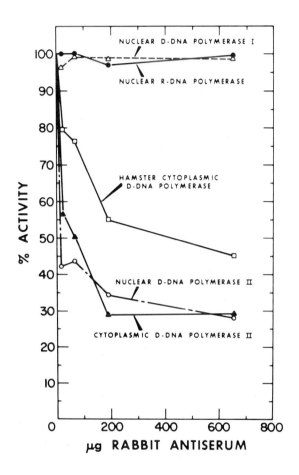

Fig. 1.   Titration of DNA polymerases against anti-serum to HeLa cell DNA polymerase α. D-DNA polymerase I and II refer to the β- and α-polymerases, respectively. R-DNA polymerase is the γ-polymerase. The preincubation of antisera with the enzymes and subsequent assay conditions have been described (8).

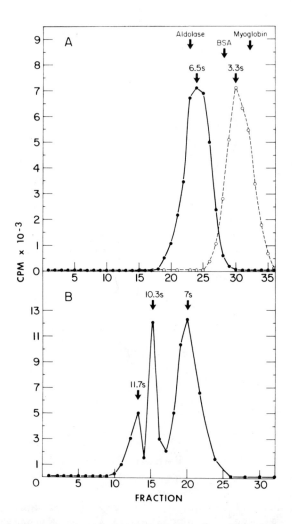

Fig. 2. Sucrose gradient analysis of the HeLa α and β polymerases. The purified enzymes were dialyzed against 0.125 M (NH₄)₂SO4 (Panel A) or 0.01 M (NH₄)₂SO4 (Panel B) before sedimentation in a 5–20% sucrose gradient for 15 hours at 37,000 rpm in a Spinco SW 50L rotor. ●----●, α-polymerase, o----o, β-polymerase.

sequences. The latter point is in agreement with some of the data of Chang and Bollum (21).

We have also examined the sedimentation behavior of purified DNA polymerases α and β at various salt concentrations. This is shown in Figure 2 in which the α polymerase was dialyzed against 0.125 M $(NH_4)_2SO_4$ (Panel A) or 0.01 M $(NH_4)_2SO_4$ (Panel B) before sedimentation in a sucrose gradient containing the equivalent salt concentration. In 0.125 M salt, DNA polymerase α shows one peak at 6.5 S but no activity is seen in the 3-4 S where the β polymerase migrates. In low salt (0.01 M) the α-polymerase shows multiple forms, and discrete peaks are seen at 10.3 and 11.7 S. Thus, the α-polymerase can be observed to aggregate to high molecular weight forms in low salt but cannot be observed to break down to units the size of the β polymerase (3-4 S). Essentially similar results have been obtained by Byrnes et al. (7) with a rabbit α-polymerase. DNA polymerase β also shows aggregation phenomena and can easily be confused with the α-polymerase if sedimentation coefficients are used as criteria for identifying the enzymes. This has been shown by Wang, Sedwick and Korn (12) who found that DNA polymerase β from KB cells exists as an aggregate form in low salt (10 mM) and essentially as a monomer in high salt (150 mM). These studies measured the aggregate and monomer forms by exclusion or retention on Sephadex G-200 gel filtration. The aggregate form, which was excluded from the column, would have an apparent molecular weight greater than 250,000. Recently, Hecht (23) has also obtained evidence which shows that DNA polymerase β can exist as aggregates which sediment at 6-8 S. The common conclusion from these studies is that DNA polymerases α, β and γ are distinct entities which are not related and that there is no interconversion between the α and β polymerases. However, because of the conflict in data, this point has not been fully resolved.

Possible Role of DNA Polymerase α in DNA Replication.
Studies with resting, dividing, or synchronized cells have suggested that DNA polymerase α is associated with DNA synthesis. L cells, when shifted from a stationary to a dividing state show a rise in α polymerase levels with no apparent change in DNA polymerase β activity (24). Regenerating rat liver also shows a marked rise in DNA

polymerase α levels (25). To examine this phenomenon in more detail, we have measured the levels of the α, β and γ polymerases during the S phase in synchronized HeLa cells obtained after a double thymidine block.

These results are summarized in Figure 3 (cf 26). DNA synthesis (as measured by ^{3}H-thymidine incorporation) in the S phase of synchronized HeLa cells continues for 8-10 hours, reaching a peak about 5 hours after release of the thymidine block. During this period there is a rapid early rise in the activity of the γ polymerase (R-DNA polymerase) in both the nucleus and cytoplasm. This activity doubles within 2 hours and then rapidly declines. Cytoplasmic DNA polymerase α (cytoplasmic D-DNA polymerase) shows a steady rise throughout the S period. In these experiments the levels of DNA polymerase α or β in the nucleus did not change significantly. Chiu and Baril (27) have recently extended these studies in synchronized HeLa cells and studied the levels of DNA polymerase α and β during a complete cell cycle. Their results are summarized in Figure 4 in which they measured DNA synthesis and mitosis (4A) and the nuclear level of DNA polymerase α and β following release of a thymidine block (4B). They found that DNA polymerase α levels start to rise in late G_1 and fall at the end of S, whilst the β polymerase levels remain unchanged. This rise in α-polymerase levels was blocked in the presence of cycloheximide but not affected by hydroxyurea, a known inhibitor of DNA synthesis. Our results and those of Chiu and Baril, support the idea that DNA polymerase α is closely allied to DNA replication. The early rise and fall of DNA polymerase γ in the S phase which we observe is difficult to interpret though it is tempting to speculate that it is related to the initiation or early steps of DNA replication.

RNA Primed DNA Synthesis. There have been a number of attempts to demonstrate that RNA priming is involved in the initiation of DNA synthesis in eukaryotic cells. In vivo, polyoma DNA covalently bound to RNA has been demonstrated (28) and Keller (29) demonstrated, in vitro, that KB cell DNA polymerase α copied a single-stranded circular phage DNA if an RNA primer was hydrogen bonded to the DNA. We have examined the HeLa cell DNA polymerases α, β and γ for their ability to utilize a template bearing a natural RNA template (30). The outline for this

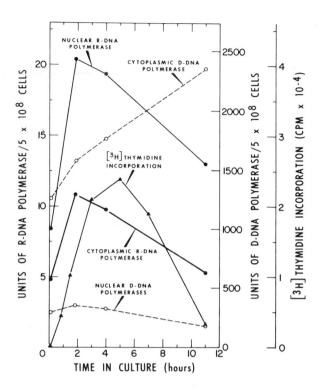

Fig. 3. DNA polymerase levels in the S-phase of synchronized HeLa S-3 cells. Cytoplasmic D-DNA polymerase refers to the α-polymerase. Nuclear D-DNA polymerases are the combined activity of the α- and β-polymerases in the nucleus. R-DNA polymerase is the previous term for DNA polymerase γ.

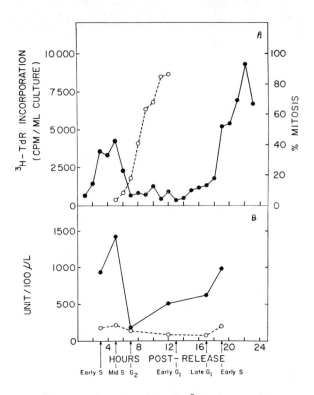

Fig. 4. Panel A - Rate of [^{3}H]-thymidine incorporation and degee of mitosis in synchronized HeLa cells.
o----o, mitosis; ●----●, [^{3}H]-thymidine incorporation.
Panel B - Variation of nuclear DNA polymerase activity
during the cell cycle of synchronized HeLa cells. o----o,
β-polymerase; ●----●, α-polymerase. These results are of
Chiu and Baril (1975).

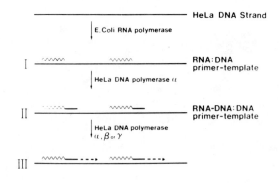

Fig. 5. Scheme for the in vitro synthesis of DNA by
HeLa cell DNA polymerases (30).

experiment is shown in Figure 5. Using a single-stranded
HeLa DNA template, a complementary RNA primer is formed
with E. coli RNA polymerase to form Structure I. This
RNA primed combination was tested as a template for the
three HeLa DNA polymerases and was found to be utilized
only by the α polymerase (Table 2). Table 2 shows the
response of HeLa cell DNA polymerases α, β and γ to a
partial RNA:DNA hybrid (Structure I) prepared as described
in "Methods." When the enzymes are added to this template
which contains small RNA pieces synthesized by the RNA
polymerase in situ on a single-stranded DNA, or double-
stranded DNA, DNA polymerases β and γ show little copying
of the DNA and are not significantly stimulated by RNA
synthesis. However, DNA polymerase α was active in this
system and inhibition of RNA synthesis by omission of
either 1 or all 4 ribonucleotides or by omission of RNA
polymerase reduced DNA synthesis by DNA polymerase α by
up to 20 fold. This would indicate that the stimulation
of DNA synthesis required RNA synthesis. As expected all
four deoxyribonucleoside triphosphates were required for
DNA synthesis and neither RNA or DNA was made in the
absence of a DNA template. We conclude that Structure I
is converted to Structure II (an RNA-DNA:DNA duplex) by
the α polymerase and Table 2 also shows that this latter
primer-template can now be extended by either the α, β or
γ polymerases to form Structure III. This study would
imply that, if RNA priming is involved in HeLa DNA replication, the α-polymerase is at least necessary for the
initiation of DNA synthesis and that further extension of
the DNA chain can be carried out by any of the three HeLa
polymerases.

Effect of Histones and Binding Proteins. Nuclear DNA
exists as a complex structure (chromatin) in which there
are many protein and RNA molecules bound to the DNA. We
have tested some of the known histones and DNA binding
proteins for their effect on the ability of DNA polymerase
α, β or γ to copy a gapped duplex (activated) DNA template.
Table 3 shows the action of histones f_1, f_2a_2, f_3 or a
complete histone mixture on the HeLa cell DNA polymerases.
All the histones tested show an inhibitory effect on the
utilization of the DNA template by each of the enzymes.
This could be attributed to the association of the histones

75

TABLE 2

RESPONSE OF HELA CELL DNA POLYMERASES TO RNA- AND DNA-PRIMED TEMPLATES

Primer-Template*	[3H]dNTP Incorporation (pmoles) by		
	DNA Polymerase α	DNA Polymerase β	DNA Polymerase γ
a. RNA-primed single-stranded HeLa DNA template (Structure I)	76.0	24.3	7.2
Without ribonucleoside triphosphates	6.9	24.0	7.1
b. RNA-primed double-stranded HeLa DNA template	35.5	21.0	3.3
Without ribonucleoside triphosphates	5.28	26.0	3.0
c. DNA-primed single-stranded HeLa DNA template (Structure II)	108.0	84.0	18.9
Without ribonucleoside triphosphates	6.9	24.0	7.1
Without deoxyribonucleoside triphosphates (d)	--	23.0	6.8

*The primer-template structures are defined in Figure 5.

LEGEND TO TABLE 2

(a) Fifty µl reaction mixtures contained 100 µg/ml of the
     RNA:DNA hybrid (Structure I) prepared as described
     in "Methods" and 0.62, 0.93 and 0.25 units of DNA
     polymerases α, β and γ respectively. The incubations
     were carried out at 37°C for 30 min with 40 µM [^{3}H]
     deoxyribonucleotides (specific activity 62 cpm/pmole).
     The single-stranded DNA used in these studies (a, b,
     and c) had a molecular weight of 700,000.

(b) Fifty µl reaction mixtures contained 100 µg/ml of the
     RNA:DNA hybrid prepared as described in "Methods"
     except that HeLa native DNA was used.

(c) Fifty µl reaction mixtures contained 100 µg/ml of the
     RNA:DNA hybrid (Structure I) formed as in (a) and
     0.62 units of DNA polymerase α which converts
     Structure I into Structure II. The polymerization
     reaction was carried out at 37°C for 30 min in the
     presence of 40 µM unlabeled deoxyribonucleotides and
     DNA polymerase α was then inactivated by heating
     7 min at 60°C. To each reaction mixture was added
     [^{3}H]dATP (final specific activity 62 cpm/pmole
     average) and 0.62, 0.93 and 0.25 units of DNA
     polymerases α, β and γ respectively to permit synthe-
     sis of Structure III. After 30 min at 37°C the acid
     precipitable counts were determined. The control
     reaction, lacking ribonucleoside triphosphates to
     prevent the initial formation of Structure I, was
     carried out in exactly the same manner.

(d) In this control reaction a RNA:DNA hybrid was in-
     cubated with DNA polymerase α in the absence of
     deoxynucleoside triphosphates. After heat inactiva-
     tion of the enzyme the desired DNA polymerase and
     [^{3}H] deoxynucleoside triphosphates were added and
     the mixture incubated at 37° for 30 min.

TABLE 3

EFFECT OF HISTONES ON THE ACTIVITY OF
HELA DNA POLYMERASES

DNA Polymerase	Histones added	Histone/DNA* Weight Ratio	Activity in %
α	-	-	100.0
	f_1	2	29.5
	f_2a_2	2	7.5
	f_3	2	60.5
	total histones	2	34.5
β	-	-	100.0
	f_1	2	32.8
	f_2a_2	2	0.6
	f_3	2	20.1
	total histones	2	32.7
γ	-	-	100.0
	f_1	2	29.7
	f_2a_2	2	2.7
	f_3	2	7.1
	total histones	2	30.9

*The concentration of activated salmon sperm DNA in the
assay was 50 μg/ml.

with the DNA template which might then interfere with the
movement of DNA polymerases along the DNA. This is most
likely an oversimplified view since each DNA polymerase
differs in its response to one or more of the histones.
Histone f_3 shows the greatest variation since it inhibits
the α polymerase 40%, the β polymerase 80% and the γ
polymerase 90% under the conditions tested.

We have also examined the action of DNA binding proteins obtained either from calf thymus (31) or murine cells
(32) upon the three HeLa DNA polymerases. Both the calf
thymus binding protein and the murine cell binding protein,
P-8, stimulate the activity of the α-polymerase with an
activated DNA template, but do not affect the β or γ
polymerases (Figure 6). It is interesting that the α
polymerase, which is felt to be the most likely candidate
to be involved in DNA replication is the enzyme which
responds to these proteins which bind to single-stranded
DNA. Herrick and Alberts (35) have reported a similar
finding with the α polymerase of calf thymus and it would
be interesting to determine if these binding proteins can
interact directly with the α-polymerase as well as affecting the structure of the DNA template.

DISCUSSION

It is apparent that our understanding of the role of
eukaryotic DNA polymerases in the replication of DNA is
rudimentary. Studies of the gross events of DNA replication (33, 34) have suggested that there are a large number
of replicons (units of DNA replication) in eukaryotic cells
and that DNA synthesis in the chromosome is discontinuous.
Thus, there may be an initiation step involving an RNA
primer and sequential formation of DNA segments which
eventually join to form the mature DNA molecule. If this
process requires several separate events then more than
one DNA polymerase could be implicated in this overall
operation. The possible ordering of the cellular DNA
polymerases in this hypothetical plan and in DNA repair
reactions remains an important goal of these studies.

Eukaryotic DNA polymerases must function, in vivo,
as part of a large replication complex which contain many
proteins such as histones or other DNA binding proteins.
The interaction of these proteins with the replicating DNA
template and appropriate DNA polymerase may be critical in

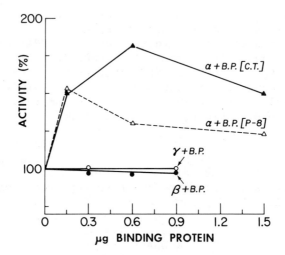

Fig. 6. Effect of DNA binding proteins on HeLa cell DNA polymerases. Assays contained 1 µg of activated DNA template and incubations were for 30 min at 37°. α, β and γ refer to the HeLa DNA polymerases. B.P.[C.T.] and B.P. [P-8] are the calf thymus (35) and murine DNA binding proteins (32), respectively. B.P. refers to either of the binding proteins.

the replication process. The data in this paper show that histones or DNA binding proteins can exert unique inhibitory or stimulatory effects with the HeLa cell DNA polymerases. These effects differ, in some cases, from one polymerase to another. One can safely assume that understanding DNA polymerase function will require further knowledge of the putative DNA replication complex and its associated sibling proteins.

REFERENCES

1. Bollum, F.J. J. Biol. Chem. 235, 2399 (1960).
2. Chang, L.M.S. and Bollum, F.J. J. Biol. Chem. 248, 3398 (1973).
3. De Recundo, A.M., Lepesant, J.A., Dichot, O., Grasset, L., Rossegnol, J.M. and Cazillis, M. J. Biol. Chem. 248, 131 (1973).
4. Holmes, A.M., Hesslwood, I.P. and Johnston, I.R. Eur. J. Biochem. 43 487 (1974).
5. Sedwick, W.D., Wang, T.S.F. and Korn, D. J. Biol. Chem. 247, 5026 (1972).
6. Weissbach, A., Schlabach, A., Fridlender, B. and Bolden, A. Nature New Biology 231, 167 (1971).
7. Byrnes, J.J., Downey, K.M. and So, A.G. Biochemistry 12, 4378 (1973).
8. Spadari, S. and Weissbach, A. J. Biol. Chem. 249, 2991 (1974).
9. Chang, L.M.S. and Bollum, F.J. J. Biol. Chem. 246, 5835 (1971).
10. Haines, M.E., Holmes, A.N. and Johnston, I.R. FEBS Letters 17, 63 (1971).
11. Berger, H., Jr., Huang, R.C.C. and Irwin, J.L. J. Biol. Chem. 246, 7275 (1971).
12. Wang, T.S.F., Sedwick, W.D. and Korn, D. J. Biol. Chem. (submitted 1975).
13. Hecht, N.B. Biochim. Biophys. Acta, in press (1975).
14. Chang, L.M.S. and Bollum, F.J. Biochemistry 11, 1264 (1972).
15. Fridlender, B., Fry, M., Bolden, A. and Weissbach, A. Proc. Nat. Acad. Sci. USA, 69, 452 (1972).
16. Bolden, A., Fry, M., Muller, R., Citarella, R. and Weissbach, A. Arch. Biochem. Biophys. 153, 26 (1972).
17. McCaffrey, R., Smoler, D.F. and Baltimore, D. Proc. Nat. Acad. Sci. USA, 70, 521 (1974).

18. Lewis, B.J., Abrell, J.W., Smith, R.G. and Gallo, R.C. Science 183, 867 (1974).
19. Yoshida, S., Ando, T. and Kondo, T. Biochem. Biophys. Res. Commun. 60, 1193 (1974).
20. Spadari, S. and Weissbach, A. J. Biol. Chem. 249, 5800 (1974).
21. Chang, L.M.S. and Bollum, F.J. Science 175, 1116 (1972).
22. Hecht, N.B. Biochim. Biophys. Acta 312, 471 (1973).
23. Hecht, N.B. Biochim. Biophys. Acta, in press (1975).
24. Chang, L.M.S. and Bollum, F.J. J. Biol. Chem. 248, 3398 (1973).
25. Baril, E.F., Jenkins, M.D., Brown, O.E., Laszlo, J. and Morris, H.P. Cancer Research 33, 1187 (1973).
26. Spadari, S. and Weissbach, A. J. Mol. Biol. 86, 11 (1974).
27. Chiu, R.W. and Baril, E.F. J. Biol. Chem., in press (1975).
28. Pigiet, V., Eliasson, R. and Reichard, P. J. Mol. Biol. 84, 197 (1974).
29. Keller, W. Proc. Nat. Acad. Sci. USA, 69, 1560 (1972).
30. Spadari, S. and Weissbach, A. Proc. Nat. Acad. Sci. USA, in press (1975).
31. Herrick, G. Dissertation, Princeton University (1973).
32. Salas, J. and Green, H. Nature New Biology 229, 165 (1971).
33. Huberman, J.A. and Horwitz, H. Cold Spring Harbor Symp. Quant. Biol. 38, 233 (1973).
34. Taylor, J.H., Adams, A.G. and Kenek, M.B. Chromosoma 41, 361 (1973).
35. Herrick, G. and Alberts, B. Fed. Proc. 32, 497 (Abs.) (1973).

ASSAY AND PARTIAL PURIFICATION OF THE DNA UNTWISTING ACTIVITY FROM RAT LIVER

James J. Champoux and Joyce M. Durnford

Department of Microbiology
School of Medicine
University of Washington
Seattle, Washington 98195

ABSTRACT. A DNA filter binding assay and a fluorometric assay have been developed for the DNA untwisting enzyme. Using these assays, the activity has been partially purified from rat liver nuclei.

INTRODUCTION

We have previously shown that both positive and negative superhelical turns are removed from closed circular DNA's when they are incubated with a nuclear extract from mouse embryo cells (1). We termed this activity the DNA untwisting enzyme. A similar activity, called ω, had been isolated from E. coli and shown to be capable of the partial removal of negative, but not positive superhelical turns (2). In both cases it was clear that the reaction carried out by the activity involved the introduction of a transient break into one of the strands of the duplex DNA. The presence of the break would permit the winding of one strand of the helix relative to the other, if there existed some internal force in the molecule (e.g. superhelical turns) to drive the winding. It has been suggested that the DNA untwisting enzyme might act in the cell to provide the swivel for DNA replication (1, 2).

Activities similar to the DNA untwisting enzyme have now been isolated from KB cells (DNA relaxing activity) (3), fertilized Drosophila eggs (4) and both HeLa and L cells (superhelix relaxation protein) (5, 6). In this

83

report we describe two rapid assays for the DNA untwisting activity and the partial purification of the enzyme
from rat liver nuclei.

RESULTS

DNA filter binding assay. The substrate for the
reaction is the naturally occurring form of SV40 DNA.
This closed circular DNA has approximately 40 negative
superhelical turns, if one assumes a value of 26^0 for
the unwinding angle of ethidium bromide (7). The product
of the reaction is the completely relaxed closed circular
form of the DNA. Vinograd et al. (8) and more recently
Wang (7) have shown that during alkaline titration of the
DNA a fraction of the bases in closed circular DNA's
containing negative superhelical turns exhibits an early
helix-coil transition. This means that there exists a pH
value at which the superhelical molecules are partially
single-stranded while the relaxed closed circles remain
completely double-stranded. Since single-stranded DNA
binds to nitrocellulose filters (9), we reasoned that at
such a pH value, the superhelical DNA, but not the relaxed
DNA, might bind to the filter.

Figure 1 shows the binding of the two forms of the
DNA as a function of the pH of the sodium phosphate buffer
used for the filtration. As predicted, the superhelical
DNA does bind to the filter at lower pH values than the
relaxed closed circular DNA. The broken vertical line
shows the pH chosen as our standard filtration condition
for discriminating between the superhelical substrate and
the relaxed product of the untwisting reaction.

The activity is quantitated according to the following procedure. Serial dilutions of the fraction to be
assayed are added to the DNA and the mixtures incubated
under the standard reaction conditions (1) for 10 minutes.
The high pH buffer is added and the solutions filtered.
The percent of the DNA bound is plotted on an inverted
scale as a function of the log of the total dilution
factor (Figure 2). The dilution required to give binding
midway between that of the substrate and the product
serves as a measure of the activity. As illustrated in

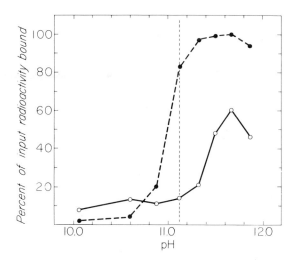

Figure 1.   Binding of superhelical and relaxed closed circular DNA to nitrocellulose filters as a function of the pH of the filtration buffer.   The superhelical DNA (closed symbols) is the naturally occurring form of SV40 DNA labeled with ^{3}H.   The procedure for preparing the relaxed form of ^{14}C-labeled SV40 DNA (open symbols) has been detailed elsewhere (14).   The two DNA preparations were mixed with 2.5 ml of the sodium phosphate buffers having the indicated pH's.   Nitrocellulose filters (13 mm, type B6, Schleicher and Schuell, Inc.) were soaked in distilled water prior to use and prewashed with the same buffer as used for the filtration.   After filtering the DNA sample, the filter was again washed with the filtration buffer, dried and counted in a toluene based scintillation fluid.   The composition of the buffer indicated by the broken vertical line is 0.39 M Na_3PO_4, 0.21 M NaH_2PO_4, 1 mM EDTA, pH 11.1.

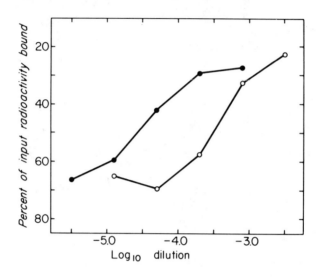

Figure 2. The DNA filter binding assay. Serial
dilutions of the fraction to be assayed were incubated
with ^{3}H-labeled SV40 DNA under the standard reaction
conditions (1) for 10 minutes at 37°C (0.10 ml reaction
mixture). After addition of 2.5 ml of the standard
filtration buffer, the samples were filtered and counted
as described in the legend to Figure 1. The percent of
the DNA bound to the filter is plotted on an inverted
scale against the log of the dilution factor. The
closed symbols correspond to the assay of the crude
nuclear extract given in Table 1. The open symbols
represent the assay of the pooled DEAE cellulose fractions
(see Table 1).

Figure 2, two fractions can be compared within the same experiment and the activity of one relative to the other determined from the displacement between the two curves.

The filter assay has the disadvantage that multiple determinations are required to quantitate a single fraction. However, the presence of DNA in the extract or the addition of as much as 50 μg of protein per assay does not interfere with the differential binding of the substrate and product DNA's. We have found that nicked DNA binds to the same extent as the relaxed product. Therefore, one must verify for each new system that no endonucleases are active under the conditions of the untwisting reaction.

Fluorometric assay. The binding of the dye, ethidium bromide, to DNA results in a complex which has a characteristic fluorescence (10). Paoletti et al. (11) previously made use of the difference in the amount of ethidium bromide bound to covalently closed and nicked circular DNA's (12) to develop fluorometric assays for endonucleases and ligases. We in turn rely on the differential binding of ethidium bromide to closed circular DNA's containing different numbers of superhelical turns (13) to fluorometrically assay the untwisting enzyme.

We have previously shown that positive superhelical turns, introduced into a closed circular DNA by the appropriate concentration of ethidium bromide, are removed by the DNA untwisting activity (1). In this case, the product of the reaction is relaxed under the conditions of the assay (i.e. in the presence of the dye), but contains more negative superhelical turns than the substrate when the two are compared in the absence of ethidium bromide. Since dye binding is directly proportional to the number of negative superhelical turns in the DNA (determined in the absence of dye) (13), the product of the reaction will have more bound ethidium than the substrate. By carrying out the reaction in a cuvette in the fluorimeter, the course of the reaction can be monitored by following the fluorescent increase as a function of time.

The recorder tracing for a typical reaction is given
in Figure 3. Addition of a second aliquot of the enzyme
at the time indicated by the arrow resulted in essentially
no further fluorescent increase, indicating that the
reaction had gone to completion. The overall increase in
fluorescence occurring on conversion of the substrate to
product under these conditions is 30 percent of the
initial value.

The time required for half the overall fluorescent
increase can be used as a measure of the rate of the
reaction. Figure 4 shows the relationship between the
reaction rate, so defined, and the amount of enzyme added
to the reaction. Knowing the dose-response for the assay
means that a single determination in the range of half
times shown, suffices to quantitate the activity.

This assay suffers from the disadvantage that any
DNA present in the fraction to be assayed will bind dye
and by contributing to the observed fluorescence, possibly
obscure the small increase due to the untwisting reaction.
For this reason the filter assay is used for assaying the
crude extracts containing DNA and the fluorometric assay
for the later stages of the purification after the DNA
has been removed.

Partial purification of the rat liver enzyme. In our
initial attempts to purify the enzyme from rat liver
nuclei we have found that greater than 90 percent of the
activity in extracts made by sonicating rat liver nuclei
in low salt (.02 M Tris) (14) is associated with a
particulate fraction containing most of the DNA. Attempts
to release the activity from the DNA prior to centrifugation have not been successful. It has proven equally
difficult to solubilize the activity in the pellet and
subsequently separate it from the DNA. Therefore, in
the purification presented here we are dealing only with
the soluble fraction present after centrifugation of the
crude nuclear extract.

This fraction is first applied to a DEAE cellulose
column and eluted with 0.20 M KCl. This separates the
activity from the DNA remaining in the crude extract.

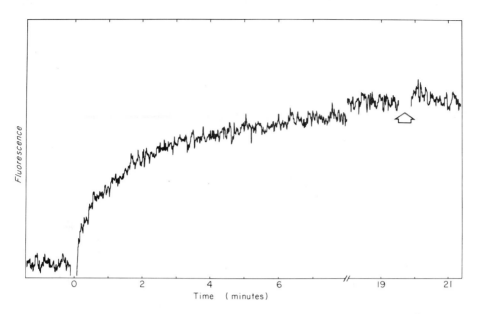

Figure 3. The fluorometric assay. The DNA untwisting reaction was carried out under the standard reaction conditions (1) in a cuvette in an Aminco-Bowman fluorimeter. The wavelength for excitation was 340 nm and the wavelength of emission was 586 nm. The time course of the fluorescent increase was monitored by a recorder attached to the fluorimeter. The arrow indicates the time at which a second aliquot of enzyme was added to the reaction.

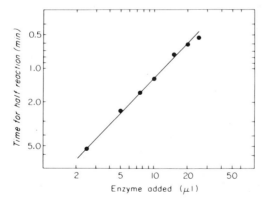

Figure 4. Reaction rate as a function of enzyme concentration for the fluorometric assay. The fluorometric assay was carried out as described in Figure 3 with various amounts of enzyme added. The log of the time for half the overall fluorescent increase is plotted against the log of the enzyme concentration.

TABLE 1

Fraction	Protein (mg)	Total Activity (Relative Units)		Specific Activity (Relative Units/mg)	Yield (%)
		Filter Assay	Fluor. Assay		
Nuclear extract	277	1.0[a]	–	1.00	100
Supernatant (after centri-fugation of extract)	105	0.066	–	0.17	6.6
Pooled DEAE cellulose	47	0.052	0.052[b]	0.31	5.2
Pooled phosphocellulose	0.69	–	0.012	4.8	1.2
Pooled carboxymethyl Sephadex (C-50)	0.05	–	0.006	33	0.6

(a) The assay for this fraction (total volume 45 ml) is given in Figure 2.

(b) This value was arbitrarily set equal to that found in the filter assay.

In agreement with the results of Keller and Wendel (3) who studied the activity from KB cells, we find that the activity can be successfully fractionated by chromatography on phosphocellulose and carboxymethyl Sephadex (C-50). As shown in Figure 5, the phosphocellulose step is particularly effective since the activity does not elute until after most of the protein.

Table 1 summarizes our purification results to date. The overall yield of activity is very low, but most of the loss occurs in the first step when the particulate material is removed by centrifugation. At present we have no way of determining whether the activity detectable in the pellet is the same as that which we have carried through the other purification steps. Overall, the activity has been purified approximately 33-fold. Disregarding the first step, the purification is 190-fold from the extract supernatant through the carboxymethyl Sephadex chromatography.

DISCUSSION

The reaction carried out by the untwisting activity results in the conversion of superhelical DNA to relaxed closed circular DNA. This reaction must involve the introduction of a discontinuity into one of the strands of duplex DNA. The lifetime of the break must be sufficiently long to allow the helix to unwind or wind up by minimally one turn, providing there exists some driving force for the winding. The break is then sealed. We presume this same nicking-sealing cycle occurs with the relaxed product, or for that matter with linear DNA, but this reaction has not yet been demonstrated.

The purification results presented here, together with the results available from other systems (2, 3, 4, 6) indicate that a single protein or tight complex of proteins carries out the nicking-sealing sequence of reactions. In the case of eucaryotic cells, the activities which remove negative and positive superhelical turns co-purify, suggesting that both reactions are carried out by the same protein.

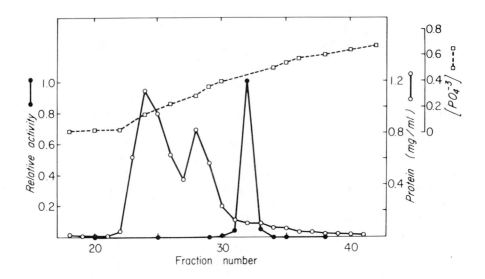

Figure 5. Phosphocellulose chromatography of the
DNA untwisting enzyme. The pooled DEAE cellulose fractions were dialyzed against 0.02 M potassium phosphate
buffer pH 7.2, containing 1 mM EDTA, 0.5 mM dithiothretol,
and 10% glycerol and applied to a phosphocellulose column
previously equilibrated with the same buffer. The column
was eluted with a phosphate gradient which ran from
0.02 M to 0.70 M. Protein (the open circles) was
determined as before (1) after precipitation of the
fractions with trichloroacetic acid. The activity (closed
circles) was determined using the fluorometric assay (see
text). The phosphate concentration (open squares) was
determined by measuring the conductivity of the fractions
using a Radiometer CDM3 conductivity meter.

Since the substrate DNA is being subjected to cycles of nicking and sealing we would predict that during the course of the reaction there should exist a steady state level of the nicked intermediate. Currently, work is underway to identify and isolate the nicked intermediate as a first step towards understanding the mechanism of the reaction carried out by this enzyme.

ACKNOWLEDGMENTS

This investigation was supported by National Institutes of Health Research Grant CA 13883 from the National Cancer Institute. J. M. D. is a Predoctoral Trainee of NIH supported by Training Grant 5-T01-GM0051.

REFERENCES

1. Champoux, J. J. and Dulbecco, R., Proc. Nat. Acad. Sci., Wash. 69, 143 (1972).
2. Wang, J. C., J. Mol. Biol. 55, 523 (1971).
3. Keller, W. and Wendel, I., Cold Spring Harbor Symp. on Quant. Biol., 39, in press (1974).
4. Baase, W. A. and Wang, J. C., Biochemistry 13, 4299 (1974).
5. Vosberg, H. P., Grossman, L. I., and Vinograd, J., Fed. Proc., Abst. Fed. Amer. Soc. Exp. Biol. 33, 1356, Abstr. No. 751 (1974).
6. Vosberg, H. P. and Vinograd, J., this volume.
7. Wang, J. C., J. Mol. Biol. 89, 783 (1974).
8. Vinograd, J., Lebowitz, J., and Watson, R., J. Mol. Biol. 33, 173 (1968).
9. Gillespie, D. and Spiegelman, S., J. Mol. Biol. 12, 829 (1965).
10. LePecq, J. B. and Paoletti, C., J. Mol. Biol. 27, 87 (1967).
11. Paoletti, C., LePecq, J. B., and Lehman, I. R., J. Mol. Biol. 55, 75 (1971).
12. Radloff, R., Bauer, W., and Vinograd, J., Proc. Nat. Acad. Sci., Wash. 57, 1514 (1967).
13. Bauer, W. and Vinograd, J., J. Mol. Biol. 54, 281 (1970).
14. Champoux, J. J. and McConaughy, B. L., Biochemistry 14, 307 (1975).

ISOLATION AND PROPERTIES OF A NICKING - CLOSING PROTEIN FROM MAMMALIAN NUCLEI

Hans-Peter Vosberg and Jerome Vinograd
Division of Biology and Division of Chemistry
and Chemical Engineering
California Institute of Technology
Pasadena, California 91125

SUMMARY. An enzyme able to release superhelical turns in closed circular DNA, called "relaxation protein", has been isolated and partially purified from the nuclei of mouse L cells. It is predominantly located in the nuclei, tightly associated with DNA and is apparently a major constituent among the nuclear non-histone proteins. It relaxes, without need for a cofactor, negatively and positively supercoiled DNA. The optimal conditions at $37°$ are 0.2 M NaCl and 10 mM potassium phosphate or Tris-HCl buffer, pH 7.0. Relaxation protein acts enzymatically, one enzyme molecule relaxes on the average at least 8 PM2 DNA I molecules in two hours. The reaction is roughly first order during the first 30 min at $37°$. The relaxation activity is measured by a fluorometric assay which uses the fluorescence enhancement of ethidium bromide bound to DNA and the fact that the binding affinity of the dye to closed circular

DNA depends on the superhelix density of the DNA. The
biological function of the enzyme is not yet known, but its
in vitro reaction with superhelical DNA suggests a role
as a nicking-closing device active on replicating DNA
ahead of the replication fork. Alternatively any nuclear
process which is either accompanied by or requires a
change in the duplex winding number of the DNA may be
affected by this enzyme.

INTRODUCTION

The nicking-closing protein described here is
tentatively called "relaxation protein", as in a previous
communication (1). The terms "relaxation" and "relaxation protein" have been used previously for other types
of enzymatic events in which supercoiled DNA serves as
a substrate. Col E1 and other plasmid DNAs have been
demonstrated to exist in the form of supercoiled DNA-
protein complexes, from which the DNA is released by
treatment with various detergents or pronase as a nicked
or "relaxed" circle. The protein involved in the conversion of superhelical DNA into the "relaxed" nicked
form, presumably a nuclease, has been called "relaxation protein" (2). Studies of the reaction mechanism of
E. coli ligase have revealed that this enzyme catalyses
an AMP dependent conversion of supercoiled DNA into a
non-supercoiled covalently closed form (3). The first
step in this reaction, the nicking of the supercoiled
closed circle, is understood to be the endonucleolytic
reversal of the main reaction of ligase. The product

95

formed has been called "relaxed" DNA.

A different reaction has been described by Wang (1971) for the ω protein of E. coli. This protein in the absence of a cofactor releases the negative supercoils of closed circular DNA (4). While the mechanism of action of the ω protein is not known, it is clear that a nicking - swivelling - closing sequence must be involved. This type of "relaxation" is also performed by the "relaxation protein" obtained from the nuclei of cultured LA9 mouse cells in our studies. Champoux and Dulbecco (1972) reported the presence of such an activity in the crude extracts of cultivated embryonic mouse cells (5). Recent reports suggest that proteins able to relax super-helical DNA in vitro are of widespread occurrence. They are referred to under different names such as untwisting enzyme (5), relaxing activity (6), ω, not only for the E. coli protein but also for the corresponding proteins from Drosophila melanogaster eggs and from calf thymus (7, 8) and "relaxation protein". Most probably these different proteins from eukaryotes are functionally similar if not identical.

We previously developed a procedure for the purification of the relaxation protein from mouse L cell nuclei (1). While the protein was apparently homogeneous in gel electrophoresis experiments, the yield of protein was low and the amounts were inadequate for studies of the properties and the function of the enzyme. We therefore revised the extraction and purification procedure and have obtained preparations of larger amounts of relaxation

protein. A first account of the new method, as far as it
has been worked out, is presented here. A more detailed
description will be published elsewhere (9). We are now
able to provide evidence that the relaxation protein acts
enzymatically rather than stoichiometrically. We finally
discuss the alternative sets of events that can lead to the
complete relaxation of a DNA molecule. The possible
in vivo role of the enzyme is discussed.

MATERIALS AND METHODS

Cell growth: LA9 cells were grown in suspension
culture in Dulbecco's modification of Eagle's medium
supplemented with 8% calf serum. Spinners were inocu-
lated at 8×10^4 cells/ml and harvested after 72 hours at
approximately 8×10^5 cells/ml.

DNA: PM2 DNA isolated from phage particles was used
as a substrate for the relaxation protein. The phage
stocks were prepared according to the procedure of
Espejo et al. (10).

Enzymes: E. coli alkaline phosphatase was from
Worthington.

Other materials: Polyethylene glycol 6000 (PEG) was
obtained from Union Carbide. Biogel HTP (hydroxyapatite)
was purchased from Bio-Rad. DEAE cellulose (type
DE 52) was from Whatman (England). Sephadex G 150
was purchased from Pharmacia (Sweden). Agarose was
from Marine Colloids Inc. and CsCl was from Reliable

Chemical Company. Ethidium bromide and propidium
diiodide were obtained from Calbiochem. Salmon sperm
DNA and horse heart cytochrome C were from Sigma.
Bovine serum albumin, fraction V (crystalline) was from
Miles Laboratories.

Standard reaction conditions: For the centrifugation
analysis and for the fluorometric assay, 1 μg of PM2
DNA I was incubated in a total volume of 250 μl containing 10 mM potassium phosphate, pH 7.0, 0.2 M NaCl,
0.2 mM EDTA, 1 mM spermidine, 50 μg/ml bovine
serum albumin and relaxation protein in 1 to 15 μl. For
gel electrophoresis in 0.8% agarose (w/v) slab gels
the total reaction volume was 50 μl containing 100 ng of
PM2 DNA I and all other components in the above concentrations.

Assay methods: Three different assay methods have
been used for the analysis of the products of the relaxation process and for monitoring the activity during the
purification. Two of the three methods are based on the
fact that the binding affinity of superhelical DNA for the
intercalating dye, ethidium bromide, decreases with the
number of negative superhelical turns in the dye-free
DNA (11). This result has been used in buoyant density
centrifugation in CsCl containing ethidium bromide. The
relaxed products bind less dye than the native DNA and
the resulting DNA-dye complex bands in a position of
higher density than the complex formed with native DNA.
Nicked DNA binds more dye than native DNA and is there-

fore easily distinguishable from both the native and the relaxed DNA. The centrifugation was done in a Beckman SW 50.1 rotor at 35,000 rev/min for at least 24 hours and the DNA bands were visualized as fluorescent bands in long wave length UV light and photographed using Kodak Royal Pan film No. 4141 with an optical system previously described (12). The centrifugation assay has been used for preliminary kinetic studies, for the evaluation of the superhelix density of PM2 DNA relaxed in the presence of ethidium bromide and finally for the titration of relaxing activity in the early steps of the purification (from fraction 0 to fraction II). These fractions contained debris and/or significant amounts of DNA from the nuclei and were therefore less suitable for the fluorometric assay (see below). For the purpose of titration, the distribution of native and relaxed DNA in the centrifuge tubes was determined densitometrically as described previously (1).

Because this method is time consuming, we developed a fluorometric procedure based on the fluorescence enhancement of ethidium bromide that occurs upon binding of the dye to double stranded DNA (13). The relaxation of the superhelical DNA results in a reduction of the amount dye bound and causes a decrease in the total fluorescence. In samples containing constant concentrations of DNA and dye, the fluorescence decrease is indicative of the reduction in superhelix density. Endonucleolytic activity under the given conditions increased the fluorescence. The details of this technique have been published elsewhere (1). The fluorescence method was first used by Paoletti et al. (14) to follow the nicking of closed circular DNA by endonucleases and the closing of open circles by DNA ligase. An alternative procedure using this method for the measurement of the relaxation activity has been independently developed by Morgan and Pulleyblank (15).

This fluorometric assay is rapid and sensitive enough for the quantitation of the relaxing activity. It has been used during the purification for the titration of the activity from fraction III on. The fluorescence decrease was measured as a function of the amount of relaxation protein present during a 30 minute incubation at 37° (Fig. 1). The relaxation reaction is taken to be complete when a further increase of the amount of relaxation protein did not change the fluorescence. Because of the logarithmic shape of such a reaction curve it is difficult to determine the amount of relaxation protein just suffi-cient for a complete reaction. The unit of the activity is therefore defined as the amount of relaxation protein that produces 50% of the maximum fluorescence decrease under the above conditions. Product analysis in a CsCl-ethidium bromide gradient of the DNA relaxed with one unit of relaxation protein exhibited an overall bimodal distribution of either unreacted or fully relaxed DNA.

The third assay method uses the fact that the migration of closed circular DNA in agarose gels depends on the superhelix density of a given DNA. Native DNA migrates faster than relaxed and nicked circular DNA (16).

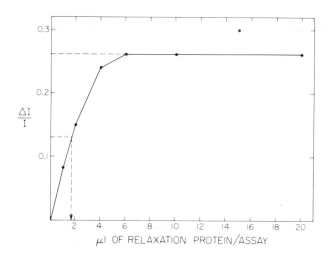

FIGURE 1   The change in the fluorescence enhancement
as a result of the incubation of PM2 DNA with varying
volumes of a solution containing relaxation protein. The
standard incubation conditions are given in the text. I is
the fluorescence of a sample containing 1 μg of unrelaxed
PM2 DNA and ΔI is the change in the fluorescence observed after incubation with the indicated volumes for
30 min at 37°. The fluorescence values were determined
after diluting the 250 μl incubation mixture into a volume
of 2.75 ml containing 0.5 μg/ml ethidium bromide, 10 mM
Tris-HCl, pH 7.6 and 0.5 mM EDTA. The excitation
wave length was 365 nm and the emission was monitored
at 584 nm at maximal amplification in a Hitachi Perkin-
Elmer MPF 2A spectrofluorometer. The relaxation protein (fraction V) was diluted with standard buffer 1 : 800.

This procedure has been extensively used by Keller (6). During these studies, this method was used for monitoring relaxation activity in fractions of column eluates and for the analysis of the distribution of superhelix densities in extensively reacted DNA.

RESULTS

Purification of Relaxation Protein

The standard buffer was 0.02 M potassium phosphate, pH 7.0 and 1 mM β-mercaptoethanol. All other potassium phosphate buffers used here were at pH 7.0 and contained 1 mM β-mercaptoethanol. All steps in the procedure were carried out at 0 to 4°.

A previously published purification procedure resulted in a preparation which produced a single band in an SDS polyacrylamide gel. The yield was less than 1% of the activity present in the crude nuclear extracts, or only approximately 50 μg protein from a 5 liter culture containing 4×10^9 cells. The improved procedure, as far as it has been worked out, is presented in Table 1.

Nuclear extracts; Cells in middle log phase were collected by centrifugation, washed and disrupted in a Dounce homogenizer. The nuclei were isolated by a low speed centrifugation (2.5 min at 2500 rev/min at 4°) and washed in standard buffer until the supernatant was no longer turbid.

It was obvious from our previous studies that the relaxation protein is relatively tightly bound to chromosomal DNA, and that incomplete dissociation of the

protein from the DNA resulted in significant losses during
the first spin down of the crude extract. The isolated
nuclei suspended in standard buffer were disrupted in a
Sorvall Omnimix, then supplemented with KCl to a final
concentration of 1 M, and sonicated with a Branson
Sonifier for 2.5 min at step 4. Salt treatment (up to 2.5 M
NaCl) without sonication, or sonication at 20 mM NaCl or
KCl resulted in lower activities in the nuclear extracts.
In combination, neither a further increase of the salt
concentration nor an extension of the sonication time
yielded a significant increase of the activity. The above
result suggests that the combination of high salt and
sonication solubilizes essentially all of the relaxation
protein present in the nuclei.

Precipitation with polyethylene glycol: The sonicate
(fraction 0) was centrifuged for 5 hours at 20,000 rev/min
(in a Spinco rotor, type 30). The activity was almost
completely recovered in the supernatant (fraction I). 5%
PEG (w/v) was dissolved into fraction I. After 20 min
of gentle stirring the resulting precipitate was removed in
a Sorvall centrifuge at 10,000 rev/min. The supernatant,
fraction II, contained 90% of the activity. The bulk of
the chromosomal DNA and approximately 40% of the proteins were removed in this step.

Hydroxyapatite I: Fraction II was diluted with an equal
volume of standard buffer to give a final concentration of
2.5% PEG and 0.5 M KCl. The solution was immediately
loaded onto a hydroxyapatite column (2.5 cm × 7.0 cm),

previously equilibrated against standard buffer containing
0.5 M KCl. PEG did not prevent binding of the relaxation
protein to hydroxyapatite. After loading, the column was
washed with 10 bed volumes of 0.5 M KCl (in standard
buffer) and then with 10 bed volumes 0.2 M potassium
phosphate. No elutable DNA remained on the column. The
relaxation protein was eluted between 0.45 and 0.65 M in
a linear gradient of 0.2 to 1.0 M potassium phosphate.
The fractions containing activity (Fig. 2) were pooled to
form fraction III.

DEAE cellulose: Fraction III was diluted to 0.05 potassium phosphate by addition of 9 volumes of 1 mM β-mer-
captoethanol and loaded onto a DEAE cellulose column
(2 cm × 6 cm) which was equilibrated against 0.05 M
potassium phosphate. The column was then developed
with 0.05 M potassium phosphate. Approximately 80%
of the recovered activity passed through unretarded
together with 30% of the total protein in fraction III. The
remaining 20% of the activity was retained and then eluted
with a linear gradient at about 0.2 M potassium phosphate.
The first portion was taken as fraction IV and purified
further. It is not known yet whether the minor portion of
the activity which bound to DEAE cellulose is identical
with the activity in the major portion. In the previous
experiments (1) all of the activity recovered from DEAE
cellulose was initially bound to the ion exchange resin.
We tentatively ascribe the discrepancy to the presence of
significant amounts of nuclear DNA in the earlier samples.
It is possible that the DNA was bound to the adsorbent,

and the relaxation protein was bound to the DNA. On the other hand the different binding behavior might have been caused by the change from a KCl to a potassium phosphate gradient.

Hydroxyapatite II: Fraction IV in 1100 ml was immediately loaded onto a small hydroxyapatite column (bed volume 1 ml) previously equilibrated against 0.05 M potassium phosphate. Approximately 50% of the protein present in fraction IV ran through the column, while the activity was retained. The column was washed with 20 ml of 0.2 M potassium phosphate, again with retention of the activity. The relaxation protein was subsequently released in a one step elution with 0.5 M potassium phosphate and collected in a volume of 6 to 7 ml. This volume was fraction V and had a specific activity of ca. 1×10^6 units/mg (see Table 1).

Fraction V contained 3 mg of partially purified relaxation protein with approximately 15% of the activity present in fraction 0. The SDS gel electrophoresis revealed a pattern of 3 or 4 protein bands, indicating, that further purification is necessary. As compared to the previous purification procedure the present one offers three major advantages: the relaxation protein is apparently completely solubilized by treatment of the nuclei with high salt and sonication. DNA is effectively removed in the early purification steps without losing more than 50% of the activity, and the overall recovery, at least up to the present level of purification, is high (15%).

TABLE I

PURIFICATION OF RELAXATION PROTEIN*

Fraction	Vol. (ml)	Total Protein (mg)	Total Activity (units)	Spec. Activity (units/mg)	Purification
0 Sonicated nuclei	102	880	20.4×10^6	2.32×10^4	x 1
I Supernatant of soni-cated nuclei	100	580	20×10^6	3.45×10^4	x 1.5
II PEG	98	350	18×10^6	5.14×10^4	x 2.2
III Hydroxyapatite I	110	20.9	10.8×10^6	5.16×10^5	x 22.2
IV DEAE cellulose	1100	7.3	5.5×10^6	7.5×10^5	x 32.3
V Hydroxyapatite II	6.3	2.96	2.96×10^6	1×10^6	x 43.1

* Prepared from 1.4×10^{10} cells

TABLE 1   The activity in fraction I - II was determined by the buoyant separation method. The activity in fraction III - V was determined by the fluorometric method. Protein determinations were carried out according to the Lowry procedure (27).

The recovered 15% of the initial activity is contained in 0.3 to 0.4% of the protein present in fraction 0. Under the assumption that there are no stimulating factors in the crude extract we conclude that fraction V contains one sixth of the relaxation protein present in the nuclei. If only 20% of the fraction V protein were relaxation protein, it would account for ca. 0.5% of the total nuclear protein or ca. 1% of the non-histone protein. The results in Table 1 and independent experiments (Wells and Vosberg, unpublished) indicate that less than 1000 log phase LA9 cells contain 1 unit of activity. The protein content of a corresponding number of nuclei is 30 to 45 ng. If we take 75,000 for the molecular weight of the relaxation protein (see below), we calculate that each nucleus contains minimally one to two million molecules of relaxation protein.

Further evidence for the relaxation protein being a major nuclear protein component comes from the low increase of specific activity during the purification. The previously observed increase was ca. 27 fold (1). The present method leads to a ca. 43 fold purification, an improvement that is probably due to more effective solubilization in the first step. A plausible explanation for the low increase in both methods is the presence of relaxation protein in high concentrations in the nuclei.

Some Properties of the Relaxation Protein

The relaxation protein requires for its optimal activity 0.2 M NaCl. Mg^{++} or other divalent cations are

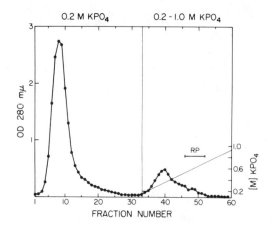

FIGURE 2   Hydroxyapatite chromatography I. Fraction
II was loaded onto the column as described in the text and
washed with 10 bed volumes of 0.5 M KCl in standard
buffer (not shown in the figure). The optical density profile at 280 nm is shown for the subsequent wash with 0.2
M potassium phosphate and for the linear gradient from
0.2 to 1.0 M potassium phosphate (all buffers at pH 7.0).
13 ml fractions were collected and the relaxation activity
was monitored using 0.8% agarose slab gels. Fractions
46 – 52 contained the activity as indicated (RP) and were
combined to form fraction III.

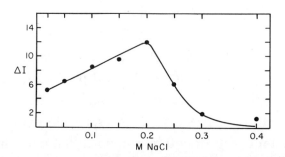

FIGURE 3   Salt optimum for the relaxation protein.
Incubations were performed under standard reaction conditions with 2 μg of native PM2 DNA and 10 units of fraction V in a total volume of 510 μl. Samples of 250 μl
were withdrawn before and after a 30 min incubation and
added to 2.5 ml of ethidium bromide solution (1 μg/ml).
The ordinate shows the change in fluorescence enhancement. The shape of the curve, but not the optimum, depends on the number of units used in the assay, cf.
figure 1.

not required. Replacement of the NaCl by 0.04 M $MgCl_2$ is partially effective. 0.2 M KCl or 0.2 M NH_4Cl replace NaCl without loss of activity. The salt dependence for the reaction as followed in the fluorescence assay is presented in Fig. 3. Relaxation protein is active over a broad range of pH between 6.5 and 8.5 as shown in Fig. 4. No activity was observed at pH 4.5.

The relaxation activity on PM2 DNA is partially inhibited by the presence of equal amounts of either double stranded or denatured salmon sperm DNA. The latter was more inhibitory (72%) than the former (35%). Inhibition by p-chloromercuribenzoate (1 mM) indicates the involvement of an SH group in the relaxation process.

In conformation with previously published results on crude extracts (5), we have found that the relaxation protein in fraction V does not contain a conventional ligase and also that there is no apparent requirement for any added cofactor. Keller has found similar results in his purified material (6).

A tentative value for the molecular weight, 75,000 to 80,000, of the native protein was found by gel filtration with Sephadex G 150 (Fig. 5). The relaxation protein was eluted after alkaline phosphatase (molecular weight 86,000) in a volume that corresponds to a molecular weight in the above range. This result was obtained in standard buffer containing 0 or 40 mM KCl. In the presence of 0.5 M KCl or NaCl however the relaxation activity eluted consistently before the alkaline phosphatase. The previous purification resulted in one visible band in an SDS polyacrylamide gel with the most purified

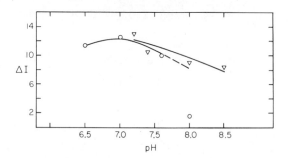

FIGURE 4   pH optimum for the relaxation protein. Incubations and analyses were essentially as described in
the legend of figure 3, except that 4 units of fraction V
were added in each assay. The solutions contained potassium phosphate, O, or Tris-HCl, Δ, buffers at indicated values of pH.

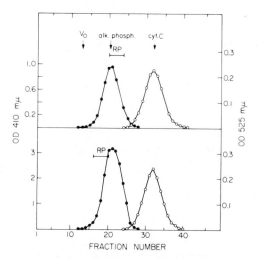

FIGURE 5   Gel filtration with Sephadex G 150. The
elution profiles of relaxation protein in standard buffer
containing 40 mM KCl (upper panel) and 0.5 M KCl (lower panel) are shown. In both cases a total of ca.
50,000 units of fraction V together with 0.05 units (upper
panel) and 0.15 units (lower panel) of alkaline phosphatase and cytochrome C were loaded onto a 2.5 cm x 45 cm
column. 5 ml fractions were collected and the relaxation
activity (indicated as RP) was analysed using agarose
gel electrophoresis. Cytochrome C was monitored directly at 525 nm and alkaline phosphatase after incubation with 1 mM p-nitrophenylphosphate in 1 M Tris-HCl,
pH 8.0, for 45 min at room temperature by measuring
the absorbance at 410 nm. V_0 is indicated by an arrow.
●, OD 410 nm; O, OD 525 nm.

fraction corresponding to a molecular weight of ca. 37,000. This latter result combined with the gel filtration in the absence of KCl suggests that the native protein contains two subunits of similar molecular weights.

Extensive kinetic studies have not yet been performed. The fluorometric procedure is rapid and sensitive enough to follow the time course of the reaction. Product analysis with ^{3}H labeled PM2 DNA in buoyant centrifugation in the presence of propidium diiodide revealed that under standard conditions with 1 μg of DNA the reaction is roughly first order for the first 30 minutes at 37°, the time chosen for the standard assay.

The Relaxation Protein is an Enzyme

With the high specific activity of relaxation protein obtained, it was possible to follow the reaction under conditions in which the number of DNA molecules was in excess over the number of relaxation protein molecules. The calculation of the latter number was based on the protein concentration, as determined by the Lowry procedure, and the assumptions that all the protein present in a fraction V is relaxation protein with a molecular weight of 75,000. The reaction was followed in a series of assays which contained 1 μg PM2 DNA I (molecular weight 6 x 10^6) and one unit of relaxation protein having a specific activity of 6.9 x 10^5 units/mg. Each assay contained a minimum 8 fold excess of DNA molecules over relaxation protein molecules. After incubation for different times the reaction was found to be complete within two hours as judged by the gradual disappearance of DNA banding at the position of native PM2-DNA and the appearance of DNA in the density position of relaxed DNA in a CsCl-propidium diiodide density gradient. Each protein molecule therefore reacted on the average with eight DNA molecules. From this result we conclude that the relaxation protein has an enzymatic mode of action.

The value of eight total relaxations per protein molecule is a low estimate since we know that the assumption of protein homogeneity in fraction V is incorrect. The number of relaxation protein molecules present in the assays was lower than that used in the calculation and the number of total relaxations will have to be corrected upward by a factor depending on the relative amount of impurity in fraction V.

Reaction With Positive Superhelical DNA

Champoux and Dulbecco reported that the relaxation protein acts on both negative and positive superhelical DNA (5). This is in contrast to the ω protein from E. coli which does not act on positive superhelical DNA as shown by Wang (4). We have examined this property with the previously purified relaxation protein and have confirmed the original observation by Champoux and Dulbecco. In addition we have estimated the extent of positive supercoiling present when the relaxation protein introduced the first obligatory chain break into the DNA molecule. Positive supercoils in these experiments were formed by the intercalation of ethidium bromide and the tests for the relaxation were made in the presence of dye.

Addition of ethidium bromide to the native PM2 DNA reduces the duplex winding number and the negative super-coiling. At a critical dye concentration the negative supercoiling disappears. Further addition of dye generates positively supercoiled DNA-dye complexes. The relaxation protein acts on these complexes removing all superhelical turns in the presence of dye and creating a relaxed DNA-dye complex, which binds more dye than the native DNA that formed the original complex. Removal of the dye after complete reaction results in a DNA with a higher negative superhelix density than the native substrate. In summarizing the results which have been published in detail elsewhere(1), we have calculated that the highest value of σ obtained ($\sigma = -0.098$) corresponds to a positive superhelix density of $\sigma = +0.023$ in the substrate complex*). We note that this positive value for σ is relatively small as compared with the absolute value of the superhelix density of the final reaction product after removal of the dye.

Mechanism of Action

A temporary chain break and a rejoining after rotation of the duplex around the intact strand opposite

---

*) All values of σ in these studies were originally based on an unwinding angle of 12° for the binding of ethidium bromide. Recent results (17, 18) suggests an unwinding angle higher by at least a factor of two. All values for σ therefore have to be corrected by the same factor.

the chain break are required for the relaxation process. The obligate nicked intermediate has not yet been demonstrated. At the present time it is not known whether the reaction proceeds stepwise, releasing only one superhelical turn at a time and then closing or whether all turns are removed after the hydrolytic event. From the fact that under standard conditions we never see a gradual shift of the whole DNA population to the position of completely relaxed DNA, we conclude that the relaxation protein has a preference for finishing the relaxation on one DNA molecule before it reacts with another one.

With respect to the structure of the extensively reacted DNA, we suggested previously on the basis of density centrifugation analysis that the final reaction products do not contain superhelical turns anymore. Agarose gel electrophoresis however has a higher resolving power for the separation of DNAs on the basis of superhelix density, as has been shown by Keller (6). Product analysis on agarose gels with extensively reacted DNA initially confirmed the notion that the product has a superhelical density of zero. Recent work in progress here (Tang, unpublished) has shown that the previously used standard conditions for agarose gel electrophoresis (19) are not effective in resolving a mixture of DNA molecules which have low superhelical densities and are heterogeneous with respect to the number of superhelical turns. The addition of 5 mM $MgCl_2$ or 5 mM Mg acetate to the standard buffer produces a multiple band pattern for DNA which migrated in a single band under standard conditions. Thus it is possible to demonstrate that the final products of the relaxation process are not homogeneous with respect to their superhelix density. It has not yet been determined whether the mean superhelix density is zero under the reaction conditions. Some supercoiling is generated by the shift from reaction conditions to analysis conditions. A possible explanation for superhelix density heterogeneity is the thermally induced fluctuation of the chain conformation in the randomly coiled polymer at the time of ring closure.

DISCUSSION

In principal there are three ways for the reaction to proceed:

a. Single turn mechanism: the removal of superhelical turns occurs stepwise via a repetition of a set of successive nicking, swivelling and closing events with each set removing one turn.

b. Single molecule mechanism: the relaxation occurs in a single step via one nicking event, followed by complete relaxation and one final closing event.

c. Abortive single molecule mechanism: each step removes several turns by a sequence of a nicking event, swivelling and a closing event.

As described in a previous section we have observed that an essentially bimodal distribution of reacted and non-reacted DNA occured under our standard assay conditions with one unit of relaxation protein. These experiments were performed in substrate (DNA molecules) excess of approximately 10 to 1. This result initially sug-

115

gests that mechanism b is operative. On the other hand
mechanism a with the repetition occuring rapidly relative to the initiation of the first nick could have produced
the same result. Furthermore mechanism c could occur
simultaneously with a or b , but less frequently. It is
important to note that we did not observe a gradual shift
of a single buoyant band during the above kinetic experiments. The gradual shift would be expected from a and
from c if limited swivelling occured and in both cases
the enzyme were released after each step. Since it is
likely that in b the enzyme remains on the DNA during
the relaxation process, we come to the conclusion that
all three mechanisms are likely to involve a continuous
association of an enzyme molecule with a DNA molecule
during its relaxation.

Relaxation protein is located predominantly in the
nucleus. Approximately 92% of the total cellular activity
is associated with the nuclei, 7% is found in the cytoplasmic supernatant of isolated nuclei, and less than 1% in
purified (20) mitochondria. Isolated rat liver chromatin
contains significant amounts of relaxation activity (Vosberg and Gottesfeldt, unpublished). The isolation of
chromatin together with the tightly bound relaxation protein has been used as an initial step in the purification of
relaxation protein from calf thymus (8). From the previous and present purification data, we tentatively conclude that the relaxation protein constitutes about at
least 1% of the non-histone nuclear proteins.

No physiological function has been identified for

either the relaxation protein or the ω protein from <u>E. coli</u>.
The relaxation activity can obviously replace the
successive action of an endonuclease and a ligase in
promoting transient swivels in duplex DNA. Such transient swivels are considered to be necessary in the
replication of closed circular DNA because of the presence
of closed circular replicative intermediates (21, 22).
Transient swivels would also eliminate the need for the
rapid rotation of long stretches of chromosomal parental
DNA ahead of the advancing replication fork.

We have investigated the amount of activity in cells
in various stages of the cell cycle in order to estimate,
whether the activity is related to DNA synthesis. For this
purpose, we separated ^{14}C thymidine labeled LA9 log
phase cells according to their size by zonal centifugation
(23). The DNA replication activity of the cells was
followed by pulse labeling with ^{3}H thymidine. The ratio
of ^{14}C to ^{3}H label in the cells after separation was used to
locate the position of the separated cells in the cell cycle.
Preliminary results indicate that the cells exhibit the
highest relaxation activity (approximately 0.6 units per
10^{3} cells) late in the G1 phase before DNA synthesis
starts (Wells and Vosberg, unpublished). This result
does not necessarily link the activity and DNA synthesis,
but it does not contradict the notion that the relaxation
protein plays a part in DNA replication. The decline of
the activity during the S phase to approximately 30% of
the peak activity could indicate that the relaxation
protein is either inactivated or used up by the replication
events.

117

Alternatively the relaxation activity could function during transcription by introducing transient swivels in the DNA ahead of the RNA synthesis sites (24).

Recently published evidence (25, 26) suggests that changes in the superhelix density in closed circular SV40 DNA occur upon the removal of the histones from SV40 "mini chromosomes". This result indicates that the free DNA in the DNA-histone complex is relaxed perhaps as a result of the relaxation protein action and suggests that the structure and/or organisation of the histone bound DNA is responsible for the formation of the superhelical turns upon DNA purification.

In DNA replication and probably in transcription, relaxation activity on positive supercoils is required. It is not known whether condensation or decondensation of DNA in eukaryotes in vivo is accompanied by the occurrence of either negative or positive supercoils. The fact that the ω protein from E. coli acts only on negative supercoils suggests that this type of reaction has a specific functional significance.

ACKNOWLDEGEMENT

This investigation was supported in part by the U. S. Public Health Service grant CA08014 from the National Cancer Institute and grant GM15327 from the National Institute of General Medical Sciences. This is contribution no. 5098 from the Division of Chemistry and Chemical Engineering.

REFERENCES

1. Vosberg, H. P., Grossman, L. I. and Vinograd, J. Eur. J. Biochem., in press (1975).
2. Blair, D. G., Clewell, D. B., Sheratt, D. J. and Helinski, D. R. Proc. Nat. Acad. Sci. (Wash.) 68, 210 (1971).
3. Modrich, P., Lehman, I. R. and Wang, J. C. J. Biol. Chem. 247, 6370 (1972).
4. Wang, J. C. J. Mol. Biol. 55, 523 (1971).
5. Champoux, J. J. and Dulbecco, R. Proc. Nat. Acad. Sci. (Wash.) 69, 143 (1972).
6. Keller, W. and Wendel, I. Cold Spring Harbor Symp. Quant. Biol., Vol. 39, 199 (1974).
7. Baase, W. A. and Wang, J. C. Biochemistry 13, 4299 (1974).
8. Pulleyblank, D. E. and Morgan, A. R. (in preparation).
9. Vosberg, H. P. and Vinograd, J. (in preparation).
10. Espejo, R. T., Canelo, E. S. and Sinsheimer, R. L. Proc. Nat. Acad. Sci. (Wash.) 63, 1164 (1969).
11. Bauer, W. and Vinograd, J. J. Mol. Biol. 54, 281 (1970).
12. Watson, R., Bauer, W. and Vinograd, J. Analyt. Biochem. 44, 200 (1971).
13. LePecq, J. B. and Paoletti, C. J. Mol. Biol. 27, 87 (1967).
14. Paoletti, C., LePecq, J. B. and Lehman, I. R. J. Mol. Biol. 55, 75 (1971).
15. Morgan, A. R. and Pulleyblank, D. E. Biochem. Biophys. Res. Comm. 61, 396 (1974).

16. Thorne, H. V. J. Mol. Biol. <u>24</u>, 203 (1967).

17. Wang, J. C. J. Mol. Biol. <u>89</u>, 783 (1974).

18. Pulleyblank, D. E. and Morgan, A. R. J. Mol. Biol.
    <u>91</u>, 1 (1975).

19. Hayward, G. S. and Smith, M. G. J. Mol. Biol. <u>63</u>,
    383 (1972).

20. Tibetts, C. J. B. and Vinograd, J. J. Biol. Chem.
    <u>248</u>, 3367 (1973).

21. Kasamatsu, H. , Robberson, D. L. and Vinograd, J.
    Proc. Nat. Acad. Sci. (Wash.) <u>68</u>, 2252 (1971).

22. Sebring, E. D. , Kelly Jr. , T. J. , Thoren, M. M.
    and Salzman, N. P. J. Virol. <u>8</u>, 478 (1971).

23. Wells, J. R. Exptl. Cell Res. <u>85</u>, 278 (1974).

24. Wang, J. C. in: DNA Synthesis in Vitro (R. D. Wells
    and R. B. Inman, eds.), 163, University Park Press,
    Baltimore, Md. (1973).

25. Griffith, J. Science <u>187</u>, 1202 (1975).

26. Germond, J. E. , Hirt, B. , Oudet, P. , Gross-Bellard,
    M. and Chambon, P. Proc. Nat. Acad. Sci. (Wash.),
    in press (1975).

27. Lowry, O. H. , Rosebrough, N. J. , Farr, A. L. and
    Randall, R. J.   J. Biol. Chem. <u>193</u>, 265 (1951).

Part II
Replicon Complexes
and Their Stabilization

RNA MOLECULES ATTACHED TO DNA IN ISOLATED
BACTERIAL NUCLEOIDS: THEIR POSSIBLE ROLE
IN STABILIZING THE CONDENSED CHROMOSOME

Ralph M. Hecht and David Pettijohn
Department of Biophysics and Genetics
University of Colorado Medical Center
Denver, Colorado 80220

ABSTRACT. The studies described here investigated the mode
of binding of RNA molecules to the isolated nucleoid of
Escherichia coli. Membrane free nucleoids with ^{3}H-labeled
RNA components were reacted briefly with RNA'ase A. The
average molecular weight of the attached nascent RNA chains
was reduced to less than one fourth the initial size;
however, very little of the RNA was released from the
nucleoid. Even when the state of condensation of the
nucleoid was further relaxed by intercalation of ethidium
bromide, the RNA fragments remained bound to the nucleoid.
The results suggest that many of the nascent RNA chains
are bound to the DNA of the nucleoid at multiple sites.

When the DNA of the nucleoid was completely unfolded,
fragmented and deproteinized, 60-100 nascent RNA chains per
nucleoid equivalent of DNA remained bound to the DNA.
These RNA chains were released from the DNA by heating at
96°C. Several properties of the bound RNA and its mode of
attachment to the DNA are consistent with the idea that
these RNA molecules are those involved in stabilizing the

condensed state of the DNA in the folded chromosome;
however, the data do not prove this interpretation.

INTRODUCTION

The condensed state of the packaged DNA in nucleoids
isolated from <u>Escherichia coli</u> is stabilized by RNA molecules attached to the nucleoid (1,2,3). When the nucleoids
are treated with RNA'ase the condensed chromosome unfolds
and acquires the usual properties of extended high-molecu-
lar weight double-helical DNA. Also the DNA is unstable in
its state of condensation when nucleoids are released from
cells grown in the presence of inhibitors of RNA synthesis
(3,4,5). Different hypotheses have been advanced to
explain how nucleoid bound RNA molecules can stabilize the
chromosome structure (2,3), but the actual mechanism by
which the interaction occurs is unknown. It is known that
most (and perhaps all) of the nucleoid attached RNA molecules are nascent chains, bound to the DNA via their associated RNA polymerase molecules (6,7). Using the most
gentle conditions of purification, essentially all of the
nascent RNA of the bacterial cell is obtained with the
isolated nucleoid and the associated RNA polymerase molecules are distributed among the genes as they were in the
cell.

To restrain a DNA fold or segregate a domain of DNA
supercoiling in the condensed chromosome, binding of an
RNA molecule at more than one site on the DNA would be
required (3). The nascent RNA chains of the nucleoid are
known only to be associated with the DNA at one site,

their 3' end which is involved in the ternary transcription
complex. In the research to be described here, we have
designed experiments to determine if any RNA molecules in
the nucleoid are multiply bound to the packaged DNA at
sites other than their 3' end. It is demonstrated that
such a class of bound RNA molecules does exist and that
this RNA fraction has properties expected of the nucleoid
stabilizing RNA.

METHODS

Bacteria: Nucleoids and nucleic acids were isolated from
Escherichia coli strain D-10 (RNAase I⁻), cultured as
previously described (6).

Nucleoid Isolation and Conditions of Radioactive Labeling:
The experiments utilized the membrane-free nucleoid isolated as before (3,8). The standard isolation buffer-
solution present in sucrose gradients during purification
steps and during resedimentation studies contained: 0.01M
Tris-HCl (pH 8.1 at 25°C), 1.0M NaCl, 1mM β-mercaptoetha-
nol and 1mM EDTA. Labeling of the DNA or RNA of the
nucleoid was accomplished by culturing the cells in the
presence of respectively (^{14}C-methyl) thymidine (0.2 to
1.0 µc/ml, 10-50 mC/mM, for 20-40 min) or 5 -^{3}H-uridine
(5-50 µc/ml, 20-50 C/mM, for 30-90 sec).

Agarose Column Chromatography: DNA and its bound RNA were
separated from free RNA on columns of agarose A150 (Bio
Rad). The sample (0.5 ml) containing the mixture of

124

nucleic acids was applied to a 50 cm x 1.8 cm^2 column which
was equilibrated with a solution containing 0.1M NaCl,
0.01M Tris (pH 7.6), 1mM EDTA and 0.02% sodium azide. The
flow rate was adjusted to 0.05 to 0.15 ml/min.

Purification of Total Nucleic Acids: This technique was
essentially as before (6). In brief, the cells were lysed
with lysozyme and sodium dodecyl sulfate (SDS) and finally
the lysate was incubated 20 min at 30°C in the presence of
0.5% SDS. The DNA was sheared by gently passing the lysate
through a 22 gauge hypodermic needle three times and the
lysate was deproteinized by phenol extraction.

RESULTS

One approach to determining the sites of attachment
of nascent RNA molecules to the nucleoid utilized RNAase
to release portions of the RNA chains. If the only site
of attachment of the RNA was at its 3' end, cleavage of
the RNA should release from the nucleoid that portion of
the RNA chain between the break and the 5' end. Conversely, if there were multiple sites of attachment of the RNA
to the nucleoid, much less RNA would be released when the
chain was cut with RNAase.

Nucleoids were isolated from cells which had been
labeled for 90 sec with 5 -^{3}H-uridine. This period of
labeling should have been sufficient to label the entire
sequence of nascent RNA molecules of length about 4000
bases or less, since the RNA chain growth rate in cells
cultured under these conditions is about 40-60

nucleotides/sec (9,10). Moreover, only a very small
fraction of the total nucleoid RNA has a chain length in
excess of 4000 bases (7)(see also Fig 2). The purified
nucleoids were incubated briefly with RNA'ase A, the
RNA'ase was inactivated and the nucleoids were sedimented
on sucrose gradients to separate them from any released
RNA (Fig 1). The conditions of incubation with the RNAase
were selected so that only partial unfolding of the chromosome occurred (3). The slightly reduced sedimentation
rate of the RNAase treated nucleoid compared to the
untreated control was typical of the partially unfolded
chromosome. As shown previously (1,3,8), higher concentrations of RNAase or longer incubation times with the RNAase
are required to complete the unfolding process as manifested by transitions in specific viscosity and sedimentation rate or as observed by electron microscopy. As
described in Figure 1 less than half of the total labeled
RNA was released from the nucleoid by the RNAase treatment.
Some of the RNA sedimenting near the top of the gradient is
attributable to the increased amounts of more extensively
unfolded DNA. The amount of RNA bound per nucleoid at the
peak of the sedimentation profile was reduced only slightly
from the untreated control as indicated by the ratio
^{3}H-RNA:^{14}C-DNA - 23 for the control and 20 for the RNAase
treated nucleoid.

RNA molecules from the RNAase treated nucleoids were
released from the DNA by heating in the presence of sodium
dodecyl sulfate and sedimented on sucrose gradients with
^{32}P labeled rRNA internal markers (Fig 2). The ^{14}C labeled
DNA sedimented to the bottom of the tube and was recovered

126

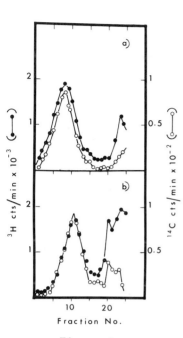

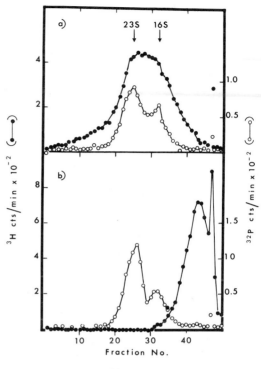

Figure 1   Figure 2

Fig. 1.   Sedimentation of isolated nucleoids before and after treatment with RNAase.   Purified membrane-free nucleoids with ^{3}H labeled RNA components and ^{14}C labeled DNA were incubated 15 min at 9°C in standard isolation buffer containing A) no RNAase or B) 5 µg/ml RNAase A. The solutions were chilled to 0°C and diethylpyrocarbonate was added to inactivate the RNAase as described previously (3); the solutions were then layered on separate sucrose gradients and centrifuged 50 min at 19,000 rpm at 4°C in an SW50 rotor.   0—0 ^{14}C-DNA;   ●—● ^{3}H-RNA.

Fig. 2.   Sedimentation of free RNA chains isolated from nucleoids.   Peak fractions of the nucleoid bands from the two sucrose gradients described in Fig. 1 were pooled separately and heated for 15 min at 70°C in the presence of 0.5% SDS.   Solutions were chilled to 20°C and purified ^{32}P labeled 23S and 16S rRNA markers were added.   The mixtures were layered on separate sucrose gradients containing 0.1M NaCl, 0.3% SDS and 1mM EDTA and centrifuged 5.5 hrs at 34,000 rpm and 22°C in an SW50 rotor.   A) RNA isolated from control nucleoids incubated without RNAase B) RNA isolated from the nucleoids treated with RNAase. ●—● ^{3}H-RNA; 0—0 ^{32}P-rRNA markers.

in the pellet (data not shown). The weight average sedimentation rate of the RNA from the RNAase treated nucleoids
was about 8S and from the control about 20S. Thus the
molecular weight of the RNA was reduced at least 4 fold
without the RNA fragments being released from the nucleoid.
This finding suggests that there is more than one site of
attachment to the nucleoid per RNA molecule.

The method of labeling the RNA chains of Figure 1
introduces a possible error because the specific activity
of the RNA precursor pools varies during the course of the
labeling. It is likely that the specific activity of
portions of a nascent RNA chain will differ depending on
whether the region is nearer the 3' or 5' end of the chain.
The amount of radioactivity released from the nucleoid
after RNA'ase treatment may, therefore, not accurately
indicate the amount of released RNA. This possible deviation would be greatest for the longest nascent RNA chains.
Our studies of precursor pool specific activity during the
90 sec pulse labeling indicate that the 5' half of a chain
4000 bases in length could differ in specific activity by
no more than a factor of 2 from the 3' half of the chain.
This result is also predictable from prior studies of
precursor pools during pulse-labeling (13,14). Even if the
specific activities did vary by as much as this, it would
not be possible to release more than 75% of the RNA mass
from the nucleoid and yet retain more than 85% of the
radioactivity. Thus the specific activity effect will not
in itself account for the results of Figs 1 and 2.

Is it possible that the hydrolysis of the nucleoid
associated RNA occurred after the nucleoid was separated

from the free RNA? This might occur, for example, if trace amounts of RNAase survived the inactivation and fractionation steps described in Fig 1. Fractions from the nucleoid peak of Fig 1B were mixed with ^{32}P-rRNA markers and incubated 30 min at 30°C. The sedimentation profile of both the ^{3}H-RNA and ^{32}P rRNA was identical to that shown in Fig 2B, showing that the contaminating RNA'ase activity was insignificant.

Is it possible that the RNA fragments are retained with the nucleoid because they are physically trapped in the packaged DNA and that their co-sedimentation with the nucleoid is not attributable to additional sites of binding? To investigate this possibility the state of condensation of the RNA'ase treated nucleoids has been relaxed additionally to determine if more RNA is released. Nucleoids with ^{14}C labeled DNA and ^{3}H labeled RNA were treated with RNA'ase exactly as described in Figure 1 and then sedimented in the presence of varied concentrations of ethidium bromide. It was previously shown (2,3) that ethidium bromide unwinds the supercoils in the folded chromosome and that at appropriate concentrations there is a substantial de-condensation of the packaged DNA. Under our conditions the maximal de-condensation, as revealed by a minimum in the sedimentation rate, was observed at an ethidium bromide concentration of 2 µg/ml (3). As shown in Table 1 below, the amount of bound RNA per nucleoid indicated by the ^{3}H-RNA:^{14}C-DNA ratio was the same in the presence of zero or 2 µg/ml ethidium bromide. As further evidence it will be demonstrated below that RNA molecules

129

Table 1	EFFECT OF ETHIDIUM BROMIDE ON THE AMOUNT OF BOUND RNA IN RNAase TREATED NUCLEOIDS	
		Ratio (cts/min) ^{3}H-RNA:^{14}C-DNA
RNAase treated nucleoids (0 ethidium br.)		6.4
RNAase treated nucleoids (2 µg/ml ethidium br.)		7.1

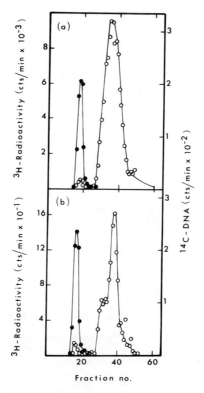

Fig. 3. Fractionation of DNA bound RNA on agarose columns. A bacterial culture was incubated for 40 min in the presence of ^{14}C-thymidine, then half the culture was removed and incubated for 3.5 min in the presence of 200 µg/ml rifampicin. Both cultures were then pulse-labeled for 30 seconds with ^{3}H-uridine (25 µC/ml; 28 C/mM). Total nucleic acids were purified from both cultures and applied to agarose columns as described in Methods. A) Untreated culture B) Rifampicin treated culture. ●—● ^{14}C-DNA; O—O ^{3}H-RNA.

bound to the DNA can be detected even after complete
unfolding of the nucleoid.

DNA Bound RNA Molecules: To determine if any of the putative RNA-DNA interactions described above can survive
disruption of the nucleoid, the chromosomes were unfolded
with strong ionic detergents, sheared, deproteinized and
the DNA and free RNA were separated on agarose columns
(Fig 3). About 2% of the total pulse labeled ^{3}H-RNA
eluted with the ^{14}C labeled DNA. This RNA remained associated with the DNA even after days of storage. When
analysed by equilibrium sedimentation in CsCl density
gradients, more than 95% of the isolated bound ^{3}H-RNA
banded with the DNA; however, after heating to 96°C the
RNA disassociated and banded in the CsCl gradient near
the bottom of the gradient where free RNA is expected
(Fig 4). Thus it appears that these DNA bound RNA molecules are attached to the DNA via non-covalent bonds.

As shown in Fig 3 the amount of DNA bound RNA was
greatly reduced when the cells were incubated briefly with
rifampicin prior to isolation of the RNA-DNA complex. The
total amount of label incorporated into RNA during the
30 sec pulse with 5 -^{3}H-uridine was reduced to 1% of the
control culture which did not receive rifampicin. Likewise
the amount of label in DNA bound RNA was reduced to 1% of
that in the control, showing that the synthesis of the DNA
bound RNA molecules is as sensitive to rifampicin as total
RNA synthesis. Also since the DNA bound RNA was labeled
even with very short pulse-labels it appears that at least
some of these chains are nascent RNAs.

131

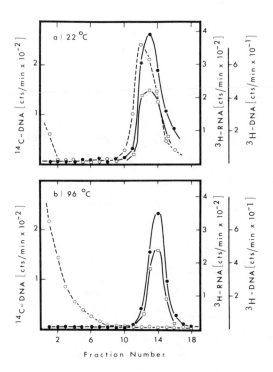

Fig. 4. Analysis of the DNA bound RNA in equilibrium CsCl gradients. Equal aliquots from the peak of the DNA band shown in Fig. 3A were incubated for 5 min at A) 22°C or B) 96°C in solutions containing 1.0M NaCl, 60 µg polyU carrier, 30 µg salmon DNA carrier, 5 mM trisodium citrate and 0.02M sodium phosphate (pH 7.5). CsCl was added to a final density of 1.75 gm/ml and the solutions were centrifuged 50 hrs at 36,000 rpm and 22°C in an SW50 rotor. Fractions were collected and ^{3}H radioactivity in RNA and DNA were determined by an alkaline hydrolysis assay. □—□, ^{3}H-DNA; ●—● ^{14}C-DNA; O==O ^{3}H-RNA.

When RNA chains were pulse-labeled in vivo with
5 -^{3}H-uridine as in Fig 3, a small fraction of the incorporated label appeared in DNA, predominantly in deoxycytidine residues. Although the amount of label in DNA was
usually less than 1% of that in RNA, this amount became
significant when the DNA was purified free of all RNA
except the DNA bound RNA chains. To distinguish between
the ^{3}H-label in RNA and DNA, an alkaline hydrolysis assay
was utilized (see for example Fig 4). This assay was
complicated by the fact that in alkali proton exchange can
occur with ^{3}H incorporated into DNA (11,12). Details of
the procedure which was developed to hydrolyze the RNA
while avoiding DNA ^{3}H-exchange will be described elsewhere.

In Fig 4A it can be seen that DNA fragments with bound
^{3}H-RNA banded at a density 1 to ½ fractions heavier than
the main DNA band, while the ^{3}H incorporated in DNA co-
banded with the main ^{14}C-DNA band. The observed density
shift of the DNA-RNA complexes was slight suggesting that
the mass of the bound RNA chain was small compared to the
mass of the DNA fragment (5-20 x 10^6 daltons).

The properties of this DNA bound RNA and its mode of
attachment to the DNA have been studied in some detail.
These experiments, which will be published elsewhere, are
beyond the scope of this brief paper. It will suffice here
to summarize several results from this more detailed
research. Pulse-chase experiments and the effects of
rifampicin on the synthesis of the DNA bound RNA (Fig 3)
suggested that the RNA molecules are nascent chains
unstable in vivo in their association with the DNA. The
RNA molecules are heterodisperse in size and have a weight

average length of 1200 nucleotides per chain. There are
60-100 RNA chains bound per nucleoid equivalent of DNA when
isolated as in Figures 3 and 4. Sensitive assays employing
radioactive labeled proteins failed to detect any protein
involved in the RNA-DNA linkage. Isopycnic analysis of
density labeled DNA molecules showed that the RNA molecules
were bound at sites distributed over the chromosome and
that they were not preferentially attached at either DNA
which has just been replicated or DNA that is about to be
replicated. End labeling studies and studies of RNAase
resistance of the bound RNA chains suggested that sites of
attachment of the RNA were not preferentially at the 3'
end. Hybridization competition analyses showed that at
least some of the bound RNA molecules have the sequences
of known mRNA's or known stable RNA species.

DISCUSSION

The properties of the DNA bound RNA described here are
similar to those of the nucleoid associated RNA molecules
which stabilize the condensed state of the DNA. The nucleoid stabilizing RNA is either lost from the nucleoid or
loses its capacity to stabilize the nucleoid after cells
are grown with rifampicin; the DNA bound RNA is released
from the DNA when cells are grown with rifampicin.
Nucleoid stabilizing RNA in its bound state is sensitive
to cleavage by RNA'ase A; the DNA bound RNA can also be
cleaved by this enzyme. The number of strongly bound RNA
molecules per nucleoid equivalent of DNA (60-100) is in
agreement with the estimates of the number of nucleoid

stabilizing RNA molecules per nucleoid if one assumes one
such RNA molecule per domain of supercoiling (3). It might
be expected that nucleoid stabilizing RNA molecules would
be bound to the DNA by at least two separate sites; evidence was summarized here that the DNA bound RNA molecules
have a binding site not involving preferentially the 3' end
and since these RNA molecules are nascent chains their 3'
end must normally be bound to the DNA via the ternary
transcription complex. It therefore appears that the DNA
bound RNA fraction would have a least two separate sites of
attachment to the chromosome. Indeed, it was demonstrated
that many of the nascent RNA chains of the nucleoid may
have multiple sites of attachment to the DNA, since after
cleavage of the nascent RNA many fragments remain attached
to the nucleoid.

Although the comparisons described above are compatible with the idea that the DNA bound RNA molecules are
those which stabilize the folded chromosome, this interpretation is not established.

It should be emphasized that the results summarized
here suggest that many different kinds of nascent RNA
chains have the potential of binding at multiple sites on
the DNA of the condensed chromosome. The stabilization of
the chromosome by bound RNA molecules may therefore be a
general property of different RNA species and not a function of a specific RNA. The physical-chemical basis for
this RNA-DNA interaction remains unknown.

ACKNOWLEDGEMENT

This research was supported by U. S. Public Health Service grant No. GM 18243-04 and by U. S. National Science Foundation grant No. GB 43358. We thank Ms. Eileen Miles for expert technical assistance. This is contribution No. 625 from the Department of Biophysics and Genetics, University of Colorado Medical Center.

REFERENCES

1. Stonington, O. G. and Pettijohn, D. E. Proc. Nat. Acad. Sci., U.S.A. 68, 6 (1971).
2. Worcel, A. and Burgi, E. J. Mol. Biol. 71, 127 (1972).
3. Pettijohn, D. E. and Hecht, R. Cold Spring Harbor Symp. Quant. Biol. 38, 31 (1973).
4. Dworsky, P. and Schaechter, M. J. Bacteriol. 116, 1363 (1973).
5. Jones, N. C. and Donachie, W. D. Nature 251, 252 (1974).
6. Pettijohn, D. E., Clarkson, K., Kossman, C. R. and Stonington, O. G. J. Mol. Biol. 52, 281 (1970).
7. Pettijohn, D. E., Stonington, O. G. and Kossman, C. R. Nature 228, 235 (1970).
8. Pettijohn, D. E., Hecht, R. M., Stonington, O. G. and Stamato, T. D. In DNA Synthesis in vitro, ed. by Wells and Inman, Baltimore: University Press, 145 (1973).
9. Manor, H., Goodman, D. and Stent, G. J. Mol. Biol. 39, 1 (1969).

10. Bremer, H. S. and Yuan, D.   J. Mol. Biol. $\underline{38}$, 163 (1968).

11. Fink, R. M.   Arch. Biochem. Biophys. $\underline{107}$, 493 (1964).

12. Evans, E. A., Sheppard, H. C. and Turner, J. C.   J. of Labelled Compounds $\underline{6}$, 76 (1970).

13. McCarthy, B. and Britten, R.   Biophysical J. $\underline{2}$, 35 (1962).

14. Salser, W., Janin, J. and Levinthal, C.   J. Mol. Biol. $\underline{31}$, 237 (1968).

VISCOMETRIC ANALYSIS OF CONFORMATIONAL TRANSITIONS IN THE ESCHERICHIA COLI CHROMOSOME

Karl Drlica and A. Worcel
Department of Biochemical Sciences
Princeton University
Princeton, New Jersey 08540

ABSTRACT. The effects of physical, chemical and enzymatic treatments on purified folded chromosomes from Escherichia coli were monitored by viscometry. The following chromosome conformation states were clearly distinguishable: a) folded, supercoiled, b) ethidium bromide relaxed, c) partially unfolded by ribonuclease, and d) completely unfolded chromosomes. The results confirm earlier sedimentation studies indicating that chromosomal DNA is supercoiled, and in addition, they show that supercoiling is lost after treatment with ribonuclease. Extensive incubation with pancreatic ribonuclease A unfolded the chromosomes only partially; complete unfolding required incubation at mild temperatures (60-70°C). Solutions of chromosomes partially unfolded by ribonuclease became even more viscous when protease VI was included in the incubation mixture, suggesting that proteins are involved in maintaining the chromosome structure.

INTRODUCTION

The DNA inside bacterial cells is organized into highly compact structures, called nucleoids or folded chromosomes, which can be purified by sucrose density gradient centrifugation following gentle lysis of the cells (1). The overall dimensions of the chromosomes change little upon isolation (2), indicating that the structural integrity of the chromosome has been preserved. Consequently, these folded chromosomes provide an excellent system for studying the molecular interactions involved in DNA packaging in bacteria.

Previous studies have shown that by varying the temperature of cell lysis, folded chromosomes can be isolated which are either attached to, or free from membrane fragments (3,4). Since the protein composition of the membrane free chromosomes is considerably less complex than that of the membrane attached ones, our structural analyses

have concentrated on the former. The chemical composition
of the membrane-free chromosome is 60% DNA, arranged in
about 50-100 independently supercoiled loops (5,6,7), 30%
RNA, and 10% protein, 90% of which is core RNA polymerase.

We have dissected the folded chromosomes with combinations of chemical, physical and enzymatic agents, and the
effect of each agent on the conformation of the chromosome
was monitored by viscometry. Four levels of structural
complexity have been clearly distinguished: a) folded,
supercoiled chromosomes; b) ethidium bromide-relaxed chromosomes; c) ribonuclease-partially unfolded chromosomes;
and d) completely unfolded chromosomes.

MATERIALS AND METHODS

Source of Reagents and Chemical Determinations

Ethidium bromide, spermidine trihydrochloride, and protease VI (from Streptomyces griseus) were purchased from
Sigma Biochemicals. Pancreatic ribonuclease A (E.C. 2.7.7.
16) was from Worthington Biochemicals. Preparations of
pancreatic ribonuclease were incubated at 100°C for 10 min
to inactivate contaminating nucleases. Chromosome concentrations are expressed in terms of DNA concentrations determined from measurements of absorbance at 260 nm. These
calculations used 6740 as the molar extinction coefficient
of E. coli DNA and 0.7 as the fraction of the chromosome
absorbance at 260 nm due to DNA (3).

Preparative Isolation of Folded Chromosomes

Escherichia coli cells growing exponentially in M9 medium were gently lysed by the method of Stonington and
Pettijohn (1) as modified by Worcel and Burgi (5). The
bacterial strain used was H560 (pol A1, endo I, thy); similar results were also obtained with strain K12 DG75 (thy,
leu) (4). The cells were lysed at 22°C, and the cell debris was removed from the lysate by centrifugation at 3,000
x g for 5 min. The membrane-free chromosomes were then
separated from the lysate by centrifugation (17,000 rpm, 40
min, 4°C, Beckman SW 25.1 rotor) into 10-30% w/v sucrose
density gradients containing 1 M NaCl, 0.01 M EDTA and
0.025 M sodium morpholinopropane sulfonic acid (MOPS buffer), pH 8.1. Approximately 20 fractions were collected
from the bottom of each centrifuge tube, and the chromosome
containing fractions were identified by measurements of ab-

139

sorbance at 260 nm. Absorbance was measured with a Zeiss
spectrophotometer from 50 µl aliquots from each fraction
diluted five fold in water. The remaining volume of the
peak 3 or 4 fractions of membrane-free chromosomes were
pooled for subsequent analyses. The ultraviolet absorption
spectrum of these chromosomes had a peak at 260 nm and a
trough at 230 nm with 260/230 and 260/280 ratios of 1.85 to
1.91, respectively. Chromosomes were stable for several
hours if kept on ice, and chromosomes were used the same
day they were isolated.

Viscosity Determinations

Chromosomes isolated from the preparative gradients
were diluted and incubated as described in subsequent sections. Viscosities were determined with a modified Gill-
Thompson rotating cartesian diver viscometer (8) in which
the driving torque was supplied by magnets attached to a
magnetic stirrer commercially available from Beckman Instru-
ments. Chromosomes were placed in a stator tube (13.2 mm
diameter), incubated, and then cooled on ice. Next, a cartesian diver, called a rotor, was carefully placed in each
chromosome solution. The stator was then placed in a water
jacket mounted between the magnets. Each rotor (6.5 mm
diameter) contained an aluminum drive ring, and the length
of the rotor was greater than 85% of the height of the DNA
solution in the stator. The vertical, submerged position
of the rotor was maintained manually by applying positive
or negative pressure with a syringe. The rotor was marked
at 90° intervals and the outer chamber of the viscometer at
10° intervals. Unless otherwise specified, all measurements were conducted at 1.5°C in 0.125 M sucrose, 1 M NaCl,
0.01 M EDTA and 0.025 M MOPS buffer pH 8.1. The angular
velocity of the rotor in the buffer alone was 20 seconds
per revolution, and the average shear stress in the standard assay was 8×10^{-3} dynes/cm^2 (calculated by the method
of Zimm and Crothers, (9)).

Handling of the DNA

To obtain reproducible results, many precautions had to
be taken to minimize DNA breakdown. The folded chromosomes
were pipetted at less than 0.1 ml/sec with plastic 1 ml
disposable pipettes having 1.5 mm diameter tip openings.
Mixing was accomplished by gently inverting the viscometer
stator tubes. Once the chromosomes were relaxed or un-

140

folded, they became extremely fragile and could be neither pipetted nor mixed. Consequently, all physical, chemical and enzymatic treatments were carried out directly in the stator tubes.

RESULTS

Response of Chromosomes to Shear Stress

When the rotation of the rotor exerted shear stress on a solution of chromosomes which had been relaxed or unfolded, the viscosity of the solution first increased, reached a peak, and then decreased. These viscosity changes are reflected in changes in the angular velocity of the rotor (plotted as period of rotation) and are illustrated in Fig. 1. Fig. 1a shows that the bell shaped viscosity response occurred at an average shear stress as low as 3.3×10^{-4} dynes/cm^2 (as a point of reference the specific viscosity of T2 phage DNA is independent of shear stress below 1.1×10^{-3} dynes/cm^2 (9)). The initial slowing of the rotor may be due to shear dependent aggregation (10) and/or stretching of the long DNA molecules. Part of the aggregation occurring with completely unfolded chromosomes (75°C, 10 min) may have been mediated by proteins since extensive digestion with proteinase K (0.5 mg/ml, 50°C, 16 hr) reduced the magnitude of the viscometric response. We could not rule out the possibility that nucleases also acted during incubation with proteinase K, but viscoelastic measurements indicated that unit length, linear E. coli DNA molecules were present after incubation (Drlica and Klotz, unpublished observations). When the rotating magnetic field was removed during this initial stage and the DNA allowed to relax, the solution exhibited the usual bell-shaped cycle when the field was reapplied.

The subsequent drop in viscosity may have been due to either a winding of the DNA molecules around the rotor (11, 12) or to a shear induced breakdown of the DNA. Either explanation is consistent with the observation that the peak viscosity decreased as the average shear stress was increased (Fig. 1a). When the rotating magnetic field was removed, the DNA allowed to relax, and the field then reapplied, the maximum viscosity of the rerun depended on the extent of initial shear; the rerun maximum decreased as the extent of initial shear increased (Fig. 1b). If the initial shear lasted long enough, the rerun showed no viscomet-

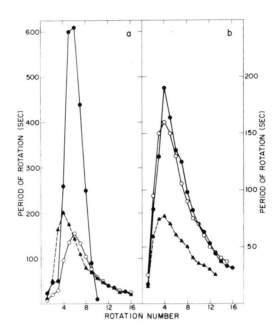

Fig. 1.   Effect of shear stress on the viscometric response of unfolded chromosomes.

a)   Chromosomes were unfolded by incubation at 73°C for 10
min and were then cooled to 1.5°C. A rotor was placed in
each solution, a rotating magnetic field was applied and
the period of each revolution of the rotor was measured at
an average shear stress of 0.33×10^{-3} dynes/cm^2 (●—●), 4.6
$\times 10^{-3}$ dynes/cm^2 (▲--▲), or 11×10^{-3} dynes/cm^2 (O—O).
Different values of shear stress were obtained by varying
the size of the rotor (4.5, 6.5, 6.5 mm diameter) and the
size of the aluminum drive ring so that the period of rotation in buffer was 280, 32 and 14 sec. Rotation periods of
the buffer alone were subtracted from all data. The DNA
concentration was 0.035 µg/ml.
b)   Chromosomes were treated as in a) and subjected to an
average shear stress of 8×10^{-3} dynes/cm^2 for 1 (●—●), 3
(O—O) or 7 (▲--▲) revolutions of the rotor. The rotating
magnetic field was removed for 30 minutes and then reapplied. The period of each rotation of the rerun was then
measured. Rotation periods of the buffer alone were subtracted from all data. The DNA concentration was 0.04
µg/ml.

ric response and the viscosity was identical to that of buffer lacking chromosomes. It appears that two processes occur, a reversible stretching and aggregation as well as an irreversible breakdown or spooling of the DNA. The former predominated during rotor deceleration and the latter during acceleration.

The shape of the viscosity response curve was similar for the three chromosomal conformations studied (Fig. 2), but quantitative distinctions could be made by the chromosome concentrations required to produce the viscometric response. In the standard viscometric determination the rotating magnetic field was applied, and measurement of the average angular velocity began after the rotor had completed one revolution. The number of revolutions was determined during a 5 min interval and related directly to specific viscosity (η_{sp}^*) by

$$\eta_{sp}^* = \frac{\eta^* - \eta_o}{\eta_o} = \left(\frac{1}{r} - \frac{1}{r_o}\right) \bigg/ \frac{1}{r_o} = \frac{r_o}{r} - 1$$

where η^* and η_o are the viscosities of the sample and buffer, respectively, and r and r_o are the number of rotor revolutions in the sample and buffer, respectively, during the 5 min interval. It is important to note that the angular velocity did not reach a steady state; consequently, the viscosity parameters are denoted with asterisks as η_{sp}^* and η_{red}^* to distinguish them from the usual parameters determined at a constant rotor velocity. The time interval was selected so that the measurement took place only on the predominantly reversible, ascending portion of the viscosity response curves. A moderate average shear stress (8 x 10^{-3} dynes/cm^2) was used routinely because it gave a broad range of η_{sp}^* values (about 50 fold) within a reasonably short time period. Under these conditions η_{sp}^* was a nearly linear function of DNA concentration (Fig. 3) and the reduced viscosity, $\eta_{red}^* = \eta_{sp}^*/[DNA]$, was a function of chromosome conformation (see results, Fig. 11). Linearity of the η_{sp}^* versus concentration curve (Fig. 3) was lost at low DNA concentrations.

Chromosome Relaxation by Ethidium Bromide

Intercalation of ethidium bromide causes the sedimentation coefficient of folded chromosomes to first decrease

143

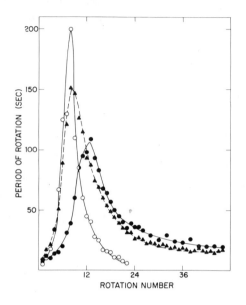

Fig. 2.   Viscometric response of relaxed and unfolded
chromosomes.
     Chromosomes were incubated with ethidium bromide (2
µg/ml, 0°C, 5 min) (●—●), pancreatic ribonuclease A (8.5
µg/ml, 37°C, 10 min) (▲--▲) or at 75°C for 10 min (0—0).
DNA concentrations were 15.6, 0.75 and 0.02 µg/ml for each
of the respective treatments. Viscometric measurements
were made as described in Fig. 1.

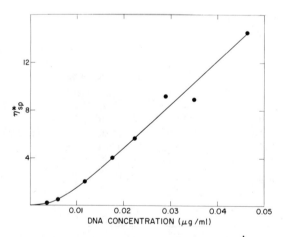

Fig. 3.   DNA concentration dependence of η_{sp}^{*} .
     Chromosomes were unfolded by incubation at 75°C for 10
min and then cooled to 1.5°C. η_{sp}^{*} was determined by the
standard assay described in Methods.

from 1500 S to a minimum of about 900 S at a dye concentration of 2 µg/ml, and then to increase to slightly more than 1500 S as the concentration of ethidium bromide is raised above 2 µg/ml (5). This dye-induced relaxation and the re-introduction of DNA supercoiling can also be measured viscometrically as shown by the solid circles in Fig. 4. When the concentration of ethidium bromide in the chromosome solution was increased, the values of η_{sp}^{*}/[DNA] increased and reached a maximum at 2 µg/ml of ethidium bromide. At higher ethidium bromide concentrations the viscosity of the chromosome solution decreased, and above 4 µg/ml of ethidium bromide the chromosome viscosity was indistinguishable from that of the buffer. Values of η_{sp}^{*} for chromosomes fully relaxed by ethidium bromide were determined at various chromosome concentrations from the maxima of curves similar to the solid circles in Fig. 4. These values, along with similar determinations for ribonuclease- and thermally-unfolded chromosomes described later, have been plotted against DNA concentration in Fig. 11. The slope of the line, which corresponds to η_{red}^{*} (ethidium bromide), is about twenty times higher than the upper estimate for η_{red}^{*} of the supercoiled folded chromosomes (the viscosity of supercoiled folded chromosomes was indistinguishable from that of the buffer up to 23 µg DNA/ml, the highest concentration tested).

Chromosome Unfolding by Ribonuclease

Previous studies indicated that chromosomes incubated with pancreatic ribonuclease unfold (1,5). When monitored by viscometry, the maximum unfolding induced by pancreatic ribonuclease A occurred at enzyme concentrations greater than 10 µg/ml if incubations were carried out in 1 M NaCl at 37°C for 10 minutes (Fig. 5). The DNA concentration dependency of η_{sp}^{*} (RNase) was determined after maximum unfolding induced by ribonuclease and is shown in Fig. 11. The DNA in chromosomes unfolded by ribonuclease appears to have lost its supercoiled conformation since η_{sp}^{*} was insensitive to ethidium bromide. After ribonuclease treatment, the plot of η_{sp}^{*} versus ethidium bromide concentration gave a flat line instead of the usual bell shaped curve (open circles, Fig. 4).

The ordinates in Fig. 4 are expressed as η_{sp}^{*}/[DNA] to emphasize the viscosity differences between the ethidium bromide-relaxed and the ribonuclease-unfolded chromosomes.

145

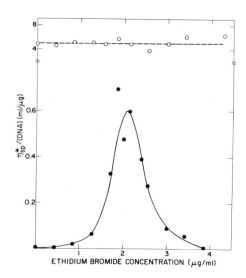

Fig. 4. <u>Chromosome relaxation by ethidium bromide.</u>
η_{sp}^{*}/[DNA] was determined at various concentrations of
ethidium bromide for folded chromosomes (●—●) and for chromosomes unfolded by incubation with pancreatic ribonuclease
A (8.5 μg/ml, 37°, 10 min) (0—0). DNA concentrations were
3.6 μg/ml and 0.79 μg/ml respectively.

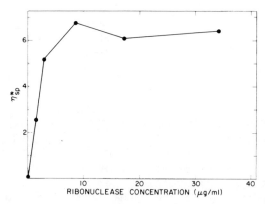

Fig. 5. <u>Chromosome unfolding by pancreatic ribonuclease A.</u>
Chromosomes were incubated with varying concentrations
of pancreatic ribonuclease A at 35°C for 10 min, and η_{sp}^{*}
was determined as described in the text. The DNA concentration was 0.68 μg/ml.

146

At 2 μg/ml ethidium bromide, both types of chromosomes have lost their superhelices, but the higher viscosity of the RNase-unfolded chromosomes indicates that the chromosomes must have lost additional tertiary conformational constraints after the RNase treatment.

Chromosome Unfolding by Temperature

Incubation of chromosomes at mild temperatures also produces unfolding (1). Viscometric measurements show that η_{sp}^* increases about 30-fold between 40° and 50°C. A narrow plateau is reached at about 48° followed by another 10-fold increase in η_{sp}^* as the incubation temperature is raised. These two temperature transitions are plotted in Fig. 6 where the ordinate is expressed as $\eta_{sp}^*/[DNA]$ to illustrate the relative magnitudes of the two transitions. The range of the assay for η_{sp}^* was inadequate to measure both transitions using a single chromosome concentration, and the two curves do not smoothly join because linearity between η_{sp}^* and DNA concentration is lost at low values of η_{sp}^* (see Fig. 3).

The high temperature transition occurred over a narrow temperature range, and in 1 M NaCl the transition midpoint is 60.3°C ± 0.4°C (S.D.). The transition did not arise from generalized DNA duplex denaturation because in 1 M Na$^+$ the T_m of E. coli DNA is above 95°C (14). Moreover, no increase in hyperchromicity occurs in the range of the viscometric transition. Chromosome unfolding by temperature is irreversible. The transition is complete in about 10 min, and little or no change in η_{sp}^* occurs with incubations up to 60 min. At higher temperatures (above 80°C) η_{sp}^* decreases markedly, probably reflecting DNA chain breakage.

Aggregation of Unfolded Chromosomes during Viscometry

The viscometric behavior of chromosomes unfolded by 60-70°C treatments differed from that expected of pure DNA. The most notable difference was the low chromosome concentrations that could be detected; viscometric measurements were made at DNA concentrations as low as 0.01 μg/ml. In addition, the viscosity dependence on DNA concentration was very high (see Fig. 11); η_{red}^* was more than 100 times greater than expected for DNA the size of the E. coli genome (13). Furthermore, the viscometric detection of chromosomes required a critical DNA concentration, a feature which can be seen in plots such as Fig. 3. These observa-

147

tions suggested that the chromosomes were aggregating during the viscometric measurements and that the conformational states of the chromosome described in the previous sections differ markedly in their aggregative properties.

To further test the aggregation hypothesis, the viscosity of fully extended chromosomes was measured at various sodium ion concentrations. Ross and Scruggs (15) showed that the viscosity of purified DNA decreases as the salt concentration is raised, probably due to compaction occurring as the salt dampens intramolecular electrostatic repulsive forces. Viscosity due to aggregation would behave the opposite way since salt would also dampen intermolecular electrostatic repulsive forces and lead to increased aggregation and increased viscosity. The solid circles in Fig. 7 clearly show that viscosity increased as the salt concentration was raised, indicating that an aggregative phenomenon was being measured by the standard viscometric assay.

The viscosity of partially unfolded chromosomes, such as those unfolded by pancreatic ribonuclease A (open circles, Fig. 7), was independent of the salt concentration. Since the ordinate in Fig. 7 is expressed as η_{sp}^{*}/[DNA], quantitative differences can be seen between RNase- and thermally- (80°C) unfolded chromosomes. Extrapolation to zero salt concentration indicates that there is only a two or three fold difference between the η_{sp}^{*}/[DNA] of the two chromosomal conformations.

Salt Dependence of the 60-70°C Thermal Transition

The salt dependency of a thermal conformational transition reflects the type of macromolecular interactions responsible for maintaining the conformation. For example, the thermal stability of nucleic acid:nucleic acid interactions increases monophasically with increasing ionic strength (14) while the thermal stability of a protein:nucleic acid interaction stabilized primarily by electrostatic forces would probably decrease as the salt concentration is raised. In order to determine which type is disrupted in the 60-70°C chromosomal unfolding, thermal transition curves were measured viscometrically at various sodium ion concentrations. Two sets of curves from different chromosomal preparations are shown in Figure 8a, b[1] and

---

[1] The raw measurements of η_{sp}^{*} have been converted to

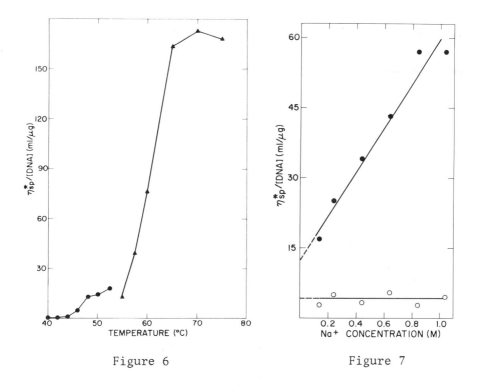

Figure 6   Figure 7

Fig. 6.   Thermal unfolding of chromosomes measured by vis-
          cometry.
     Folded chromosomes were incubated at various tempera-
tures for 10 min and η_{sp}^{*} was determined as described in the
text.   Viscosity is expressed as η_{sp}^{*} divided by DNA concentrations which were 0.095 µg/ml (▲——▲) and 0.76 µg/ml (●—●).

Fig. 7.   Effect of Na^{+} concentration on η_{sp}^{*} of unfolded
          chromosomes.
     Chromosomes were incubated with 8.5 µg/ml pancreatic
ribonuclease A for 10 min at 37°C (O—O) or 80° (●—●) at
various Na^{+} concentrations and assayed by the standard viscometric assay described in the Methods.   Chromosome concentrations were 1.4 and 0.14 µg DNA/ml for the 37°C and
80°C treatment, respectively.

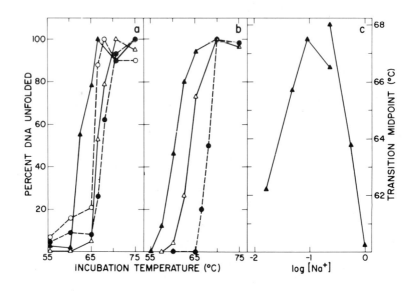

Fig. 8.   Effect of Na$^+$ concentration on the 60-70°C thermal unfolding of chromosomes.
     Solutions of folded chromosomes containing various con-
centrations of Na$^+$ were incubated at various temperatures
for 10 min, and values of n^*_{sp} were determined for each solution as described in the methods. The values of n^*_{sp} were
then normalized as described in the text[1] and are expressed
as percent DNA unfolded. The DNA concentration was 0.07
µg/ml.
a)   Sodium ion concentrations were 0.017 $\underline{M}$ (▲–▲); 0.05 $\underline{M}$
(0--0); 0.09 $\underline{M}$ (●–●); and 0.24 $\underline{M}$ (Δ–Δ).
b)   Sodium ion concentrations were 0.24 $\underline{M}$ (●--●); 0.6 $\underline{M}$
(Δ—Δ); and 1 $\underline{M}$ (▲—▲).
c)   The temperature of each transition midpoint was determined from Fig. 8a, b and is plotted against log [Na$^+$].

the thermal midpoint of each curve is plotted in Fig. 8c to illustrate the biphasic relationship of transition temperature and salt concentration. Increasing the Na^+ concentration at low ionic strengths increased the stability of the chromosome, presumably due to salt-screening of electrostatic repulsive forces between the DNA phosphates. Above 0.2 $\underline{M}$ Na^+ increasing the salt concentration reduced chromosome stability, perhaps by weakening electrostatic DNA:protein and/or DNA:polycation interactions which may stabilize chromosomal folding.

Polycations such as spermidine can replace the monovalent cations normally needed to maintain chromosome stability in vitro (16) and stabilize chromosomes against thermal unfolding (17). As shown in Fig. 9, the addition of spermidine raises the transition temperature of the 60-70°C transition. Thus at least part of the stabilizing effect of spermidine is directed at the interactions disrupted during the 60-70°C transition.

Effect of Protease

Completely folded, supercoiled chromosomes are unaffected by incubation with proteases (1,5). However, chromosomes unfolded by ribonuclease treatment become even more viscous when protease VI is included in the incubation mixture (Fig. 10). A decrease in viscosity is seen at high protease concentrations, perhaps due to the digestion of ribonuclease before it can degrade all the RNA molecules

---

percent unfolded DNA to correct for the slight nonlinearity in the relationship between n^*_{sp} and DNA concentration (see Fig. 3) and for the effect of the salt concentration on the viscosity of completely unfolded chromosomes (see Fig. 7). Values of n^*_{sp} for each salt concentration were first measured after incubation at various temperatures and at a fixed DNA concentration. The temperature at which the chromosomes were completely unfolded, usually 75°C, was then used to unfold at various DNA concentrations, and the values of n^*_{sp} obtained were plotted against DNA concentrations to give a calibration curve similar to Fig. 3. These calibration curves were then used to convert the values of n^*_{sp} initially obtained at different temperatures at a given DNA concentration to percent unfolded DNA. For each salt concentration the maximum viscosity value was equated to 100% unfolded DNA.

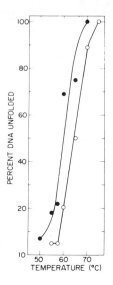

Fig. 9.   Effect of spermidine on the 60-70°C thermal un-
           folding of chromosomes.
     Folded chromosomes in 1 M̲ NaCl, 0.025 M̲ MOPS pH 8,
0.01 M̲ EDTA were incubated for 10 min at various temperatures and the percent DNA unfolded was determined as described in the legend for Fig. 8. Chromosome solutions
(0.08 µg DNA/ml) contained only the buffer described above
(●--●) or also contained 3 mM spermidine trihydrochloride
(0--0).

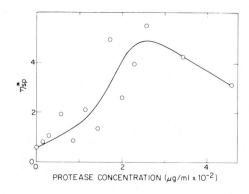

Fig. 10.   Effect of Protease VI on chromosomes unfolded by
           ribonuclease.
     Chromosomes in 0.57 M̲ NaCl were incubated for 20 min
at 36°C with 28 µg/ml pancreatic ribonuclease A and various
amounts of protease VI. Viscosity was determined as described in the methods. The DNA concentration was 0.22 µg/
ml.

involved in chromosomal folding. Because of the fragility
of the ribonuclease-unfolded chromosomes, protease could
not be added after ribonuclease digestion (DNA shearing
could not be avoided during mixing, see Methods). The maximum viscosity obtained by the combination of the two enzymes is about five times above the level seen with ribonuclease alone.

DISCUSSION

Solutions containing high molecular weight DNA exhibit
two reponses to shear stress. Steady state shear stresses
distort the shape of DNA molecules and orient them along the
lines of flow; the result is non-Newtonian flow producing
lower viscosities at higher shear stresses (8,9). If high
molecular weight DNA solutions are subjected to a sudden
increase in shear stress, the viscosity increases, probably
due to aggregative effects associated with the reorienta-
tion and entangling of the molecules (10,18). We have
shown that purified bacterial chromosomes, which contain
60-70% DNA by weight, respond to the application of shear
stress in a way characteristic of the compactness of the
chromosome. The more compact structures, as judged by sedimentation analysis (5; Drlica and Worcel, unpublished observations), have a much smaller viscometric response.
This is most evident from the differing chromosome concentrations required to produce a response and the DNA concentration dependence of the response (Fig. 11). A standard
set of assay conditions was selected which gave an almost
linear viscometric response to chromosome concentration,
thus making it possible to establish quantitative relationships among the various chromosomal conformations.

Folded chromosomes were observed undergoing three distinct transitions, suggesting that there are at least four
levels of tertiary complexity. The first level or organization, in which the DNA is folded and supercoiled, is the
most compact one. The folded and supercoiled chromosomes
have a sedimentation coefficient of about 1500 S and exhibit no viscometric response at the highest concentrations
tested. It has been previously shown (5) that the DNA is
organized into about 50 independently supercoiled loops.
Treatment of the chromosome with 2 µg/ml ethidium bromide
relaxes the supercoiling in such a way that the sedimentation coefficient decreases to about 900 S (5). Likewise,

153

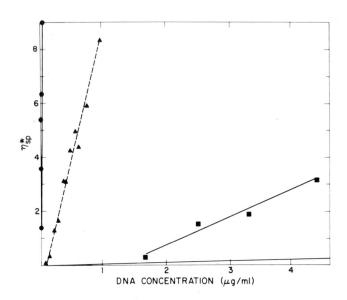

Fig. 11. Effect of chromosome conformation on the DNA concentration dependence of η_{sp}^*.
η_{sp}^* was determined at various DNA concentrations using chromosomes that were kept at 0°C (——), relaxed by incubation with ethidium bromide for 10 min at 0°C (■—■), unfolded by incubation with pancreatic ribonuclease A (8.5 µg/ml, 37°C, 10 min) (▲--▲), or unfolded by incubation at 75°C for 10 min (●—●). The values of η_{sp}^* for the ethidium bromide relaxed particles were derived from the peaks of η_{sp}^* versus ethidium bromide concentration curves similar to Fig. 4.

at 2 µg/ml ethidium bromide a viscometric response can be observed in which n_{sp}^* is at least 20 times greater than that of the folded and supercoiled chromosome in the absence of ethidium bromide. Analogous results have been reported for supercoiled viral DNA by Revet, Schmir and Vinograd (19); their bell shaped viscosity response curve of SV40 DNA versus ethidium bromide concentration is similar to our Fig. 4.

There are about 10,000 negative superhelices in the folded and supercoiled E. coli chromosome (about one negative supertwist per 400 base pairs (5)). Ethidium bromide intercalation gradually eliminates the negative superhelices, and at the equivalence point (20,21), i.e., at about 2 µg/ml ethidium bromide and 3 µg/ml chromosomal DNA, the superhelical content of the chromosomal DNA is equal to zero. Thus, the ethidium bromide relaxation represents the first conformational transition of the folded chromosome, from a folded and supercoiled structure to a folded and relaxed one (5). The extent of the increase in n_{red}^* (over 20-fold) and decrease in the sedimentation coefficient (2-fold) represents the specific contribution of the supercoiling to DNA compaction.

The second conformational transition is observed after ribonuclease treatment of the folded and supercoiled chromosomes. Incubation with ribonuclease causes a 3-fold decrease in the sedimentation coefficient (from 1500 S to about 500 S; Drlica and Worcel, unpublished observations) and a 200-fold increase in n_{red}^*. Since ribonuclease-treated chromosomes no longer exhibit the bell shaped viscosity response curve to ethidium bromide concentration (Fig. 4), we conclude that the DNA is no longer supercoiled. Both the ribonuclease-treated chromosomes and the ethidium bromide-relaxed ones have lost their superhelices, and yet the ribonuclease treated chromosomes have a 10-fold greater n_{red}^* and a 2-fold lower sedimentation coefficient than the chromosomes relaxed by 2 µg/ml ethidium bromide. Therefore the ribonuclease-treated chromosomes must have lost additional tertiary conformational constraints. Those conformational constraints may be the 50 or so DNA loops per chromosome which are eliminated by the ribonuclease treatment, allowing the rotation of the DNA chains around the few DNA nicks (probably occurring during isolation) to relax completely the chromosome (5,6). Thus, the ribonuclease-induced unfolding represents the second conformational transition of the folded chromosome, from a folded and supercoiled structure to an unfolded and relaxed one. The extent of the increase in the η^*_{red} (over 200-fold) and decrease in sedimentation coefficient (over 3-fold) represents the contribution of the superhelices plus the DNA loops to chromosomal compacting.

Two thermal transitions have been detected viscometrically; in 1 $\underline{M}$ NaCl the lower one takes place at about 45° and the upper at about 60-70°. The values of η^*_{red} after incubation at 50° is similar to that following ribonuclease treatment, and the value of η^*_{sp} following incubation in either ribonuclease or at 50° is independent of Na^+ concentration. Thus, the low temperature transition may be identical with the ribonuclease-induced transition. On the other hand, η^*_{sp} following the 60-70°C transition is highly dependent on Na^+ concentration, suggesting that a qualitative change has occurred in the aggregative properties. Moreover, the values of η^*_{red} (in 1 $\underline{M}$ NaCl) is 20-30 times greater than that of the RNase-treated chromosomes, although at zero NaCl concentration the η^*_{red} of the thermally (80°) treated chromosomes is only 2-3 times greater than that of the RNase-unfolded chromosomes. The lower η^*_{red} of the RNase treated chromosomes suggests that they still retain some tertiary structure; this residual structure is completely eliminated after the 60-70°C treatment.

The 60-70°C transition therefore represents the third conformational change in the folded chromosome, leading probably to a fully extended DNA molecule. This final transition may be due to disruption of protein-DNA stabilizing interactions; although completely folded, supercoiled chromosomes are resistant to protease attack, chromosomes partially unfolded by ribonuclease unfold even more when treated with protease VI, indicating that proteins may be involved in maintaining the chromosomal architecture. Additional support for this concept comes from the Na^+ dependence of the thermal midpoint of the 60-70°C transition. At low ionic strength increasing the Na^+ concentration stabilizes the chromosomes; however, above 0.2 $\underline{M}$ increasing the Na^+ destabilizes the chromosomes, suggesting that ionic forces are being weakened.

In spite of the fact that the viscometric measurements could not be conducted at a constant rotor speed, the data presented above is qualitatively significant, and the interpretation of the data is consistent with the previous sedimentation and electron microscopic studies (5,22). Indeed, even with the above mentioned technical limitations, viscometry appears to be the most sensitive tool currently available for the study of the conformational transitions in a giant macromolecule the size and complexity of the Escherichia coli chromosome.

ACKNOWLEDGEMENTS

We thank Dr. B. Zimm, Dr. L. Klotz and our colleagues at Princeton for many stimulating discussions. The work was supported by grants from the National Science Foundation and the American Cancer Society.

REFERENCES

1. Stonington, O.G. and Pettijohn, D.E. Proc. Nat. Acad. Sic. USA. 68, 6 (1971).
2. Hecht, R.M., Taggart, R.T. and Pettijohn, D.R. Nature 253, 60 (1975).
3. Pettijohn, D.E., Hecht, R.M., Stonington, O.G. and Stamato, T.D. in DNA Synthesis in vitro. eds., R.O. Wells and R.B. Inman, University Park Press, Baltimore pp. 145-162 (1973).
4. Worcel, A. and Burgi, E. J. Mol. Biol. 82, 91 (1974).
5. Worcel, A. and Burgi, E. J. Mol. Biol. 71, 127 (1972).
6. Pettijohn, D.R. and Hecht, R.M. Cold Spring Harbor Symp. Quant. Biol. 38, 31 (1973).
7. Kavenoff, R. and Ryder, O. J. Mol. Biol. (in press).
8. Gill, S.J. and Thompson, D.S. Proc. Nat. Acad. Sci. USA. 57, 562 (1967).
9. Zimm, B. and Crothers, D. Proc. Nat. Acad. Sci. USA. 48, 905 (1962).
10. Thompson, D., Hays, J. and Gill, S. Biopolymers 7, 571 (1969).
11. Shafer, R.D. Biophys. Chem. 2, 185 (1974).
12. Shafer, R.H., Laiken, N. and Zimm, B. Biophys. Chem. 2, 180 (1974).
13. Klotz, L. and Zimm, B. J. Mol. Biol. 72, 779 (1972).
14. Marmur, J. and Doty, P. J. Mol. Biol. 5, 109 (1962).
15. Ross, P.D. and Scruggs, R.L. Biopolymers 6, 1005 (1968).
16. Kornberg, T., Lockwood, A. and Worcel, A. Proc. Nat.

Acad. Sci. USA. <u>71</u>, 3189 (1974).

17. Flink, I. and Pettijohn, D. Nature <u>253</u>, 62 (1975).
18. Massa, D.J. Biopolymers <u>12</u>, 1071 (1973).
19. Revet, B., Schmir, M. and Vinograd, J. Nature New Biol. <u>229</u>, 10 (1971).
20. Bauer, W.R. and Vinograd, J. J. Mol. Biol. <u>33</u>, 141 (1968).
21. Crawford, L.V. and Waring, M.J. J. Mol. Biol. <u>25</u>, 23 (1967).
22. Delius, H. and Worcel, A. J. Mol. Biol. <u>82</u>, 107 (1974).

PROPERTIES OF MEMBRANE-ASSOCIATED FOLDED CHROMOSOMES OF *Escherichia coli* DURING THE DNA REPLICATION CYCLE

Oliver A. Ryder*, Ruth Kavenoff[†],
and Douglas W. Smith*
*Department of Biology and
[†]Department of Chemistry
University of California, San Diego
La Jolla, California 92037

ABSTRACT. Membrane-associated folded chromosomes from several *E. coli* strains have been examined. Utilizing a modification of the procedure of Stonington and Pettijohn (14), quantitative yields of membrane-associated folded chromosomes may be obtained. Folded chromosomes remain associated with the cell envelope during their replication and after completion of residual synthesis under conditions which prohibit reinitiation of new rounds of chromosome replication. Membrane-associated folded chromosomes become more rapidly sedimenting as replication proceeds in the absence of reinitiation. Sedimentation analysis of folded chromosomes prepared from amino acid-starved cultures or from a *dnaC* initiation mutant suggests that a unique structure is associated with completion or near completion of rounds of chromosome replication. Before reinitiation of chromosome replication occurs after restoring required amino acids to amino acid-starved cells or after lowering the temperature in a thermosensitive *dnaC* mutant, sedimentation velocities of the membrane-associated folded chromosomes decrease substantially. The decrease in sedimentation velocity does not depend on renewed DNA synthesis, but does require the activity of at least the *dnaC* gene product.

Electron microscopy of membrane-associated folded chromosomes from exponentially growing and amino acid-starved *E. coli* 15 TAU-bar, utilizing aqueous spreading techniques, has revealed a spectrum of structures ranging from condensed structures with no DNA fibers visible to extended structures with DNA fibers. The structures with free DNA most probably arise from the condensed structures, possibly during the spreading procedure. In the extended structure, loops of DNA radiate from residual envelope, the loops frequently appear supercoiled, and the number and length of loops approximates previous estimates for the length of the genome obtained from physical and biochemical data. Additionally, the average number of loops present in chromosomes from amino acid-starved cells appears to be somewhat larger than the average number of loops present in chromosomes from exponentially growing cells.

INTRODUCTION

The series of events which results in the duplication of the bacterial chromosome is closely regulated, such that during the normal growth of bacteria the DNA content of a cell is doubled once each generation. This regulation of chromosome duplication is integrally fused with the cell division cycle, resulting in the precise transmission of the genetic information to successive generations. The overall process of chromosome duplication results in a semi-conservative replication of cellular DNA and the replicated chromosomes are subsequently segregated into daughter cells.

The process of chromosome replication has been conveniently divided into the processes of initiation of chromosome replication, elongation of the growing DNA chains, and termination of replication.

The mechanism which controls the timing of initiation of chromosome replication is of major importance in the process of cell multiplication because the rate of cell division appears to be governed by the frequency of initiation of rounds of chromosome replication, whereas the rate of elongation appears to be fairly constant over a wide range of growth rates (1,2). It is not surprising then, that initiation is a closely regulated process. Conditional lethal mutants in the process of initiation of rounds of chromosome replication have been operationally defined to be those mutants which, under the restrictive conditions, achieve a level of residual DNA synthesis corresponding to completion of existing rounds of replication in the absence of any reinitiation. Mutants in the *dnaA* and *dnaC* loci fall into this operational category.

When protein synthesis is inhibited by removal of a required amino acid from the growth medium of an amino acid auxotroph, DNA synthesis continues for a period of time.

The extent of residual synthesis which occurs in the
absence of required amino acids corresponds to the
amount of synthesis expected if replicating chromosomes
complete rounds of replication in progress but cannot initiate new rounds of replication (3). For these reasons,
Maaløe and Hanawalt (3) postulated that protein synthesis
was necessary to initiate, but not to sustain, chromosome
replication. Detailed genetic and biochemical analyses
have suggested that in the absence of required amino acids,
chromosomes are replicated to a fixed region of the chromosome up to, or close to, the terminus of replication (4-8).

   Once initiation of chromosome replication has occurred,
elongation proceeds to the terminus of replication by a
mechanism which has yet to be completely described but
which involves a pattern of discontinuous synthesis (9).
(The products of the *dnaB, polC, dna G, dnaH, dnaI, polA* and
lig loci may be involved in the elongation process.)

   Very little is known about the end product of the
chromosome replication sequence for *E. coli* or for other
replicons. This is due, in part, to the fact that small,
easily isolatable replicons are replicated very quickly,
and therefore, the end products of replication may be
rapidly processed. Nonetheless, studies of the replication of small, circular DNA molecules of both prokaryotic
and eukaryotic origin have demonstrated the production of
multiple circular forms (circular oligomers consisting of
two of more genomes tandemly linked, and catenated forms
where the circular forms are interlocked). The production
of these molecules has been ascribed both to replication
and recombination (10-13).

   Analysis of pieces of DNA obtained after bacterial
lysis has provided considerable insight into the mechanism
of chromosome replication. However, investigation of some
aspects of chromosome organization might best be performed
on intact isolated chromosomes. Questions concerning the
nature of the organization of the chromsome *in vivo*, the
characterization of molecular factors responsible for maintaining the native chromosomal conformation, the supramole-
cular structure of intermediates in the process of chromosome duplication, and the conformation of chromosomes at
the stages of initiation and termination of chromosome
replication can be most directly investigated only by
isolating intact chromosomes.

   The description by Stonington and Pettijohn (14) of

a procedure which allows the isolation of the DNA of *E. coli* in a non-viscous, rapidly sedimenting form has provided the methodology needed to initiate the studies on bacterial chromosome organization and conformation mentioned above. Analysis of the DNA-containing complexes obtained by the procedure of Stonington and Pettijohn (14) has demonstrated that a portion of the cell envelope is associated with the bacterial DNA when lysis occurs at 0°-4°C (15-17) and that, as the temperature at which the lysis is performed is raised, a decreasing proportion of the DNA complexes isolated possess associated membrane (15-19). The DNA complexes which are isolated at lower temperatures and include a portion of the bacterial envelope are referred to as membrane-associated folded chromosomes, membrane-associated nucleoids, and/or membrane-attached folded chromosomes. The DNA complexes which are more slowly sedimenting and which lack extensive amounts of bacterial envelope components are referred to as folded chromosomes, membrane-free folded chromosomes, and/or nucleoids.

Membrane-associated folded chromosomes isolated from exponentially growing *E. coli* have been characterized by their sedimentation velocity, 3000-4000S (14-19), by the presence of high levels of cosedimenting protein (15-19), by their affinity for M-bands (16), and by electron microscopy (20-22).

Folded chromosomes have been shown to be organized into loops or folds which are stabilized by RNA molecules (14,15,23,24). Folded chromosomes possess a superhelical density similar to other naturally occurring supercoiled DNA molecules (23,24), but unlike other supercoiled molecules a single nick in a DNA strand is insufficient to totally destroy the superhelical properties of the molecule (23,24).

It is of interest to examine the properties of membrane-associated folded chromosomes corresponding to the events of initiation, elongation, and termination of DNA replication. Analysis of these stages of chromosome replication utilizing amino acid starvation of auxotrophic strains and a termosensitive mutant defective in initiation has been conducted. These studies demonstrate that folded chromosomes remain associated with portions of the cell envelope during the DNA replication cycle and become more rapidly sedimenting as chromosomes proceed to completion of rounds of replication in the absence of

reinitiation. A unique chromosome structure may exist
after completion of a round of replication. Additionally,
these studies suggest that the *dnaC* gene product participates in a process which results in an alteration of the
sedimentation properties of membrane-associated folded
chromosomes before reinitiation of DNA synthesis.

METHODS

Bacterial Strains and Growth Conditions: *Escherichia coli*
15 TAU-bar *thy⁻ ura⁻ arg⁻ pro⁻ met⁻ trp⁻* (32), *E. coli* K12
CT28-4 *dnaC* F⁻ *thy⁻ arg⁻ leu⁻ thr⁻ pro⁻ bio⁻ str*ᴿ *T6ˢ lac⁻ ara⁻*
xyl⁻ man⁻ (a *his⁺*, more thermostable derivative of CT28-3b
(26)),and *E. coli* K12 DG75 F⁻ *leu⁻ thy⁻* (5) were used.
E. coli CT28-4 was provided by C. I. Davern. TG₀ medium
(25) was supplemented when necessary with 4 μg/ml thymine,
20 μg/ml uracil, and 40 μg/ml L-arginine, L-leucine, L-
methionine, L-proline, L-threonine, and L-tryptophan.
Radioactive labelling was performed with 1 μCi/ml ¹⁴C-
thymine (New England Nuclear) or with 2.5 μCi/ml ³H-thymine
(Schwartz), and with 4 μCi/ml ³H-arginine (New England
Nuclear) or with 4 μCi/ml ³H-leucine (Schwartz). When labeling with ³H-leucine or ³H-arginine, cold leucine or arginine was present at 15 μg/ml instead of 40 μg/ml. TAU-bar
and DG75 were grown with aeration at 37°C. CT28-4 was
grown with aeration at 37°C (permissive temperature) and
shifted to 42°C (restrictive temperature) as indicated.
Transfer of media was accomplished by rapid filtration (3).
When cultures of TAU-bar were deprived of required amino
acids, 20 μg/ml L-methionine was added to the medium.

Preparation of Membrane-Associated Folded Chromosomes:
Membrane-associated folded chromosomes were prepared by a
modification of the method of Stonington and Pettijohn
(14) as previously described (17,22,25). 10 ml aliquots
of cell cultures were quickly cooled to 0°C in a -80°C
bath, harvested by centrifugation (8 min, 8000 x g, 0°C),
and the cell pellet drained of any residual medium. Pelleted cells were resuspended in 0.20 ml solution A (0.10 M
Tris-Cl pH 8.0, 0.01 M NaCl, 0.01 M NaN₃, 20% [w/v] sucrose
[special density gradient sucrose, Mann]). 50 μl of freshly prepared solution B was added (4 mg/ml egg white lysozyme [Worthington] in 0.12 M Tris-Cl pH 8.0, 0.05 M Na-
EDTA); the solution was gently mixed and incubated at 0°C
for 3 minutes. 0.25 ml lysis mixture (1% w/v Brij-58, 0.4%
w/v Na-deoxycholate, 0.01 M Na-EDTA, 2.0 M NaCl, sterilized

by filtration through a 0.45 μ membrane filter [Millipore HA]) was added. The lysate was mixed carefully and incubated at 4°C for 30 minutes. ^{32}P–T4 bacteriophage were added as indicated as a sedimentation marker and 0.45 ml of the lysate was layered onto a 5 ml linear 10%–30% sucrose (density gradient grade sucrose, Mann) gradient containing 0.10 M Tris–Cl pH 8.0, 1 mM Na–EDTA, 1.0 M NaCl, 1 mM 2–mercaptoethanol. Centrifugation was performed at 2°C in the Spinco SW50.1 rotor as indicated in the figure legends.

Collection of sucrose gradients and analysis for trichloroacetic acid insoluble radioactivity were performed as described (17).

Chloramphenicol (Sigma) and nalidixic acid (Calbiochem) were freshly prepared before dilution to concentrations sufficient to inhibit protein and DNA synthesis respectively.

Electron Microscopy: The Kleinschmidt and Zahn spreading technique (29,30) was modified to minimize shear to membrane–associated folded chromosomes (22).

RESULTS

A.   Preparation of Membrane–Associated Folded Chromosomes from Exponentially Growing Cells: We have obtained consistent and reproducible results utilizing the isolation procedure of Stonington and Pettijohn (14) for the preparation of membrane–associated folded chromosomes at 0°–4°C; however, the total yields of chromosomes are on the order of 30% to 50%. In order to exclude ambiguities in interpretation of results when a minority of the total DNA is isolated, attempts to increase the total yield of chromosomes have been made.

Following the suggestion of Dr. David Pettijohn, the effect of the 4000 x g centrifugation in the Stonington and Pettijohn (14) procedure was examined.

When membrane–associated folded chromosomes are prepared at 0°–4°C from exponentially growing E. coli 15 TAU–bar, the sedimentation profile of membrane–associated folded chromosomes obtained upon sucrose gradient centrifugation of the supernatant of the 4000 x g centrifugation is nearly identical to the sedimentation profile obtained when a crude lysate is subjected to sucrose gradient centrifugation (Figure 1). The more slowly sedimenting material which lacks appreciable amounts of cosedimenting protein represents folded chromosomes released from the cell envelope (17); a small proportion of membrane-free folded chromosomes are commonly observed in sucrose gradient displays of lysates prepared at 0°-4°C. Figure 1 illustrates data obtained from cells growing in medium supplemented with naturally occurring L-amino acids; essentially identical results are obtained from cells grown in minimal medium.

Yields of membrane-associated folded chromosomes are significantly higher when the 4000 x g centrifugation is omitted in the lysis procedure (Table 1), although the membrane-associated folded chromosomes obtained are similar or identical in behavior to those obtained when the 4000 x g centrifugation is included (Figure 1).

Utilizing this modification of the Stonington and Pettijohn (14) technique, we have obtained quantitative yields of membrane-associated folded chromosomes from several *E. coli* strains: TAU-bar, CT28-4, MRE601 and DG75.

B. Preparation of Membrane-Associated Folded Chromosomes from Amino Acid-Starved Cells: Dworsky and Schaechter (16) have shown that inhibition of protein synthesis by a variety of antibiotics results in increased sedimentation velocities for membrane-associated folded chromosomes. We have shown (17,25) that inhibition of protein synthesis by amino acid deprivation of an auxotroph results in increased sedimentation velocities of membrane-associated folded chromosomes under conditions which allow their isolation (Table 1, Figure 2). Thus, because of the rapidly sedimenting nature of membrane-associated folded chromosomes, the 4000 x g centrifugation step must be omitted, and the time of sucrose gradient centrifugation must be decreased, in order to obtain high yields of membrane-associated folded chromosomes from amino acid-starved cells (Table 1). When ^{14}C-thymine-labeled exponentially growing cells are mixed prior to lysis with ^{3}H-thymine-labeled cells starved 60 minutes for required amino acids, centrifugation of the resulting lysate unambiguously demonstrates the sedimentation velocity differences between membrane-associated folded chromosomes obtained from amino acid-starved cells and exponentially growing cells (Figure 2); the membrane-associated folded chromosomes

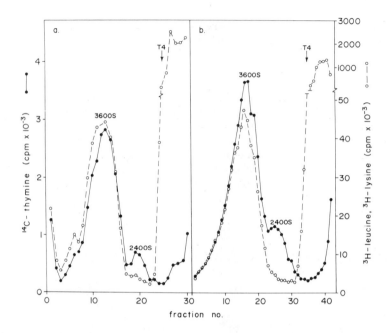

Fig. 1. Effect of 4000 x g centrifugation on the sedimentation properties of membrane-associated folded chromosomes. Growth was in media as described previously (17). Membrane-associated folded chromosomes were prepared and subjected to sucrose gradient centrifugation for 10.0 minuntes at 16,000 rpm in the SW50.1 rotor at 2°C. 1a) 4000 x g supernatant 1b) crude lysate. ^{32}P-T4 bacteriophage were added as a sedimentation marker. Copyright J. Bacteriol.

Table 1

RECOVERY OF LABELLED DNA

source of lysate	fractionation of acid-insoluble ^{14}C-thymine counts in 4000 x g centrifugation[a]		recovery of chromosomes from sucrose gradient centrifugation		
	supernatant	pellet	4000 x g centrifugation	time at 16,000 rpm	recovery of chromosomes[a]
exponentially growing cells	32%	68%	+	10 min	29%
			-	10 min	96%
			-	3 min	94%
60 min amino acid-starved cells	3%	97%	+	10 min	3%
			-	10 min	11%
			-	3 min	89%

a) expressed as percent of counts in the crude lysate

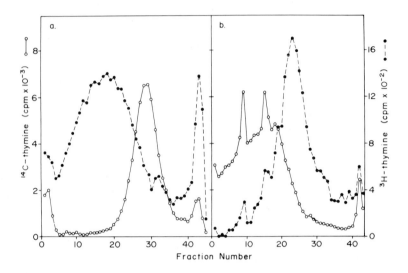

Fig. 2. Sedimentation properties of membrane-associated folded chromosomes isolated from exponentially growing and amino acid-starved TAU-bar cells. Exponentially growing cultures of *E. coli* TAU-bar were labeled with either ^{14}C-thymine or ^{3}H-thymine as described in METHODS. Cultures to be amino acid-starved were transferred to medium lacking required amino acids except methionine by rapid filtration (Maaløe and Hanawalt, 1961). Amino acid-starved cultures were mixed with exponentially growing cultures and membrane-associated folded chromosomes were prepared as described in experimental procedures. Cell densities in the lysate were $\leq 2 \times 10^{9}$ cells/ml. Centrifugation was performed at 2°C for 3.0 min at 16,000 rpm in the SW 50.1 rotor. 2a) ^{3}H-thymine-labeled cells starved for required amino acids for 60 minutes were mixed with ^{14}C-thymine-labeled exponentially growing cells. 2b) ^{14}C-thymine-labeled cells starved for required amino acids for 90 minutes were mixed with ^{3}H-thymine-labeled exponentially growing cells. Copyright Cell.

from amino acid-starved cells are more rapidly sedimenting. Assigning a sedimentation coefficient of 3600S to the peak of the sedimentation profile for membrane-associated folded chromosomes obtained from exponentially growing cells (17) yields a sedimentation coefficient of 5000-6000S for membrane-associated folded chromosomes obtained from amino acid-starved cells. Similar results have been obtained with other *E. coli* strains; Figure 2 illustrates that membrane-associated folded chromosomes obtained from amino acid-starved *E. coli* K12 DG75 are more rapidly sedimenting than membrane-associated folded chromosomes obtained from exponentially growing DG75 cells. Other investigators(18,19) have been unable to isolate membrane-associated chromosomes from amino acid-starved *E. coli* K12 DG75 cells possibly because their conditions of isolation and centrifugation would not have allowed recovery of chromosomes sedimenting as rapidly as 5000-6000S.

C.   Electron Microscopy of Membrane-Associated Folded Chromosomes: We have begun analyzing membrane-associated folded chromosomes of *E. coli* 15 TAU-bar in the electron microscope. With membrane-associated folded chromsomes from both exponentially growing (22) and amino acid-starved cultures, aqueous spreading techniques reveal two basic forms: condensed forms and expanded forms. The condensed forms resemble cells, with DNA condensed within an envelope which is often torn or partially disrupted. The expanded forms resemble lysed cells, with loops of DNA radiating from an irregularly shaped patch of residual envelope. Some loops appear supercoiled. The expanded forms may arise from condensed forms (22).

Plate 1 is an electron micrograph of an expanded membrane-associated folded chromosome from an amino acid-starved culture. This is similar in appearance to expanded membrane-associated folded chromosomes from exponentially growing cultures (22). In such extended structures, it is possible to count the approximate number of loops with a reproducibility of about $\pm$ 5%. Seven complexes from exponentially growing cultures had 65-130 loops, with the distribution shown in Figure 4 and an average of about 90 loops per structure. Three complexes from an amino acid-starved culture had 85-115 loops, with the distribution shown in Figure 4 and an average of about 105 loops per structure. For comparison, the number of supercoiled

Plate 1.   Electron micrograph of a membrane-associated folded chromosome purified from an *E. coli* 15 TAU-bar
culture starved 75 minutes for required amino acids.   Sample contained 1 mM spermine.   Grid was prepared 1 min after
spreading.

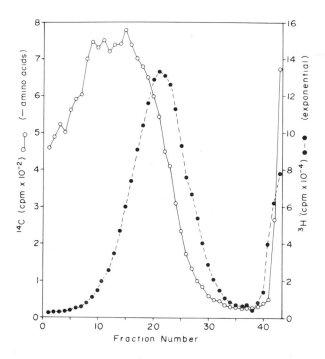

Fig. 3. Sedimentation properties of membrane-associated folded chromosomes isolated from exponentially growing and amino acid-starved DG75 cells. ^{14}C-thymine cells were grown as described in METHODS, subjected to 120 minutes of leucine deprivation, mixed with exponentially growing cells pulse-labeled 30 minutes with ^{3}H-thymidine (10 μCi/ml), and lysed together. Cell density in the lysate was $\leq 1 \times 10^9$ cells/ml.

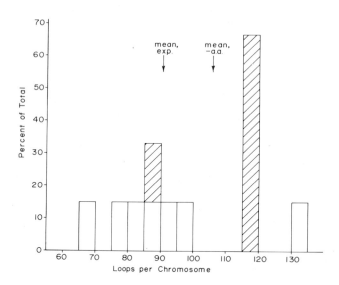

Fig. 4. Histogram of the number of loops present in membrane-associated folded chromosomes from exponentially growing cells (open bars) and amino acid-starved cells (crosshatched bars). Membrane-associated folded chromosomes from *E. coli* 15 TAU-bar were purified by sucrose gradient centrifugation and prepared for electron microscopy as described in METHODS. Repeated measurements of the number of loops present in individual chromosomes indicated a varience of $\pm$ 5%. NOTE: These data are for structures in which loops of DNA radiate from membrane patches (e.g. Plate 1). In 30% of the structures from amino acid-starved cells, envelope appeared detached from the DNA, and in these cases it was not possible to count loops. The frequency of structures with envelope detached during spreading was much lower with exponentially growing cells.

loops (per structure from exponentially growing cultures)
estimated from physical and biochemical data is: 12–80
(23) and 20–100 (24). Another possible difference between
the membrane-associated folded chromosomes from exponentially growing cells and amino acid-starved cells is that
flagella are found on almost all of the structures from
exponentially growing cultures, but they are not seen on
the structures from amino acid-starved cultures.

D. Preparation of Membrane-Associated Folded Chromosomes
from a Thermosensitive Initiation Mutant: If the increased
sedimentation velocity of membrane-associated folded chromosomes obtained from amino acid-starved cells is due to
the lack of synthesis of protein species necessary for initiation of DNA synthesis, then analysis of thermosensitive
mutants deficient in initiation at the restrictive temperature should also result in more rapidly sedimenting folded
chromosomes. Furthermore, because initiation mutants continue to synthesize protein at a normal rate under the
restrictive conditions, the effects of protein synthesis
inhibition can be separated from the inhibition of initiation of rounds of chromosome replication. For these studies we have used *E. coli* K12 CT28-4, a more thermostable
derivative of CT28-3b, the *dnaC* mutant isolated and
characterized by Schubach, Witmer and Davern (26) and
demonstrated by autoradiographic techniques to complete
rounds of chromosome replication at the restrictive temperature by Rodriguez, Dalby, and Davern (27).

The permissive temperature for the mutant is 37°C; the
restrictive temperature is 42°C. When CT28-4 is grown at
37°C and then shifted to 42°C, protein synthesis and mass
increase exponentially while DNA synthesis diminishes in
rate until synthesis ceases after 40–50 minutes (25,26).
When the temperature is reduced to 37°C, DNA synthesis
resumes at an exponential rate after 0–15 minutes (depending on the rapidity with which the temperature shift is
achieved).

Figure 5 illustrates the sedimentation properties of
membrane-associated folded chromosomes isolated from CT28-4
grown at the permissive temperature (Figure 5a) and subsequently shifted to nonpermissive temperature, aliquots
being taken at 20 minute intervals (Figures 5b-e). The
sedimentation coefficient for the peak fraction from exponentially growing CT28-4 (Figure 5a) is 3600S relative to

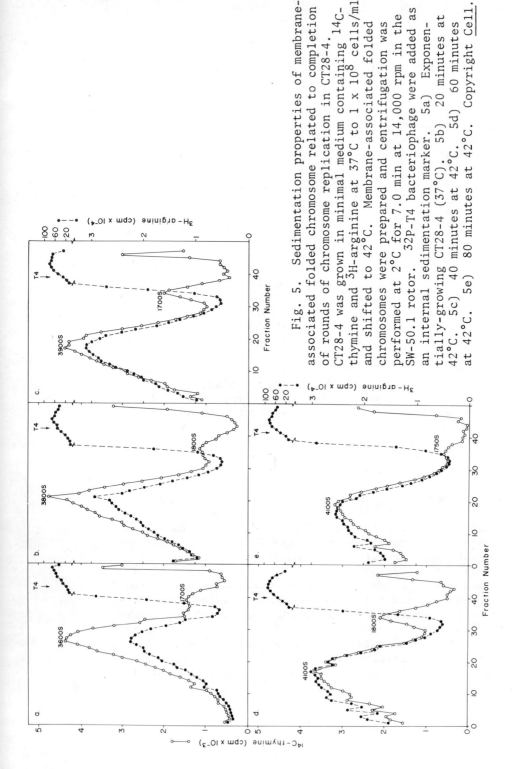

Fig. 5. Sedimentation properties of membrane-associated folded chromosome related to completion of rounds of chromosome replication in CT28-4. CT28-4 was grown in minimal medium containing 14C-thymine and 3H-arginine at 37°C to 1 x 10⁸ cells/ml and shifted to 42°C. Membrane-associated folded chromosomes were prepared and centrifugation was performed at 2°C for 7.0 min at 14,000 rpm in the SW-50.1 rotor. 32P-T4 bacteriophage were added as an internal sedimentation marker. 5a) Exponentially-growing CT28-4 (37°C). 5b) 20 minutes at 42°C. 5c) 40 minutes at 42°C. 5d) 60 minutes at 42°C. 5e) 80 minutes at 42°C. Copyright Cell.

marker bacteriophage T4 (1025S, ref. 28). With increasing
time at the restrictive temperature (Figure 5b,c), the
isolated membrane-associated folded chromosomes become
more rapidly sedimenting until residual DNA synthesis has
ceased (40-60 minutes). After cessation of DNA synthesis
at 42°C the sedimentation velocities of the chromosomes
are not altered (Figure 5d,e). The increase in sedimentation velocity is not due to an artifact of the higher
temperature but is apprently related to *dnaC* gene product
inactivation because no increase in sedimentation velocity
of membrane-associated folded chromosomes is observed if
a thermoresistant revertant of CT28-4 is treated for 80
minutes at 42°C (data not shown).

These data confirm the results obtained with membrane-
associated folded chromosomes from amino acid-starved cells;
as chromosomes proceed to termination of a round of replication in the absence of reinitiation, membrane-associated
folded chromosomes become more rapidly sedimenting (17,25).

E.   Properties of Membrane-Associated Folded Chromosomes
at Initiation of Chromosome Replication: When amino acids
are resupplied to a culture of *E. coli* TAU-bar which has
completed residual synthesis in the absence of required
amino acids, DNA synthesis does not resume until 10-20
minutes after restoration of amino acids (Figure 6).
When amino acids are restored to an amino acid-starved
culture, the sedimentation properties of membrane-associated folded chromosomes are altered; they revert to the
range of sedimentation velocities observed for chromosomes
obtained from exponentially growing cells (Figure 7a).
Figure 6 illustrates the kinetics for the sedimentation
velocity decrease of terminalized membrane-associated
folded chromosomes related to the initiation of DNA synthesis. It is clear that the sedimentation velocity decrease
observed tends to preceed the initiation of rounds of DNA
synthesis. To further examine the relationship between
the sedimentation velocity decrease of terminalized chromosomes and initiation of DNA synthesis, cultures of *E.
coli* TAU-bar which had been starved for amino acids were
resupplied with amino acids in the absence of thymine so
that protein synthesis might resume but deoxyribonucleotide
incorporation would be inhibited (Figure 7b). A 30 minute
incubation in medium containing required amino acids but
lacking required thymine allows a substantial proportion

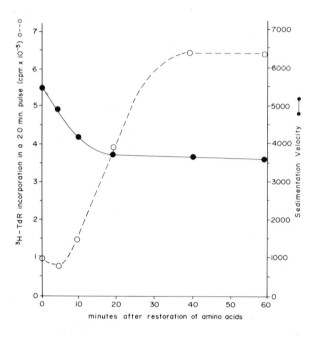

Fig. 6.   Sedimentation velocities of membrane-associated folded chromosomes and initiation of DNA synthesis
after restoring amino acids to amino acid-starved cells.
E. coli 15 TAU-bar was grown to 1 x 10^8 cells/ml in minimal medium containing 1.0 μCi/ml ^{14}C-thymine and starved
90 minutes for required amino acids except L-methionine.
10 ml aliquots (containing approximately 1 x 10^9 cells)
were removed at intervals and pulse-labeled with ^{3}H-thymi-
dine (5 μCi/ml) for 2.0 minutes; pulse labeling was terminated by addition of KCN and NaN$_3$ (0.01 M final concentration of each) and quick chilling of the culture.   Membrane-
associated folded chromosomes were prepared and acid-insoluble isotope incorporation was determined.   Sucrose gradient centrifugation was performed on the crude lysates at
2°C in the SW50.1 rotor at 16,000 rpm for 4.0 minutes.
Sedimentation velocities are estimated on the basis of
position of the complexes in sucrose gradients relative to
the position in an identical gradient of membrane-associated folded chromosomes isolated from exponentially growing cells.

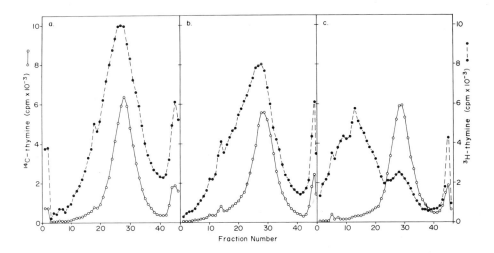

Fig. 7.   Conversion of sedimentation properties of
terminalized membrane-associated folded chromosomes in the
absence of DNA and protein synthesis. Exponentially growing cultures were labeled with either ^{14}C-thymine or ^{3}H-
thymine as described in experimental procedures. The ^{3}H-
thymine-labeled culture was transferred to medium lacking
required amino acids except methionine for 90 minutes.
Prior to lysis, ^{14}C-thymine-labeled exponentially growing
cells were added to the ^{3}H-thymine-labeled cells. Cell
density in the lysate was 2 x 10^{9} cells/ml. Centrifugation of lysates was performed at 2°C for 3.0 minutes at
16,000 rpm in the SW50.1 rotor. 7a) Required amino acids
were resupplied to the ^{3}H-thymine-labeled culture for 30
minutes before preparation of membrane-associated folded
chromosomes. 7b) The ^{3}H-thymine-labeled cells were incubated with required amino acids in the absence of thymine
for 30 minutes before preparation of membrane-associated
folded chromosomes. 7c) ^{3}H-thymine-labeled cells were
incubated for 30 minutes in medium containing 200 μg/ml
chloramphenicol and required amino acids but lacking thymine prior to preparation of membrane-associated folded
chromosomes. Copyright Cell.

of the chromosomes to be converted to complexes with sedimentation velocities similar to those observed for chromosomes obtained from exponentially growing cells (Figure 7b). The decrease in sedimentation velocity after restoration of required amino acids therefore does not depend on renewed DNA synthesis.

If 200 µg/ml chloramphenicol is added to an amino acid-starved culture 5 minutes prior to readdition of required amino acids, the bulk of the membrane-associated folded chromosomes prepared after a 30 minute incubation do not exhibit any reduction in sedimentation velocity (Figure 7c). Thus, the reduction in sedimentation velocity of membrane-associated folded chromosomes related to the readdition of amino acids to a starved culture requires protein synthesis.

In order to examine the properties of chromosomes at initiation of DNA replication in CT28-4, cultures of the mutant were allowed to complete rounds of replication at the restrictive temperature and then were shifted to the permissive temperature to allow synchronous initiation (26). Figure 8a again illustrates the sedimentation properties of membrane-associated folded chromosomes from cells having completed rounds of replication at the restrictive temperature. When the temperature is shifted down to 37°C for 10 minutes (Figure 8b), the sedimentation velocity of the membrane-associated folded chromosomes decreases. DNA synthesis resumes 10-20 minutes after reduction of temperature under our conditions of temperature shift (25). By 20 minutes after lowering the temperature, the majority of the membrane-associated folded chromosomes sediment at velocities similar to those of membrane-associated folded chromosomes obtained from exponentially growing cells (Figure 8c). Some cells of CT28-4 do not reinitiate DNA synthesis upon temperature reduction (R.L. Rodriguez, personal communication); this is probably reflected in the continued presence of some chromosomes with sedimentation velocities similar to terminalized chromosomes (Figures 8b,c).

Figure 8d demonstrates that inhibition of DNA synthesis by thymine deprivation does not deter the decrease in sedimentation velocity of membrane-associated folded chromosomes related to the initiation of DNA replication in CT 28-4. This is similar to the result obtained with *E. coli* 15 TAU-bar (Figure 7b). If the culture of CT28-4 is

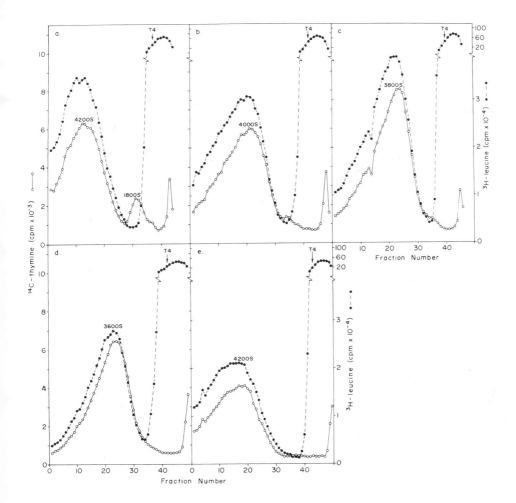

Fig. 8.   Sedimentation properties of membrane-associated folded chromosomes related to release of the thermosensitive block in initiation of chromosome replication in
CT28-4. CT28-4 was grown in minimal medium containing [14]C-
thymine and ^{3}H-leucine at 37°C to approximately 7 x 10[7]
cells/ml and shifted to 42°C for 60 minutes. Aliquots of
the culture were either returned to 37°C or were transferred
to 42°C medium lacking thymine and returned to 37°C. Membrane-associated folded chromosomes were prepared and
centrifuged at 2°C for 7.0 minutes at 14,000 rpm in the
SW50.1 rotor. ^{32}P-T4 bacteriophage were added as a
sedimentation marker. 8a) 60 minutes 42°C. 8b) 60
minutes 42°C, 10 minutes 37°C. 8c) 60 minutes 42°C, 20
minutes 37°C. 8d) 60 minutes 42°C, 20 minutes without
thymine at 37°C. 8e) 60 minutes 42°C, 20 minutes without
thymine at 42°C. Copyright Cell.

held at the restrictive temperature in the absence of
thymine (Figure 8e), no sedimentation velocity decrease
of the complexes is observed, thus suggesting that it is
not thymine deprivation but rather the activity of the
dnaC gene product which induces the reduction in sedimentation velocity.

To further examine the effect of DNA synthesis inhibition on the sedimentation velocity decrease of chromosomes observed upon return of CT28-4 to the permissive
temperature, 30 µg/ml nalidixic acid was added to a culture of CT28-4 before reducing the temperature. After
20 minutes at 37°C (Figure 9c), the sedimentation velocities of membrane-associated folded chromosomes from the
nalidixic acid-treated cells are similar to those of the
control culture which received no nalidixic acid before
temperature reduction (Figure 9b). In both cases the 20
minute treatment at the permissive temperature was sufficient to result in the decrease of the sedimentation velocity of membrane-associated chromosomes compared to the
velocity of membrane-associated chromosomes from cells
which were not returned to 37°C (Figure 9a). Nalidixic
acid treatment apparently has no observable effects on the
sedimentation properties of membrane-associated folded
chromosomes (16, our unpublished data). These observations,
further demonstrate that DNA synthesis is not one of the
events required for the decrease in sedimentation velocity
observed after terminalized chromosomes are returned to
the permissive temperature.

The thermolabile defect in initiation of chromosome
replication in CT28-4 is reversible upon return to the
permissive temperature; thus, the majority of the cells
within a culture of CT28-4 can initiate rounds of replication in the presence of chloramphenicol after allowing
completion of rounds of replication at 42°C (26). Consistent with this behavior, the membrane-associated folded
chromosomes decrease in sedimentation velocity after return
to 37°C, even if 200 µg/ml chloramphenicol is added to the
culture before the temperature reduction (Figure 9d).

If the temperature is again raised to 42°C after
allowing initiation to occur by a 20 minute incubation at
37°C, chromosomes continue a synchronous round of replication (26), and should again yield the rapidly sedimenting
chromosomes related to completion or near completion of
rounds of replication. Figure 9e illustrates that this

180

prediction is confirmed.

DISCUSSION

The results presented here and elsewhere (17,22,25) using *E. coli* strains grown in minimal glucose medium demonstrate that completion or near completion of rounds of chromosome replication in the absence of reinitiation permits isolation of membrane-associated folded chromosomes which sediment more rapidly than membrane-associated folded chromosomes isolated from exponentially growing cells (Figures 2,3,5). The very rapidly sedimenting membrane-associated folded chromosomes that may be isolated when chromosome initiation has been inhibited by amino acid deprivation or by mutation possess unique sedimentation and ultrastructural properties and might therefore represent a unique chromosomal structure. The molecular arrangement of this apparently unique structure remains to be elucidated, but might possibly consist of two chromosomes which are isolated as a single unit.

Electron microscopy of membrane-associated folded chromosomes purified from exponentially growing and amino acid-starved cells shows certain similarities in structure. In both cases, compact and extended structures are observed. Attempts to quantitate the number of loops in well-spread complexes suggest that the number of DNA loops present in chromosome complexes from amino acid-starved cell approaches the upper boundary in loop number present in chromosome complexes from exponentially growing cells (Figure 4). Preliminary results suggest that, as replication proceeds, the size of individual loops remains approximately constant, but the total number of loops increases as chromosomes proceed to completion of a round of replication. This hypothesis is presented schematically in Figure 10.

After restoring defective function(s) by resupplying amino acids to amino acid-starved cells (Figure 7a) or by returning to the non-restrictive temperature in a thermosensitive initiation mutant (Figures 8c,9b), membrane-associated folded chromosomes undergo a transition resulting in a reduction of the sedimentation velocities of the complexes to the range observed for chromosomes purified from exponentially growing cells. The sedimentation velocity reduction of membrane-associated folded chromosomes related to restoration of protein synthesis in amino acid-

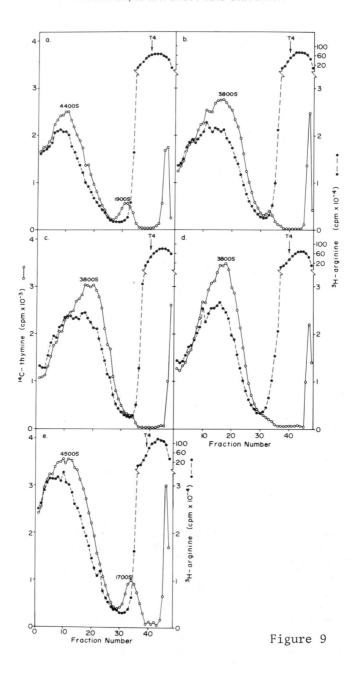

Figure 9

Fig. 9. Effect of inhibition of protein and DNA synthesis on sedimentation velocity changes associated with release of the thermosensitive block in CT28-4. CT28-4 was grown in minimal medium containing ^{14}C-thymine and ^{3}H-arginine at 37°C to approximately 7×10^{7} cells/ml and shifted to 42°C for 60 minutes. Before shifting the cultures back to 37°C inhibitors were added to some of the cultures. Membrane-associated folded chromosomes were prepared and centrifuged at 2°C for 7.0 minutes at 15,000 rpm in the SW50.1 rotor. ^{32}P-T4 bacteriophage were added as a sedimentation marker. 9a) 60 minutes 42°C. 9b) 60 minutes 42°C, 20 minutes 37°C. 9c) 60 minutes 42°C, 20 minutes at 37° in the presence of 30 µg/ml nalidixic acid. 9d) 60 minutes 42°C, 20 minutes at 37°C in the presence of 200 µg/ml chloramphenicol. 9e) 60 minutes 42°C, 20 minutes 37°C, 40 minutes 42°C. Copyright Cell.

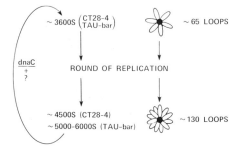

Fig. 10. A schematic model for properties of membrane-associated folded chromosomes during rounds of replication.

183

starved cells and with restoration of *dnaC* gene product
function in CT28-4 appears not to be related to the actual
initiation of deoxyribonucleotide incorporation. This is
supported by observations that the sedimentation velocity
reduction occurs when DNA synthesis is inhibited by thymine
deprivation (Figures 7b,8d,9c) or by nalidixic acid treatment (Figure 9c). The reduction in sedimentation velocity
of the membrane-associated folded chromosomes does appear to
require protein synthesis in cells which are deprived of
required amino acids (Figure 7c). The *dnaC* gene product
must be among the species involved since the reduction in
sedimentation velocity depends on reduction of the temperature in the thermosensitive CT28-4 mutant (Figure 8a,
8b, 9a, 9b). This evidence demonstrates that the active
dnaC gene product is capable of leading to a reduction in
the sedimentation velocity of terminalized membrane-associated folded chromosomes independently of the initiation
of DNA synthesis. Therefore, it would appear that some
processing of completed chromosomes - altering their sedimentation properties - precedes the onset of DNA synthesis,
and that at least the *dnaC* gene product is involved in
this processing.

ACKNOWLEDGEMENTS

We thank D.E. Pettijohn, E.P. Geiduschek, and R.L.
Rodriguez for helpful advice, B.H. Zimm and K. Tokuyasu
for facilities and encouragement, C.Barna for technical
assistance, and D. Williams for typing the manuscript.
This work was supported by grants from the American Cancer
Society, U.S. Public Health Service, Anna Fuller Foundation,
and the W.H.O. International Agency for Research on Cancer.

REFERENCES

1. Maaløe, O. and Kjeldgaard, N.O. Control of Macromole-
    cular Synthesis. Benjamin, New York. (1966).
2. Helmstetter, C.E., Cooper, S., Pierucci, O., and
    Revelas, E. Cold Spring Harbor Symp. Quant. Biol.
    33:809 (1968).
3. Maaløe, O.and Hanawalt, P.C. J. Mol. Biol. 3:144
    (1961).
4. Lark, K.G., Repko, T., and Hoffman, E.J. Biochim.
    Biophys. Acta. 190:88 (1963).

5. Wolf, B., Pato, M.L., Ward, C.B., and Glaser, D.A. Cold Spring Harbor Symp. Quant. Biol. 33: 575 (1968).

6. Caro, L.G. and Berg, C.M. J. Mol. Biol. 45:325 (1969).

7. Cerda-Olmedo, E., Hanawalt, P.C., and Guerola, N. J. Mol. Biol. 33:705 (1968).

8. Marunouchi, T. and Messer, W. J. Mol. Biol. 78:211 (1973).

9. Okazaki, R., Okazaki, T., Sakabe, R., Sugimoto, K., and Sugino, A. Proc. Nat. Acad. Sci. U.S. 59:598 (1968).

10. Jaenisch, R., Hofschneider, P.H., and Preuss, A. Biochim. Biophys. Acta. 190:88 (1969).

11. Meinke, W. and Goldstein, D.A. J. Mol. Biol. 61:543 (1971).

12. Kupersztoch-Portnoy, Y.M. Ph.D. Thesis. University of California, San Diego (1974).

13. Benbow, B.M., Eisenburg, M., and Sinsheimer, R.I. Nature New Biol. 237:141 (1972).

14. Stonington, O.G. and Pettijohn, D.E. Proc. Nat. Acad. Sci. U.S. 68:6 (1971).

15. Pettijohn, D.E., Hecht, R.M., Stonington, O.G., and Stomato, T.D. DNA Synthesis in vitro. R.D. Wells and R.D. Innman (eds.). University Park Press, Baltimore, p. 145 (1973).

16. Dworsky, P. and Schaechter, M. J. Bacteriol. 116: 1364 (1973).

17. Ryder, O.A. and Smith, D.W. J. Bacteriol. 120:1356 (1974).

18. Worcel, A., Burgi, E., Robinton, J., and Carlson, C.L. Cold Spring Harbor Symp. Quant. Biol. 38:43 (1973).

19. Worcel, A. and Burgi, E. J. Mol. Biol. 82:91 (1974).

20. Delius, H. and Worcel, A. Cold Spring Harbor Symp. Quant. Biol. 38:53 (1974).

21. Delius, H. and Worcel, A. J. Mol. Biol. 82:107 (1974).

22. Kavenoff, R. and Ryder, O.A. In press: Chromosoma (1975).

23. Worcel, A. and Burgi, E. J. Mol. Biol. 72:127 (1972).

24. Pettijohn, D.E. and Hecht, R. Cold Spring Harbor Symp. Quant. Biol. 38:31 (1973).

25. Ryder, O.A. and Smith, D.W. Cell 4: 337 (1975).

26. Schubach, W.H., Whitmer, J.D., and Davern, C.I. J. Mol. Biol. 74:205 (1975).

27.   Rodriguez, R.L., Dalby, M.S., and Davern C.I. J. Mol. Biol. 74:599 (1973).

28.   Cummings, D.J. Virology 23:408 (1964).

29.   Kleinschmidt, A. Methods in Enzymology. S. Colowick and N. Kaplan (eds.). Academic Press, New York. Vol. 12, part B, p. 361 (1968).

30.   Davis, R., Simon, M. and Davidson, N. Methods in Enzymology. L. Grossman and K. Moldave (eds.). Academic Press, New York. Vol. 21, part D, p. 413 (1971).

31.   Cairns, J. Cold Spring Harbor Symp. Quant. Biol. 28: 43 (1963).

32.   Hanawalt, P.C. Nature 198:286 (1963).

MEMBRANE-DNA COMPLEX IN BACILLUS SUBTILIS

Sumi Imada, Lynn E. Carroll and Noboru Sueoka
Department of Molecular, Cellular
and Developmental Biology
University of Colorado
Boulder, Colorado 80302

ABSTRACT. A DNA-membrane complex (M2) from Bacillus
subtilis has previously been separated from the major
membrane fraction (M1) by a CsCl-sucrose double gradient
centrifugation. The M2 DNA has been shown to be enriched
in genetic markers which map close to the replication
origin and terminus. The complex has also been shown to
possess a characteristic set of proteins. Here, we report:
(a) The complex originated from the membrane attached
folded chromosome. (b) The complex (M2) is disrupted by
detergents such as sarkosyl, Brij 58, Tween 80 and sodium
deoxycholate, while RNase and phospholipase A do not disrupt the complex. (c) Iodination of intact cells with ^{125}I
shows a preferential labeling of the complex suggesting
that at least part of the complex is exposed on the
external membrane surface. (d) Possible models of the DNA-
membrane complex are proposed.

INTRODUCTION

An attachment of the bacterial chromosome to the cell
membrane, as first proposed by Jacob, Brenner and Cuzin
(1), suggests a number of interesting possibilities for
chromosome organization and replication. First, it permits
a firm localization of the chromosome in the cell and provides a basis for mechanical separation of the two daughter
chromosomes (1). Secondly, DNA replication initiates and
proceeds in a special metabolic environment.

Substantial evidence has been accumulated on the
association of the bacterial chromosome with the cell
envelope (2). We have been focusing our attention on the
topological nature of chromosome-membrane association.

We present here further results of our work on the chromosome-membrane complex in <u>Bacillus</u> <u>subtilis</u>.

MATERIALS AND METHODS

<u>Strains</u>: <u>B</u>. <u>subtilis</u> 168TT (<u>thy</u> <u>trp</u>) was used for the major part of the study, and 168 <u>leu</u>8-<u>metB</u>5-<u>purA</u>16 was used as the transformation recipient.

<u>Culture media and radioactive labeling</u>: <u>B</u>. <u>subtilis</u> 168TT was labeled with ^{35}S-sulfate and ^{3}H-thymine in GM-11 medium (3) to which 50 µg/ml of tryptophan and 2 µg/ml of thymine were added, with magnesium chloride and ammonium chloride replacing magnesium sulfate and ammonium sulfate, respectively. Bott's & Wilson's Transformation Medium (4) was used with supplements of adenine, histidine and tryptophan (50 µg/ml of each), leucine and methionine (100 µg/ml of each) added where required. The transformation selection plates were made with 2% agar, C medium (5), 100 µg/ml of required amino acids and 50 µg/ml of required base.

<u>Cell lysis procedure</u>: <u>B</u>. <u>subtilis</u> 168TT was grown in the above medium to Klett-Summerson Colorimeter unit 40 in the presence of appropriate radioactive isotopes. The culture was chilled and cells were harvested by centrifugation at 5,000 rpm for 10 min at 4°C. The drained cell pellet was frozen in a dry ice-ethanol bath and stored frozen at -80°C. The cell pellet of a 10 ml culture was thawed and resuspended in 0.3 ml of Tris (50mM)-EDTA (0.1M)-sucrose (0.5M), pH 8.0 plus 0.1 ml of 10 mg/ml of egg white lysozyme (Worthington Co.) and 0.04 ml of 0.1 M of KCN. The mixture was kept at 0°C for 60 min and diluted by rapid addition of 1.06 ml of Tris (50mM)-EDTA (0.1M)-sucrose (0.15M), pH 8.0 buffer. The protoplasts were burst by this dilution although the lysates were still turbid. Further incubation at 30°C was performed in some cases as indicated. The lysate was stirred with a Vortex mixer in a tube of 13 mm inner diameter for 1 min at the highest speed setting to shear the DNA. The lysate was kept on ice except as noted.

<u>CsCl-sucrose double gradient</u>: A double linear gradient of CsCl (0.5 molal to 5 molal) and sucrose (10-20% w/w) was formed as described in the previous paper (6). The sample

was layered on top of the gradient and centrifuged in an
SW 50.1 rotor at 35,000 rpm for 30 min at 4°C.

Sucrose step centrifugation: Sucrose step gradients were
prepared by layering 3.0 ml of 20% sucrose over 1.0 ml of
62% sucrose. Sucrose solutions (w/w) were prepared in
1/10 NET (NET: 0.1M NaCl, 10mM EDTA, 10mM Tris-Cl, pH 8.0
(7)). Samples were centrifuged for 60 min at 35,000 rpm
in an SW 50.1 rotor at 4°C.

External cell labeling with ^{125}I: Iodination of B. subtilis intact cells was performed according to the method
of Hubbard and Cohn (8). Exponentially growing cells were
harvested by centrifugation (6,000 rpm, 5 min at 4°C) and
washed with phosphate-buffered saline (PBS pH 7.4: one
liter of PBS contains 8 gm of NaCl, 0.2 gm of KCl, 1.15 gm
of Na_2HPO_4 and 0.2 gm of KH_2PO_4). The cells were resuspended in PBS at a concentration of ~5x10^9 cells/ml and
iodinated at 10°C for 30 min. The reaction mixture contained 5 mM glucose, 3.6 units of glucose oxidase (Sigma
Chem. Co.), 200-300 μCi of carrier-free Na^{125}I (New England Nuclear Co.), 2.7 mM KI and 20 μg/ml of lactoperoxidase (Calbiochem). The reaction was stopped by dilution
with 5 volumes of PBS containing 1 mM phenylmethyl
sulfonyl fluoride (PMSF) and 0.01 mM sodiumthiosulfate.
The cells were then washed twice with PBS. The washed
pellet was frozen in a dry ice-ethanol bath and stored at
-80°C.

SDS-polyacrylamide gel electrophoresis: All samples for
the gel analysis were dialyzed against distilled water,
lyophilized, and redissolved in SDS-polyacrylamide gel
electrophoresis sample buffer (9). The protein samples
were heated for 3 min at 95°C in the buffer. The gel
electrophoresis in 10% acrylamide in the presence of 0.1%
sodium dodecyl sulfate was performed according to Laemmli
(9).

RESULTS

DNA-membrane complex: In a previous paper (6) it was shown
that a fraction containing membrane-bound DNA and minor
amounts of membrane protein could be separated from the
major portion of the B. subtilis membrane fraction.

This paper presents additional data obtained when the previously described lysis conditions (6) were modified (see Materials and Methods).

When the bacteria were treated with lysozyme at 0°C, then gently disrupted by osmotic shock, the resulting lysate was free of cells, but not viscous in contrast to the lysate obtained with the unmodified procedure. Upon CsCl-sucrose centrifugation most (>90%) of the DNA was recovered in a single peak. Approximately 10 to 15% of the total ^{35}S-labeled protein was present in this peak. If the lysates were agitated by Vortex before CsCl-sucrose centrifugation, the profiles (Fig. 1a) were identical to those obtained with the non-vortexed preparation. The DNA of the complex is apparently not sensitive to shearing force, suggesting that it could be tightly folded. Thus, the complex might be compared to the folded chromosome isolated from E. coli (10,16).

The complex was heat-sensitive. The gently opened cells (Fig. 1a) were incubated at 30°C for 10 min (Fig. 1b) 20 min (Fig. 1c), and for 40 min (Fig. 1d). The lysates were then vortexed and centrifuged on CsCl-sucrose double gradients. During the incubation at 30°C, the DNA became shear-sensitive and even after short incubation (10 min) approximately 80% of the total DNA could be removed from the complex by the vortex treatment as seen in Figure 1.

After heat incubation, the centrifugation profiles of the lysate resembled those reported in the previous publication (6). The complex separated into two fractions, one of which corresponded to M2 with more DNA and less protein, and the other to M1 carrying less DNA and more protein.

Marker ratio: Since purA16 is located near the origin, leu8 approximately midway between origin and terminus, and metB5 near the terminus of the B. subtilis chromosome, the ratios purA16$^+$/leu8$^+$ and metB5$^+$/leu8$^+$ provide useful information concerning the portion of the chromosome remaining in the complex following heat-treatment. The fractions between the two arrows in each gradient (Fig. 1a through 1d) were pooled and the numbers of transformants for the markers purA16$^+$, leu8$^+$ and metB5$^+$ were examined. The marker ratios of the complex increased as the percentage of total DNA in the complex decreased. Thus, the enrichment of origin and terminus markers in the DNA-membrane increased with the time of 30°C treatment of the

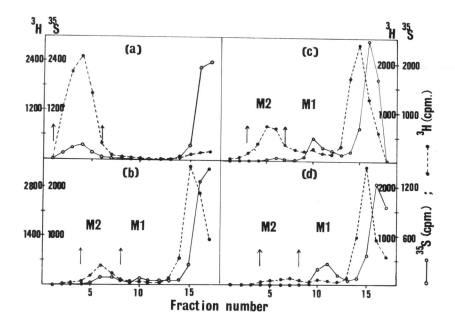

Fig. 1. Fractionation of cell lysate by CsCl-sucrose double gradient cnetrifugation. B. subtilis 168TT was labeled with 20 μCi/ml of ^{3}H-thymine and 5 μCi of ^{35}S-sulfate and the pellet obtained from 10 ml of culture was treated with lysozyme as described in Materials and Methods. After the protoplasts were burst by dilution, the sample (a) was taken and kept at 0°C. The rest of the lysate was incubated at 30°C for periods of 10 min (b), 20 min (c), and 40 min (d). After the incubation periods, samples were kept at 0°C. Samples (a) through (d) were agitated by Vortex mixer and centrifuged on CsCl-sucrose double gradients. Seventeen or eighteen fractions were collected from each gradient with polystaltic pump. An aliquot of each fraction was used for measurement of TCA precipitable counts of ^{3}H and ^{35}S. The fractions between the two arrows in each gradient were pooled and used for determination of marker ratios, the results of which are shown in Table 1. o——o ^{35}S (protein); •----• ^{3}H (DNA)

TABLE 1

ENRICHMENT OF puraA16 AND metB5 MARKERS RELATIVE TO leu8

DNA from gradients in Fig. 1	$puraA16^+/leu8^+$			$metB5^+/leu8^+$		
	Transformants/Plate		Ratio	Transformants/Plate		Ratio
	$puraA16^+$	$leu8^+$	$puraA16^+/leu8^+$	$metB5^+$	$leu8^+$	$metB5^+/leu8^+$
(a)	2799	734	3.81	2361	4967	0.48
(b)	3323	752	4.42	2271	3668	0.62
(c)	6932	1962	3.53	4183	5141	0.81
(d)	2225	386	5.76	1273	1489	0.85

The fractions between the arrows were pooled from each of the gradients shown in Fig. 1, and the pooled sample was dialyzed against 0.1 M potassium phosphate buffer (pH 7.0). The transformation was performed for the three markers by using 0.5 ml of competent culture and 0.2 ml of a sample which had been appropriately diluted to give a statistically meaningful number of transformants. As the transformation efficiency of $puraA16$ is higher than that of $metB5$, the ratios of $puraA16^+/leu8^+$ and $metB5^+/leu8^+$ were obtained using two different concentrations of DNA. The transformants were scored by plating 0.1 ml of the transformation mixture on each selection plate. The numbers shown in this table are averages of two plates.

192

lysate. The low ratio of purA16$^+$/leu8$^+$ of sample (c) was probably caused by the use of a DNA amount exceeding the linear transformation range.

Proteins in M2 complex: The protein in the M2 fractions was analyzed by polyacrylamide gel electrophoresis. As shown in Figure 2, the protein species recovered in the M2 fraction slots (b)-(d) were distinct from that of the original DNA-membrane complex which is shown in slot (a). Although the protein patterns of slots (b)-(d) were generally similar to each other, some of the proteins, the molecular weights of which are shown in Figure 2, disappeared or decreased from the M2 fraction with the increasing duration of 30°C incubation, and they were recovered in the M1 fraction. The protein patterns in slots (c) and (d) were reproduced in over five different preparations and represent the typical protein pattern of the M2 fraction. These results suggest that the M1 and M2 fractions in the previous paper (6) were derived from a complex of tightly folded chromosome and membrane, and the M2 fraction was a specific part of the chromosome complexed with specific membrane proteins.

RNase, phospholipase and detergent treatment of M2 complex: The M2 fraction was isolated from the lysate incubated at 30°C for 20 min. The pooled fraction was dialyzed against NNET (NNET: 50 mM NaCl, 20 mM EDTA, 10 mM Tris-Cl, pH 8.0), and its sensitivity to the following enzymes and detergents was tested: RNase (50 µg/ml), phospholipase A (20 µg/ml), Triton X100 (1%), sarkosyl (1%), Brij 58 (1%), Tween 80 (1%), and sodium deoxycholate (1%). The dialyzed M2 fraction was incubated with an enzyme or a detergent at 30°C for 10 min and then recentrifuged on a CsCl-sucrose double gradient. The resulting patterns were compared with those of M2 fraction treated the same way in the absence of detergent or enzyme.

Approximately 40% of the DNA and 60-70% of the protein was recovered at the M2 position after recentrifugation of the control M2 fraction.

Treatment with RNase or phospholipase A had no apparent effect on the M2 fraction. On the other hand, the M2 fraction was sensitive to all detergents employed in these experiments except Triton X100. Thus, detergents seem to solubilize the membrane and destroy the M2

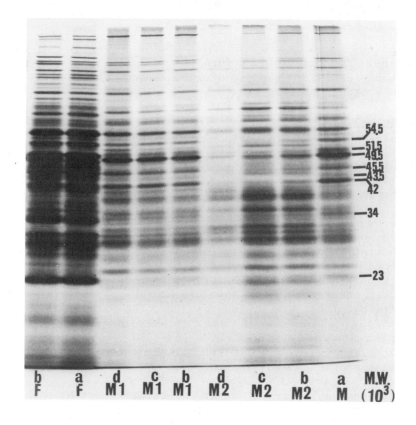

b   a   d   c   b   d   c   b   a   M.W.
F   F   M1 M1 M1 M2 M2 M2 M   (10³)

Fig. 2. Sodium dodecyl sulfate polyacrylamide gel
analysis of proteins. Strain 168TT was labeled with
20 µCi/ml of ^{3}H-thymine and 100 µCi/ml of ^{35}S-sulfate, and
the cells from 10 ml of culture were lysed and the lysates
were centrifuged as described in Figure 1. Profiles
similar to those shown in Figure 1 were obtained and the
fractions corresponding to those found between the two
arrows were pooled, dialyzed against distilled water, and
lyophilized. Gel electrophoresis in 10% acrylamide in the
presence of 0.1% of SDS was performed (9). An autoradiogram was made on Kodak SB54 X-ray film. From right to
left: the DNA membrane complex from the non-heated lysate,
the M2 fractions from lysates heated 10 min, 20 min, and
40 min, the M1 fractions of the same preparations, and the
top fractions of gradients similar to those shown in
Figure 1a,b. Marker proteins: bovine serum albumin
(MW 67,000), ovalbumin (44,500), carboxypeptidase B
(34,300), trypsin (23,500), myoglobin (17,200) and lysozyme
(14,400) were electrophoresed on the same slab gel and used
for calculation of molecular weight of proteins.

fraction; although some fraction of DNA (10-30% of M2 DNA
or 3-8% of total DNA) and a small amount of protein (approximately 5-10% of M2 protein) sedimented to the bottom of
the gradient. This complex has a higher density than M2
presumably due to a higher ratio of DNA to protein than
that seen in the original M2. A preliminary experiment
showed that protein species in these fractions are more or
less similar to the original M2 proteins. The DNA in this
complex is yet to be analyzed.

Surface protein iodination: Cells growing exponentially
in a media containing 10 μCi/ml of ^{35}S-sulfate were
harvested by centrifugation and the outside of the cells
were labeled with ^{125}I as described in Materials and
Methods. With this method, only proteins on the external
surface of the membrane are labeled in mammalian cells (8).
The labeled cells were lysed and the membrane was separated
from free proteins by a sucrose step gradient centrifugation in which the membrane sedimented to the interface,
while free proteins remained on top of the gradient. As
shown in Figure 3a, the membrane fraction was preferentially labeled with ^{125}I. The lysate of iodinated cells was
also examined by CsCl-sucrose double gradient (Fig. 3b).
The lysate was prepared exactly the same way as the samples
for Figure 1c and Figure 3a, i.e., the osmotically shocked
cells were further incubated at 30°C for 20 min and the
lysate was subjected to agitation with Vortex for 1 min.
The ^{125}I recovered in M2 and M1 fractions was approximately 70% of the total TCA precipitable ^{125}I counts in the
lysate, and the M2 fraction was preferentially labeled.
Further studies of the M2 fraction of iodinated cells are
in progress.

DISCUSSION

Our previous work on the membrane attachment of the
B. subtilis chromosome gave the following results: 1. The
membrane fraction obtained from exponentially growing cells
is enriched for the genetic markers close to the replication origin and the terminus when compared with internal
markers (13,14). 2. A pulse of ^{3}H-thymidine at the replication origin remains in the membrane fraction after a
chase by non-radioactive thymidine (13,14). 3. The enrichment of the origin and terminal markers in the membrane

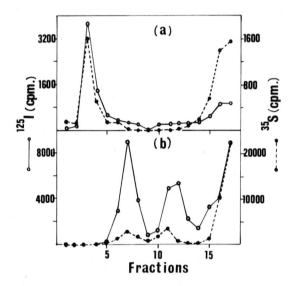

Fig. 3.   Fractionation of the lysate obtained from cells externally labeled with ^{125}I.   Strain 168TT was grown in the presence of 10 μCi/ml of ^{35}S-sulfate.   The intact cells obtained from 5 ml of culture were washed and iodinated as described in Materials and Methods and then lysed.   The lysate was incubated at 30°C for 20 min and agitated with a Vortex mixer.   An aliquot of the lysate was centrifuged on a sucrose step gradient (a), and the rest was centrifuged on a CsCl-sucrose double gradient (b).

o———o ^{125}I (iodinated protein)

•----• ^{35}S (total protein)

fraction are found consistently in various growth conditions with different cell generation times (14). 4. All
four origins of a chromosome during synchronous dicotomous
replication are associated with the membrane, i.e., two
successive initiations result in four copies of the replication origin and one copy of the terminus per chromosome
(15). 5. The DNA-membrane complex can be separated from
the major portion of the membrane. The complex displays
unique sets of proteins in acrylamide gel electrophoresis
(6). 6. The possibility that the DNA-membrane complexes
are artifacts of the lysate preparation is unlikely (6).
    The following information emerges from this work:
1. The two membrane subfractions (M1 and M2) are derived
from a much larger DNA-protein complex. When the cell
lysate is prepared at 0°C practically all DNA is recovered
as a rapidly sedimenting fraction with 10 to 15% of the
total protein (Fig. 1a). The complex may correspond to the
nuclear body preparations in Escherichia coli originally
reported by Stonington and Pettijohn (10). The fact that
this large complex contains all membrane fraction is indicated by the absence of M1 fraction in CsCl-sucrose double
gradient centrifugation (Fig. 1a). Upon incubation at
30°C, M1 and M2 fractions are generated from the complex.
2. The M2 complex may indeed be a part of the membrane,
since although it contains some unique proteins, its overall protein pattern is similar to that of M1 (major
membrane fraction). The present method of lysate preparation gives highly reproducible M2 protein patterns in gel
electrophoresis, even though more protein bands are
observed with our previous method (6). This point requires
further investigation with more sensitive methods of
fractionation. 3. Some proteins of the M2 complex may be
exposed on the outside of the membrane. The proteins
labeled with ^{125}I by iodinating intact cells with glucose
oxidase and lactoperoxidase are most likely those which
are in the membrane and partially exposed on the outside
of the membrane (8). The M2 proteins are iodinated most
readily and show highest specific activity among the three
fractions, M1, M2 and free protein (Fig. 3b). These findings also support a possible in vivo location of the M2
complex in the membrane. In our isolation method of the
DNA-membrane complex, the cell wall is enzymatically
digested prior to the preparation of the lysate. Therefore,
if any association of the complex to the cell wall exists,

as has been suggested for E. coli (16,17,18), it is lost
during protoplast preparation. However, the present
results indicate that some proteins of the M2 complex are
exposed on the membrane surface and, therefore, it is
possible that some of the M2 proteins might be attached to
cell wall components. This point is particularly worth
examination.

<u>Models to be tested</u>: We present the following models which
reflect our current idea of chromosome arrangement and
which will direct future experiments. A model which is
consistent with available facts is shown in Figure 4. The
models incorporate the possibility that the replication
forks are attached to the membrane (7,13,19).
    A unique feature of the model is that the initiation
complex (IC) is attached to the cell wall, while the
replication complex (RC) and the termination complex (TC)
are not, although all three are associated with the
membrane. The cell wall attachment should localize the
chromosome and aid in separating the two daughter chromosomes (16,17), while the RC and the TC are movable in the
membrane.
    An interesting alternative to this model is one in
which the TC attaches to the cell wall, while the other
two float in the membrane. This alternative model is also
consistent with available data and may provide a somewhat
better explanation of the constant interval (20 min)
between termination of replication and cell division, as
observed by Cooper and Helmstetter (20). In this model,
the two daughter chromosomes cannot be pulled apart until
the terminus replicates.

ACKNOWLEDGMENTS

    We are grateful to Dr. E. Milewski and Mr. R. J. Hye
for their critical reading of the manuscript.
    This work was supported by NSF GB-40090X and NIH
GM-20352-03.

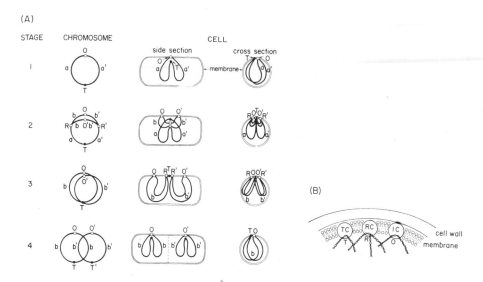

Fig. 4. A model of chromosome attachment to the cell envelope. This model is defined as shown in the two sets of diagrams, A and B.

A: Replication stages of the chromosome and association of the chromosome with the membrane.

0,0': Replication origins.

R,R': Replication forks.

T,T': Replication termini.

a,a': Left and right halves of the nonreplicated chromosome.

b,b': Left and right halves of the replicated part of the chromosome.

The terminus is assumed, for the sake of convenience, to be located equidistant bidirectionally from the origin since in B. subtilis the location of the terminus is still unresolved (21,22).

B: Association of chromosome with the membrane at initiation, replication and termination complexes. Note that in this model the initiation complex (IC) is not only embedded in the membrane but is also attached to the cell wall. In contrast, the replication complex (RC) and the termination complex (TC) are not attached to the cell wall, permitting them greater mobility on the membrane.

REFERENCES

1. Jacob, F., Brenner, S. and Cuzin, F. Cold Spring Harbor Symp. Quant. Biol. 28, 329 (1963).
2. Goulian, M. Ann. Rev. Biochem. 40, 855 (1971).
3. Kennett, R. H. and Sueoka, N. J. Mol. Biol. 60, 31, (1971).
4. Bott, K. F. and Wilson, G. A. J. Bacteriol. 94, 562, 1967).
5. Spizizen, J. Proc. Nat. Acad. Sci. USA 44, 1072 (1958).
6. Sueoka, N. and Hammers, J. M. Proc. Nat. Acad. Sci. 71, 4787 (1974).
7. Smith, D. W. and Hanawalt, P. C. Biochim. Biophys. Acta 149, 519 (1967).
8. Hubbard, A. L. and Cohn, Z. A. J. Cell Biol. 55, 390 (1972).
9. Laemmli, U. K. Nature 227, 680 (1970).
10. Stonington, G. O. and Pettijohn, D. E. Proc. Nat. Acad. Sci. 68, 6 (1971).
11. O'Sullivan, M. A. and Sueoka, N. J. Mol. Biol. 27, 349 (1967).
12. Yoshikawa, H. and Sueoka, N. Proc. Nat. Acad. Sci. USA 49, 559 (1963).
13. Sueoka, N. and Quinn, W. G. Cold Spring Harbor Symp. Quant. Biol. 33, 695 (1968).
14. Sueoka, N., Bishop, R. J., Harford, N., Kennett, R. H., Quinn, W. G. and O'Sullivan, A. In Genetics of Industrial Microorganisms, eds., Vanet, Z., Hostatek, Z. and Cudlin, J. (Academia, Prague) p. 73 (1973).
15. O'Sullivan, M. A. and Sueoka, N. J. Mol. Biol. 69, 237 (1972).
16. Worcel, A. and Burgi, E. J. Mol. Biol. 82, 91 (1974).
17. Ryter, A., Hirota, Y. and Schwarz, U. J. Mol. Biol. 78, 185 (1973).
18. Olsen, W. L., Heidrich, H. G., Hannig, K. and Hofschneider, P. H. J. Bacteriol. 118, 646 (1973).
19. Ganesan, A. T. and Lederberg, J. Biochem. Biophys. Res. Commun. 18, 824 (1965).
20. Cooper, S. and Helmstetter, C. E. J. Mol. Biol. 31, 519 (1968).
21. Wake, R. G. J. Mol. Biol. 77, 569 (1973).
22. O'Sullivan, A., Howard, K. and Sueoka, N. J. Mol. Biol. 91, 15 (1975).

THE UNIT CHROMOSOMAL FIBER:
EVIDENCE FOR ITS UNIVERSAL NATURE

Jack D. Griffith

Department of Biochemistry
Stanford University School of Medicine
Stanford, California 94305

ABSTRACT. DNA in eukaryotic cells is condensed into unit fibers 110Å in diameter. Several parameters of the fibers: a 7-fold contraction of the enclosed DNA duplex, a 200 base pair substructure, and a partial melting have been measured. By gentle disruption of prokaryotic cells their DNA is seen also to exist in apparently the same unit fiber.

The fine structure of the eukaryotic chromosome is being dissected in many laboratories. It has been shown that in the higher cell DNA exists in an elemental unit fiber roughly 110Å in diameter (1). The unit fiber structure confers several unique properties on the DNA. The duplex is foreshortened 7-fold as a chain of 200 base pair beads and is melted by about 10% (2). We see evidence of this melting as supercoiling. When such condensed DNA becomes covalently closed to form a duplex circle, it will appear topologically relaxed as long as it is maintained in that form. If the factors generating the condensation and partial melting are disrupted, the DNA will renature and the strain of the melting will spread uniformly over the DNA and appear as a topological twisting or supercoiling of the circle. In higher cells the formation of the unit fiber is believed to be dependent on the presence of histone proteins present in an equal mass with the DNA. This report will suggest that a similar condensation is present in the DNA of prokaryotic cells where the histones are absent.

The Unit Fiber in Eukaryotic Cells

Interphase chromosomes visualized by electron microscopy appear as a network of fibers with a mean diameter of 110Å and a beaded substructure (Figure 1). This is even more clearly seen if the minor histone F_1 is removed (3). Several parameters of the unit fiber have been measured with the aid of a unique viral probe. During a lytic infection of cultured monkey cells by the papova virus SV40, the progeny DNA is found in the nucleus of the infected cell in the form of a nucleoprotein complex. The proteins of the complex comprise the same complement of histones as are found in monkey cell chromatin (but lack histone F_1). Visualization of the complexes in physiologic salt solutions by electron microscopy (Figure 2a) showed them to be short 2000Å loops of the unit fiber. Comparison of this length with that of the DNA purified from the complexes and examined under identical conditions provided an estimate of 7:1 for the packing ratio of DNA in the unit fiber. In lower ionic strength (Figure 2b) 21 beads each 110Å in diameter joined by 130Å duplex bridges were seen. Because the total number of base pairs in the SV40 circle is known (5, 100), each bead must contain 170 to 200 base pairs of DNA and each bridge 30 (2). These structures appear topologically relaxed which is to be expected since they represent the state of the DNA in the cell. Removal of the histones generated supertwisted DNA circles. If the SV40 circle contains 42 supertwists (4), then one base pair must be melted for each pitch of the DNA helix. Numerous studies from other laboratories are in agreement with these results (5, 6, 7).

A Unit Fiber in Prokaryotic Cells

The chromosome of E. coli is folded into a structure containing 50 to 100 loops (see the accompanying papers of this volume). The DNA of each loop was found to be supercoiled with a superhelix density (a measure of unwinding) similar to that of small circular DNAs. Because strong agents (1 M NaCl, deoxycholate) were used to extract the folded chromosomes, it was not surprising that proteins or other factors would have been lost from the DNA during the extraction giving rise to the supertwisted state. Little protein in fact was found associated

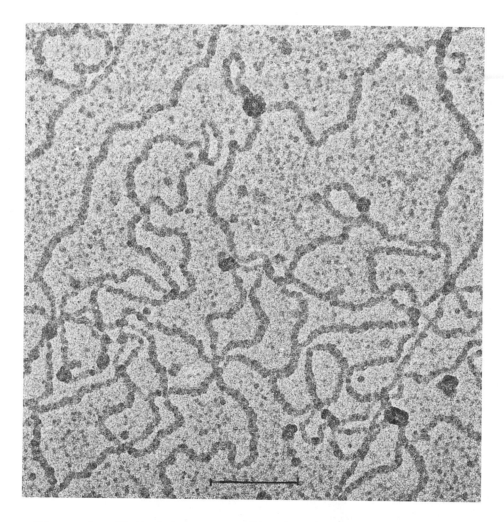

Figure 1   Unit chromosome fibers of monkey cells.
   Interphase nuclei of cultured monkey cells were gently disrupted in 0. 15 M NaCl, 0. 01 M Tris, pH 7. 5 with
0. 25% Triton X-100.   The chromatin fibers were purified
by sucrose velocity sedimentation fixed and prepared for
high resolution electron microscopy (10).   The unit chromosomal fibers depicted here have a diameter of 100Å to
120Å and exhibit a beaded substructure.
Bar equals 1000Å.

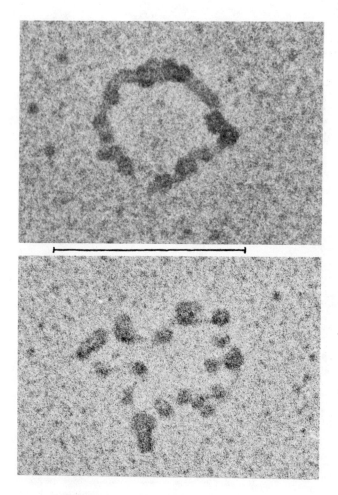

Figure 2   SV40 miniature chromosomes
    Newly replicated SV40 DNA in infected monkey cell
nuclei can be isolated as a DNA-histone complex identical
in its basic composition and structure to the unit chromosomal fibers of the host cell (2).   This figure shows an
example (a) of such a complex prepared and fixed for high
resolution electron microscopy in 0. 15 M NaCl, and (b)
fixed in 0. 015 M NaCl to show the beaded substructure
more clearly.   DNA purified from these topologically
relaxed structures appears supercoiled and has a 7-fold
greater circumference than the structures in (a).
Bar equals 1000.Å.

with the folded chromosome (8).

To investigate the state of DNA within the prokaryotic cell, very gentle lysis and preparative procedures for electron microscopy have been developed. These techniques and results will be reported elsewhere (9). A typical micrograph showing a stationary phase <u>E. coli</u> cell prepared by these methods is in Figure 3. The DNA exists as a fiber 115Å to 130Å (mean of 120Å) in diameter, arranged in loops (very roughly 50 to 100) which are topologically relaxed and appear to be attached to the cell envelope. A beaded repeat 115Å to 135Å in length is seen along the fiber (Figure 4).

In work to be reported (9), evidence will be given from aldehyde fixation studies that relatively little protein is associated with the DNA in the 120Å fiber. Measurement of the entire length of an apparently whole chromosome free of the cell evelope indicated a roughly 10-fold packing ratio of the DNA in the 120Å fiber. A more precise measurement of this ratio using phage lambda in the manner in which SV40 was applied in eukaryotic cells appears to give a packing ratio of 7:1. The lambda circle is seen as a relaxed 2.5 micron loop of the 120Å fiber. Each 115Å to 135Å repeat along the fiber would contain about 200 base pairs of DNA.

Structure of the Basic Fibers

DNA in the eukaryotic nucleus is present in a fundamental unit fiber 110Å in diameter. The DNA is condensed 7-fold in a chain of 200 base pair units and is melted by about 10%, measured by electron microscopy and the superhelix density. In the eukaryotic cell the presence of the histone proteins is thought to be required for this structure (5).

The initial studies described here indicate that DNA in the prokaryotic cell is also present in a fundamental fiber with parameters remarkably similar to those in the eukaryotic cells. In prokaryotic cells there is no evidence for histone-like proteins in amounts required to form a one-to-one complex with the DNA as do the histones of higher cells. Rather, it is thought that prokaryotic DNA is complexed with Mg^{++} and polyamines. The greater lability of the prokaryotic fibers to disruption is consistent with the condensing factors being small cations.

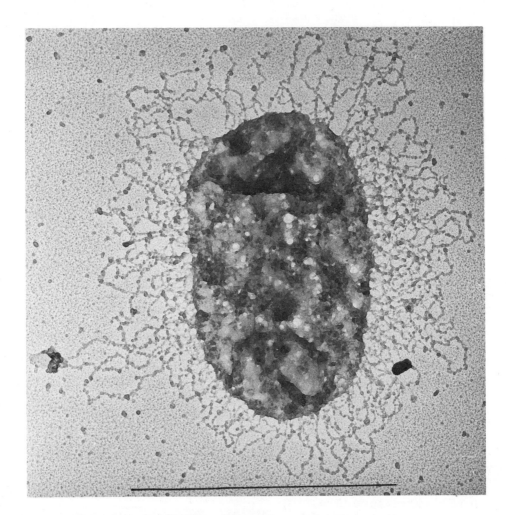

Figure 3   Gently disrupted E. coli cells.
    Stationary phase E. coli cells were disrupted in
0. 15 M NaCl, 0. 01 M Tris, pH 7. 5, directly on the
microscopic supporting film by a method previously described (11).   The DNA is in the form of a fiber 120Å in
diameter.   Shear, exposure to high salt solutions or ionic
detergents disrupt these fibers exposing 20Å DNA filaments.
Bar equals 1 micron.

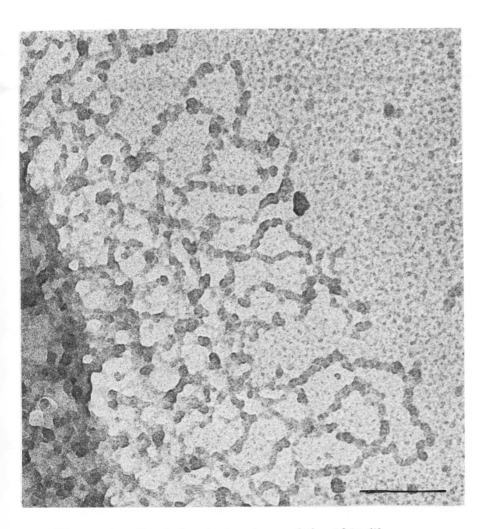

Figure 4   Beaded substructure of the 120Å fiber.
    A higher magnification of a portion of the cell in
Figure 3 shows more clearly the 115Å to 135Å beaded
substructure of the 120Å prokaryotic fiber.
Bar equals 1000Å.

The fundamental question to be answered is how the
DNA duplex is folded or twisted in these two unit fibers.
It is possible that although the basic parameters are
nearly identical, the folding is quite different in each. It
seems more likely, however, that the folding is the same
and the template for folding resides in the nature of the
DNA helix itself. It may be that a variety of molecules
which neutralize the charge of the phosphate groups and
exclude a certain amount of water of hydration will induce
the duplex to assume this basic condensed state. The
histones may stabilize this state to a much greater degree than the condensing factors of prokaryotic cells and
provide a mechanism for even further coiling into metaphase structures. A precise elucidation of this folding
may be most easily accomplished in the prokaryotic
fibers.

References

1.   Bram, S. and Ris, H. (1971) J. Mol. Biol. 55, 325.
2.   Griffith, J. D. (1975) Science 187, 1202.
3.   Oudet, P. , Gross-Bellard, M. and Chambon, P.
     Proc. Nat. Acad. Sci. USA (in press).
4.   Wang, J. personal communication.
5.   Germond, J. E. , Hirt, B. , Oudet, P. , Gross-Bellard,
     M. and Chambon, P. Cell (in press).
6.   Kornberg, R. D. (1974) Science 184, 868.
7.   Clark, R. J. and Felsenfeld, G. (1974) Biochemistry
     13, 3622.
8.   Stonington, O. G. and Pettijohn, D. E. (1971) Proc.
     Nat. Acad. Sci. USA 68, 6.
9.   Griffith, J. in preparation.
10. Griffith, J. D. in Methods in Cell Biology (D. M. Pres-
     cott, ed.) Academic Press, New York, 1973, vol. 7
     129.
11. Pratt, D. , Laws, P. and Griffith, J. D. (1974)
     J. Mol. Biol. 82, 425.

Part III
Systems for Replication of DNA *in vitro*

MULTIENZYME SYSTEMS IN THE REPLICATION
OF ØX174 AND G4 DNAs

Jean-Pierre Bouché, Randy Schekman*, Joel Weiner,
Kasper Zechel** and Arthur Kornberg
Department of Biochemistry
Stanford University
Stanford, California 94305

ABSTRACT. Eight purified proteins are required for the conversion of ØX174 from single-stranded circular DNA (SS) to RF II (a duplex with a small gap). These are: products of genes dnaB, dnaC, dnaE (DNA polymerase III*) and dnaG, the DNA-unwinding protein, copolymerase III*, protein i and protein n. In addition to the DNA-unwinding protein, protein n may play a structural role as judged by a six-fold condensation of the viral DNA into a beaded structure visualized by electron microscopy. The conversion can be separated into a state of initiation and a stage of DNA synthesis. Addition of the ribonucleoside triphosphates during the first stage, but not during the second stage, stimulates DNA synthesis, suggesting the synthesis of a primer RNA.

Evidence for the role of dnaG protein as an RNA polymerase in the synthesis of such a primer has been obtained with the DNA of phage G4 (a ØX-like particle) as template. Conversion of G4 DNA to RF II requires only dnaG protein, the DNA unwinding protein, DNA polymerase III* and copolymerase III*. The dnaG protein transcribes the complex of G4 SS plus the DNA-unwinding protein to yield RNA fragments about 20 nucleotides in length. After replication, one of these fragments is covalently linked to the synthetic complementary strand at or near a unique initiating 5'-end of the DNA.

Present address:
*Department of Biology, University of California at San Diego, La Jolla, California 92037
**Max-Planck-Institut für biophysikalische Chemie, D-3400 Göttingen, Postfach 968, Germany

INTRODUCTION

Involvement of multiple enzymes in the replication of the DNA in E. coli has been indicated by both genetic and biochemical evidence. Several loci (dnaA, C, H, I, and P (1-4)) affect the initiation of a round of chromosome replication; others (dnaB, E, G and Z (1, 5)) affect the growth of chromosome chains. Still others alter the processing of nascent DNA fragments into the long DNA polymer. These are: polA 1 (affecting the DNA polymerase I), polAex (affecting the 5' → 3' exonuclease of DNA polymerase I), lig (affecting the DNA ligase) and dnaS (6-9). DNA polymerases, contrary to RNA polymerases, do not start polynucleotide chains in vitro. Involvement of RNA polymerase has been established in studies of replication of phage M13 DNA in vitro (10, 11) and an RNA-DNA covalent linkage has been presented in a number of systems (12). Studies of the initiation of nascent DNA fragments (12, 13) and of conversion of the DNA of phage ØX174 from SS to RF (14, 15) provided evidence for an RNA polymerase other than the principal RNA polymerase of E. coli.

Analysis of the rate of joining of short DNA pieces in various mutants suggested that gene dnaG might code for this function (16, 17).

Formation of an RNA-DNA covalent linkage requires removal of the RNA fragment, gap filling and ligation of the DNA pieces. A nuclease specific for RNA-DNA hybrids (RNAse H) is found in E. coli (18) as well as in higher organisms. However, the ability of DNA polymerase I to perform both the gap filling and the degradation of the RNA piece by its 5' → 3' exonuclease was shown in the case of M13 DNA replication (15), and the properties of DNA polymerase I mutants (6, 19) make it a likely candidate for these two functions in vivo.

Another aspect concerns the structural form of the DNA to be replicated. Replication of a double-stranded DNA requires strand separation and, although no mutant has been identified for this function in E. coli, this may be achieved by DNA unwinding proteins (20). The existence of phage-specific unwinding proteins (20, 21) suggests that the DNA-binding, "structural" proteins play an important role in specification of the particular multienzyme system which replicate a particular DNA.

Two multienzyme systems will be described in this
paper: one supports the conversion of the single-strand
of ØX174 to form RF II, and the other that of phage G4
(a ØX-like particle (22)). Both rely exclusively on a set
of bacterial soluble enzymes, and are entirely resistant
to the inhibitors of RNA polymerase (10, 23).

MATERIALS AND METHODS

ØX174 and G4 DNAs were purified according to published procedures. E. coli H560 (F$^+$, pol Al$^-$, Endo I$^-$)
was grown and cell extracts were prepared (24) as later
modified (25). Extracts from E. coli NY73 RIF^S (dnaG3,
thy$^-$, pol Al, leu$^-$, met E$^-$, str^R) for assay and purification of the DNA-unwinding protein and dnaG protein
using G4 DNA as template are described elsewhere (26,
27). DNA polymerase III*, copolymerase III* and the
holoenzyme form of DNA polymerase III were purified
fractions (24, 25).

A detailed account of the purification procedures of
dnaB protein, dnaC protein, protein i and protein n is to
be published (28).

RESULTS

Resolution of ØX174 multienzyme system. A combination
of two approaches, complementation and total fractionation, was used to resolve and identify the components of
the ØX174 system. The flow diagram in Fig. 1 indicates
the major steps used to segregate the major components.
The purification procedure for a component was reexam-
ined, starting in each case with an independent procedure
from the first step to optimize yield and purity.

The dnaC protein was separated from other essential
components by ammonium sulfate fractionation and assayed by complementation of a mutant (dnaC) extract (28).
After further purification by phosphocellulose chromatography, this fraction serves in rifampicin-resistant synthesis by the mutant extract (data not shown). Requirement for dnaC protein was the touchstone for authentic
reconstitution in subsequent fractionation. The other
components were divided into two groups by passage
through a DNA-cellulose column (Fig. 1). At this stage,
three fractions were needed for ØX174 DNA synthesis:

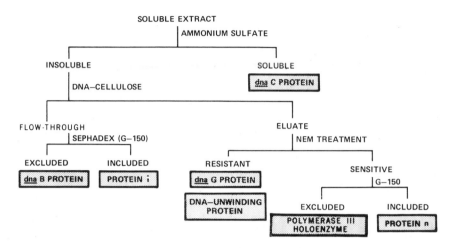

Figure 1   Scheme for resolution of the ØX174 enzymes.

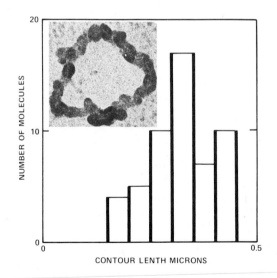

Figure 2   Contour length distribution of the ØX174-protein n complex.   The average length is 0. 32 microns. A molecule of ØX174 single-stranded DNA complexed by protein n is shown in the inset.

the fraction unadsorbed to DNA-cellulose, the fraction bound to and eluted from it, and the dnaC protein fraction.

The fraction unadsorbed to DNA-cellulose was further divided by chromatography on Sephadex G-150 into an excluded and an included fraction. Each of these was then purified by further chromatography to yield, respectively, dnaB protein (identified by complementation with a dnaB mutant extract) and protein n (not identified with any of the dnaA, B, C, E or G gene products). At this stage, the components required for activity were DNA-cellulose binding fraction, dnaB protein, protein i, and dnaC protein.

The fraction adsorbed to, and eluted from, DNA-cellulose was dissected by sensitivity to the sulfhydryl-blocking agent N-ethylmaleimide (NEM) and further resolved by chromatography on Sephadex G-150. Excluded from the gel was the NEM-sensitive, DNA polymerase III holoenzyme (a complex containing DNA polymerase III[*] and copolymerase III[*] activities (25). Included in the gel were three essential components. Two were NEM-resistant and separable into dnaG (identified by complementation (27)) and the DNA-unwinding protein (26); the other was an NEM-sensitive protein (protein n), corresponding to none of the dnaA, B, C, E or G gene products.

Each of the seven proteins was purified and was required in a reconstituted reaction (Table I). There was at least a 5-fold decrease of DNA synthesis when any one of the purified proteins was missing. Based on current states of purification, there may be, per cell, in the order of 100 to 300 molecules each of the dnaE protein, dnaG protein, protein i, protein n and DNA unwinding protein.

Binding of protein n to ØX174 DNA. Protein n was purified to homogeneity. It is sensitive to N-ethylmaleimide, and has a molecular weight of 82,000 as judged by electrophoresis in an acrylamide-sodium dodecyl sulfate gel. It aggregates at low ionic strength (at neutral pH) at concentrations as low as 0.1 mg/ml. In the electron microscope these aggregates appeared, by the Griffith technique (29), as long twisted filaments of proteins (data not shown). Mixing with single-strands of ØX174 resulted in condensation of the DNA in a beadlike structure (Fig. 2). These beads numbered about 25 per chromosome, cor-

TABLE 1

Omitted item	Reconstitution (% relative to complete reaction)	
	ØX174	G4
None (complete)	100	100
dnaB protein	21	150
dnaC protein	2	150
DNA polymerase III holoenzyme	7	12
dnaG protein	7	13
Protein i	12	130
Protein n	4	150
Spermidine	2	75
DNA unwinding protein	15	140
DNA unwinding protein + spermidine	1	12

Requirements of the reconstituted ØX174 and G4 reactions. Experimental conditions are described in reference 32.

responding to about 200 nucleotides per bead. The contour length of the chromosome averaged 0. 32 microns (Fig. 2), which is 6- to 7-fold less than the contour length of free ∅X174 single strands.

When protein n was added in subsaturating amounts to ∅X174 single-stranded circular DNA, two classes of DNA molecules were formed and separated by sedimentation in a sucrose gradient. One sedimented at the position of free single strands; the other sedimented faster to the same position as DNA saturated with protein n. This indicates that binding of protein to DNA is cooperative (data now shown). Protein n did not bind to the DNA of SV40 form I (a twisted duplex molecule), except at one site, presumably the SI-sensitive region of the molecule (data not shown).

Binding to single-stranded DNA suggests that protein n is involved in an early step of the conversion of ∅X174 DNA to RF II molecules. Whether DNA polymerase III holoenzyme can elongate the DNA of the strand complementary to the highly condensed DNA-protein n complex has not been determined.

Resolution of G4 multienzymatic system. Extracts of E. coli prepared by gentle lysis are efficient in the replication of the single-stranded DNA of phage G4, a ∅X174-like particle (23). Even in the absence of spermidine, replication of G4 DNA is very effective and rapid; under the same conditions ∅X174, S13 and G14 DNAs are poorly replicated (Fig. 3). The product of the reaction is a full-length linear molecule; conversion to form RF II is not inhibited by rifampicin, an inhibitor of the RNA polymerase (Fig. 3).

Inhibition of cell extracts with NEM could be reversed by addition of DNA polymerase III holoenzyme, indicating that no other NEM-sensitive protein, such as dnaC protein or protein n, was required for the conversion. Gel filtration through Sephadex G-150 resolved the G4 replication system into three components: DNA polymerase III holoenzyme, DNA unwinding protein and a third component presumed to be responsible for the priming of DNA replication.

The priming component was purified by taking advantage of its ability to complement dnaG mutant extracts supplemented with DNA-unwinding protein for the repli-

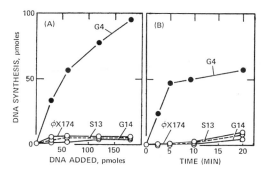

Figure 3   Comparison of phage DNAs as templates for replication in the presence of rifampicin and in the absense of spermidine.   Incubations were as described in reference 23, using an extract of gently lysed cells as source of enzymes, with the indicated amounts of DNA for 15 min in (A) and for indicated times with 60 pmoles of DNA in (B)..

TABLE 2

Component omitted	Nucleotide incorporation (residues/circle)			
	labeled nucleoside triphosphate			
	ATP	GTP	UTP	CTP
None (Complete)	0. 65	3. 35	0. 48	0. 26
G4 DNA	NT	0. 28	0. 02	<0. 01
DNA unwinding protein	0. 07	0. 42	0. 08	0. 08
Unlabeled rNTPs	0. 09	0. 41	0. 01	0. 03
dnaG protein	<0. 01	0. 35	0. 04	0. 02

Requirements of the ribonucleoside-triphosphate incorporation reaction catalyzed by dnaG protein.   The DNA of phage G4 is used as a template unless omitted.   Experimental conditions are described in reference 27. NT = not tested.

217

cation of G4 DNA. When large amounts of mutant extracts
are used, DNA-unwinding protein alone restores full
activity to the extract (26). However, small amounts of
mutant extract require in addition an NEM-resistant protein having the function of the priming protein. The
priming protein was purified to about 50% homogeneity.
It was identified with the product of gene dnaG by the
following criteria: resistance to NEM (30), similar
molecular weight (60,000-daltons) as judged from velocity sedimentation data (30), ability to complement a dnaG
mutant extract, and thermolability when purified from a
dnaG strain (30). Comparison of electrophoretic patterns
in acrylamide-sodium dodecyl sulfate gels suggest a
monomeric structure (27).

The complete reaction could be reconstituted with
only three purified proteins: the DNA-unwinding protein,
dnaG protein and DNA polymerase III holoenzyme (Table I).

dnaG protein synthesizes a primer RNA. DNA synthesis,
either with an extract of gently lysed cells or a mixture
of partially purified fractions, was shown to be greatly
stimulated, not only by ATP (required for the action of
copolymerase III[*] (31)) but also by GTP and UTP (23).
This suggested the synthesis of an RNA primer. Using
labeled rNTPs and an extensively purified preparation of
protein dnaG, RNA synthesis was indicated by gel filtration (Table 2). The RNA synthesis required G4 DNA,
DNA-unwinding protein, dnaG protein, rNTPs (Table 2)
and $MgCl_2$ (near 3 mM). The half-saturating concentration of rNTPs was about 10^{-6} M for GTP, CTP and
UTP, and 10^{-5} for ATP. Some GTP incorporation could
be observed in the absence of one of the other three
rNTPs; however, only the RNA-DNA hybrid synthesized
in the presence of all four rNTPs was a good substrate
for subsequent replication (Table 3). This suggests that
incomplete RNA chains formed in the absence of ATP,
CTP or UTP (or presumably GTP) are not recognized as
primers by the DNA polymerase III holoenzyme. The
RNA synthesized in the presence of the four rNTPs is
fairly homogenous in size (Fig. 4). Its size was estimated to about 20 nucleotides by electrophoresis in acryl-
amide-urea (data not shown).

In view of the requirement for dnaG protein in ØX174
conversion from SS to RF and of the binding of protein n

TABLE 3

First Stage		Second Stage
Unlabeled rNTP added	[^{32}P] GTP incorporation (residues/circle)	Deoxynucleotide incorporation (pmoles)
ATP, CTP, UTP	2.96	80
CTP, UTP	0.13	5
ATP, UTP	1.66	10
ATP, CTP	0.56	18
Controls		9
		12

RNA priming required for DNA synthesis.
Incubation mixtures in the first stage contained G4 DNA,
DNA unwinding protein, dnaG protein [α-^{32}P]-labeled
GTP and non-labeled rNTPs as indicated. [^{32}P]-GTP
incorporation was measured after filtration on a Bio-Gel
A-15M column. To aliquots of the excluded fraction containing 200 pmoles of DNA (as residues) were added
DNA polymerase III holoenzyme, [^{3}H]-dNTPs and ATP,
for a second step of DNA synthesis. Controls were only
incubated in the second stage, with further addition of
either DNA unwinding protein or DNA unwinding protein
plus dnaG protein. Experimental conditions are described in reference 27.

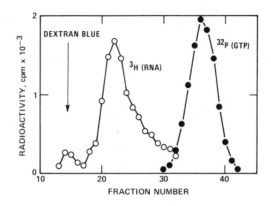

Figure 4 Gel filtration of primer-RNA on Sephadex G-75. [^{3}H] GTP-labeled primed single-stranded G4 DNA (15,500 cpm, 230 µl) was mixed with [^{32}P] GTP (12,500 cpm) and dextran blue, made 7 M in urea, heated at 100° for 5 min and filtered on a 7 ml-column of Sephadex G-75 equilibrated with 20 mM phosphate buffer, pH 7.5, 1 mM EDTA and 7 M urea.

TABLE 4

Binding protein added to DNA	Incorporation (GTP/circle)		
	G4	ØX	M13
None	0.33	0.40	0.51
Protein n	2.7	3.05	3.3
DNA unwinding protein	2.3	0.44	0.58
DNA unwinding protein + protein n	3.0	2.4	2.4

Incorporation of ^{32}P-labeled GTP into RNA with G4, ØX174 and M13 DNAs as templates. The assay was as described in reference 27, except that RNA synthesis was measured by adsorption to DEAE paper circles, following a modification of the procedure of Brutlag and Kornberg (33).

to single-stranded DNA, RNA synthesis was qualitatively examined with ØX174 or M13 DNAs, and in the presence of protein n (Table 4). In the presence of protein n, RNA synthesis occurs whether G4, ØX or M13 DNAs are used. In contrast, RNA synthesis in the presence of DNA-unwinding protein takes place only with G4 DNA. Moreover, DNA-unwinding protein appears to compete with the protein n-dependent RNA synthesis with ØX and M13 DNAs.

Evidence for covalent linkage of the primer RNA. G4 single-strands were converted to RF II molecules using ^{32}P-labeled dNTPs, and then filling the remaining gap from the 3' end of the complementary strand using DNA polymerase I and ^{3}H-labeled dNTPs. The product was treated with endonuclease Eco R I and sedimented in an alkaline sucrose gradient (Fig. 5). The ^{32}P label was recovered in two peaks, containing 95% and about 5% of the radioactivity, indicating that the Eco R I site is at 5% from the gap (Fig. 5). Since most of the ^{3}H-labeled 3'-end was found in the fast-sedimenting peak, the position of the Eco R I site is established near the 5'-side of the gap (Fig. 5).

In a separate experiment with purified proteins of the G4 system, G4 DNA was first primed with [^{32}P]-labeled rNTPs, dna G protein and DNA-unwinding protein. The primed single-strands were replicated with [^{3}H]-labeled dNTPs by DNA polymerase III* and copolymerase III*. An aliquot of the product was treated with Eco R I and then treated and untreated aliquots were sedimented in formamide-sucrose gradients (Fig. 6). Before Eco R I treatment, the RNA is covalently attached to DNA molecules sedimenting, as expected, as full-length molecules (Fig. 6a). After Eco R I treatment (Fig. 6b), a small peak of DNA, similar to the one seen in Fig. 5, was generated. All the RNA was found cosedimenting with this peak indicating that it is covalently attached to the DNA at or near the unique 5'-end of the complementary strands.

DISCUSSION

Properties of the ØX174, M13 (11) and G4 replicative enzyme systems are summarized in Table 5. This table suggests that the major difference between the systems

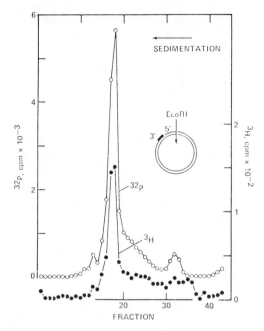

Figure 5   Location of Eco R1 cleavage site near the
5'-end of the complementary strand of G4 DNA.   The
inset illustrates the position of the gap relative to the
Eco R1 site.   The experimental procedure is described
in reference 23.

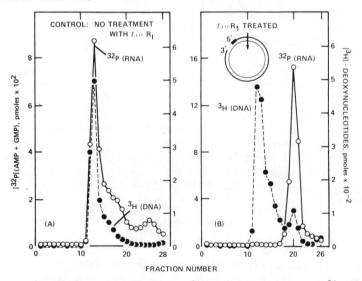

Figure 6   Sedimentation in a linear sucrose gradient in
98% formamide of primed-replicated G4 DNA.   The experimental conditions are described in reference 27.

TABLE 5

Item	Function	Template ØX174	G4	M13
Initiation				
Unique origin		-	+	+
Rifampicin-sensitive		-	-	+
RNA polymerase III	RNA synthesis	-	-	+
dnaG protein	RNA synthesis	+	+	-
Protein n	DNA binding	+	-	-
dnaB, dnaC, protein i	?	+	-	-
Initiation and Elongation				
DNA unwinding protein	DNA binding	+	+	+
spermidine	?	+	-	-
Elongation				
DNA polymerase III holoenzyme	DNA synthesis	+	+	+
Termination				
DNA polymerase I	DNA synthesis	+	+	+
DNA ligase	DNA ligation	+	+	+

Summary of the properties of the in vitro replication of circular viral DNAs from phages ØX174, G4 and M13.

resides in the mechanism of initiation rather than in DNA
elongation or termination. Initiation depends on the synthesis of a primer RNA catalyzed by RNA polymerase III
in the case of M13, and by dnaG protein in the case of G4.
In the case of ØX174, evidence for an RNA-DNA covalent
linkage (14), requirement for dnaG protein and stimulation of DNA synthesis in the first stage of a two-stage
reaction by rNTPs (32) suggest strongly a similar mechanism of initiation, although the functions of dnaB, dnaC
proteins and protein i remain to be established. The
apparent lack of requirement for dnaB protein differen-
tiates the replication of G4 DNA and the formation of
short DNA fragments at the moving fork of the E. coli
chromosome. Progress in the dissection of ØX174
mechanism of initiation depends on obtaining the many
protein components of this system in a more purified
form. ØX174 replication, by its requirement for dnaG
protein, dnaB protein, and the protein product of the
initiation-controlling dnaC gene, is of key interest for
insights into E. coli chromosome initiation and replication.

ACKNOWLEDGEMENT

This work was supported in part by grants from the
National Institutes of Health and the National Science
Foundation. Jean-Pierre Bouche is supported by the
Centre National de la Recherche Scientifique; J. H. Weiner
is a Damon Runyon Memorial Fund for Cancer Research
Fellow; R. Schekman was a graduate student; and K.
Zechel a fellow of the Deutsche Forschungsgemeinschaft.

REFERENCES

1. Wechsler, J. A. and Gross, J. D. Mol. Gen. Genet.
   113, 273 (1971).
2. Sakai, H., Hashimoto, S. and Komano, T.
   J. Bacteriol. 119, 811 (1974).
3. Beyersmann, D., Messer, W. and Schlicht, M.
   J. Bacteriol. 118, 783 (1974).
4. Wada, C. and Yura, T. Genetics 77, 199 (1974).
5. Filip, C. C., Allen, J. S., Gustafson, R. A., Allen,
   R. G. and Walker, J. R. J. Bacteriol. 119, 443
   (1974).

6.   Kuempel, P. L. and Veomett, G. E.   Biochem.
     Biophys. Res. Commun. 41, 973 (1970)
7.   Konrad, E. B. and Lehman, I. R.   Proc. Nat. Acad.
     Sci. USA 71, 2048 (1974).
8.   Pauling, C. and Hamm. L.   Proc. Nat. Acad. Sci.
     USA 64, 1195 (1969).
9.   Konrad, E. B. and Lehman, I. R.   Proc. Nat. Acad.
     Sci. USA (in press).
10. Wickner, W., Brutlag, D., Schekman, R. and
     Kornberg, A. Proc. Nat. Acad. Sci. USA 69, 965
     (1972).
11. Geider, K. and Kornberg, A.   J. Biol. Chem. 249,
     3999 (1974).
12. Reichard, P., Eliasson, R. and Söderman, G. Proc.
     Nat. Acad. Sci. USA 71, 4901 (1974).
13. Sugino, A. and Okazaki, R. Proc. Nat. Acad. Sci.
     USA 70, 88 (1973).
14. Schekman, R., Wickner, W., Westergaard, O.,
     Brutlag, D., Geider, K., Bertsch, L. L. and Korn-
     berg, A. Proc. Nat. Acad. Sci. USA 69, 2691
     (1972).
15. Westergaard, O., Brutlag, D. and Kornberg, A.
     J. Biol. Chem. 248, 1361 (1973).
16. Lark, K. G. Nature New Biol. 240, 237 (1972).
17. Olivera, B. M., Lark, K. G., Herrmann, R. and
     Bonhoeffer, F. in DNA Synthesis In Vitro (Wells, R.
     and Inman, R., eds.) University Park Press, Mary-
     land, pp. 215 (1973).
18. Berkower, I., Leis, J. and Hurwitz, J. J. Biol.
     Chem. 248, 5914 (1973).
19. Konrad, E. B., Modrich, P. and Lehman, I. R.
     J. Mol. Biol. 90, 115 (1974).
20. Alberts, B. M. and Frey, L. Nature 227, 1313
     (1970).
21. Alberts, B., Frey, L. and Delius, H. J. Mol. Biol.
     68, 139 (1972).
22. Godson, N. G. Virology 58, 272 (1974).
23. Zechel, K., Bouche, J. P. and Kornberg, A. J. Biol.
     Chem. (in press).
24. Wickner, W., Schekman, R., Geider, K. and
     Kornberg, A. Proc. Nat. Acad. Sci. USA 70, 1764
     (1973).

25. Wickner, W. and Kornberg, A. J. Biol. Chem.
    249, 6244 (1974).
26. Weiner, J. H. , Bertsch, L. L. and Kornberg, A.
    J. Biol. Chem. 250, 1972 (1975).
27. Bouche, J. P. , Zechel, K. and Kornberg A. J. Biol.
    Chem. (in press).
28. Schekman, R. , Weiner, J. H. , Weiner, A. and
    Kornberg, A. J. Biol. Chem. (in press).
29. Griffith, J. D. in Methods in Cell Biology (David
    Prescott, ed.) Academic Press, N. Y. , Vol VII
    pp 129 (1973).
30. Wickner, S. , Wright, M. and Hurwitz, J. Proc.
    Nat. Acad. Sci. USA 70, 1613 (1973).
31. Wickner, W. and Kornberg, A. Proc. Nat. Acad.
    Sci. USA 70, 3679 (1973).
32. Schekman, R. , Weiner, A. and Kornberg, A.
    Science 186, 987 (1974).
33. Brutlag, D. and Kornberg, A. J. Biol. Chem. 247,
    241 (1972).

IN VITRO SYNTHESIS OF DNA

Sue Wickner[*] and Jerard Hurwitz[†]

[*] Viral Carcinogenesis Branch, National Cancer Institute,
National Institutes of Health, Bethesda, Maryland 20014

[†] Department of Developmental Biology and Cancer, Division
of Biological Sciences, Albert Einstein College of Medicine, Bronx, New York 10461

ABSTRACT $\phi\chi174$ DNA-dependent DNA synthesis is catalyzed
in vitro by the combination of 10 purified protein preparations, ATP, 4 dNTPs and Mg^{2+}. The reaction has been resolved into 2 steps; the $\phi\chi174$ DNA-protein complex formed in
the first step (dependent upon ATP, $\phi\chi174$ DNA and 6 protein
components) has been isolated by gel filtration and supports DNA synthesis following the addition of dNTPs and the
remaining protein components. Partial reactions involving
proteins required in the first step include: (a) ribonucleoside triphosphatase activity of dnaB gene product which
is stimulated by single-stranded DNA, (b) ATP-dependent
complex formation by dnaB and dnaC(D) gene products, and
(c) $\phi\chi174$ DNA-dependent ATPase activity associated with DNA
replication factor Y.

INTRODUCTION

T hree types of single-stranded circular phage DNAs
are converted to duplex DNA *in vitro* by purified *Escherichia coli* proteins. Table 1 summarizes the requirements we
observe for DNA synthesis by these systems. The most complex system and the one which will be discussed further below, uses $\phi\chi174$ DNA or $\phi\chi$ahb DNA.[1] DNA synthesis requires
ATP, dNTPs, and 10 protein preparations isolated from uninfected *E. coli* (Table 2 and reference 2). Four of these
proteins, dnaB, dnaC(D), dnaE, and dnaG gene products, are
required for the replication of the *E. coli* chromosome as

---

[1] Phage $\phi\chi$ahb is described in reference 1.

TABLE 1

Requirements for in vitro DNA-dependent DNA synthesizing systems

	φχ174 φχahb	ST-1 φχtb	fd	Primed single-stranded DNA
Specific DNA	+	+	+	−
ATP	+	?	+	ATP or dATP
UTP, CTP, GTP	−	?	+	−
dATP, dCTP, dGTP, dTTP .	+	+	+	+
dnaB, dnaC(D) gene products, factors X, Y, and Z	+	−	−	−
dnaG gene product	+	+	−	−
DNA binding protein . . .	+	+	+	−
DNA polymerase III, elongation factors I and II	+	+	+	+
RNA polymerase	−	−	+	−

TABLE 2

Protein requirements for ϕχ174 DNA-dependent DNA synthesis

Additions	dTMP incorporated	
	Expt. 1	Expt. 2
	(pmol/30 min)	
Complete reconstituted ϕχ174 DNA system	38.5	19.7
- dnaB gene product	0.3	-
- dnaC(D) gene product	0.3	-
- dnaG gene product	< 0.2	-
- DNA binding protein	3.3	-
- DNA polymerase III and DNA elongation factor II	< 0.2	-
- DNA elongation factor I	2.0	-
- replication factor X	1.1	< 0.2
- replication factor Y	-	1.2
- replication factor Z	-	0.9

Each assay (0.025 ml) contained 20 mM Tris-HCl (pH 7.5), 10 mM MgCl$_2$, 4 mM dithiothreitol, 1 mM ATP, 0.04 mM each of dATP, dCTP, dGTP, and [^{3}H]TTP (500-1,000 cpm/pmol), 10 µg/ml of rifampicin, 200 pmol ϕχ174 DNA, 0.5 mg/ml bovine serum albumin, dnaB gene product (0.05 U, 0.06 µg), dnaG gene product (0.08 U, 0.06 µg), dnaC(D) gene product (0.03 U, 0.1 µg), DNA elongation factor I (0.2 U, 0.04 µg), *E. coli* DNA binding protein (0.2 U, 0.25 µg), DNA replication factor X (0.05 U, 0.06 µg), DNA replication factor Z (0.05 U, 0.1 µg), DNA polymerase III (0.2 U containing 0.04 U DNA elongation factor II, 0.2 µg), and DNA replication factor Y (0.02 U). All proteins with the exception of DNA replication factors Y and Z were isolated as previously described (2). Procedures for the isolation of factors Y and Z will be published elsewhere. After incubation at 30° for 30 minutes, acid insoluble radioactivity was measured.

well and are defined by mutants temperature sensitive for DNA synthesis. The dnaE gene product is DNA polymerase III (3, 4). The dnaB gene product has an associated ribonucleoside triphosphatase activity which is stimulated by single-stranded DNA (5). No enzymatic activity is as yet associated with dnaC(D) gene product; however, we will show below that *in vitro* the purified protein interacts physically and functionally with dnaB gene product (6). It has been suggested that dnaG gene product has RNA polymerase activity (7). The other 6 proteins required for *in vitro* φχ174 DNA-dependent DNA synthesis are as yet undefined by genetic loci. Three of these have been characterized in other reactions. DNA binding protein interacts physically with some proteins in the absence of DNA as well as affecting the activities of some enzymes by binding to single-stranded DNA (8, 9). DNA elongation factors I and II are required for DNA synthesis catalyzed by DNA polymerase III in the presence of long single-stranded DNA primed with RNA or DNA (10, 11). The other 3 proteins involved in the φχ174 DNA synthesizing system are referred to as DNA replication factors X, Y, and Z and are defined by their requirement in this system (2). To date, DNA replication factors X and Z have no known enzymatic activities; we will discuss below φχ174 DNA-dependent ATPase activity associated with DNA replication factor Y.

The second DNA synthesizing system we have studied uses φχtB or ST-1 DNA[2] and requires dnaG gene product, DNA binding protein, DNA elongation factors I and II, and DNA polymerase III. The requirement for dnaG in this system has been shown in the following ways: (a) dnaG complementing activity in the φχ174 system (14) and stimulation of DNA synthesis dependent on φχtB or ST-1 DNA plus DNA binding protein, DNA polymerase III, and DNA elongation factors I and II cosedimented through glycerol gradients and comigrated on native polyacrylamide gel electrophoresis; and (b) φχtB and ST-1 DNA-dependent activity was thermolabile when dnaG gene product isolated from dnaG temperature sensitive cells was used as compared to activity when wild-type dnaG gene product was used.

---

[2] Phage φχtB and ST-1 are described in references 12 and 13, respectively.

The third DNA synthesizing system uses fd DNA and requires NTPs, RNA polymerase, DNA binding protein, DNA
polymerase III, and DNA elongation factors I and II (10,
11).

Three common proteins in these 3 specific DNA synthesizing systems are DNA polymerase III and DNA elongation
factors I and II. The combination of these proteins will
elongate any DNA or RNA primed single-stranded DNA with no
template specificity (10, 11). A requirement for ATP or
dATP can be seen when the primer template is poly(dA) ·
oligo(dT); no ATP requirement can be seen when the template requires that dATP be present for incorporation.

RESULTS

The results presented below concern our studies with
the reconstituted φχ174 DNA synthesizing system. Some
partial reactions involving components of this system have
been examined independent of DNA synthesis:

*Resolution of φχ174 DNA synthesizing reaction into 2
steps*. Incubation of the required proteins, ATP, and φχ-
174 DNA in the absence of dNTPs (step 1) increased the
initial rate of DNA synthesis following addition of dNTPs
(step 2) (2). In addition to ATP and φχ174 DNA, rapid
initial rate of DNA synthesis in the second step depended
on the presence of dnaB and dnaC(D) gene products, DNA
binding protein, and DNA replication factors X, Y, and Z;
the stimulation was not seen when any of these components
was added during step 2 only. In contrast to these components, DNA polymerase III, DNA elongation factors I and
II, and dnaG gene product were not required during the
first step; the initial rate of dNMP incorporation was
the same whether they were present during step 1 and step
2 or during step 2 only. The φχ174 DNA-protein complex
produced in the first step was isolated by gel filtration
(BioGel A-5m agarose). DNA synthesis by this complex required the addition of dNTPs, dnaG gene product, DNA
polymerase III, and DNA elongation factors I and II.

Interaction of dnaB and dnaC(D) gene products. Both
dnaB and dnaC(D) gene products have been isolated by complementation assays in which protein preparations from

wild-type cells stimulate $\phi\chi174$ DNA-dependent DNA synthesis in heat inactivated crude extracts of dna temperature sensitive cells (5, 15, 16). These dna gene products have been isolated from both wild-type and temperature sensitive strains and the thermolability of the temperature sensitive protein demonstrated in the $\phi\chi174$ DNA-dependent complementation system. The dnaC(D) gene product[3] has a molecular weight of 25,000 and is N-ethylmaleimide sensitive; as yet no enzymatic activity has been found associated with it (16). The dnaB gene product has a molecular weight of 250,000 and is N-ethylmaleimide insensitive (15). The purified dnaB gene product has ribonucleoside triphosphatase activity which is stimulated by single-stranded DNA. These triphosphatase activities and dnaB complementing activity copurify over the last 20-fold of a 40,000-fold purification procedure (16). NTPs are hydrolyzed, not dNTPs, and the products are P_i and NDPs. We have found these 2 proteins interact physically and functionally *in vitro* (6).

The dnaC(D) gene product was detected (as measured by stimulation in the $\phi\chi174$ DNA-dependent complementation assay) physically associated with the dnaB gene product (also measured by the $\phi\chi174$ DNA-dependent complementation assay) when the 2 purified proteins were mixed and subjected to gel filtration in the presence of ATP (Fig. 1A). Fig. 1B shows that in the presence of ATP and in the absence of dnaB gene product, dnaC(D) eluted as expected for a protein with a molecular weight of about 25,000. Fig. 1C shows that dnaB gene product in the presence of ATP and absence of dnaC(D) gene product eluted as expected for a protein with a molecular weight of about 250,000. Since the elution profile of dnaB gene product was not significantly affected by its interaction with dnaC(D) gene product, probably only one or a few dnaC(D) molecules were bound to one native dnaB molecule.

When dnaB and dnaC gene products were mixed in the absence of ATP and filtered through a column in the absence of ATP, the 2 proteins were not associated with each

---

[3] The dnaC and dnaD gene products are identical *in vitro* (16) and *in vivo* (17) and will be referred to as dnaC(D) gene product.

232

other (Fig. 1D). Furthermore, most of the dnaB activity
was excluded from the column; the nature of this aggregation has not been studied, nor has it been demonstrated
that it was catalyzed exclusively by the dnaC(D) gene product. In the absence of ATP, dnaB and dnaC(D) gene products alone eluted as they had in the presence of ATP.
Fig. 1E shows that dnaB and dnaC(D) activities were not
associated with each other if they were incubated together
for 30 minutes at 30° in the presence of ATP and then filtered through the agarose column in the absence of ATP.
Thus ATP was required for detecting the physical complex of
dnaB and dnaC(D) gene products. The nucleotide requirement was specified for ATP; other NTPs and dNTPs did not
satisfy this requirement.

In the experiment shown in Fig. 1A, half of the
dnaC(D) activity recovered was associated with dnaB gene
product. With half the amount of dnaC(D) gene product and
the same amount of dnaB gene product, all of the dnaC(D)
activity was associated with dnaB activity (Fig. 1F). Thus
the amount of complex depended on the ratio of dnaB to
dnaC(D) gene product.

In addition to interacting physically, the dnaB and
dnaC(D) gene products mutually affected one another in reactions involving ATP (6): (a) the dnaB DNA-independent
ATPase activity was inhibited by dnaC(D) gene product and
(b) the dnaC(D) complementing activity was protected from
N-ethylmaleimide inactivation by the combination of ATP
and dnaB gene product. This protection specifically required ATP; other NTPs and dNTPs did not satisfy the requirement.

*φχ174 DNA-dependent ATPase of DNA replication factor
Y.* As mentioned above, DNA replication factor Y is defined by its requirement in the reconstituted φχ174 DNA
synthesizing system. We have found this factor to contain
DNA-dependent ATPase activity. The association of DNA-
dependent ATPase activity with factor Y was shown in the
following ways: (a) the 2 activities copurified over 3
consecutive purification steps with a constant ratio of
factor Y to ATPase; (b) they comigrated on native polyacrylamide gel electrophoresis; (c) both activities were
heat inactivated at the same rate; and (d) both showed

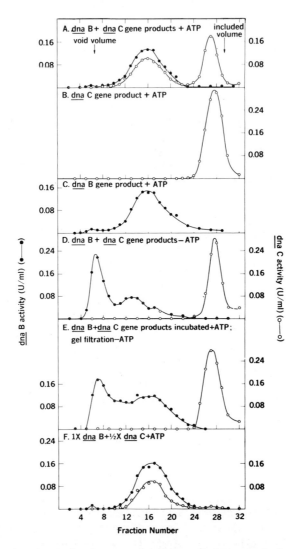

FIG. 1. *Physical association of <u>dnaB</u> and <u>dnaC(D)</u> gene products.*

FIG. 1. *Bio-Gel A-5m agarose gel filtration of* dnaB, dnaC(D), *and the combination of gene products*. (A) A reaction (0.075 ml) containing dnaB (1.2 U), dnaC(D) (1.5 U), 0.5 mM ATP, 1.0 mM $MgCl_2$, 0.05 mg/ml bovine serum albumin, 0.03 M potassium phosphate, pH 6.0, and 10 mM dithiothreitol was applied to a 0.5 cm x 20 cm column of BioGel A-5m agarose equilibrated with 0.5 mM ATP, 1.0 mM $MgCl_2$, 0.05 mg/ml bovine serum albumin, 0.03 M potassium phosphate, pH 6.0, and 10 mM dithiothreitol at room temperature. The column was developed with the same buffer at room temperature and 0.1 ml fractions were collected in tubes on ice containing 1 nmol ATP. Fractions were assayed as previously described for dnaB (●) (15) and dnaC(D) (O) complementing activities (16). (B) A reaction minus dnaB gene product was filtered through an A-5m agarose column as described in part A. (C) A reaction minus dnaC(D) gene product was filtered as described in (A). (D) A reaction minus ATP was filtered as described in (A) but with ATP omitted from the column buffer. (E) A reaction containing dnaB (1.2 U), dnaC(D) (1.5 U), 0.5 mM ATP, 1 mM $MgCl_2$, 0.03 M potassium phosphate, pH 6.0, 10 mM dithiothreitol, and 0.05 mg/ml bovine serum albumin was incubated 30 minutes at 30°. The reaction was then filtered as described in (A) but with ATP omitted from the column buffer. (F) A reaction as described in (A) containing dnaB (1.2 U) and dnaC(D) (0.75 U) was filtered as described in (A).

TABLE 3

Properties of DNA-dependent ATPase associated with factor Y

Additions	ATP or dATP hydrolysis	
	P_i	ADP or dADP
	(nmol/30 min)	
Complete	10.0	10.6
- φχ174 DNA	< 0.1	< 0.2
- ATP + dATP	-	9.4
- Mg^{2+}	< 0.1	-

Each assay (0.04 ml) contained 50 mM Tris-HCl (pH 8.0), 1 mM dithiothreitol, 20 mM KCl, 0.1 mg/ml bovine serum albumin, 10 μg/ml rifampicin, 0.7 mM $MgCl_2$, 0.7 mM [α- or γ-^{32}P]ATP or [α-^{32}P]dATP (10-50 cpm/pmol), 200 pmol φχ174 DNA, and DNA replication factor Y (0.02 U). After 30 minutes at 30°, ^{32}P_i was determined.

identical patterns of N-ethylmaleimide sensitivity. The properties of the ATPase are shown in Table 3. Factor Y hydrolyzed ATP or dATP producing P_i and ADP or dADP; other NTPs and dNTPs were not hydrolyzed. The reaction was stimulated 50- to 100-fold by $\phi\chi174$ or $\phi\chi$ahb DNA. It was not stimulated (0- to 5-fold) by other single-stranded DNA ($\phi\chi$-tB, ST-1, fd, heat-denatured *E. coli* DNA, poly dT, poly dC, poly dA, poly dG), by double-stranded DNA ($\phi\chi174$ RFI, *E. coli* DNA), or by RNA (Qβ, tRNA, poly A). Thus the 2 DNAs which are replicated by the $\phi\chi174$ system of purified components stimulate the factor Y DNA-dependent ATPase.

DISCUSSION

It is now possible to study detailed mechanisms of initiation of DNA synthesis and elongation catalyzed by this system. The reaction has been resolved into 2 steps. In the first step, *E. coli* DNA binding protein, dnaB and dnaC(D) gene products, and DNA replication factors X, Y, and Z interact with each other, ATP and $\phi\chi174$ DNA in reactions prior to dNMP incorporation. The $\phi\chi174$ DNA-protein complex formed supports DNA synthesis upon the addition of DNA polymerase III, DNA elongation factors I and II, dnaG gene product, and dNTPs. DNA polymerase III and elongation factors I and II probably function in DNA elongation in the second step as they do in elongation of RNA or DNA primed single-stranded DNA. The role of dnaG gene product in the second step is unclear.

So far several ATP-dependent reactions involving components of this system have been described which may contribute to the ATP requirement for DNA synthesis: (a) dnaB gene product contains ATPase (NTPase) activity which is stimulated by single-stranded DNA; (b) complex formation by dnaB and dnaC(D) gene products specifically requires ATP; (c) replication factor Y contains $\phi\chi174$ DNA-dependent ATPase (dATPase) activity; and (d) elongation of primed single-stranded DNA by DNA polymerase III and DNA elongation factors I and II requires ATP (dATP). These activities are sufficient to account for the specific ATP requirement of the system; however, there may be still other ATP-dependent partial reactions.

236

Although the purified $\phi\chi174$ DNA synthesizing system is specific for $\phi\chi174$ and $\phi\chi$ahb DNA using the assay conditions described, many of the proteins involved in this system are also involved in other DNA replicating systems, including those summarized in Table 1 and *E. coli* chromosome replication. It is not known if the *in vitro* reactions involving these proteins also occur *in vivo*. However, if dnaB and dnaC(D) gene products interact physiologically, then either (a) dnaB gene product would be implicated in initiation of chromosome replication since dnaC(D) is required for this process (18), or (b) dnaC(D) gene product would be implicated in chromosome elongation since dnaB is required for elongation (19), or (c) both proteins would be implicated in both chromosome initiation and elongation. Consistent with the possibility that these two proteins function jointly, some dnaC(D) mutants stop DNA synthesis immediately at the elevated temperature (20). This suggests that dnaC(D), like dnaB, is involved in chromosome elongation. If the interaction of dnaB and dnaC(D) gene products and the accompanying inhibition of ATPase are physiologic reactions, then the hydrolysis of ATP by dnaB in the absence of dnaC(D) is possibly an uncoupled reaction. Similarly, the $\phi\chi174$ DNA-dependent ATP hydrolysis by replication factor Y is possibly an uncoupled reaction; other proteins may be required to couple this ATP hydrolysis to other reactions involved in DNA synthesis. Thus, while some of the partial reactions catalyzed by these proteins *in vitro* may be specific to various *in vitro* DNA synthesizing systems, others may be more general to all DNA replication.

ACKNOWLEDGMENTS

This research was carried out while S. W. was a guest scientist in Dr. W. Parks' laboratory. The authors are indebted to Dr. Parks for his hospitality. The work was supported in part by grants from the National Institutes of Health, National Science Foundation, and the American Cancer Society. S. W. is a National Institutes of Health postdoctoral fellow.

REFERENCES

1. Vito, C. C., Primrose, S. B., and Dowell, C. E.,
    J. Virol. 15, 281-287 (1975).
2. Wickner, S., and Hurwitz, J., Proc. Nat. Acad. Sci.
    U.S.A. 71, 4120-4124 (1974).
3. Gefter, M. L., Hirota, Y., Kornberg, T., Wechsler,
    J., and Barnoux, C., Proc. Nat. Acad. Sci. U.S.A.
    68, 3150-3153 (1971).
4. Nüsslein, V., Otto, B., Bonhoeffer, F., and Schaller,
    H., Nature New Biol. 234, 285-286 (1971).
5. Wickner, S., Wright, M., and Hurwitz, J., Proc. Nat.
    Acad. Sci. U.S.A. 71, 783-787 (1974).
6. Wickner, S., and Hurwitz, J., Proc. Nat. Acad. Sci.
    U.S.A., in press.
7. Schekman, R., Weiner, A., and Kornberg, A., Science
    186, 987-993 (1974).
8. Sigal, N., Delius, H., Kornberg, T., Gefter, M., and
    3537-3541 (1972).
9. Molineux, I., Friedman, S., and Gefter, M., J. Biol.
    Chem. 249, 6090-6098 (1974).
10. Hurwitz, J., Wickner, S., and Wright, M., Biochem.
    Biophys. Res. Commun. 51, 257-267 (1973).
11. Hurwitz, J., and Wickner, S., Proc. Nat. Acad. Sci.
    U.S.A. 71, 6-10 (1974).
12. Bradley, D. E., Can. J. Microbiol. 16, 965-971 (1970).
13. Bone, D. R., and Dowell, C. E., Virology 52, 319-329
    (1973).
14. Wickner, S., Wright, M., and Hurwitz, J., Proc. Nat.
    Acad. Sci. U.S.A. 70, 1613-1618 (1973).
15. Wright, M., Wickner, S., and Hurwitz, J., Proc. Nat.
    Acad. Sci. U.S.A. 70, 3120-3124 (1973).
16. Wickner, S., Berkower, I., Wright, M., and Hurwitz,
    J., Proc. Nat. Acad. Sci. U.S.A. 70, 2369-2373
    (1973).
17. Wechsler, J. A., in DNA Synthesis In Vitro (Wells, R.
    D., and Inman, R. B., Eds.), University Park Press,
    Baltimore, Maryland, pp. 375-383 (1973).
18. Carl, P. L., Mol. Gen. Genet. 109, 107-122 (1970).
19. Kohiyama, M., Cold Spring Harbor Symp. Quant. Biol.
    33, 317-324 (1968).
20. Wechsler, J. A., J. Bacteriol. 121, 594-599 (1975).

IN VITRO SYNTHESIS OF BACTERIOPHAGE T7 DNA

W. E. Masker, D. C. Hinkle, D. Mark,
P. A. Modrich, and C. C. Richardson
Department of Biological Chemistry
Harvard Medical School
Boston, Massachusetts 02115

The virulent bacteriophage T7 contains a 26×10^6 dalton
linear duplex DNA molecule with coding capacity for approximately 30 proteins. Genetic analysis of phage T7 has
shown that the products of genes 1, 2, 3, 4, 5, and 6 are
required for viral DNA synthesis in vivo, and several of
these products have been purified and characterized (1).
Recently cell-free systems that support replication of T7
DNA in vitro have been developed (2,3). Incorporation of
deoxyribonucleotides into T7 DNA in vitro is dependent on
the products of phage genes 4 and 5, but not those of genes
2, 3, or 6. This requirement has provided a complementation assay for the purification of the product of genes 4
and 5 (4). In addition in vivo T7 DNA replication is completely blocked in Escherichia coli tsnC mutants (6).
Extracts of these mutants are also unable to support in
vitro T7 DNA synthesis; and this requirement has made possible the purification and characterization of the protein
that is lacking or altered in tsnC mutants (6,7). Recent
results indicate that the TSNC protein serves as a subunit
for the T7 DNA polymerase (7) and that the product of the
tsnC gene is identical to E. coli thioredoxin (8).

Alkaline zone sedimentation of the DNA synthesized
in vitro by extracts of T7-infected polA hosts indicates
that most of the product consists of short DNA fragments.
However, similar extracts of T7-infected $polA^+$ hosts produce larger DNA molecules, some of which have the same
molecular weight as intact T7 chromosomes. Similar results
have been obtained by adding purified DNA polymerase I to
extracts of T7-infected polA cells. Isopycnic gradient
analysis shows that there is no covalent linkage between
the high molecular weight DNA synthesized in vitro and the
exogenous template DNA (9).

A spheroplast system (10) has been used to test the
product of the in vitro system for biological activity.
Contributions to the infectivity caused by endogenous DNA
present in the assay mixtures were eliminated by performing

phage infections with T7 carrying amber mutations in essential genes and then performing transfection assays with suppressor-free spheroplasts and indicator bacteria. In order to eliminate infectivity due to exogenous template DNA, wild type T7 [^{13}C^{15}N]DNA was used as a template and then separated from the newly synthesized (normal density) product by isopycnic centrifugation. The results of these experiments show that the fidelity of in vitro DNA synthesis in this system is sufficiently accurate to produce biologically active DNA (11).

REFERENCES

(1)   Studier, F. W. (1972) Science 176: 367-376.
(2)   Strätling, W., Ferdinand, F. J., Krause, E., and Knippers, R. (1973) Eur. J. Biochem. 38: 160-169.
(3)   Hinkle, D. C. and Richardson, C. C. (1974) J. Biol. Chem. 249: 2974-2984.
(4)   Hinkle, D. C. and Richardson, C. C., submitted to J. Biol. Chem.
(5)   Chamberlin, M. (1974) J. Virol. 14: 509-516.
(6)   Modrich, P. A. and Richardson, C. C., submitted to J. Biol. Chem.
(7)   Modrich, P. A. and Richardson, C. C., submitted to J. Biol. Chem.
(8)   Mark, D. and Modrich, P. (1975) Fed. Proc. 34: 639.
(9)   Masker, W. E. and Richardson, C. C., manuscript in preparation.
(10) Henner, W. D., Kleber, I., and Benzinger, R. (1973) J. Virol. 12: 741-747.
(11) Masker, W. E. and Richardson, C. C., manuscript in preparation.

RECONSTRUCTION OF THE T4 BACTERIOPHAGE DNA REPLICATION APPARATUS FROM PURIFIED COMPONENTS

Bruce Alberts, C. Fred Morris, David Mace, Navin Sinha, Michael Bittner and Larry Moran

Department of Biochemical Sciences
Princeton University
Princeton, New Jersey 08540

ABSTRACT: The protein products of T4 genes 41, 43 (T4 DNA polymerase), 45, 44 and 62 have been purified to near homogeneity, using individual "complementation assays" which measure their ability to stimulate DNA synthesis in an appropriate mutant-infected crude lysate. The gene 44 and 62 proteins are copurified as a tight complex which displays DNA-dependent ATPase activity. In a reaction requiring 45 protein, this complex speeds up <u>in vitro</u> polymerization by T4 DNA polymerase on primed single-stranded templates to near <u>in vivo</u> rates. Further addition of gene 32 and gene 41 proteins allows two additional reactions:

> (1) Extensive synthesis by strand-displacement on double-stranded DNA templates (as expected for synthesis on the "leading side" of a replication fork),
>
> (2) <u>De novo</u> DNA chain initiation on single-stranded DNA templates, dependent on the presence or ribo C, G and U nucleoside triphosphates (as expected for RNA-priming of Okazaki pieces on the "lagging side" of the replication fork).

With the complete system operating on circular fd DNA templates, the products visualized in the electron microscope resemble classical rolling circles, with discontinuous DNA synthesis clearly evident on the long, unbranched double-stranded tails.

INTRODUCTION: Since 1971, efforts in this laboratory have been directed towards achieving <u>in vitro</u> reconstruction of the T4 bacteriophage DNA replication apparatus from purified components. With this communication, we report our first real success.

The work to be described was inspired by the elegant genetic studies begun by R.H. Epstein, R. Edgar and colleagues, which indicated that at least six major T4 gene

products are necessary for in vivo replication fork movement (1-3). Fig. 1A presents a highly schematized version
of the current T4 genetic map (4), emphasizing the six "replication proteins," henceforth designated as the gene 32,
gene 41, gene 43, gene 44, gene 45, and gene 62 proteins,
respectively. Of these gene products, only the proteins
corresponding to gene 43 (T4 DNA polymerase (5-10)) and
gene 32 (DNA-unwinding protein (11-13)) had been purified
when our work began.

Isolation of the Replication Proteins   In our current view,
each of the above replication proteins constitutes part of
a large multienzyme complex or "machine" for replicating
DNA; thus, individual catalytic functions need not exist
for them outside of this complex, and some may even have a
purely structural role. Nevertheless, by quantitating the
small amounts of DNA synthesis observed in a concentrated
crude extract, and exploiting the T4 mutants isolated by
the geneticists, these proteins can be detected via an "in
vitro complementation assay" (14,15). With this assay, we
have been able to purify the four previously unpurified replication proteins to homogeneity in active form. Fig. 1B
shows a SDS-polyacrylamide gel electrophoretic analysis of
our present preparations of the above six T4 proteins.
Using purification procedures to be described elsewhere
[and starting with cells infected with overproducing T4
mutants isolated by J. Wiberg and J. Karam (16,17)], we can
obtain at least 10 mg quantities of each protein from 100 g
of infected cells. The major protein bands in Fig. 1B have
been identified as the indicated gene products by standard
protein double-label experiments, using mixed radioactive
amber-mutant and wild-type-infected extracts as the starting material for a complete isolation. On further fractionation by sucrose gradient sedimentation, each protein band
comigrates with the appropriate activity in the in vitro
complementation assay, and with the activities to be described below.

   As previously reported (14), the gene 44 and 62 proteins are isolated as a tight complex (180,000 daltons)
which appears to contain four molecules of 44-protein
(34,000 daltons each) and two molecules of 62-protein
(20,000 daltons each). The other four proteins contain
only a single type of subunit, although the gene 32-protein
appears to exist mainly as a 6-8 S aggregate and the gene

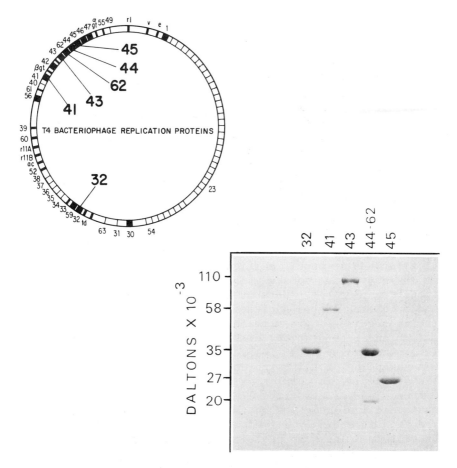

Figure 1. (A). The genetic map of T4 bacteriophage with
replication genes emphasized (1,4). The broadest black
segments indicate the relative locations of mutationally
identified genes with a major effect on DNA replication at
37°C. Mutants in genes 32, 41, 43, 44, 45 or 62 are unique
in synthesizing little or no DNA, even though all four
deoxyribonucleotide triphosphate precursors are present
(42).

(B). Polyacrylamide gel electrophoresis of the purified
T4 replication proteins. The separating gel was made with
12.5% acrylamide and the stacking gel with 6% acrylamide,
in buffers containing 0.1% sodium dodecyl sulfate (43).
The protein samples were diluted from their concentrated
stocks directly into SDS-mercaptoethanol sample buffer
and boiled for 2 minutes at 100°C just prior to electrophoresis. To estimate purity, the relative intensities of
the Coomassie-blue stained protein bands shown were determined with a Joyce-Loebl densitometer.

45-protein as a dimer. The gene 41-protein preparation is
the least pure (being about 85% a single species), whereas
all of the others appear to be on the order of 99% homogeneous. Tested on supercoiled PM2 native DNA or fd DNA single-strands, the addition of all these proteins at the concentrations at which they are routinely used reveals no
detectible endonuclease activity. However, it should be
noted that even our best preparations of T4 DNA polymerase
exhibit low levels of DNA ligase activity (as do the preparations of others (10)).

Partial Reactions   The most striking in vitro activities
of the T4 replication proteins require that all six of them
be present simultaneously. But since analysis of individual functions in such a multicomponent system is difficult,
it is instructive to study in detail as many "partial reactions" as possible. Here we include the DNA polymerase (7)
and proof-reading exonuclease (18) activities of 43-protein
(which require single-stranded DNA templates), the denaturation (and renaturation) of DNA catalyzed by 32-protein
(11), and the specific interaction of this polymerase and
DNA-unwinding protein with each other (13). Additionally,
the 44-62 protein complex is an ATPase which can hydrolyze
either ribo- or deoxyribo-ATP to the corresponding diphosphate and inorganic phosphate. As shown in Fig. 2A, this
reaction is strongly stimulated by DNA and by addition of
45-protein. The same proteins stimulate T4 DNA polymerase
utilization of long single-stranded DNA templates (Fig. 2B)
and enhanced stimulation is obtained upon rATP addition.
However, as illustrated in Fig. 3, no polymerase stimulation is seen when very short, single-stranded DNA molecules
are used as templates.
    Analysis by classical methods reveals that all the DNA
product in Fig. 2B and 3 is template-linked in both stimulated and control reactions and that no new primer sites
are being generated. Therefore, considering the results
in Fig. 3, it seemed likely that the 44-62 and 45 proteins
interact with T4 DNA polymerase to increase its average
"sticking-distance," or processiveness. That this will
give stimulation of net synthesis only on long template
strands is clear from the model shown in Fig. 4. This
model assumes that initiation of synthesis by polymerase is
rate-limiting under our conditions and that the polymerase
then continues along the DNA template for some distance be-

244

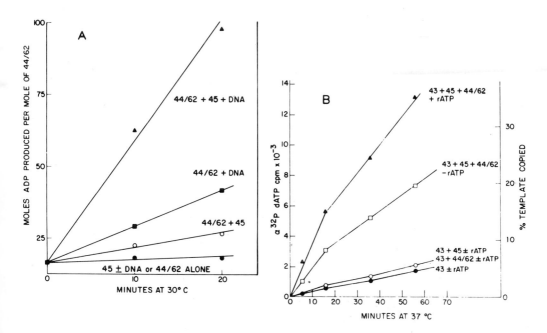

Figure 2. (A). Time course for the hydrolysis of rATP by
44-62 protein. The complete reaction contained 0.25 mM
[^{3}H]-rATP, 10 mM Tris-Cl pH 7.4, 25 mM KCl, 5 mM MgCl$_2$,
5 mM 2-mercaptoethanol, 400 µg/ml BSA, 7.5 µg/ml calf-
thymus single-stranded DNA, 10 µg/ml 45 protein, and
30 µg/ml 44-62 protein. Aliquots were spotted directly on
to PEI plates, overlayered with a mixture of carrier AMP,
ADP and ATP, and then developed in 1 M LiCl. The spots
seen under U.V. light were cut out and counted.

(B). Time course for DNA synthesis on T4 single-stranded
DNA. The reaction contained 10 mM Tris-Cl (pH 7.4), 25 mM
KCl, 5 mM MgCl$_2$, 5 mM 2-mercaptoethanol, 40 µM each
dNTP (including [α^{32}P]-dATP), 250 µg/ml BSA, 5 µg/ml T4
single-stranded DNA, 0.1 µg/ml T4 DNA polymerase, and
where indicated: 5 µg/ml 45 protein, 30 µg/ml 44-62
protein, and 0.25 mM rATP. Aliquots were removed at the
indicated times and assayed for acid-insoluble ^{32}P.

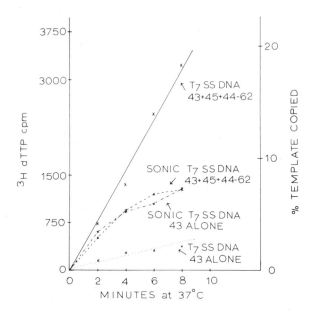

Figure 3. Time course for DNA synthesis on T7 single-stranded DNA molecules of different size. All reactions contained 10 mM Tris-Cl pH 7.4, 25 mM KCl and 2 mM MgCl$_2$, 0.5 mM dithiothreitol, 0.5 mM rATP, 50 μM each dNTP, 75 μg/ml BSA and 0.7 μg/ml T4 DNA polymerase. In addition, as indicated, the reactions contained as template either 5 μg/ml intact T7 single-stranded DNA or 5 μg/ml sonicated T7 single-stranded DNA, plus 138 μg/ml 45 protein and 85 μg/ml 44-62 protein.

The relatively rapid rates seen with the "sonic DNA" template are due to its enrichment for 3'-OH terminal priming sites, which allows more frequent polymerase starts.

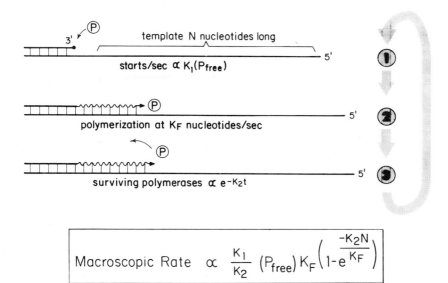

$$\text{Macroscopic Rate} \propto \frac{K_1}{K_2}(P_{free})K_F\left(1 - e^{\frac{-K_2N}{K_F}}\right)$$

Figure 4. A proposed microscopic DNA polymerase cycle.
The T4 DNA polymerase is viewed as starting on a 3'OH
primed single stranded template with a rate proportional
to the concentration of such 3'OH termini. Polymerization
then proceeds with a uniform rate (K_F)until terminated by
one of two alternate paths. The first, direct dissociation from the middle of a template chain, is typically
first order, while the second path is due to polymerases
running off the template end. On very large templates
the latter pathway is insignificant under our reaction
conditions. The total rate of incorporation ("macroscopic
rate") is determined by the number of polymerases bound
to a primer end (i.e., one that can be extended by the
addition of at least one more deoxyribonucleotide) multi-
plied by the rate of movement of polymerase, K_F. The
final expression is seen to include an exponential term
which reflects the relative rates of the above two termination processes.

fore falling off. It is clear that increasing K_F (the
microscopic polymerization rate) and/or decreasing K_2 (the
falling off rate) will increase the average polymerase
"sticking-distance". This in turn will stimulate the macroscopic rate of DNA synthesis, providing that the template
length, N, is longer than the typical sticking-distance for
polymerase alone.

Our attention was originally directed to the question
of DNA polymerase processiveness by Dr. Lucy Chang, who
found that calf DNA polymerases work processively for less
than 5 seconds (her limit of resolution) before falling off
(19). To explain the differential stimulation of the macroscopic synthesis rate for long templates observed (Fig.
3), calculations from the model in Fig. 4 demand a sticking
length of at least 1,500 nucleotides for T4 DNA polymerase
alone. This led us into a detailed investigation aimed at
determining K_F and K_2 in the T4 system.

Fig. 5 is a histogram of the length of double-stranded
DNA product made during 5 minutes of incubation, as measured by electron miscroscopy. Under the conditions used,
an average of only one template molecule in five had a
polymerase initiation. Thus, Fig. 5 may be viewed as a
display of microscopic "sticking lengths" in the stimulated
and control reactions. Note that the sticking-length for
polymerase alone clearly exceeds 1,500 nucleotides, as required to explain the results in Fig. 3. Although the
double-stranded DNA lengths are quite heterogeneous in the
stimulated reaction, a clear and striking increase in average product length is seen (apparently only about half of
the polymerase molecules are affected by the 44-62 and 45
protein additions). Assuming the same number of polymerase
initiations in the stimulated and control reactions (as in
Fig. 4), we calculate a 4.5 fold stimulation in macroscopic
synthesis rate from the histogram in Fig. 5, in agreement
with the 3.9-fold increase actually observed.

Our interpretation of the experiment in Fig. 5 is in
agreement with direct measurements of the rate of polymerization in the stimulated and unstimulated reactions obtained with 2 to 20 second incubations, to be described in
detail elsewhere (D.M. and B.A., in preparation). Here,
using electron microscopy or physical sizing of S_1 nuc-
lease-resistant product as independent measures of product
size, we estimate for T4 DNA polymerase alone a microscopic
polymerization rate (K_F) of ∿250 nucleotides/sec and an

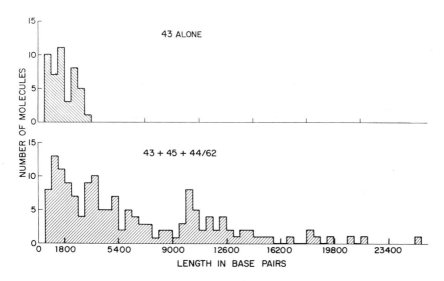

Figure 5. The lengths of double-stranded regions in
in vitro DNA polymerase reaction products as measured by
electron microscopy. The reactions were carried out
essentially as described in Fig. 2B, except that 2 mM
$MgCl_2$ was present and 5 μg/ml single-stranded T7 DNA was
used as template. For the lower panel, 7.5 μg/ml 45
protein and 9 μg/ml 44-62 protein was added. The reactions
were stopped at 5 min by adding an equal volume of cold
0.2 M EDTA, and spread by the Inman procedure (24).

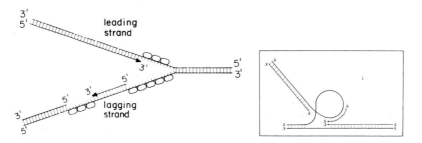

Figure 6. Schematic representation of a general DNA replication fork with DNA-unwinding protein binding to the single-
stranded DNA exposed by fork movement. The insert illustrates an alternative visualization of the same replication fork, in which DNA synthesis on the "leading" and
"lagging" strands is coupled. In this latter form, one
can easily rationalize the existence of discontinuous
synthesis on both sides of a fork, as observed
(27; see text).

off-rate (K_2) of $\sim$0.1 sec^{-1} under our conditions (see Figs. 2B and 3). Addition of 44-62 and 45 proteins increases the K_F to 800-900 nucleotides/sec, without a major effect on K_2.

Note that these values are much greater than the 10 nucleotides/sec commonly suggested for K_F from standard macroscopic rate measurements (13). This means that T4 DNA polymerase runs at close to the in vivo rate of polymerization (estimated from the data of R. Werner (20) at 1,000 nucleotides/sec in our simple in vitro system, even without the further stimulation expected from addition of 32-protein (13). We conclude, therefore, that the other pro-teins of the replication apparatus need only decrease K_2 (by tying down polymerase) to account for the observed speed of in vivo fork movement.

In other experiments, we have been able to demonstrate tight binding of 44-62 protein, but not 45 protein, to sin-gle-stranded DNA in a reaction requiring 37° incubation (unpublished results of M. Davies). For this reason, our tentative view of the DNA polymerase interaction is that schematically illustrated: 44-62 is viewed as holding onto both the template strand and the DNA polymerase, and its ATP hydrolysis serves to generate protein conformations which help move the DNA polymerase more quickly along the DNA.

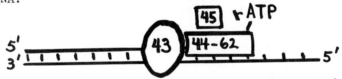

Modeling Suspected Polymerization Processes at the Replica-tion Fork. Based upon work in many laboratories, a schema-tic representation of a replication fork can be drawn (Fig. 6). Because of the anti-parallel orientation of strands in the DNA double-helix, a discontinuous mode of DNA synthesis is necessary on at least one template strand if all of the DNA is to be made with a proof-reading, 5' → 3' DNA poly-merase (21). Biochemical analysis of initial DNA products (22, 23) and electron microscopy of forks (24-25) indicate lengths of 1 to 2 x 10^3 nucleotides between adjacent poly-merase starts on the "lagging side" of the fork. On the "leading side" of the fork, synthesis also appears to be discontinuous (26,27). The restarts on the leading strand, however, need not lead to the large single-stranded gaps observed in the lagging strand intermediates. This single-

stranded DNA is almost certain to be covered with tightly
bound DNA-unwinding protein. In addition, we propose that
32-protein has a helix-opening role in leading strand synthesis. [Opening of the helix ahead of the fork seems essential, since double-helical templates cannot be copied by
the T4 DNA polymerase directly (8)].

The model illustrated in Fig. 6 assumes that two DNA
polymerase molecules are working at any one time in the
fork. The 44-62 ATPase would be expected to be traveling
along with one or both of these. (Note that these two
polymerization sites could be physically linked together,
as suggested in the Insert to Fig. 6). In addition, because T4 DNA polymerase is unable to start chains de novo,
it seems likely [by analogy with observations in other
systems (28-30)] that discontinuous synthesis is primed by
a special piece of RNA made by the replication proteins on
single-stranded template DNA generated at the fork. After
this RNA serves as a primer for the DNA polymerase, it must
be erased by another enzyme and replaced by DNA. We feel
that this mechanism makes good biological sense, since it
can be argued that the enzyme that synthesizes the primer
will necessarily make a relatively inaccurate copy; thus,
the choice of RNA (rather than DNA) as primer automatically marks these sequences as "bad copy" to be removed (for
details, see Ref. 21).

In the simplest view, one replication protein (e.g.,
gene 41 protein?) might function solely to synthesize an
RNA primer. To test this idea, we set out to study the
predicted reactions on the two sides of the replication
fork in Fig. 6 separately. For lagging strand synthesis
one might expect to obtain ribonucleoside triphosphate dependent de novo chain initiations on a T4 single-stranded
DNA template (see also article by Bouche and Kornberg, this
volume):

For leading strand synthesis one might anticipate rapid
strand displacement beginning from a nick on a T4 DNA
duplex.

As will be demonstrated, the T4 replication proteins cata-

251

lyze both of the above model reactions in an efficient manner. Although we initially expected that different combinations of replication proteins would be required, both the
leading and lagging-strand reactions turn out to require
all six of the T4 proteins introduced in Fig. 1 for maximal
DNA synthesis.

A "Lagging-Strand" Model Reaction.   Incorporation data
strongly suggesting RNA-primed DNA chain starts in vitro
are presented in Fig. 7. Using alkali-denatured T4 single-
strands as a template one obtains extensive DNA synthesis
which requires the presence of all six of the replication
proteins. As shown, this synthesis is completely dependent
upon the addition of a mixture of rCTP, rGTP and rUTP. Alkaline sucrose gradient sedimentation reveals that the small
amount of product made in the absence of either rCTP, rGTP
and rUTP or 32 protein cosediments with the template (as
expected for template priming by fold-back at the 3' end).
In contrast, essentially all of the product made at early
times in the complete reaction sediments with a mean chain
length much smaller than the template (data not shown).
These results suggest that RNA-primed de novo initiation of
new DNA chains is occurring. This reaction proceeds under
conditions which approximate intracellular concentrations
of salts and with protein concentrations which are not
greater than those found in vivo (see Fig. 7, heading).
Therefore, we believe that it is biologically significant.
However, the sites at which the presumed primer is being
initiated are not restricted to T4 DNA. Shown in Table 1
are results obtained with T4, fd, T7 and synthetic DNA
single-strands. We find that the ribo-dependent reaction
similar to that shown in Fig. 7 proceeds on many DNA templates (e.g., T4 and fd) but that not all DNAs are active
(e.g., T7 and synthetic DNAs). Thus, we conclude that
under our conditions, the T4 replication proteins are recognizing some special sequence or structure in this
priming reaction. To explain why all six T4 proteins are
required, we suggest that a special multienzyme complex
is formed between them, and that this is needed to generate
the correct environment for the postulated synthesis and
utilization of the primer RNA.
     Note that the regular DNA-dependent RNA polymerase
utilized for transcription is not involved in our reactions:
we are unable to detect the activity of this enzyme in any

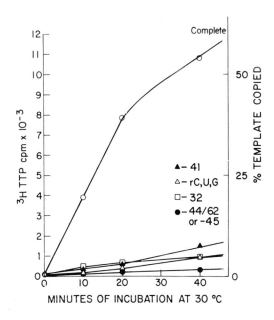

Figure 7. Time course of ribonucleotide-dependent DNA synthesis on a single-stranded T4 DNA template. The "complete" reaction components consist of 43 protein 1 μg/ml, 41 protein 80 μg/ml, 44-62 protein 8 μg/ml, 45 protein 25 μg/ml, 32 protein 125 μg/ml, dNTP's 0.1 mM each, rATP 0.5 mM, nuclease-free BSA 50 μg/ml, $MgSO_4$ 5 mM, KCl 25 mM, Tris-Cl (pH 7.4) 20 mM, dithiothreitol 0.5 mM, alkali-denatured (44) T4 single-stranded DNA 10 μg/ml, and rCTP, rUTP, rGTP 0.1 mM each. [³H]-TTP was used to label DNA products made. For the incomplete reactions, only the specified component was deleted. At the specified times, aliquots were removed and incorporation into acid insoluble material measured.

253

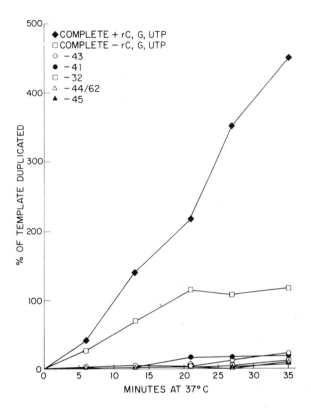

Figure 8. Time course of synthesis on a double-stranded
T4 DNA template. The complete reaction components were
essentially the same as those specified in Fig. 7, the
major changes being addition of K$^+$ to 67 mM, substitution of 4 µg/ml intact T4 double-stranded DNA as
template, and removal of rCTP, rGTP and rUTP. For
incomplete reactions, only the specified component was
deleted. Note that only the one reaction indicated
contained the three ribonucleoside triphosphates.

of our purified proteins; moreover, the ribonucleoside triphosphate requirement for DNA synthesis is retained even
after addition of high levels of rifampicin (10 µg/ml) or
antibody to T4 modified RNA polymerase.

A "Leading-Strand" Model Reaction   The T4 DNA polymerase
by itself will not utilize nicked double-helical DNAs as a
template, as it is unable to copy base-paired regions (8).
However, in 1974 Nancy Nossal made the important observation that addition of high concentrations of 32 protein
allows a limited amount of DNA synthesis to begin on such
templates (31). Most of the product DNA made is reversibly
denaturable, re-forming hairpin-type double-helical structures after heating and quick cooling (31). This is the
expected result if branch migration (32) repeatedly inter-
rupts the polymerization process (33,34), as diagrammed
below:

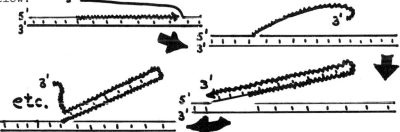

When additional T4 replication proteins are added to this
system, such interruptions are apparently prevented. One
observes, instead, the rapid and efficient strand-displace-
ment reaction expected for leading strand synthesis. All
double-stranded DNAs thus far tested function as templates
for this synthesis, including the DNA from bacteriophages
T4, T7, lambda, PM2 and G4; E. coli; SV40 virus; and monkey
cells. The protein requirements for extensive DNA synthesis are illustrated in Fig. 8, where double-stranded T4 DNA
has been used as the template. Once again, all 6 T4 replication proteins are required; note that an amount of product equal to the input template is made within 20 min at
37°. Alkaline sucrose gradient analysis demonstrates that
long continuous DNA strands are being synthesized, and test
of the S_1 nuclease susceptibility of product and template
revealsthat more than 99% of each is fully denaturable at
all extents of reaction (data not shown).
    Further addition to this system of a mixture of rCTP,

rGTP and rUTP stimulates the observed rate of DNA synthesis
between 2- and 5-fold, depending on the exact conditions
and template used (e.g., ~2-fold in Fig. 8). Clearly, part
of this stimulatory effect is due to the ribo-dependent
lagging strand synthesis which ensues (see below), but the
fact that the effect can be greater than 2-fold leaves open
the possibility that another ribo-dependent reaction is additionally involved.

Complete Reactions   When leading and lagging strand reactions proceed simultaneously, they should generate a replication fork resembling that described earlier in Fig. 6.
This is readily seen when circular templates such as PM2
or fd bacteriophage DNA are used. We shall demonstrate
that, following de novo chain initiation on a fd template,
polymerization proceeds around the circular, single-stranded DNA molecules to yield a circular double-helix with one
strand-specific interruption. Further synthesis on some of
these double-stranded circles induces strand displacement
and generates a "rolling circle" (35) mode of DNA synthesis.
Because the initial template is exclusively (+) strand, the
32-protein coated, single-stranded regions (formed as intermediates on these rolling circle tails) are located exclusively on the (-) strand. These regions therefore cannot
renature with each other to confuse the analysis. [In contrast, renaturation between rolling circle tails can be a
problem when the initial template is a circular double-
stranded DNA molecule (such as PM2 or SV40)].

Reaction on Single-Stranded fd DNA Templates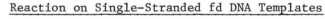

The only components used in the "complete" reaction on
fd DNA templates are the 6 replication proteins, rNTPs,
dNTPs, bovine serum albumin (BSA), Mg^{++} and salts. If any
of the T4 proteins (44/62, 43, 41 or 45) is left out of the
reaction mixture, one finds less than 1% the amount of DNA
synthesis seen in the complete reaction. Therefore, each
is clearly essential. As can be seen in Fig. 9, there is
also a dramatic decrease in synthesis when rUTP, rGTP and
rCTP are omitted. [The rATP is not omitted, since a role
for it in stimulation of DNA synthesis with proteins 43,
44/62 and 45 has already been established (Fig. 2B)]. This

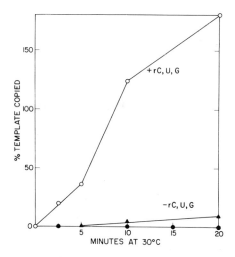

Figure 9. Time course of the complete reaction using [³H]-labeled fd single-stranded circular DNA template. The complete reaction components were essentially the same as those designated in Fig. 7. [α ³²P]-dGTP was used to label DNA products made. The +rUTP, +rGTP. +rCTP labeled data represents a time course of the complete reaction and the -rUTP, -rGTP, -rCTP data represents the time course observed if these components are omitted.

TABLE 1

Template	% DNA made relative to template	
	+ { rUTP, rGTP, rCTP	− { rUTP, rGTP, rCTP
	4' at 37°C	
Fd	101%	3%
T₇ - H strand	5%	4%
T₇ - L strand	5%	7%
	20' at 30°C	
SS T4 DNA	35%	0.6%
Poly dI	0.7%	2%
Poly dG	1%	0.6%
Poly (dI, dT)	0	0

For conditions used, see Fig. 9.

strong requirement for ribonucleotides suggests that an RNA
primer is being made. However, in preliminary attempts, it
has not been possible to detect a significant fraction of
radioactive ribonucleotides incorporated into the DNA product. We suspect that RNA primer is both being made and degraded in our system, perhaps by a RNAse activity intrinsic
to the purified proteins.

Since synthesis continues far past the level equivalent to a complete copying of the fd template (Fig. 9), this
system must do more than simply convert the initial template to a circular double-helical DNA molecule. To examine this reaction in more detail, ^{3}H-labeled fd DNA was
used as template and ^{32}P-labeled deoxyribonucleoside triphosphates incorporated. Template & product could then be
separately analyzed by sedimenting the reaction mixture
through alkaline sucrose gradients. The results are shown
in Fig. 10. As expected, the sedimentation value of the
^{3}H-template remains unchanged during the course of the reaction. Since it never cosediments with the bulk of the
product DNA, it cannot be covalently attached to it. The
average strand length of the ^{32}P-labeled product, however,
is seen to grow with time. The product made during the
first 2.5 minutes is the size of a linear fd genome or
smaller (Fig. 10a). By 10 minutes, a small amount of product is seen sedimenting faster than fd circular template
(Fig. 10b), and by 20 minutes almost half the product (45%
in this experiment) is found to sediment faster than the
template strands (Fig. 10c). These results are those expected if, following de novo initiation of (−) strand synthesis on the fd circular template [(+) strand], this (−)
strand grows to multiple fd lengths by continuous (−)
strand displacement as the circular (+) strand template
"rolls" (see schematic diagram in text, above). This basic
conclusion is supported by separate experiments in which
DNA-DNA hybridization has been used to determine the strandedness of the largest and smallest DNA products isolated
from alkaline sucrose gradients (unpublished results of
C.F.M.).

Direct S$_1$ nuclease digestion (in 0.2% SDS, pH 4.6) of
the DNA products obtained at various times during the
course of reaction in Fig. 9 revealed that while only 35%
of the ^{3}H template is resistant to S$_1$ nuclease digestion at
2.5 minutes, 70% is resistant at 5 minutes and nearly 100%
at 10 minutes. As might be expected, the initial ^{32}P pro-

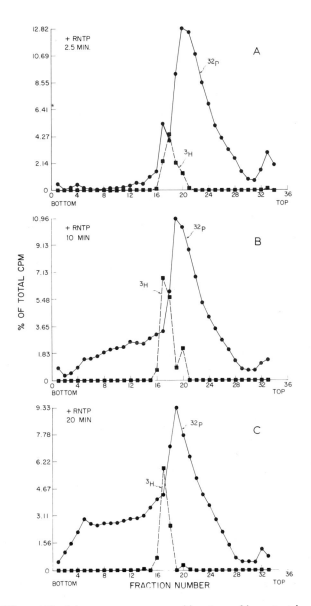

Figure 10. Alkaline sucrose gradient sedimentation of
the reaction products made on a [³H]-labeled fd DNA
template, as described in Fig. 9. At the indicated times,
aliquots were removed from the reaction mix into an
equal volume of cold 20 mM Na₂EDTA. Samples were then
layered on 5-20% alkaline sucrose gradients (0.8 M NaCl,
5 mM EDTA, 0.2 M NaOH) with 60% sucrose shelves and
centrifuged for 3 hours at 50,000 rpm, in a Spinco SW
50.1 rotor (4°C). Fractions were collected from the
bottom onto glass fiber filters, and washed and counted
by standard techniques. Counts were plotted by computer
after overlap corrections. (The fraction of [³H]-template
counts has been adjusted to 10% of their real value
in order to allow the product counts to be better
visualized).

duct made is almost completely resistant (92%) to S_1 digestion. At later times, the fraction of resistant product
decreases, but reaches an apparent limit where 75% of it is
still double-stranded. Since product DNA would be continually displaced as single-strands from a rolling circle, we
conclude that most of it must be used as a template for
synthesis of its complementary strand after it enters the
tail.

Electron Microscopy of the Reaction Products   For electron
microscopy, aliquots were removed from reactions on an fd
DNA template and examined after cytochrome spreading on a
formamide hypophase, using the modified Kleinschmidt procedure developed by R. Inman (36). This type of analysis
is not strictly quantitative, since a substantial fraction
of DNA molecules are not well enough spread to be scored,
and thus a special minor class of product can easily be
overlooked. Moreover, the shear forces generated by
spreading appear to fragment some of the longer DNA molecules (especially those containing substantial single-
stranded interruptions). Therefore, although   simple linear double-stranded products are seen, it is not clear that
they were present as such before spreading.

      Despite these qualifications, it is clear that a major
class of product DNA has precisely the geometry predicted
from the sedimentation analysis described earlier (Fig. 10).
As illustrated in Fig. 11, double-stranded DNA circles of
fd length are the predominant structures seen at early
times in the reaction. On further incubation, long double-
stranded DNA tails are seen attached to some of these circles (Figs. 12 and 13). As expected from the rolling circle model, the DNA at the point of attachment of circle and
tail is nearly always single-stranded. Moreover, double-
and single-stranded regions frequently alternate in this
region, as in the example shown in Fig. 14. We therefore
conclude that the (+) strand complement made on the single-
stranded tail is synthesized by a discontinuous mechanism,
as predicted (see lagging strand in Fig. 6).

      Similar structures to those shown in Figs. 12 and 13
are seen whether fd or G4 bacteriophage DNAs (single-stranded), or PM2 bacteriophage or SV40 virus DNAs (double-stranded) provide the initial circular template. In all cases,
although an amount of DNA equivalent to multiple copies of
the template is made in the first 30 min of incubation,

only a small percentage of double-stranded circular molecules appear as circles with attached tails. In a typical electron microscopic field, many of these tails are greater than 50 microns in length (10^8 daltons). Thus, the mechanism of polymerization at these replication forks appears to be highly processive, with a calculated rate of polymerization per active rolling circle of at least 600 nucleotides/ sec at 37°. Note that this is close to the rate at which the T4 replication fork moves in vivo (20).

The Fidelity of Synthesis   The most sensitive test for fidelity in this system would be to quantitate the biological activity of the DNA product relative to that of the input template. While these studies are not yet completed, we have determined that all (>99%) of the product DNA made is denatured by 3 minutes of heating at 100° followed by rapid cooling, as judged by its becoming sensitive to S_1 endonuclease degradation. Thus, our products do not contain the hairpin inversions found when either E. coli DNA polymerase I (33,34) or T4 DNA polymerase plus 32-protein (31) utilize double-stranded DNA as template. Moreover, Fig. 15 shows a sedimentation analysis with and without R_I restriction endonuclease treatment of double-stranded DNA synthesized in vitro using single-stranded DNA from G4 phage as template. As expected for faithful copying, the fast-sedimenting rolling circle products made on either G4 or SV40 virus DNA templates [which contain a single EcoR$_I$ restriction site (37,38)] are cut to unit-length linear double-helices by this enzyme treatment (SV40 data not shown). Finally, analysis by CsCl gradient equilibrium sedimentation of products made with deoxybromouridine triphosphate replacing TTP reveal a classical semiconservative pattern of DNA synthesis (unpublished results, N.S.). We conclude therefore that double-helical DNA templates are copied with a reasonable degree of fidelity in our in vitro system.

Future Directions   After a T4 bacteriophage DNA molecule enters an E. coli cell, it synthesizes replication proteins and these are then directed so as to selectively replicate only the T4 genomes present. The E. coli DNA is not used as template, even though it is in great excess over phage DNA. This remains true for T4 mutant infections in which this host DNA is not significantly degraded (39). It therefore seems likely that special DNA sequences or structures

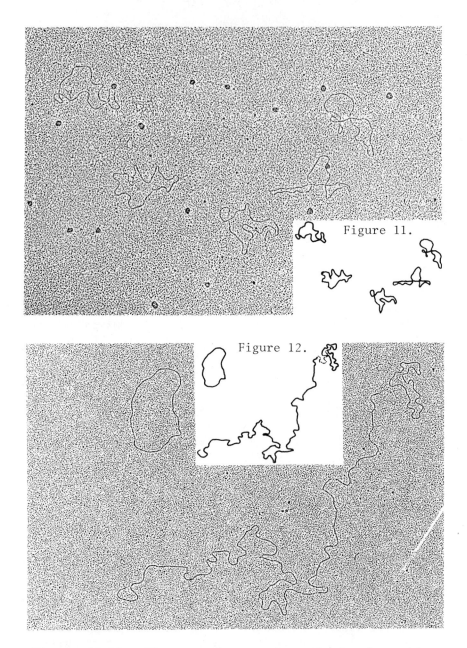

Figure 11.

Figure 12.

Figures 11-14. Electron microscopic analysis of reaction
products. From the reaction described in Fig. 9, 5 μl of
diluted sample was added to 5 μl of freshly prepared
buffer [(4 ml H_2O, 0.3 ml 1M Na_2CO_3, 0.4 ml Na_2EDTA 0.13M);
the high pH of this buffer is designed to eliminate 32
protein binding to single-stranded DNA]. 10 μl of
formamide is then added and gently mixed, followed by

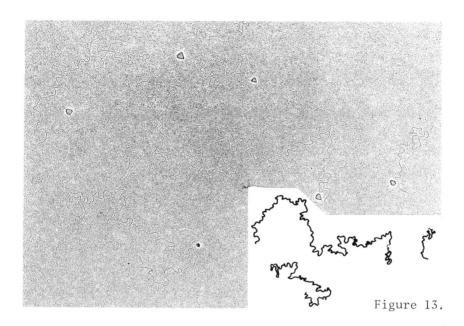

Figure 13.

Figure 14.

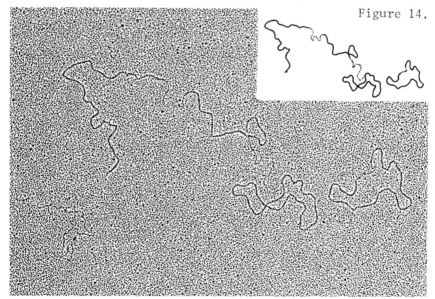

addition of 2 μl of 0.1% cytochrome C. The DNA is then
spread, platinum shadowed, and placed on copper grids
for examination in the electron microscope, essentially
as described by Inman (24). Fig. 11 is from the 5 min
time point, while the remaining micrographs are from
the 10 and 20 min time points.

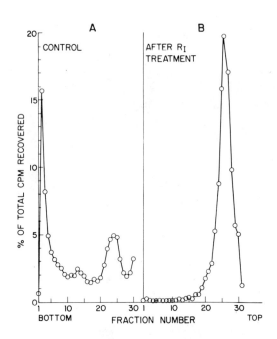

Figure 15. Sedimentation analysis in neutral sucrose gradients of DNA synthesized in vitro using single-stranded G4 phage DNA as template under conditions similar to those described in Fig. 8. The reaction mixture contained 0.2 mM each rCTP, rGTP, and rUTP; and α^{32}P-dGTP was used to label the in vitro product. After incubation at 37°C for 30 minutes a 10 μl aliquot was withdrawn [for part (a)]. Purified (non-specific endonuclease free) EcoR$_I$ restriction endonuclease was added [for part (b)]. After another 15 minutes of incubation at 37°C the reactions were stopped by adding an equal volume of 20 mM Na$_3$EDTA and the samples layered on neutral gradients (5-30% in a buffer containing 20 mM Tris-Cl, pH 8.1, 1 M NaCl, 1 mM EDTA and 0.1% Sarkosyl). A 0.3 ml cushion was made by mixing equal volumes of 80% sodium iothalamate (Angio-Conray) and 30% sucrose. The gradients were centrifuged for 90 minutes at 50,000 rpm and 4°C in a Spinco SW 50.1 rotor.

264

on the T4 genome are required for initiation of new replication forks in vivo.

In contrast, our in vitro T4 system appears to be able
to establish replication forks almost equally well on most
double helical DNA molecules. Although not yet proven, all
our results are consistent with the hypothesis that these
forks arise by strand displacement from a nick, followed by
de novo initiation of complementary strand synthesis on
this displaced single-strand. While this mode of replication fork initiation must be strongly suppressed in vivo
(possibly by special blocking proteins), it conveniently
allows us to bypass the (presumably complex) biological
pathway for T4 fork initiation, and enables the mechanism
of fork propagation to be examined in a relatively simple
in vitro system.

The characteristics of the DNA synthesis described
above make it appear that, despite its apparent non-biolog-
ical start, the propagation reaction seen in vitro closely
mimicks that found in vivo. However, several important
questions remain to be resolved:

(1) Our inference that a short RNA primer is made and
degraded in this system remains to be proven by direct isolation and characterization of such RNA.

(2) In vivo, the DNA on the leading, as well as on
the lagging strand in Fig. 5 appears to be made discontinuously, with about the same average piece size for each
(27). This is perhaps best explained by postulating some
direct coupling of the synthesis on the two strands, such
as that schematically illustrated in the Insert to Fig. 6.
In this scheme, both leading and lagging strand DNA polymerase molecules are joined (via other replication proteins). As soon as the lagging strand polymerase runs out
of single-stranded template to copy (by reaching the 5' end
of the previous Okazaki piece), the postulated structure
must rearrange. This could force interruption of both
leading and lagging strand synthesis simultaneously, with
de novo starts on each strand.

Although we find that most of the (-) strand in the fd
reaction rolls out as a continuous molecule of multiple fd
length (e.g., Fig. 10), sealing of discontinuous starts
on this leading strand could be fast enough to mask their
presence(see article by Sternglanz, this volume). Thus, to
determine whether strand coupling occurs in our in vitro
system will require that we prepare each of the replication

proteins in a completely DNA ligase-free state, a task not yet accomplished.

(3) The strongly synergistic effects of the 6 T4 replication proteins implies that they interact with each other to make a large complex of unique structure. All our attempts to isolate this structure have thus far failed, suggesting that at least some of the interactions are quite weak. Weak interactions might also explain why rather high concentrations of some of the replication proteins are required for maximal activity (see Fig. 7, heading). It is clearly essential to identify all of the interactions and the exact roles of each individual protein in the replication process. This remains as a challenging task for future research.

(4) It is not clear why nature requires so many proteins to make DNA. In particular, one might wonder about the necessity for the 44-62 DNA-dependent ATPase, a large protein of about 180,000 daltons. (Note that DNA-dependent ATPases are also found among the E. coli replication proteins: see article by Wickner and Hurwitz, this volume). From the most general point of view, the efficacy of using bound nucleotide hydrolysis to drive an ordered series of allosteric changes in protein conformation has been recognized for many years (40). For example, consider three conformational states for a protein molecule, determined by bound nucleotide, e.g.:

If hydrolysis of ATP occurs when it is bound to the protein the transition between state I and state II will be strongly biased towards II. Such hydrolysis is thereby sufficient to drive the entire conformational cycle clockwise. In this way, a protein molecule can readily do useful work (e.g., pumping small molecules unidirectionally through a membrane, walking unidirectionally down a DNA chain, etc.). Such miniature "protein machines" play important roles throughout living systems (40). In our case, ATP hydrolysis by the 44-62 complex may serve to set up the clock mechanism needed for a pre-incorporation "kinetic-proofreading" scheme (41). The result would be a substantial increase in the fidelity of the base-pair fit prior to polymerase action, analogous to the fidelity increasing

266

mechanism proposed by Hopfield for several other important biosynthetic reactions (41). Thus, much of the complexity of the replication machinery may reflect the necessity of achieving a very accurate DNA copy.

ACKNOWLEDGEMENTS

We gratefully acknowledge the skillful technical assistance of Monique Davies, Judy Goldberg, Linda McAfee and Mei Lie Wong. This work was supported by grants from the National Institutes of Health (PHS-GM 14927) and the American Cancer Society (NP 152) to Bruce Alberts. Navin Sinha and Fred Morris were supported by postdoctoral fellowships from the National Cancer Institute (fellowship #'s 5F22CA01829 and 6F22CA03266 respectively).

REFERENCES

1. Epstein, R. H., Bolle, A., Steinberg, C. M., Kellenberger, E., Boy de la Tour, E., Chevallez, R., Edgar, R.S. Susman, M., Denhardt, G. H. and Lielausis, A., Cold Spring Harbor Symp. Quant. Biol. <u>28</u>, 375 (1963).

2. Warner, H. R. and Hobbs, M. D., Virology <u>33</u>, 376 (1967).

3. Riva, S., Cascino, A. and Geiduschek, E. P., J. Mol. Biol. <u>54</u>, 85 (1970).

4. Wood, W. B. In "Survey of Genetics" (King, R. C. ed.), Plenum, N. Y. (1973).

5. DeWaard, A., Paul, A. V. and Lehman, I. R., Proc. Nat. Acad. Sci. U.S.A. <u>54</u>, 1241 (1965).

6. Warner, H. R. and Barnes, J. E.,Virology <u>28</u>, 100 (1966)

7. Englund, P. T., J. Biol. Chem. <u>246</u>, 5684 (1971).

8. Goulian, M., Lucas, Z. J. and Kornberg, A., J. Biol. Chem. <u>243</u>, 627 (1968).

9. Nossal, N. G. and Hershfield, M. S., J. Biol. Chem. <u>246</u>, 5414 (1971).

10. Lehman, I. R., Methods in Enzymology <u>29</u>, 46 (1974).

267

11. Alberts, B. and Frey, L., Nature <u>227</u>, 1313 (1970).

12. Delius, H., Mantell, N. and Alberts, B., J. Mol. Biol. <u>67</u>, 341 (1972).

13. Huberman, J., Kornberg, A. and Alberts, B., J. Mol. Biol. <u>62</u>, 39 (1971).

14. Barry, J. and Alberts, B., Proc. Nat. Acad. Sci. U.S.A. <u>69</u>, 2717 (1972).

15. Barry J., Hama-Inaba, H., Moran, L., Wiberg, J. and Alberts, B., In "DNA Synthesis In Vitro" (Wells, R. D. and Inman, R. B. eds.), Baltimore, Md. (1973).

16. Wiberg, J. S., Mendelsohn, S., Warner, V., Hercules, K. Aldrich, C. and Munro, J. L., J. Virol. <u>12</u>, 775 (1973).

17. Karam, J. D. and Bowles, M. G., J. Virol. <u>13</u>, 428 (1974).

18. Brutlag, D. and Kornberg, A., J. Biol. Chem. <u>247</u>, 241 (1972).

19. Chang, L., J. Mol. Biol. <u>93</u>, 219 (1975).

20. Werner, R., Cold Spring Harbor Symp. Quant. Biol. <u>33</u>, 501 (1968).

21. Alberts, B. In "Molecular Cytogenetics" (Hamkalo, B. and Papconstantinou, J., eds.), Plenum. N. Y. (1973).

22. Sugino, A. and Okazaki, R., J. Mol. Biol. <u>64</u>, 61 (1972).

23. Okazaki, R., Okazaki, T., Sakabe, K., Sugimoto, K., Kainuma, R., Sugino, A. and Iwatzuki, N., Cold Spring Harbor Symp. Quant. Biol., <u>33</u>, 129 (1968).

24. Inman, R. and Schnos, M., J. Mol. Biol. <u>56</u>, 319 (1971).

25. Wolfson, J. and Dressler, D., Proc. Nat. Acad. Sci. U.S.A. <u>69</u>, 2682 (1972).

26. Sugimoto, K., Okazaki, T. and Okazaki, R., Proc. Nat. Acad. Sci. U.S.A. 60, 1356 (1968).

27. Sternglanz, R., This Volume (1975).

28. Reichard, P., Eliasson, R. and Söderman, G., Proc. Nat. Acad. Sci. U.S.A. 71, 4901 (1974).

29. Wickner, W., Brutlag, D., Schekman, R. and Kornberg, A., Proc. Nat. Acad. Sci. U.S.A. 69, 965 (1972).

30. Okazaki, R., This Volume (1975).

31. Nossal, N. G., J. Biol. Chem. 249, 5668 (1974).

32. Lee, C. S., Davis, R. W. and Davidson, N., J. Mol. Biol. 48, 1 (1970).

33. Inman, R. B., Schildkraut, C. L. and Kornberg, A., J. Mol. Biol. 11, 285 (1965).

34. Masamune, Y. and Richardson, C.C., J. Biol. Chem. 246, 2692 (1971).

35. Gilbert, W. and Dressler, D., Cold Spring Harbor Symp. Quant. Biol. 33, 473 (1968).

36. Inman, R. B., Methods in Enzymology 29, 451 (1974).

37. Morrow, J. F. and Berg, P., Proc. Nat. Acad. Sci. U.S.A. 69, 3365 (1972).

38. Godson, G. N. and Boyer, H., Virology 62, 270 (1974).

39. Snustad, P. and Conray, L. M., J. Mol. Biol. 89, 663 (1974).

40. Hill, T. L., Proc. Nat. Acad. Sci. U.S.A. 64, 267 (1969).

41. Hopfield, J. J., Proc. Nat. Acad. Sci. U.S.A. 71, 4135 (1974).

42. Mathews, C. K., J. Biol. Chem. 247, 7430 (1972).

43. Laemmli, U. K., Nature 227, 680 (1970).

44. Studier, F. W., J. Mol. Biol. 41, 189 (1969).

STIMULATION OF DNA REPLICATION *IN VITRO* BY
ESCHERICHIA COLI PERIPLASMIC FACTORS

Douglas W. Smith, Catherine Kemper,
and J. Weaver Zyskind
Department of Biology
University of California, San Diego
La Jolla, California 92037

ABSTRACT. Osmotic shock fluid obtained from *E. coli* contains macromolecular factors which stimulate the ATP-dependent semiconservative mode of *in vitro* DNA synthesis. Evidence is presented that these factors come from the *E. coli*
periplasmic space. This stimulation is observed when
assayed using either toluenized cells or lysed agar-embedded
lysozyme spheroplasts as the *in vitro* DNA synthesis system.
Shock fluid obtained from a given *E. coli dna* mutant does
not stimulate *in vitro* DNA synthesis by that mutant. However, in some cases, shock fluid from one class of *dna* mutants does stimulate ATP-dependent *in vitro* DNA synthesis
by another class of *dna* mutants, in a thermosensitive reaction. Gently prepared cell extracts also stimulate ATP-
dependent *in vitro* DNA synthesis, whereas cell extracts
prepared by more severe procedures inhibit this *in vitro*
synthesis. The stimulating activity in shock fluid is
nondialyzable, partially RNase sensitive, and heat sensitive. Several factors may be present in the osmotic shock
fluid, including products of *E. coli dna* genes.

INTRODUCTION

The *E. coli* cell envelope consists of an inner
membrane, a peptidoglycan layer, and an outer membrane
(1-3). The cell permeability properties are controlled
mainly by the inner membrane, and the rigid rod shape of
the cell is maintained primarily by the murein or peptidoglycan layer, a single cross-linked mucopeptide molecule
(4). Ethylenediaminetetraacetate (EDTA) treatment, or
combined EDTA and lysozyme treatment, of *E. coli* cells dis-
turbs the cell surface so that suspension of these treated

cells in low osmotic strength medium causes the release of a select 4% to 5% of the total cell protein into the medium (osmotic shock fluid). These proteins, termed the periplasmic factors, are compartmentalized in the cell surface mainly between the inner membrane and the murein layer in the periplasmic space (5,6). Many hydrolytic enzymes, including alkaline phosphatase, DNA endonuclease I, and ribonuclease I, and transport proteins for sugar and amino acid transport are found among the periplasmic factors (5,6).

We have examined osmotic shock fluid from *E. coli* cells for factors capable of stimulating *in vitro* DNA replication. Both toluenized cells (7) and cells lysed following agar immobilization (8,9) have been used as *in vitro* DNA synthesis systems to assay the effects of these factors on *in vitro* DNA replication. These two systems are representative of the two major classes of *in vitro* DNA synthesis systems capable of ATP-dependent semiconservative DNA synthesis *in vitro* (DNA replication *in vitro*) (10,11,12). For comparative purposes, stimulation of *in vitro* DNA replication by cell extracts prepared by several methods has also been examined.

METHODS

Bacteria and Growth Conditions: The following *Escherichia coli* strains were used: W3110 *thy dna⁺ polA⁺ polB⁺* (obtained from J. Gross); p3478 *thy polA1* (from J. Gross); D110 *thy polA1 endA* (from C. Richardson); MRE601 *thy rnsA* (from D. Dütting); CRT4636 *thi his mal mtl strA polA1 dnaA46* (ts) (from Y. Hirota); CRT2667 *thi his mal strA polA1 dnaB266* (ts) (from Y. Hirota); BT1026 *thy polA1 endA polC1026* (ts) (from F. Bonhoeffer); H10265 *thy polA1 endA polC1026 polB* (from M. Gefter); and HMS83 *thy lys rha lacA polA1 polB* (from C. Richardson). A thymine auxotroph of MRE601 was derived using trimethoprim selection (19). Cells were grown as previously described (9).

In vitro DNA Synthesis Procedures: Procedures for the *in vitro* DNA synthesis systems using agar immobilized cells, including immobilization of cells, growth of immobilized cells in preincubation medium, conversion of such cells into spheroplasts using lysozyme and ethyleneglycol-bis-(β-amino ethyl ether)N,N'-tetraacetate (EGTA), spheroplast lysis, *in vitro* DNA synthesis reaction conditions, and

assay of DNA synthesized *in vitro*, were previously
described (9). The procedure for spheroplast formation
using lysozyme and EDTA was modified from that of Bazill
et al (13). Fragments from 1.0 ml agar preparation were
slowly collected onto a 45 mm Millipore HA filter, slowly
washed with 50 ml cold sucrose solution (0.50 M sucrose,
0.10 M KCl, 25 mM NaCl, 50 mM Tris-Cl pH8) containing 1 mM
EDTA, and then washed with 10 ml cold sucrose solution.
The agar fragments were suspended in 2 ml sucrose solution,
incubated for 5 min at 37°C, 2 ml cold 100 µg/ml egg white
lysozyme, in 50 mM Tris-Cl, pH 8, was added, and the suspension was incubated at room temperature for 5 min and
then at 4°C for 10 min. Toluenized cells were prepared by
a previously described (9) modification of the procedure of
Moses and Richardson (7).

Analysis of DNA: CsCl gradient analysis was as previously
described (9). For sucrose gradient analysis, agar fragments
in the termination mixture were homogenized 10 times in a
small tissue grinder and incubated at 50°C for 30 min following addition of 3 mg solid pronase. Residual agar fragments
were removed by centrifugation (6,000 rpm, 15 min, 4°C),
and the DNA was denatured by addition of 0.1 volume 2 N KOH
to the supernatant fluid. This solution was gently layered
over a 35 ml 5% to 20% linear sucrose gradient in 0.7 M
NaCl, 0.3 M NaOH preformed in an SW27 centrifuge tube. After centrifugation for 10 hrs at 25,000 rpm and 15°C, approximately 1.5 ml fractions were collected and acid-precipitable radioactivity was determined as previously described
(9). Phage T7 (^{3}H)DNA was sedimented in a separate tube as
a marker.

Osmotic Shock Fluid: Using the procedure of Neu and Heppel
(14), the indicated *E. coli* strain was grown in 1000 ml
tryptone broth (17) at 37°C to 5 x 10^8 cells/ml, harvested
(10,000 x g, 10 min, 4°C), and washed twice with 400 ml
cold 30 mM NaCl, 10 mM Tris-Cl, pH 7.4. The cell pellet
was suspended in 20 ml 33 mM Tris-Cl, pH 7.1, and 20 ml
40% sucrose, 33 mM Tris-Cl, pH 7.1, was added with rapid
stirring (solution 1). Following the addition of 0.04 ml
100 mM Na-EDTA, pH 7.1, the solution was slowly stirred at
0°C for 10 min. Following centrifugation (10,000 x g,
10 min, 4°C), the well-drained cell pellet was gently
smeared over the inside of a chilled centrifuge bottle.

The cells were then rapidly suspended in 40 ml cold 0.1 mM
$MgCl_2$, stirred at 0°C for 10 min, centrifuged, and the
supernatant (the osmotic shock fluid; solution 2) was
concentrated to 1 to 2 ml overnight in dialysis tubing using
polyethylene glycol. The pellet of shocked cells was suspended in 40 ml 33 mM Tris-Cl, pH 7.1 (solution 3).
    Some of the characteristics of osmotic shock fluid
from *E. coli* MRE601 cells are shown in Table 1; the solutions are those given above. These results and others show
that about 95% of the shocked cells retain colony forming
ability, that the DNA of such cells is insensitive to added
pancreatic DNase, that approximately 5% of the total cell
protein is recovered in the shock fluid, and that this
shock fluid contains the majority of DNA endonuclease I and
alkaline phosphatase activity, essentially no β-galactosidase activity, and less than 5% of the total cell DNA, properties previously described for osmotic shock fluid (5).

Other Methods: Cell extracts were prepared as previously
described (9). Standard procedures were used for assay of
protein content (15), for alkaline phosphatase activity
(16), and for β-galactosidase activity (17). DNA content
was determined from the acid-insoluble radioactivity of
cells labeled with (^{3}H)thymidine. Relative DNA endonuclease
I activity was determined from the rate of sRNA-inhibitable
decrease in viscosity of 0.25 mg/ml salmon sperm DNA in
0.05 M Tris-Cl, pH 8, 10 mM $MgCl_2$, with or without 100 μg/ml
yeast sRNA. Inhibitor stock solutions were prepared as
previously described (9). Radioactive phage T7 DNA was
prepared according to Thomas and Abelson (18).

RESULTS

Lysozyme Agar *in vitro* DNA Synthesis Systems: When cells
were converted into spheroplasts with lysozyme immediately
after agar immobilization, they exhibited very little *in
vitro* DNA synthesis ability and no ATP-dependent synthesis
(Fig. 1; zero time). However, during growth at 36°C in a
penassay broth-sucrose-Mg^{++} preincubation medium prior to
lysozyme spheroplast formation, the cells formed small
colonies in the agar fragments and the *in vitro* DNA synthesis ability per cell increased dramatically. After an
initial lag, this ability increased nearly linearly with

273

TABLE 1

Properties of Osmotic Shock Fluid[a]

Property	Total Lysate (Solution 1) Units	Shock Fluid (Solution 2) Units Percent		Shocked Cells (Solution 3) Units Percent	
β-galactosidase	48.7	1.7	3.8	44.6	96.2
Alkaline Phosphatase	17.3	16.9	95.5	0.8	4.5
DNA endo- nuclease I	4.2	4.1	93.2	0.3	6.8
Protein	7.30	0.34	4.7	6.90	95.3
DNA	45,500	2,050	4.4	44,300	95.6

[a]Aliquots (1 ml) of each indicated solution using E. *coli*
MRE601 cells were sonicated (three 30 sec treatments, at
0°C), and appropriate dilutions used to assay each indicated property. The results are units per ml of each
solution. Standard units are used for β-galactosidase
(17) and alkaline phosphatase (16) activity; isopropyl-
thio-β-D-galactoside (IPTG) was added to 0.1 mM to the
growth medium to induce the lactose operon. Each unit of
protein is 1 mg per ml solution, using bovine serum
albumin as a standard (15). Each unit of DNA endonuclease
I activity is that which decreases the relative viscosity
50% in 60 min at 37°C. Each unit of DNA is 1 count/min
acid-insoluble radioactivity (9) per ml solution, using
cells labeled with (^{3}H)thymidine.

time, reaching a maximum at 3 to 5 hrs incubation. The ATP-dependent *in vitro* DNA synthesis was 2 to 3 fold higher than the ATP-independent synthesis (Fig. 1). Similar results are observed using either EGTA-lysozyme or EDTA-lysozyme spheroplasts.

The development of this *in vitro* DNA synthesis ability per cell was prevented by incubation in the presence of inhibitors of macromolecular synthesis (Fig. 1). When nalidixic acid, rifampin, or chloramphenicol were present during the entire preincubation period, essentially no development of *in vitro* DNA synthesis ability was observed. When added after 2 hrs preincubation, nalidixic acid caused a decrease in ATP-dependent *in vitro* DNA synthesis ability within 30 min, demonstrating that the observed *in vitro* DNA synthesis is sensitive to nalidixic acid inhibition. When added after 2 hrs preincubation, both rifampin and chloramphenicol initially prevented further increase in the ATP-dependent *in vitro* DNA synthesis ability per cell. The observed decrease in *in vitro* synthesis ability per cell following exposure to these inhibitors for 90 min may reflect their prevention of initiation of new rounds of DNA replication (reviews:11,12). These results suggest that synthesis of one of more RNA and/or protein species during the preincubation period is required for the development of the *in vitro* DNA synthesis ability per cell. Schaller *et al* (20) have shown that one or more nondialyzable components must be present at *in vivo* concentrations for replicative *in vitro* DNA synthesis to proceed. The agar immobilization procedure used here may decrease the concentration of one or more of these components below a required minimal level, so that the preincubation requirement permits the necessary resynthesis of such components.

ATP-Dependent Semiconservative *in vitro* DNA Synthesis:

The properties of the *in vitro* DNA synthesis characteristic of these lysozyme systems preincubated for 3 hrs were examined further in a density labeling experiment using CsCl gradient analysis (Fig. 2). When synthesized *in vitro* in the absence of ATP, the (^{3}H)DNA formed a single band at the buoyant density of thymine-containing *in vivo* synthesized (^{14}C)DNA (Figs. 2b and 2d). Thus, *in vitro* DNA synthesis in the absence of ATP using lysozyme spheroplasts is nonconservative. In contrast, (^{3}H)DNA synthesized *in vitro* in the presence of ATP banded mainly at the

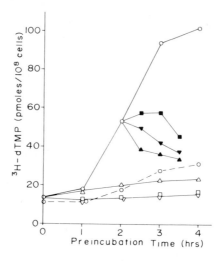

Fig. 1. Development of *in vitro* DNA Synthesis Ability during Preincubation in the Presence and Absence of Inhibitors of Macromolecular Synthesis. Agar immobilized *E. coli* W3110 cells were grown in preincubation medium for the times indicated in the presence or absence of the indicated inhibitors. Aliquots were removed, the cells were converted into EGTA-lysozyme spheroplasts, lysed, and incubated in a standard reaction mixture at 36°C for 25 min. Termination of the reaction and analysis of the DNA synthesized *in vitro* were as previously described (9). Circles: aliquots from the culture grown in the absence of inhibitors, with the reaction mixture containing (O——O) or not containing (O– —O) ATP. Other symbols: aliquots from cultures containing inhibitors, with reaction mixtures all containing ATP. Open symbols: the indicated inhibitor was present in the preincubation medium for the entire time indicated. Closed symbols: the indicated inhibitor was added 2 hrs after growth in preincubation medium was begun. Symbols: O , no inhibitors; △ , 100 µg/ml nalidixic acid added; ▽ , 200 µg/ml rifampin added; □ , 200 µg/ml chloramphenicol added. Copyright J. Bacteriol.

276

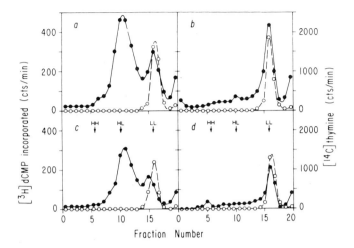

Fig. 2.   Density Labeling Analysis of DNA Synthesized *in vitro* using EGTA- and EDTA-Lysozyme Spheroplasts.
Lysed agar embedded *E. coli* W3110 EGTA-lysozyme and EDTA-
lysozyme spheroplasts preincubated for 3 hrs were incubated in reaction mixtures with and without ATP for 10
min at 36°C.   The reaction mixtures were as described (9),
except that (^{3}H)dCTP, BrdUTP, and deoxycytosine were substituted for (^{3}H)dTTP, dTTP, and thymidine, respectively.
The four reaction mixtures contained lysed EGTA-lysozyme
spheroplasts with (a) or without (b) ATP, or lysed EDTA-
lysozyme spheroplasts with (c) or without (d) ATP.   The
reaction was terminated, and the DNA was extracted and
analyzed via equilibrium sedimentation in CsCl gradients
as described (9).   The positions of *E. coli* native light
DNA (LL: thymine in both strands), hybrid DNA (HL: thymine in one strand, bromuracil in the other), and heavy
DNA (HH:bromuracil in both strands) are indicated by the
arrows shown in representative gradients, as determined
using refractive index determinations.   Symbols: O − −O,
(^{14}C)DNA; ●——●, (^{3}H)DNA.

density of *E. coli* hybrid DNA, with a smaller band near
the position of the light (^{14}C)DNA (Figs. 2a and 2c).
Thus, most of the *in vitro* DNA synthesis observed in the
presence of ATP is semiconservative. Similar results were
obtained using either EGTA-lysozyme spheroplasts (Figs. 2a
and 2b) or EDTA-lysozyme spheroplasts (Figs. 2c and 2d),
showing that both of these lysozyme systems exhibit an
ATP-dependent semiconservative mode of *in vitro* DNA synthesis.

Stimulation of *in vitro* DNA Synthesis by Added Osmotic
Shock Fluid: The effects of added osmotic shock fluid from
two *E. coli* strains deficient in DNA polymerase I on the
kinetics of *in vitro* synthesis using EGTA-lysozyme spheroplasts of cells preincubated for only 1 hr are shown in
Fig. 3. In the absence of added shock fluid (Fig. 3a), a
low rate of ATP-independent *in vitro* synthesis was observed,
typical for cells preincubated for 1 hr (see Fig. 1).
Shock fluid from *E. coli* D110, a strain deficient in DNA
endonuclease I and DNA polymerase I, stimulated the ATP-
dependent *in vitro* synthesis and inhibited the ATP-independent *in vitro* synthesis (Fig. 2b). Shock fluid from
E. coli p3478, a strain deficient in DNA polymerase I,
stimulated the ATP-dependent *in vitro* synthesis and had
little effect on the ATP-independent *in vitro* synthesis
(Fig. 2c). In both cases, ATP stimulated the resulting
in vitro synthesis about 3 fold, in a time-independent
manner.

Using shock fluid from several *E. coli* strains, a
stimulation of the ATP-dependent *in vitro* synthesis exhibited by cells preincubated for 1 hr has always been seen,
whereas little or no stimulation, and often an inhibition,
of the ATP-independent *in vitro* synthesis by such cells is
observed. Shock fluid from some *E. coli* strains also
stimulates the ATP-dependent *in vitro* synthesis exhibited
by cells preincubated for 3 hrs prior to lysozyme treatment. Thus, the level of stimulation for a saturating
amount of shock fluid varies with the *E. coli* source of the
shock fluid. Shock fluid from *E. coli* MRE601, a strain
deficient in ribonuclease I (21), is an excellent source
for the stimulating activity; the concentration dependence
of *in vitro* DNA synthesis on shock fluid from *E. coli*

278

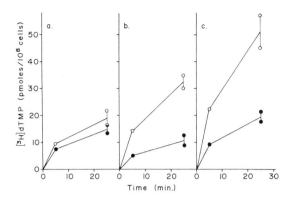

Fig. 3.   Incorporation Kinetics of Osmotic Shock
Fluid Stimulation. Lysed agar embedded *E. coli* W3110 EGTA-
lysozyme spheroplasts preincubated for 1 hr and concentrated osmotic shock fluid were incubated in standard reaction mixtures with (O—O) or without (●—●) ATP at 37°C.
The reaction mixtures contained (a) no added shock fluid;
(b) shock fluid   (90 µg protein) from *E. coli* D110, and
(c) shock fluid (90 µg protein) from *E. coli* p3478. Aliquots from the reactions were terminated at the indicated
times and analyzed for DNA synthesized *in vitro* as previously described (9). Copyright <u>J</u>. <u>Bacteriol</u>.

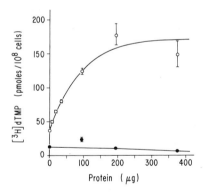

Fig. 4.   Concentration Dependence of Stimulation by
E. coli MRE601 Shock Fluid. Lysed agar embedded *E. coli*
W3110 EGTA-lysozyme spheroplasts preincubated for 1 hr
and concentrated *E. coli* MRE601 shock fluid, in the
amounts indicated, were incubated in standard reaction
mixtures with (O—O) or without (●—●) ATP for 25 min at
37°C. Termination of the reaction and determination of
DNA synthesized *in vitro* were as previously described (9).
The results are the average of duplicate experiments, presented as error bars shown. Copyright <u>J</u>. <u>Bacteriol</u>.

MRE601 is shown in Fig. 4. The ATP-independent *in vitro*
DNA synthesis showed little or no stimulation and some inhibition at high periplasmic factor concentrations. In contrast, the ATP-dependent *in vitro* DNA synthesis was strongly stimulated, with a maximum at about 200 μg protein. The
combined stimulation of the ATP-dependent *in vitro* DNA synthesis and inhibition of the ATP-independent *in vitro* DNA
synthesis resulted in a maximum ATP stimulation of 15-fold.

Periplasmic Factor Stimulation of *in vitro* DNA Synthesis
Exhibited by Toluenized Cells: The degree of stimulation
by periplasmic factors of *in vitro* DNA synthesis by cells
treated with toluene is dependent on the concentration of
cells used in the reaction mixture (Table 2). When high
concentrations were used, little stimulation was observed.
However, as the cell concentration was reduced, a progressively greater stimulation per cell of the ATP-dependent
synthesis was observed. At 10^8 cells per ml reaction mixture, a 2 to 3 fold stimulation of the ATP-dependent *in
vitro* DNA synthesis by osmotic shock fluid was observed,
whereas added shock fluid inhibited the ATP-independent *in
vitro* DNA synthesis (Table 2). Thus, when toluenized
cells are used as the assay system, periplasmic factor
stimulation is also specific for the ATP-dependent mode of
in vitro DNA synthesis, resulting in nearly a 100-fold ATP
stimulation. The rather low degree of stimulation of the
ATP-dependent *in vitro* synthesis is probably due to 1) lack
of free entry of macromolecular species into toluene
treated cells (22), and 2) the presence of the stimulating
species in nearly optimal concentrations in the toluenized
cells. This again shows that an assay system must initially be defective or deficient in the functional activity
under study.

Preferential Stimulation of ATP-dependent Semiconservative
in vitro DNA Synthesis: To further characterize the DNA
synthesized due to addition of osmotic shock fluid, a density labeling experiment using CsCl gradient analysis was
performed (Fig. 5). In the absence of added shock fluid,
gradient profiles similar to those of Fig. 2 were obtained
(Figs. 5A and 5B). Shock fluid from *E. coli* W3110 stimulated the ATP-dependent *in vitro* DNA synthesis about 3-fold,
with nearly an equal stimulation of both (^{3}H)DNA peaks in

TABLE 2

Stimulation of Toluenized Cells
by Osmotic Shock Fluid[a]

Cell Concentration in Reaction Mixture	Shock Fluid	(^3H)dTMP incorporation (pmoles/10^8 cells)	
		+ATP	−ATP
3×10^9 cells/ml	−	87.3	5.6
	+	97.8	2.0
7×10^8 cells/ml	−	103.4	9.9
	+	169.9	2.5
3×10^8 cells/ml	−	110.7	11.3
	+	211.5	2.3

[a]*E. coli* W3110 cells were toluene treated, incubated in standard reaction mixtures in the presence and absence of ATP, with and without 100 µg protein of osmotic shock fluid from *E. coli* MRE601, for 20 min at 37°C, and the reactions were terminated and analyzed as previously described (9). Cell concentrations were decreased as indicated by gentle dilution in TRB buffer (9) after toluene treatment. The results are each the averages of duplicate experiments.

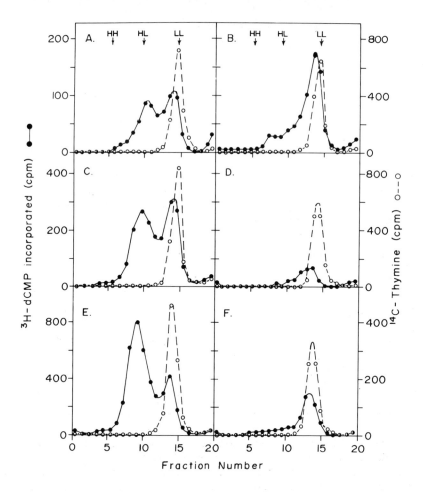

Fig. 5. Density Labeling Analysis of DNA Synthesized *in vitro* in Presence of Osmotic Shock Fluid. Lysed agar embedded *E. coli* W3110 EGTA–lysozyme spheroplasts preincubated for 3 hrs and concentrated osmotic shock fluid (100 μg protein) were incubated in reaction mixtures as described in the legend to Fig. 2 with and without ATP for 10 min at 36°C. The six reaction mixtures contained (A) ATP with no shock fluid, (B) no ATP and no shock fluid, (C) ATP with W3110 shock fluid, (D) ATP with heat treated (95°C, 10 min) W3110 shock fluid, (E) ATP with MRE601 shock fluid, and (F) MRE601 shock fluid but no ATP. The reaction was terminated, and the DNA was extracted and analyzed via equilibrium sedimentation in CsCl gradients are previously described (9). The positions of *E. coli* native light (LL) DNA, hybrid (HL) DNA, and heavy (HH) DNA are indicated (see also Fig. 2). Symbols: O–O, (^{14}C) DNA; ●——●, (^{3}H)DNA. Copyright <u>J</u>. <u>Bacteriol</u>.

the CsCl gradient (Fig. 5C), whereas essentially no stimulation of the ATP-independent *in vitro* DNA synthesis was
observed (data not shown). In contrast, addition of heat
treated *E. coli* W3110 shock fluid inhibited the ATP-dependent *in vitro* DNA synthesis, resulting in no semiconservative DNA synthesis (Fig. 5D). Thus, the stimulating activity present in osmotic shock fluid is heat-sensitive. Shock
fluid from *E. coli* MRE601 cells produced a greater total
stimulation of ATP-dependent *in vitro* DNA synthesis, with a
preferential stimulation of synthesis of hybrid density
(^{3}H)DNA (about 14-fold) compared with the light density
(^{3}H)DNA (about 6-fold) (Fig. 5E). No hybrid density DNA
was synthesized in the ATP-independent reaction (Fig. 5F),
although some stimulation of light density DNA (about 3-fold)
was observed. Thus, addition of periplasmic factors preferentially stimulates the ATP-dependent semiconservative mode
of *in vitro* DNA synthesis.

Effects of Osmotic Shock Fluid on the Size of *in vitro*
Synthesized DNA: Alkaline sucrose gradient analysis permits analysis of the size distributions of the single
strands of DNA, both the (^{14}C)DNA synthesized *in vivo* and
the (^{3}H)DNA synthesized *in vitro*. In the absence of added
osmotic shock fluid, the *in vitro* synthesized (^{3}H)DNA,
synthesized either in the presence of ATP (Fig. 6B) or in
the absence of ATP (Fig. 6A), sedimented at about 10S to
20S relative to a phage T7 DNA marker, whereas the *in vivo*
synthesized (^{14}C)DNA sedimented at about 60S, typical for
denatured sheared DNA. Thus, the *in vitro* DNA product is
synthesized primarily as short pieces of DNA, noncovalently
associated with previously *in vivo* synthesized DNA, again
mimicking DNA replication *in vivo*. Addition of osmotic
shock fluid from *E. coli* MRE601 cells to the reaction mixture resulted in a lower size distribution for the (^{14}C)DNA,
independent of the presence of ATP in the reaction mixture
(Figs. 6C and 6D), indicating that some endonuclease activity is present in the osmotic shock fluid. An increase in
very slowly sedimenting (less than 10S) *in vitro* synthesized
(^{3}H)DNA was also observed. However, some of the DNA synthesized *in vitro* in the presence of ATP (Fig. 6D) sedimented considerably more rapidly (maximum about 40S), with
a sedimentation coefficient comparable to that of the (^{14}C)
DNA. Although these effects are complicated by the

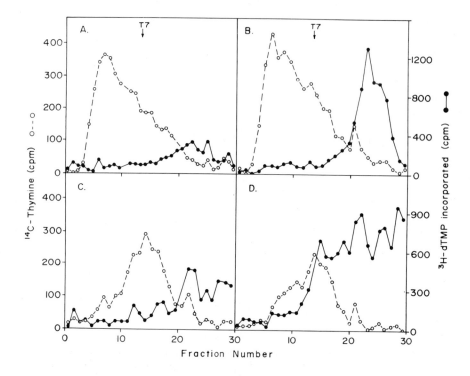

Fig. 6. Alkaline Sucrose Gradient Analysis of DNA Synthesized *in vitro* In Presence of Osmotic Shock Fluid. Lysed agar embedded *E. coli* EGTA–lysozyme spheroplasts incubated for 3 hrs and concentrated *E. coli* MRE601 osmotic shock fluid (100 μg protein) were incubated in reaction mixtures with and without ATP for 25 min at 36°C. The four reaction mixtures contained (A) no ATP and no shock fluid, (B) ATP and no shock fluid, (C) shock fluid and no ATP, and (D) ATP and shock fluid. The reaction was terminated and the DNA was extracted and analyzed via zone sedimentation in alkaline sucrose gradients as described in METHODS. Symbols: O— —O ,(^{14}C)DNA; ●——●, (^{3}H)DNA.

presence of endonuclease activity, these results indicate
that the osmotic shock fluid contains factors capable of
catalyzing the joining together of the *in vitro* synthesized
short DNA pieces. Since this increase in sedimentation
coefficient was not observed for the DNA synthesized *in
vitro* in the absence of ATP (Fig. 6C), the overall joining
reaction appears to be ATP-dependent.

Periplasmic Factors from *E. coli dna* Mutants: The specificity of the stimulation by periplasmic factors was further
examined by testing shock fluid obtained from several *E.
coli dna* mutants (23) for ability to stimulate *in vitro*
DNA synthesis by these same strains (Table 3). Shock fluid
from a given *dna* mutant did not stimulate *in vitro* synthesis by the same mutant for any of the *dna* mutants tested.
However, in some cases, shock fluid from one type of *dna*
mutant did stimulate the ATP-dependent *in vitro* synthesis
exhibited at 30°C by another type of *dna* mutant. For
example, shock fluid from *E. coli* BT1026 stimulated the
ATP-dependent *in vitro* DNA synthesis exhibited by *E. coli*
CRT2667 and CRT4636 at 30°C about 2- to 3-fold (Table 3).
This stimulation was not observed when assayed at the
restrictive temperature of 43°C, although stimulation by
osmotic shock fluid from the *dna+* strain MRE601 was similar
at both temperatures. These observations show that the
stimulation by the periplasmic factors is thermosensitive
when some thermosensitive *dna* mutants are used, providing
further evidence that these effects are physiologically
relevant to *in vivo* DNA replication. Further, the osmotic
shock fluid appears to contain some of the *dna* gene products. Stimulation by *E. coli* MRE601 shock fluid is high
in each of the mutants tested and represents a potential
source for the products of the *dna* genes in question.

Properties of *E. coli* p3478 Shock Fluid: Properties of
osmotic shock fluid obtained from *E. coli* p3478 are given
in Table 4. Omission of sRNA from the reaction mixture
somewhat decreased the osmotic shock fluid stimulated ATP-
dependent *in vitro* DNA synthesis, whereas addition of a
3-fold higher concentration of sRNA had no effect. No
effect of the presence or absence of sRNA in the reaction
mixture on the ATP-independent mode of *in vitro* DNA synthesis was observed. Since sRNA is a competitive inhibitor
of DNA endonuclease I (24), these results, together with

285

TABLE 3

Stimulation of *E. coli dna* Mutants
by Osmotic Shock Fluid[a]

Temperature of Reaction	Shock Fluid Source	(^{3}H)dTMP or (^{32}P)dAMP Incorporation (pmoles/10^9 cells)			
		W3110[b] dna^+	CRT4636 $dnaA46$	CRT2667 $dnaB266$	BT1026 $polC$
	None	190–150[c]	124–70	174–140	113–27
	CRT4636 $dnaA$	184–85	150–78	295–230	----
30°C	CRT2667 $dnaB$	205–65	142–70	213–278	----
	BT1026 $polC$	265–120	250–118	500–185	63–45
	MRE601 dna^+	1250–230	1165–365	1240–400	570–90
	None	332–230	70–150	208–250	----
43°C	BT1026 $polC$	95–27	97–89	160–207	----
	MRE601 dna^+	1207–67	854–343	1150–275	----

[a]Lysed agar embedded *E. coli* EGTA–lysozyme spheroplasts of the indicated strains, preincubated for 1 hr, and concentrated osmotic shock fluid from the indicated *E. coli* strains, were incubated in reaction mixtures with and without ATP for 25 min at 30°C or at 43°C as indicated. The reaction mixtures contained either (^{3}H)dTTP and thymidine, or alpha (^{32}P)dATP and deoxyadenosine (9), and an amount of shock fluid containing 100 to 200 μg protein. The reactions were terminated and analyzed as previously described (9).

[b]Agar embedded *E. coli* strains are given in this row.

[c]The pairs of numbers represent *in vitro* DNA synthesis in the presence and absence of ATP, respectively, and are each the averages of duplicate experiments.

286

TABLE 4

Properties of *E. coli* p3478 Shock Fluid[a]

Reaction Mixture	Shock Fluid	(^3H)dTMP incorporation (pmoles/10^8 cells)	
		+ATP	−ATP
Complete	−	19.0	16.0
Complete	+	52.0	19.0
Minus sRNA	−	19.0	16.5
Minus sRNA	+	35.0	16.0
Plus 3 x sRNA	+	53.0	16.5
Plus ribonuclease	−	19.5	17.0
Plus ribonuclease	+	41.0	16.5
Heat treated shock fluid	+	20.0	14.0

[a]Lysed agar embedded *E. coli* W3110 EGTA-lysozyme sphero-plasts preincubated for 1 hr and osmotic shock fluid from *E. coli* p3478 were incubated in standard reaction mixtures with and without ATP for 25 min at 36°C. Pancreatic ribonuclease (Worthington), prepared at 2 mg/ml in 0.1 M KCl, 10 mM MgCl$_2$, 10 mM Tris-Cl, pH 7.4, was heat-treated (95°C, 10 min) prior to use. Termination of the reaction and determination of acid-insoluble radio-activity and pmoles dTMP incorporated per 10^8 cells were as described (9). The averages of duplicate experiments are presented.

those using *E. coli endA* mutants deficient in DNA endonuclease I (cf. Fig. 3), indicate that DNA endonuclease I is apparently not involved in the stimulation by the periplasmic factors. Addition of pancreatic ribonuclease to the reaction mixture decreased the stimulation by added shock fluid about 30%, whereas heat treatment of the shock fluid prior to addition to the reaction mixture nearly completely inactivated its stimulating activity. Thus, the stimulating activity found in the osmotic shock fluid from *E. coli* p3478 is nondialyzable, heat sensitive, and partially ribonuclease sensitive.

Effects on *in vitro* DNA Synthesis Due to Added Cell Extracts: A comparison of the effects of added cell extracts with the effects of osmotic shock fluid was made to provide evidence that the stimulating factors are released from the cell surface of the bacterial cell rather than from the cytoplasm of a small percentage of lysed cells (Fig. 7). Cell extracts of *E. coli* MRE601 prepared by procedures which extensively shear the bacterial DNA (an EDTA-lysozyme-vortex mixed extract and a sonication extract) strongly inhibit the ATP-dependent *in vitro* DNA synthesis (Fig. 7, dashed curve). In contrast, cell extracts prepared by gentle procedures from which bacterial DNA is removed by centrifugation (nonsheared extracts) exhibited stimulating characteristics similar to those of osmotic shock fluid. Such cell extracts include a freeze-thaw-lysozyme extract (25,26) and a brij-lysozyme extract (13). A several fold stimulation of the ATP-dependent synthesis was observed (Fig. 7, solid curves), whereas no cell extract stimulated or inhibited the ATP-independent mode of *in vitro* DNA synthesis.

The degree of stimulation per µg protein by the nonsheared cell extracts was very similar to that observed using osmotic shock fluid (Fig. 7). If the stimulating activity present in osmotic shock fluid came from the cytoplasm of a few lysed cells, the degree of stimulation per µg protein in total cell extracts should be much greater than that observed for osmotic shock fluid. For example, if the stimulating activity came from the cytoplasm of 3% lysed cells, then the degree of stimulation per µg protein should be approximately 3-fold higher for a total cell extract than for osmotic shock fluid. Such

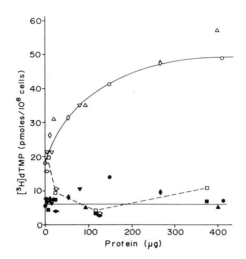

Fig. 7.   Concentration Dependence of Stimulation by
E. coli MRE601 Cell Extracts.   Lysed agar embedded *E. coli*
W3110 EGTA-lysozyme spheroplasts preincubated for 1 hr and
E. coli MRE601 cell extracts prepared as previously
described (9) were incubated in standard reaction mixtures
with (open symbols) or without (closed symbols) ATP for
4 min at 37°C.   Termination of the reaction and determination of DNA synthesized *in vitro* were as previously
described (9).   The results are the average of duplicate
experiments. Symbols: O, osmotic shock fluid;   △,
freeze-thaw-lysozyme extract; ▽, freeze-thaw-lysozyme
extraction of cell pellet resulting from osmotic shock
treatment;   ◊, brij-lysozyme extract; ◇ , EDTA-lysozyme-
vortex mixed extract; □, sonication extract. Copyright
J. Bacteriol.

a stimulation was not observed using any cell extract, thus providing evidence that the stimulating factors do not come from the cytoplasm of a few lysed cells but rather are released from the bacterial cell surface during the shock treatment.

On the other hand, if all of the stimulating factors were released by osmotic shock treatment, the degree of stimulation per µg protein for the shock fluid should be about 20-fold higher than that of a total cell extract, since the shock fluid contains only about 5% of the total cell protein. This also was not observed. The results suggest that some stimulating activity found in total cell extracts is not found in osmotic shock fluid. To test this possibility directly, osmotically shocked cells were further extracted using the freeze-thaw-lysozyme procedure. This nonsheared cell extract exhibited a stimulating activity per µg protein nearly identical with the other nonsheared extracts (Fig. 7). Thus, stimulating factors are released by minimal shear procedures which do not appear to be released by the osmotic shock procedure. Conversely, the stimulating activity found in the osmotic shock fluid does not appear to come from the cytoplasm of a few lysed cells.

DISCUSSION

Evidence is presented that macromolecular species, or factors, can be obtained from within the cell surface of *E. coli* cells which preferentially stimulate the ATP-dependent semiconservative mode of *in vitro* DNA synthesis. This stimulation has been observed in two classes of *in vitro* DNA synthesis systems: 1) in toluene-treated cells (7), in which the permeability properties of the cells are altered but the cells remain essentially intact; and 2) in agar embedded cells which have been converted into spheroplasts using EGTA and lysozyme and then osmotically lysed (9). Thus, the observed stimulation appears to be independent of the *in vitro* DNA synthesis system used to assay the stimulating activity.

Osmotic shock fluid from *E. coli dna*[+] cells contains factors capable of stimulating *in vitro* DNA synthesis by at least three classes of thermosensitive *E. coli dna* mutants (Table 3), as well as the ATP-dependent synthesis exhibited by *E. coli polA* and *polB* mutants (9). In contrast, shock fluid obtained from a given *E. coli dna* mutant does not stimulate *in vitro* DNA synthesis by the same strain. However, shock fluid from *E. coli* BT1026 stimulates the ATP-dependent *in vitro* DNA synthesis exhibited by agar-embedded *E. coli* CRT4636 and *E. coli* CRT2667 at 30°C, but not that exhibited at 43°C. Thus, cross complementation in the sense of stimulation of one type of *dna* mutant by shock fluid from another type of *dna* mutant occurs in some cases, in a thermosensitive reaction. These results imply that the observed stimulation is not a nonspecific property of the shock fluid, and is probably physiologically related to the process of DNA replication *in vivo*.

The shock fluid from *E. coli* MRE601 stimulates the ATP-dependent *in vitro* DNA synthesis by all *E. coli dna* mutants examined (Table 3), suggesting that the replication factors deficient in these mutants may be found to some extent in osmotic shock fluid from wildtype *E. coli* cells. Further, the stimulating activity found in the osmotic shock fluid is heat labile, nondialyzable, and partially ribonuclease sensitive (Table 4). These results suggest that more than one stimulating factor is present in the shock fluid. Since shock fluid from RNase I deficient cells exhibits the greatest stimulating activity, and since the activity is partially ribonuclease sensitive, one or more of the factors may be RNA species. The shock fluid also appears to contain inhibitory activity, such as endonuclease activity (Fig. 6), and the relative content of these different factors in shock fluid obtained from different *E. coli* strains may vary. Purification of the factors is required for definitive clarification of their respective activities. The functional roles of these factors in the steps found in the process of DNA replication have not been determined, although activities which stimulate the overall process of semiconservative synthesis (Fig. 5) and the joining together of the newly synthesized short DNA pieces (Fig. 6) appear to be present.

Evidence has been presented (Fig. 7) showing that the stimulating activity found in the osmotic shock fluid does not arise from the cytoplasm of a small percentage of lysed cells, but is in fact released from the bacterial cell surface during shock treatment. On the other hand, the cell extract studies (Fig. 7) also show that additional stimulating activity, not found in the osmotic shock fluid, is present in these extracts. The relative content of different replication factors in osmotic shock fluid and in nonsheared total cell extracts can best be determined by purification of the factors.

Cell extracts obtained using lysis procedures which shear the bacterial DNA inhibit rather than stimulate the ATP-dependent *in vitro* DNA synthesis. Since these extracts contain nearly all intracellular DNA and β-galactosidase activity (data not shown), they probably also contain most of the cellular stimulating activity. The observed inhibition then is probably due to the simultaneous presence of inhibiting activity. When added to the reaction mixture, purified sheared *E. coli* DNA does not by itself inhibit the ATP-dependent *in vitro* DNA synthesis (data not shown). Cell extracts prepared with minimal shear of the bacterial DNA contain only 50% to 80% of the total cell protein, 30% to 70% of the total β-galactosidase activity, and less than 5% of the total cellular DNA found in extracts containing extensively sheared DNA (data not shown). Thus, the nonsheared extracts possibly do not contain much of the putative inhibitory activities apparently present in the sheared extracts. However, kinetic experiments similar to those of Fig. 3 show a rapid decrease in the initial rate of the ATP-dependent *in vitro* DNA synthesis reaction when nonsheared extracts are included in the reaction mixture, suggesting that these extracts do contain more inhibitory factors than are present in the osmotic shock fluid. Whether the variation in relative amounts of stimulating and inhibitory factors found in these different preparations is due to intracellular compartmentalization is not known, although the nonsheared extracts would probably contain a greater fraction of intracellular "soluble" protein than protein found localized in the cell surface.

Considerable evidence suggests that the origin and growing fork(s) of a semiconservatively replicating DNA molecule are noncovalently associated with a large intracellular structure, possibly a site at the cell surface, in bacteria and in phage-infected bacteria (11,27,28).

The development of methods for separation of the inner
and outer membranes of the *E. coli* cell surface (29) has
permitted the demonstration that phage M13 parental replicative form DNA is associated with the inner membrane (30),
whereas in contrast, phage øX174 parental replicative form
DNA is associated with the outer membrane (31). The *E.
coli* nucleoid, or folded chromosome, when isolated at 4°C,
appears to be associated with a membrane fragment characterized by outer membrane markers (32). Treatment of
toluenized cells with 1% Triton-X100, a mild detergent
which preferentially alters the inner membrane of the *E.
coli* cell surface (33), increases the permeability of
these cells to large molecules, yet without disruption
of the ATP-dependent *in vitro* DNA synthetic capability of
these cells (22). These observations suggest that, at
most, only part of the inner membrane is involved in replication of the bacterial chromosome and that the outer membrane and components found between the two membranes may
also be involved in this process. In addition, replica-
tions factors are apparently lost from growing cells
during the agar immobilization procedure, even though cell
viability is maintained. These factors must be resynthesized during the preincubation growth period for the cells
to exhibit *in vitro* ATP-dependent DNA synthesis ability
(Fig. 1). It would appear likely that these factors are
released from within the bacterial cell surface during
agar immobilization. Such ideas provided the rationale
for examination of osmotic shock fluid for DNA replication
stimulating factors, and it will be of interest to determine the location and function in DNA replication of these
factors.

ACKNOWLEDGMENTS

We gratefully thank Drs. F. Bonhoeffer, J. Cairns,
D. Dutting, M. Gefter, Y. Hirota, and C.C. Richardson for
permission to use their *E. coli* strains for these studies.
This research was supported by a grant, E-625, from the
American Cancer Society.

REFERENCES

1. Tomasz, A. Nature 234:389 (1971).
2. Leive, L. (ed.) "Membranes and Walls of Bacteria". Dekkar, New York. (1973).
3. Braun, V., and Hantke, K. Ann. Rev. Biochem. 43:89 (1974).
4. Ghuysen, J.M., and Shockman, G.D. in "Membranes and Walls of Bacteria". Leive, L. (ed.). Dekkar, New York. p. 37. (1973).
5. Heppel, L.A. in "Structure and Function of Biological Membranes". Rothfield, L.I. (ed.). Academic Press, New York. p. 223. (1971).
6. Rosen, B.P., and Heppel, L.A. in "Membranes and Walls of Bacteria". Leive, L. (ed.). Dekkar, New York. p. 209. (1973).
7. Moses, R.E., and Richardson, C.C. Proc. Natl. Acad. Sci. USA. 67:674 (1970).
8. Smith, D.W., Schaller, H.E., and Bonhoeffer, F.J. Nature 226:711 (1970).
9. Boerner, P., and Smith, D.W. J. Bacteriol.,in press. (1975).
10. Klein, A., and Bonhoeffer, F. Ann. Rev. Biochem. 41: 301 (1972).
11. Smith, D.W. Prog. Biophys. Mol. Biol. 26:321 (1973).
12. Matsushita, T., and Kubitschek, H. Adv. Microb. Physiol., in press. (1975).
13. Bazill, G.W., Hall, R., and Gross, J.D. Nature New Biol. 233:281 (1971).
14. Neu, H.C., and Heppel, L.A. J. Biol. Chem. 240:3685 (1965).
15. Lowry, O.H., Rosebrough, N.J., Farr, A.L., and Randall, R.J. J. Biol. Chem. 193:265 (1951).
16. Garen, A., and Levinthal, C. Biochim. Biophys. Acta, 38:470 (1960).
17. Miller, J.H. "Experiments in Molecular Genetics". Cold Spring Harbor Laboratory, New York. (1972).
18. Thomas, C.A., Jr., and Abelson, J. in "Proc. Nuc. Acid Res." Cantoni, G.L., and Davies, D.R. (eds.). p. 553. (1966).
19. Stacey, K.A., and Simson, E. J. Bacteriol. 90:554 (1965).
20. Schaller, H., Otto, B., Nüsslein, V., Huf, J., Herrman, R., and Bonhoeffer, F. J. Mol. Biol. 63: 183

20. (1972).
21. Gesteland, R.F. J. Mol. Biol. 16:67 (1966).
22. Moses, R.E. J. Biol. Chem. 247:6031 (1972).
23. Gross, J.D. Curr. Topics Microbiol. Immun. 57:39
     (1972).
24. Lehman, I.R., Roussos, G.G., and Pratt, E.A. J.
     Biol. Chem. 237:829 (1962).
25. Schekman, R., Wickner, W., Westergaard, O., Brutlag,
     D., Geider, K., Bertsch, L.L., and Kornberg, A. Proc.
     Natl. Acad. Sci. USA. 69:2691 (1972).
26. Wickner, W., Brutlag, D., Schekman, R., and Kornberg,
     A. Proc. Natl. Acad. Sci. USA 69:965 (1972).
27. Siegel, P.J., and Schaechter, M. Ann. Rev. Microbiol.
     27:261 (1973).
28. Slater, M., and Schaechter, M. Bacteriol. Rev. 38:
     199 (1974).
29. Osborn, M.J., Gander, J.E., Parisi, E., and Carson,
     J. J. Biol. Chem. 247:3962 (1972).
30. Jazwinski, S.M., Marco, R., and Kornberg, A. Proc.
     Natl. Acad. Sci. USA 70:205 (1973).
31. Kornberg, A. "DNA Synthesis". Freeman, San Francis-
     co. p. 258 (1974).
32. Worcel, A., and Burgi, E. J. Mol. Biol. 82:91 (1974).
33. Schnaitman, C.A. J. Bacteriol. 108:545 (1971).

STUDIES ON THE IN VITRO REPLICATION
OF THE ESCHERICHIA COLI CHROMOSOME

Thomas Kornberg
Department of Biochemical Sciences
Princeton University
Princeton, New Jersey 08540

INTRODUCTION

Evidence obtained from in vivo studies of bacteriophage and bacterial chromosome replication indicates that
DNA synthesis proceeds discontinuously (1). We have recently shown that the E. coli replication apparatus can be
reconstructed in vitro and that this in vitro replication
of the E. coli chromosome also proceeds by a discontinuous
mechanism (2). In this system, the newly made DNA fragments, "Okazaki pieces", can be joined together; sealing
of these replicative intermediates to high molecular
weight DNA requires both DNA polymerase I and DNA ligase,
in agreement with other in vivo (3-7) and "quasi in vitro"
(8,9) observations. As reported below, we have further
characterized the synthesis and joining of in vitro synthesized Okazaki pieces, and find that the sealing of
these replication intermediates can be described by a
simple, first order mechanism.

MATERIALS AND METHODS

The preparation and method of assay for DNA synthesis
has been described (2).

RESULTS AND DISCUSSION

Folded E. coli chromosomes (10,11), stabilized by the
polyamine spermidine, can be replicated in a semi-conservative manner in vitro when supplemented with Mg^{++}, deoxynucleoside triphosphates, ATP, and a DNA-free, soluble
enzyme fraction (2). We have identified several factors
in this reaction in the following way. Nalidixic acid,
an inhibitor of in vivo DNA replication (12), substantially inhibited both the initial rate and final extent of DNA
synthesis (Fig. 1A). However, when the standard soluble

TABLE 1

Reactants	Incubation time in minutes	p moles nucleotide incorporated
Complete System	10	13.6
	20	19.2
Complete System + Antiserum to Un- winding Protein	10	2.4
Complete System + Antiserum to Un- winding Protein Added at 10 min	20	16.8
Complete System + Antiserum to co-Pol III*	10	4.8
Complete System + Antiserum to co-Pol III* added at 10 min	20	14.4

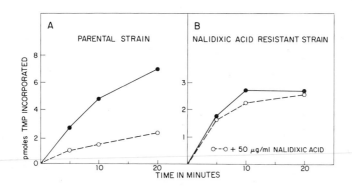

Figure 1 Effect of Nalidixic Acid on DNA synthesis.

enzyme fraction was replaced by an analogous enzyme fraction prepared from an E. coli mutant resistant in vivo to nalidixic acid (200 µg/ml), the DNA synthesis catalyzed in vitro was now resistant to nalidixic acid (Fig. 1B). Thus the nalidixic acid sensitivity of DNA replication depends in vitro, as it does in vivo, on a single mutable component. Addition of antiserum directed against the E. coli DNA unwinding protein (13) or the addition of antiserum directed against co-DNA polymerase III* (14) abolished all incorporation. Addition of either antiserum after the reaction had proceeded for 10 min did not affect DNA already synthesized (Table 1). We therefore conclude that the nalidixic acid sensitive component of the DNA replication apparatus, the DNA unwinding protein, and co-DNA polymerase III* participate in and are required for this in vitro synthesis.

The DNA synthesized in the presence of the folded chromosome and enzyme fraction remains associated with the folded chromosome under conditions of zonal sedimentation in neutral sucrose, but when denatured, it sediments slowly, as small DNA fragments. If, however, DNA polymerase I and DNA ligase are included, the reaction product sediments as high molecular weight DNA during sedimentation in alkaline sucrose (Figs. 2,3). In order to demonstrate that high molecular weight DNA synthesized in the presence of Pol I and DNA ligase is a consequence of the ligation of Okazaki pieces, rather than elongation of pre-existing primer-templates, the fate of the DNA fragments and the capacity to synthesize them at all times during the reaction was examined. Fig. 4A illustrates that after 5 min of incubation, a 0.5 min pulse of [^{3}H]-TTP results in the synthesis of small DNA fragments. Fig. 4B shows that DNA synthesized during a short pulse can be chased into high molecular weight DNA. Thus the kinetics of synthesis observed in the presence of Pol I and DNA ligase are a consequence of the continued initiation synthesis, and sealing of small DNA fragments.

The kinetics of DNA synthesis shown in Fig. 3 indicates that DNA of intermediate molecular weight does not accumulate. This result is in agreement with similar continuous label experiments performed in vivo (1). We have derived a mathematical expression which describes the molecular weight distribution as seen in vivo and in vitro. The first order expression requires only a single rate

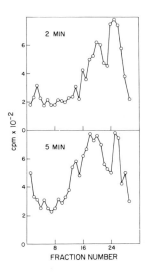

Figure 2

Alkaline sucrose sedimentation
analysis of DNA synthesized in
vitro.

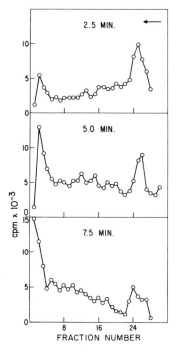

Figure 3
Alkaline sucrose sedimentation
analysis of DNA synthesized in
vitro with added DNA polymerase
I and DNA ligase.

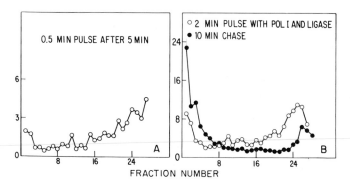

Figure 4   Alkaline sucrose sedimentation analysis of DNA
synthesized in vitro with added DNA polymerase I
and DNA ligase.   A: pulse label; B: pulse-chase.

constant, and was derived in the following way:

One side of a pulse labeled replication fork is schematically represented below in which the Okazaki pieces are synthesized sequentially from left to right.

```
                              pulse              newly
Okazaki                       labeled            synthesized
fragment                      fragment           strand

   i — 3 — 2 — 1 ⌒ 1 — 2 — 3— j —
  _____
                                        template strand
```

$(1 - e^{-kt})$ = the fraction, within a large population of such replication forks, of a specific nick (i.e., the nick #1 adjacent and to the left of the labeled piece) that has been closed.

$k \equiv$ rate constant of sealing

$t \equiv$ time since the labeled piece was completed, measured in units of time required to synthesize a single Okazaki piece.

$$\prod_{i=1}^{\ell} (1 - e^{-k(t + i - 1)})(e^{-k(t + i)}) = \text{the probability}$$

that the labeled piece is sealed to pieces that were synthesized before the pulse and thus is a member of an aggregate $\ell + 1$ pieces long.

$$\prod_{j=1}^{r} (1 - e^{-k(t - j)})(e^{-k(t - (j + 1))}) = \text{the probability}$$

ity that the labeled piece is sealed to pieces that were synthesized after the pulse and thus is a member of an aggregate $r + 1$ pieces long.

The probability that a piece will be a member of an aggregate of any size S pieces long is:

$$\sum_{\ell=2}^{S-1} [\prod_{i=1}^{\ell} (1 - e^{-k(t + i - 1)})(e^{-k(t + i)})$$

$$\prod_{j=1}^{j=S-\ell} (1 - e^{-k(t - j)})(e^{-k(t - (j + 1))})] +$$

$$\prod_{i=1}^{i=S} (1 - e^{-k(t + i - 1)})(e^{-k(2t + i - 1)}) +$$

300

$$\prod_{j=1}^{j=t-1} (1 - e^{-k(t - j)})(e^{-k(2t = 1 - (j + 1))})$$

This expression describes the fate of a given Okazaki piece; it is effectively a 'pulse-chase'. The final expression can be used to calculate the molecular weight distribution of a 'continuous label' from the sum of 'pulse-chase' distributions of equal pulse length and successively shorter chases. Calculated values for 'pulse-chase' and 'continuous label' experiments are shown in Fig. 5, in which the fraction of fragments at any aggregate size 1 - 100 pieces long (10S - 63S) is plotted against fragment size. The continuous label calculation describes an exponential with no discrete intermediates. Although the slope is determined by k and t, the shape will be unaffected by either parameter. Thus intermediate molecular weight pieces will not accumulate and this formulation is in agreement with experimental observation and strongly suggests that the sealing of Okazaki pieces in vitro and in vivo proceeds via a first order reaction mechanism determined by a single rate constant of sealing.

ACKNOWLEDGMENTS

I thank A. Worcel for his encouragement and advice and thank B. Alberts and D. Cleveland for their help in the derivations and programming. This work was supported by National Science Foundation grant No. GB-43134X.

REFERENCES

1. Okazaki, R., Okazaki, T., Sakabe, K., Sugimoto, K. and Sugino, A. Proc. Nat. Acad. Sci. 59, 598 (1968).
2. Kornberg, T., Lockwood, A. and Worcel, A. Proc. Nat. Acad. Sci. 71, 3189 (1974).
3. Kuempel, P.L. and Veomett, G.E. Biochem. Biophys. Res. Commun. 41, 973 (1970).
4. Okazaki, R., Arisawa, M. and Sugino, A. Proc. Nat. Acad. Sci. 68, 2954 (1971).
5. Konrad, E.B. and Lehman, I.R. Proc. Nat. Acad. Sci. 71, 2048 (1974).
6. Konrad, E.B., Modrich, P. and Lehman, I.R. J. Mol. Biol. 77, 519 (1973).
7. Gottesman, M.M., Hicks, M.L. and Gellert, M. J. Mol.

Biol. <u>77</u>, 531 (1973).

8. Moses, R. and Richardson, C.C. Proc. Nat. Acad. Sci.
    <u>67</u>, 674 (1970).

9. Schaller, H., Otto, B., Nusslein, V., Huf, J. Herrmann,
    R. and Bonhoeffer, F. J. Mol. Biol. <u>63</u>, 183 (1972).

10. Stonington, G.O. and Pettijohn, D.E. Proc. Nat. Acad.
    Sci. <u>68</u>, 6 (1971).

11. Worcel, A. and Burgi, E. J. Mol. Biol. <u>71</u>, 127 (1972).

12. Cook, T.M., Brown, K.G., Boyle, J.V. and Gross, W.A.
    J. Bacteriol. <u>92</u>, 1510 (1966).

13. Sigal, N., Delius, H., Kornberg, T., Gefter, M. and
    Alberts, B.M. Proc. Nat. Acad. Sci. <u>69</u>, 3537 (1972).

14. Wickner, W. and Kornberg, A. J. Biol. Chem. <u>249</u>,
    6244 (1974).

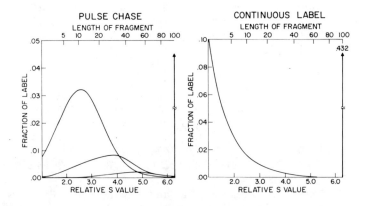

Figure 5   Calculated molecular weight distribution of
           Okazaki pieces.
           A. Pulse chase: pulse length = 1; time = 50, 75,
              100; k = .05
           B. Continuous label: pulse length = 50; time =
              50; k = .1

Part IV
Intermediates in DNA Replication

MUTANTS OF <u>ESCHERICHIA</u> <u>COLI</u> DEFECTIVE IN THE JOINING OF NASCENT DNA FRAGMENTS

E. Bruce Konrad and I. R. Lehman, Department of Biochemistry, Stanford University School of Medicine, Stanford, California 94305

ABSTRACT

A new screening procedure has been developed which has permitted the isolation of mutants in which the joining of nascent DNA fragments is retarded. In addition to DNA ligase and DNA polymerase I defectives, these mutants include a group which, during short pulses, incorporate $[^3H]$ thymidine into DNA fragments that are substantially smaller than Okazaki fragments. These small fragments can be chased into DNA of high molecular weight, and thus may be precursors in DNA replication. During longer pulses, label also appears in DNA of higher molecular weight, though at an abnormally slow rate. The mutations map at a previously undescribed locus (<u>dnaS</u>) at 72 minutes on the <u>E. coli</u> chromosome.

INTRODUCTION

We have developed a screening procedure for the isolation of mutants that are defective in the joining of nascent DNA fragments ("Okazaki fragments") (1). This screening procedure, the details of which will be reported elsewhere (2), is based on the prediction that the retarded joining of Okazaki fragments will stimulate recombination by increasing the frequency of nicks or gaps in the chromosome. The "hyper rec" mutants that we have isolated in this way include ligase defectives, DNA polymerase I defectives (3), and a third, novel class that has not heretofore been described.

This paper will summarize our current information regarding the latter group of mutants. A more detailed study will appear elsewhere (4). The characteristic feature of these mutants, which we have designated <u>dnaS</u>, is that following short pulses of $[^3H]$ thymidine, label is incorporated into nascent DNA fragments that are

304

substantially smaller than the Okazaki fragments found in wild type, or DNA polymerase I or DNA ligase mutants (5-8).

RESULTS
Isolation of dnaS mutants.   Strains carrying the dnaS mutation were isolated by screening hyper rec mutants, obtained at 30º, successively for temperature-sensitive conditional lethality at 44º, capacity to form filaments at 44º, and abnormal accumulation of label in small DNA fragments following short pulses of [^{3}H]thymidine.   Two of the mutants, dnaS1 and dnaS3, were induced by ethyl-methane sulfonate, and the third, dnaS2, was induced by nitrosoguanidine.   Although the dnaS mutants were isolated as conditionally lethal at 44º, in each case secondary mutations were responsible for the lethality, since on transfer of the dnaS mutation to a wild type background this property was lost.

The following results were obtained with a dnaS1 mutant, but qualitatively similar results were found with dnaS2 and dnaS3 mutants.

Accumulation of short DNA fragments in dnaS mutants. At 44º, the dnaS mutant incorporates [^{3}H] thymidine into DNA fragments with sedimentation coefficients in alkaline sucrose gradients progressing from 3.0S, for a 1-second pulse, to 4.4S for a 10-second pulse, indicating that they are considerably smaller in size than the 5.1S to 8.2S Okazaki fragments labeled during equivalent pulses with a wild type strain, or accumulated by strains with deficiencies in DNA ligase or polymerase I (5-8).   The [^{3}H] thymidine in these smaller fragments, like that in Okazaki fragments, appears in higher molecular weight DNA (>20S) after a 2-minute chase with unlabeled thymidine. During longer pulses, the rate of appearance of [^{3}H] thymidine in high-molecular-weight DNA is considerably retarded in the dnaS mutant in a manner analogous to strains deficient in DNA ligase or DNA polymerase I.

Other characteristics of dnaS mutants.   The levels of DNA ligase and DNA polymerase I (polymerase and 5'→3'

exonuclease activities) in extracts of the dnaS and wild
type strains are indistinguishable. Besides accumula-
ting very small fragments, these mutants differ phenotypically from DNA polymerase I and DNA ligase mutants
in their resistance to methylmethane sulfonate, their
slight sensitivity to ultraviolet irradiation, and capacity
to support plaque formation by λ phage defective in general recombination (λ red⁻) (3, 5, 6, 9, 10, 11).

    The dnaS locus is closely linked to pyrE at 72 minutes.
The hyper rec phenotype associated with the dnaS1 mutation is transferred as an early marker by Hfr strains
PK3 and KL228 but not by Hfr strain R1 (12). This local-
izes dnaS1 between the origins of PK3 and R1. Further
mapping within this interval by transduction with the generalized transducing phage P1 established very close
linkage between dnaS1 and the pyrE locus (12). We have
not determined the order of pyrE and dnaS. The linkage
of dnaS1 and of dnaS2 and dnaS3 to pyrE has been confirmed by demonstrating that the capacity to accumulate
very small fragments as well as the hyper rec character
is cotransduced with pyrE. We tentatively assign all of
these mutations to the same locus. The dnaS1 mutation
is recessive, since introduction of an episome carrying
the wild type locus (F'111) restores the wild type phenotype to a dnaS1 mutant (12).

DISCUSSION

    We have described a group of E. coli mutants (dnaS)
with a novel defect in DNA replication. During short
pulses, dnaS mutants incorporate [³H] thymidine into
DNA fragments much smaller than the Okazaki fragments
found in wild type strains, and accumulated by mutants
defective in DNA ligase or DNA polymerase I. The [³H]
thymidine can be chased into high molecular weight DNA,
thus, the dnaS fragments may be intermediates in DNA
replication. DNA replicative intermediates of comparable size have been found in dnaB mutants (13), and in
ether-treated E. coli (14). Slower incorporation of [³H]
thymidine into high molecular weight DNA by dnaS

306

mutants indicates that the very small fragments persist longer than Okazaki fragments and suggests the mutants are also defective in joining these fragments. The dnaS mutations are closely linked to pyrE at 72 minutes on the E. coli map. At least one of them is recessive.

If the very small fragments occur as intermediates in wild type E. coli, they may be converted to Okazaki fragments either via joining by an unknown ligase, or via extension by chain propagation.

Alternatively, the dnaS mutation might alter a component of the replication machinery in such a way as to shorten the length of Okazaki fragments. If this were the case, the very small fragments would represent abnormally short Okazaki fragments that are specific to these mutants. At the present time we cannot distinguish between these three possibilities.

REFERENCES

1. Okazaki, R., Okazaki, T., Sakabe, K., Sugimoto, K., Kainuma, R., Sugino, A. and Iwatsuki, N. (1968) Cold Spring Harbor Symp. Quant. Biol. 33, 129-142.
2. Konrad, E. B., in preparation.
3. Konrad, E. B. and Lehman, I. R. (1974) Proc. Nat. Acad. Sci. USA 71, 2048-2051.
4. Konrad, E. B. and Lehman, I. R., in press, Proc. Nat. Acad. Sci. USA.
5. Konrad, E. B., Modrich, P. and Lehman, I. R. (1973) J. Mol. Biol. 77, 319-529.
6. Gottesman, M., Hicks, M. and Gellert, M. (1973) J. Mol. Biol. 77, 531-547.
7. Kuempel, P. L. and Veomett, G. E. (1970) Biochem. Biophys. Res. Commun. 41, 973-980.
8. Okazaki, R., Arisawa, M. and Sugino, A. (1971) Proc. Nat. Acad. Sci. USA 68, 2954-2957.
9. Pauling, C. and Hamm, L. (1968) Proc. Nat. Acad. Sci. USA 60, 1495-1502.
10. De Lucia, P. and Cairns, J. (1969) Nature 224, 1164-1166.

11. Glickman, B. , van Sluis, C. A. , Heijneker, H. L. and Rörsch, A. (1973) Mol. Gen. Genet. 124, 69-82.
12. Low, B. (1972) Bacteriol. Rev. 36, 587-607.
13. Lark, K. G. and Wechsler, J. , personal communication.
14. Hess, U. , Dürwald, H. and Heffman-Berling, H. (1973) J. Mol. Biol. 73, 407-423.

Evidence That Both Growing DNA Chains at a Replication
Fork Are Synthesized Discontinuously

Rolf Sternglanz, Helen F. Wang, and James J. Donegan

Department of Biochemistry, State University of New York,
Stony Brook, N.Y. 11794

ABSTRACT

E. coli and B. subtilis have been labeled with short
pulses of ^{3}H-thymidine and the labeled DNA examined by
sedimentation in alkaline sucrose. In both systems, the
great majority of the DNA labeled by a short pulse is found
in the form of small DNA chains of 10S, the so-called
Okazaki pieces. The B. subtilis nascent DNA fragments hybridize with equal efficiency to the separated strands of
B. subtilis. The results suggest that both growing DNA
chains at a given replication fork are synthesized discontinuously in the case of both E. coli and B. subtilis. We
have found that the method used to terminate the pulse
determines the fraction of labeled DNA found as short DNA
fragments. Previous reports of discontinuous DNA synthesis
on only one growing DNA chain and continuous synthesis on
the other DNA chain are probably due to preferential
joining of Okazaki pieces after termination of the pulse,
on the DNA chain growing in the overall $5' \longrightarrow 3'$ direction.

INTRODUCTION

The well-known work of Okazaki and coworkers has
shown that short pulses of ^{3}H-thymidine are incorporated
into small DNA chains, the so-called Okazaki pieces, in a
variety of bacterial and bacteriophage systems (1,2).
Pulse-chase experiments have demonstrated that the Okazaki
pieces are the precursors of long DNA chains; the latter
are indistinguishable in length from the bulk of the DNA.

Thus the idea has arisen that one or both of the two growing DNA strands at a replication fork are synthesized discontinuously, and later joined together by ligase.

The question of whether one or both growing DNA chains
at a given replication fork are synthesized discontinuously
is still not settled. Okazaki et al. (1,2) reported that
label from a very short pulse appears almost exclusively
in small pieces in E. coli and T4, implying that both
strands are made discontinuously. In three bacteriophage
systems, T4, λ, and SPP1, Okazaki pieces hybridize equally
to both of the separated DNA strands (3,4,5). However,
since DNA is now known to replicate bidirectionally in
many systems, hybridization data alone cannot be
definitive in settling the question.

There have been reports in the case of E. coli (6,7)
and B. subtilis (8,9) that approximately half the label in
a short pulse is found in large DNA chains. DNA hybridization studies as well as other methods have led these
workers to conclude that only one DNA strand at a replication fork is synthesized discontinuously in the form of
Okazaki pieces, while the other strand is synthesized
continuously.

In this paper we report on pulse-labeling studies of
E. coli and B. subtilis. In both cases, the great
majority of the DNA labeled by a short pulse under our
conditions is in the form of small DNA fragments; i.e.,
Okazaki pieces. We have found that the method used to
terminate the pulse determines the fraction of labeled
DNA fragments. Previous reports of discontinuous DNA
synthesis on only one DNA chain and continuous synthesis
on the other DNA chain are probably due to preferential
joining of Okazaki pieces after termination of the pulse,
on the DNA chain growing in the overall 5'→3' direction.

METHODS

Growth, pulse-labeling and lysis of E. coli. Cells
were grown in M9 medium supplemented with 0.4% glucose and
0.1% casamino acids. The medium for E. coli 15 TAU was
supplemented with 2 μg/ml thymine, 20 μg/ml uracil,
50 μg/ml arginine, and 80 μg/ml tryptophan. The medium

for E. coli LC 434 (7) was supplemented with 2 μg/ml thymine, 25 μg/ml leucine, and 1 μg/ml thiamine. Cells were grown at 37° to 1.5 x 10⁸ cells/ml, switched to 25° and grown to 2.5 x 10⁸ cells/ml, diluted 1:1 with fresh medium and again grown at 25° to 2.5 x 10⁸ cells/ml. Five ml aliquots of the culture were pulse-labeled at 25° by the addition of 50 μl of 1 mC/ml ³H-thymidine (7 C/mmole, Nuclear Dynamics). Pulses were terminated with 20 ml of -10° acetone. The cells were pelleted, washed in 6 ml of cold 0.02 M KCN, 0.02 M EDTA, pH = 7.5, and re-pelleted. They were then resuspended in 0.1 ml of 0.02 M EDTA, pH = 7.5, and lysed by the addition of 0.5 ml of 0.3 M NaOH, 0.02 M EDTA and 50 μl of 10% sarkosyl (Geigy Industrial Chemicals). The lysates were placed in boiling water for 2 min, and cooled quickly in ice water.

Growth, pulse-labeling and lysis of B. subtilis. B. subtilis 168 thy⁻ trp⁻ (10) was grown in a minimal medium described previously (11). Cells were grown at 37° to 1.5 x 10⁸ cells/ml, switched to 25° and grown to 2 x 10⁸ cells/ml, diluted 1:1 with fresh medium and grown at 25° to 3 x 10⁸ cells/ml. Ten ml aliquots of the culture were pulse-labeled at 25° by the addition of 0.1 ml of 1 mC/ml ³H-thymidine (7 C/mmole). Pulses were terminated with 40 ml of -10° acetone. The cells were pelleted, washed in 12 ml of cold 0.02 M KCN, 0.02 M EDTA, pH = 7.5, and re-pelleted. They were then resuspended in 0.1 ml of 0.02 M EDTA, 20% sucrose. Ten μl of 1 mg/ml lysozyme (Sigma) was added and the mixture incubated at 37° for 10 min. The cells were lysed by the addition of alkali and sarkosyl exactly as described above for E. coli.

Alkaline sucrose gradient sedimentation. The lysates described above were layered on top of linear 5 to 20% (w/v) alkaline sucrose gradients containing 0.3 M NaOH, 0.5 M NaCl, 0.01 M EDTA. At the bottom of each sucrose gradient (10.8 ml) was a 1 ml CsCl shelf (ρ = 1.74 gm/cm³; 5.8 gms CsCl dissolved in 4.2 ml of 0.3 M NaOH, 0.5 M NaCl, 0.01 M EDTA). Sedimentation was in a Beckman L2-65B ultracentrifuge for 16 hr at 28,000 rpm using an SW41 rotor and cellulose nitrate tubes. Centrifuge tubes were fractionated from the top and the acid-precipitable radioactivity in each fraction determined as described previously (11), except that 17 equal fractions were collected

311

from each tube. Also, the bottom of each tube was washed
with 1 ml of 0.3 M NaOH, 0.5 M NaCl, 0.01 M EDTA and then
TCA-precipitated. This was called the 18th fraction of
each gradient.

Some gradients contained ^{32}P-labeled denatured DNA
sedimentation markers. SV40 DNA converted to the linear
form by the Eco R1 restriction endonuclease (12), a gift
of C. Mulder, and T7 DNA, a gift of F.W. Studier, have
sedimentation coefficients of 16 S and 37 S, respectively.

RESULTS

Pulse-labeling of E. coli. In order to re-examine the
question of whether one or both growing DNA strands at a
replication fork are synthesized discontinuoully in bacteria, we wanted a DNA pulse-labeling procedure designed to
minimize nuclease or ligase action after termination of
the pulse. Also, it seemed desirable to label cells at the
same temperature as had been used for growth of the cells,
rather than to shift the temperature just before the pulse.
Our procedure is summarized below; details are given in
Methods. Cells are grown exponentially at 37°, diluted
with fresh medium and then grown at 25° for about two
generations before pulse-labeling with ^{3}H-thymidine. The
pulse is terminated with -10° acetone, the cells washed in
0.02 M KCN, 0.02 M EDTA, and finally lysed with alkali and
sarkosyl. The lysates are immediately sedimented through
alkaline sucrose gradients.

Fig. 1 presents the results of pulse-labeling two E.
coli strains, B and 15 TAU, for 15 sec and 13 sec, respectively, using the procedure described above. It can be
seen that the great majority of the pulse-labeled DNA
sediments at the standard position for Okazaki pieces,
about 10 S. Specifically, the fraction of the total radioactivity in Okazaki pieces (<20 S, in fractions 2-8) is 66%
for E. coli B and 75% for E. coli 15 TAU. For both
strains, the fraction of the total radioactivity in large
DNA (>40 S, at the shelf in fractions 16-18) is only 4%.
These results are in agreement with Okazaki's work (1) and
clearly suggest that both strands are synthesized discontinuously at a given replication fork.

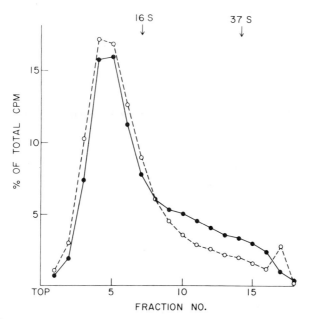

Figure 1
    Alkaline sucrose gradient sedimentation of pulse-labeled DNA from E. coli B and E. coli 15 TAU. Cells were pulse-labeled with ³H-thymidine at 25° and the pulse terminated with -10° acetone as described in Methods. E. coli B (●——●), 15 sec pulse, 7693 total CPM. E. coli 15 TAU (o----o), 13 sec pulse, 13214 total CPM.

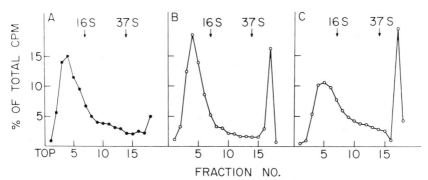

Figure 2
    Alkaline sucrose gradient sedimentation of pulse-labeled DNA from E. coli LC 434. A: Cells pulse-labeled for 15 sec at 25° and the pulse terminated with -10° acetone as described in Methods (●——●), 5343 total CPM. B and C: Cells grown at 37°, pulse-labeled at 20°, the pulse terminated and the cells washed with 10% pyridine, 1 mM KCN as described by Louarn and Bird (7). B: (o——o) 12 sec pulse, 1012 total CPM. C: (o——o) 30 sec pulse, 10084 total CPM.

Table 1

Pulse-Labeling of E. coli LC 434 Under Various Conditions[a]

Exp't.	Growth conditions[b]	Stop solution	Pulse time (sec)	Total CPM	% of total CPM in fractions 2-8 (Okazaki pieces)	% of total CPM in fractions 16-18 (large DNA)
1	Medium A, 25°	acetone	15	5,108	71%	8%
	Medium A, 25°	pyridine	15	2,931	66%	14%
2	Medium A, 25°	acetone (Fig. 2A)	15	5,343	67%	10%
	Medium A, 25°	pyridine	15	5,870	59%	20%
3	Medium B', 37°→20°	acetone	8	486	79%	5%
	Medium B', 37°→20°	pyridine	8	274	80%	16%
4	Medium A, 37°→20°	acetone	12	1,449	70%	15%
	Medium A, 37°→20°	pyridine	12	1,484	64%	19%
	Medium A, 37°→20°	pyridine	30	18,275	50%	27%
	Medium B, 37°→20°	acetone	12	1,629	68%	12%
	Medium B, 37°→20°	pyridine (Fig. 2B)	12	1,012	65%	20%
	Medium B, 37°→20°	pyridine (Fig. 2C)	30	10,084	50%	24%
Ref. 7[c]	Medium B, 37°→20°	pyridine	30	277,000	42%	29%

[a]Cells were grown, pulse-labeled, and lysed as summarized in the Table, and described in Methods. The lysates were sedimented through alkaline sucrose gradients, and the fraction of the total radioactivity in Okazaki pieces and in large DNA determined. See Fig. 1 and 2 for typical sucrose gradients.

[b]Medium A is M9 supplemented with 2 μg/ml thymine and 0.1% casamino acids. Medium B' is supplemented with 4 μg/ml thymine and 0.1% casamino acids. Medium B is supplemented with 4 μg/ml thymine and 0.5% casamino acids (7). In all cases, 25 μg/ml leucine and 1 μg/ml thiamine were also present.

[c]This row summarizes the data calculated from the alkaline sucrose gradient depicted in Fig. 2A of Louarn and Bird (7).

314

On the other hand, a recent paper by Louarn and Bird
(7) concluded that E. coli DNA is synthesized discontinuously on only one strand, at least for pol A⁺ strains. We
have examined pulse-labeled DNA from the strain used by
Louarn and Bird, E. coli K12, strain LC 434 (a generous
gift of R. Bird). Fig. 2A shows the labeling pattern for
LC 434 pulsed for 15 sec under our usual conditions. The
result is similar to that seen for E. coli B and E. coli
15 TAU; 67% of the radioactivity is in Okazaki pieces
(fractions 2-8) and only 10% is in large DNA (fractions
16-18). Louarn and Bird reported a different labeling
pattern for strain LC 434. They found that a large fraction of the DNA labeled during a 30 sec pulse at 20° sedimented as large DNA; even with a 5 sec pulse at 20° at
least 40% of the DNA sedimented with S>15 in their hands.
Their procedure differed from ours in three major ways:
1) their supplemented M9 growth medium was slightly richer
than ours (generation time was 30 min rather than 35 min
at 37°), 2) they grew the cells at 37°, shifted them to
20° for 15 min and then pulsed at 20°, and 3) they terminated pulses with 10% pyridine, 1 mM KCN and washed the
cells in the same solution.

When we use the procedure of Louarn and Bird to label
strain LC 434 we obtain results similar to theirs. Fig.
2B and 2C show 12 sec and 30 sec pulses for strain LC 434
labeled using their conditions. A significant amount of
radioactivity is seen in large DNA (fractions 16-18) for
both pulses. In order to determine whether the growth
medium, the temperature, or the solution used to terminate
the pulse (hereafter called the stop solution) was causing
the difference in the labeling pattern (contrast Fig. 2A
with 2B and 2C) we systematically varied the labeling procedure. Table 1 summarizes the results of four experiments.
The quantities to be compared are the % of the total radioactivity found in Okazaki pieces (<20 S) and in large DNA
(>40 S) under different conditions. Table 1 clearly shows
that the stop solution is an important variable. In all
cases, the fraction of the pulse-labeled DNA sedimenting
to the shelf is greater in pulses terminated with pyridine
than in comparable pulses terminated with acetone. On the
other hand, the growth medium does not seem to be a significant variable (exp't 4, Table 1; compare Medium A with
Medium B). Comparison of experiments 1 and 2 (cells grown

315

and pulsed at 25°) with experiments 3 and 4 (cells grown at 37° and pulsed at 20°) suggests that shifting the temperature from 37° to 20° before pulse-labeling also leads to a greater fraction of large DNA at the shelf. Table 1 also compares the fraction of total radioactivity in Okazaki pieces and in large DNA for a 30 sec pulse stopped with pyridine (exp't. 4) with the same pulse performed by Louarn and Bird (7). One can see that there is good agreement between their data and ours.

In conclusion, the data of Fig. 2 and Table 1 show that strain LC 434 gives results similar to other E. coli strains when pulse-labeled under our conditions; most of the radioactivity is found in Okazaki pieces (<20 S). As argued in detail in the Discussion, we believe that the acetone stop solution and the KCN-EDTA washing of the cells prevent joining of Okazaki pieces after termination of the pulse. In contrast, pyridine and KCN do allow some joining to occur.

Pulse-labeling of B. subtilis and hybridization of the nascent DNA with B. subtilis DNA separated strands. Okazaki and coworkers have reported that when B. subtilis is labeled with short pulses of ^{3}H-thymidine some of the radioactivity is found in 10 S fragments and the rest is found in large DNA (8,9). When they hybridized the labeled DNA with the separated strands of B. subtilis DNA they found that the 10 S pieces hybridized preferentially to the L strand and the large DNA hybridized preferentially to the H strand. Okazaki and coworkers therefore concluded that B. subtilis DNA is replicated discontinuously off the L template strand and continuously off the H template strand.

However, in our hands the same results are obtained with B. subtilis as with E. coli. The great majority of the radioactivity incorporated during a short pulse labels Okazaki pieces and very little radioactivity is found in large DNA. Fig. 3 shows the results of 8 sec and 15 sec pulses of B. subtilis labeled using the conditions described in this paper (25° growth and labeling, acetone stop solution, KCN-EDTA washing of the cells). These results suggest that both growing DNA chains at a replication fork can be replicated discontinuously in B. subtilis.

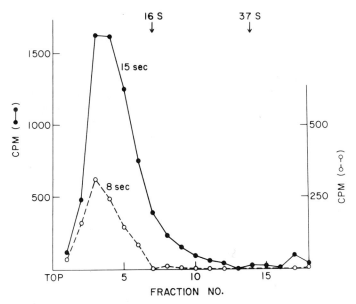

Figure 3
    Alkaline sucrose gradient sedimentation of pulse-
labeled DNA from B. subtilis 168 thy⁻ trp⁻. Cells were
pulse-labeled at 25° and the pulse terminated with -10°
acetone as described in Methods. (o----o) 8 sec pulse.
(●——●) 15 sec pulse. Note the different ordinates for
the two pulses.

                            Table 2

    Hybridization of B. subtilis Nascent DNA With the
        Separated Strands of B. subtilis DNA

| | % Hybridization to | |
Experiment	L Strand	H Strand
1	18%	18%
2	29%	30%

    Both experiments involved a 15 sec pulse at 25°. The
distribution of radioactivity was very similar to that
seen for the 15 sec pulse in Fig. 3. Fractions from the
peak of radioactivity in the sucrose gradient were pooled,
concentrated, and used for hybridization as described in
ref. 16. Experiment 1 involved the use of 80 ml of cells
concentrated to 5 ml before pulse-labeling as described
previously (11). Experiment 2 used unconcentrated cells.

$$\% \text{ hybridization} = \frac{(\text{CPM remaining on filter})(100\%)}{(\text{CPM added to filter})}$$

If both DNA chains are replicated discontinuously then the Okazaki pieces should hybridize to both of the separated strands of B. subtilis DNA. This is true even if B. subtilis replicates bidirectionally (13,14,15). Table 2 shows that indeed the Okazaki pieces do hybridize with equal efficiency to both the L and H strands of B. subtilis DNA. Possible reasons for the differences between our results and those of Okazaki and coworkers will be considered in the Discussion.

DISCUSSION

We have re-examined the question of whether one or both growing DNA chains at a replication fork are synthesized discontinuously. Under our conditions, the majority of the radioactivity incorporated into DNA during a short pulse is found in Okazaki pieces. The fraction of pulse-labeled DNA in short pieces (S<20) is about 75% for E. coli (Fig. 1 and 2A) and more than 90% for B. subtilis (Fig. 3). On the other hand, uniformly-labeled E. coli or B. subtilis DNA sediments to the shelf (fractions 16-18 of the sucrose gradients) under these conditions (data not shown).

If one of the two growing DNA strands at a replication fork is replicated continuously, then one half of the radioactivity incorporated into DNA during a short pulse should be in large DNA. Presumably the DNA will be as large as the bulk of the DNA. Our evidence is clearly inconsistent with such a model. Even workers who have suggested that one strand is replicated continuously never find as much as 50% of the pulse-label in large DNA (6-9). Instead they find a substantial amount of labeled DNA sedimenting at intermediate positions, larger than Okazaki pieces but smaller than bulk DNA. Since they find that the fraction of radioactivity in intermediate size pieces plus that in large DNA (i.e., all the DNA larger than 20S) often approaches 50%, these workers have concluded that one strand is made continuously.

In the case of E. coli, we find approximately 20-30% of the labeled DNA in the intermediate size class (greater than 20 S but less than 40 S; fractions 9-15 of Fig. 1 and

2A). This DNA could either be due to: 1) joining of smaller Okazaki pieces either before or after termination of the pulse, or 2) a rapid rate of DNA polymerization relative to initiation of Okazaki pieces on the strand growing in the overall 5'——➤3' direction. The latter case would lead to some chains longer than a single Okazaki piece, a situation known to occur during in vitro DNA replication in the cellophane disc system (17). However, we would still consider this case an example of discontinuous synthesis.

Our results show that the method of stopping the pulse affects the fraction of labeled DNA which is larger than Okazaki pieces (S>20). The use of pyridine-KCN to stop the pulse and wash the cells leads to a greater fraction of such DNA than does the use of -10° acetone as a stop solution and KCN-EDTA to wash the cells (see Fig. 2 and Table 1). Previous workers who observed about half the pulse-labeled DNA in pieces larger than 20 S (and who therefore concluded that one strand is synthesized continuously) all used pyridine and/or KCN to stop the pulse and wash the cells (6-9). Our results suggest that the pyridine-KCN stop solution allows some joining of Okazaki pieces after termination of the pulse, and that acetone and/or KCN-EDTA used in our procedure prevent much of this joining.

Louarn and Bird concluded that one strand is synthesized continuously in E. coli, and they ingeniously demonstrated that it is the strand growing in the overall 5'——➤3' direction (7). They pulse-labeled an E. coli strain lysogenic for phage λ, and hybridized the labeled DNA to the separated strands of λ DNA. Using a pol A⁺ strain, they found that a short pulse labeled both short and large DNA fragments; the former hybridized preferentially to the l strand and the latter to the r strand of λ DNA. Knowing the orientation and position of the integrated λ DNA in the E. coli chromosome, they concluded that E. coli DNA is replicated discontinuously on the strand growing in the overall 3'——➤5' direction and is replicated continuously on the strand growing in the overall 5'——➤3' direction. However, in the case of a pol A⁻ strain, they found mainly small DNA after a short pulse, and concluded that both strands are replicated discontinuously in such mutants.

The results of Louarn and Bird taken in conjunction with our results using the same pol A$^+$ strain (Fig. 2 and Table 1) lead us to conclude that both strands are replicated discontinuously, and that there is preferential joining of Okazaki pieces on the strand growing in the overall 5'→3' direction under their experimental conditions. A possible reason why there is preferential joining on the strand growing in the overall 5'→3' direction is that there may be a greater probability of two Okazaki pieces being immediately adjacent to each other on that side of the replication fork.

Additional support for this point of view comes from Louarn and Birds results with the pol A$^-$ strain. Since the only known defect of pol A$^-$ strains in DNA replication is in the joining of Okazaki pieces (18), it seems reasonable to conclude that the difference between pol A$^+$ and pol A$^-$ strains found by them (7) is simply due to the difference in the ability of the two strains to join Okazaki pieces. Okazaki and coworkers have come to a similar conclusion in the case of phage P2 (see this Volume).

Our conclusion that both growing DNA strands at a replication fork are replicated discontinuously in the case of E. coli and B. subtilis is based on two assumptions: 1) ^{3}H-thymidine labels both growing DNA chains with approximately equal efficiency, and 2) there is no unusual fragility of the newly replicated DNA.

We thank Aviva Mukamel and Judith P. Baird for technical assistance and the American Cancer Society and the National Institutes of Health for financial support.

REFERENCES

1. Okazaki, R., Okazaki, T., Sakabe, K., Sugimoto, K., and Sugino, A., Proc. Nat. Acad. Sci. U.S.A. 59, 598 (1968).

2. Okazaki, R., Okazaki, T., Sakabe, K., Sugimoto, K., Kainuma, R., Sugino, A., and Iwatsuki, N., Cold Spring Harbor Symp. Quant. Biol. 33, 129 (1968).

320

3. Sugimoto, K., Okazaki, T., Imae, Y., and Okazaki, R., Proc. Nat. Acad. Sci. U.S.A. 63, 1343 (1969).

4. Ginsberg, B. and Hurwitz, J., J. Mol. Biol. 52, 265 (1970).

5. Polsinelli, M., Milanesi, G., and Ganesan, A., Science 166, 243 (1969).

6. Iyer, V. and Lark, K., Proc. Nat. Acad. Sci. U.S.A. 67, 629 (1970).

7. Louarn, J. and Bird, R., Proc. Nat. Acad. Sci. U.S.A. 71, 329 (1974).

8. Okazaki, R., Sugimoto, K., Okazaki, T., Imae, Y., and Sugino, A., Nature 228, 223 (1970).

9. Kainuma, R. and Okazaki, R., J. Jap. Biochem. Soc. 42, 464 (1970).

10. Wilson, M., Farmer, J., and Rothman, F., J. Bact. 92, 186 (1966).

11. Wang, H. and Sternglanz, R., Nature 248, 147 (1974).

12. Mulder, C. and Delius, H., Proc. Nat. Acad. Sci. U.S.A. 69, 3215 (1972).

13. Wake, R., J. Mol. Biol. 77, 569 (1973).

14. Lepesant-Kejzlarova, J., Lepesant, J., Walle, J., Billault, A., and Dedonder, R., J. Bact. 121, 823 (1975).

15. Harford, N., J. Bact. 121, 835 (1975).

16. Wang, H., Ph.D. Thesis, State University of New York at Stony Brook (1973).

17. Olivera, B. and Bonhoeffer, F., Nature New Bio. 240, 233 (1972).

18. Kuempel, P. and Veomett, G., Biochem. Biophys. Res. Comm. 41, 973 (1970).

DNA SYNTHESIS IN HUMAN LYMPHOCYTES

Ben Y. Tseng, Wolfgang Oertel*, Richard M. Fox+
and Mehran Goulian
Department of Medicine
University of California, San Diego
La Jolla, California 92037

ABSTRACT: Three in vitro systems for the study of DNA
synthesis in a human lymphocyte cell line are described.
DEAE-Dextran treated cells take up nucleotides but in many
other features resemble intact cells, whereas the two
other systems employ broken cells. The characteristics
of synthesis and the nature of the products are described.
In all three systems nascent DNA sediments in a 4S range,
which by polyacrylamide gel analysis is approximately 120
nucleotides in length. The three systems have been used
to examine the mechanism of DNA synthesis in the human
cell line. Nascent DNA is the precursor to high molecular
weight DNA and extends from 100 up to several hundred nucleotides before attaching to the growing daughter strand.
The technique of nearest neighbor transfer of incorporated
^{32}P-labeled deoxynucleotides provides evidence for RNA in
the nascent DNA. There is one RNA-DNA junction per nascent chain but the RNA sequence is not specific. Removal
of the RNA is correlated with the ability to synthesize
DNA. The evidence suggests that the RNA serves a primer
function for the synthesis of nascent DNA.

INTRODUCTION

Recent studies on the mechanism of DNA synthesis,
carried out in a number of laboratories, have provided a
remarkable number of important details (1, this volume).

---

* Present address: Institut für Mikrobiologie, der
Universität Würzburg, West Germany
+ Present address: Monash University Medical School,
Melbourne, Victoria, Australia

Eukaryote DNA synthesis is not as clearly defined. Replication at the level of chromosomes occurs by initiation of multiple replicons of 10-30 μ which grow bidirectionally (2,3). The distance between replicon origins changes dependent upon cell type (3) or the length of S-phase (4). At the replicon level, bidirectional growing forks consist of discontinuous nascent segments of DNA on one or both parental strands. Sizes that have been reported for the animal cell and viral "Okazaki" fragments range from 4S up to 10S (5-13), although it is unclear whether the differences reflect changes between cells or technical procedures.

To facilitate study of the mechanism of synthesis in eukaryote cells, three in vitro systems have been examined for their requirements and the products of synthesis. The cell line employed is a permanent human lymphoblast grown in suspension culture. One system permeabilizes the cell to nucleotides in the presence of DEAE-Dextran. Cells remain intact and exclude trypan blue stain although recovery of growth does not occur. A lysate, or broken cell, system utilizes a non-ionic detergent to disrupt the plasma membrane but leaves the nuclear membrane intact. The third or nuclear system, analogous to previous systems used by other groups (14-17), is a suspension of nuclei washed free of cytosol material.

RESULTS AND DISCUSSION

DEAE-Dextran Treated Cells: DEAE-Dextran facilitates the uptake of nucleotides into the cell allowing direct incorporation into DNA. The apparent initial rate of DNA synthesis is 20-40% the in vivo rate and may be an underestimation due to dilution of label by intracellular nucleotide pools. Incorporation is essentially linear for 10 min and thereafter gradually diminishes in rate up to 60 min. Several lines of evidence show that access to the cell of extracellular nucleotides is limited; one example of this is difficulty in saturating the system with deoxynucleoside triphosphates. Continued de novo synthesis of deoxynucleotides was indicated by the fact that hydroxyurea which inhibits ribonucleotide reductase (18), enhanced the effect of omitting one deoxynucleotide substrate. These properties show the relatively intact nature of the DEAE-Dextran treated cell, and the latter has been employed to

complement studies with the broken cell systems. The early
products of DNA synthesis in DEAE-Dextran cells are found
in a fraction sedimenting at 4S in alkaline sucrose gradients, as also observed with the broken cell preparations;
the discontinuous nascent DNA will be discussed further in
the other systems.

Whole Cell Lysate: The whole cell lysate used in these
experiments is prepared from a concentrated cell suspension
treated with 0.25% NP-40, which results in solubilization
of the plasma membrane but not the nuclear membrane. Incorporation is approximately 10% of the calculated in vivo
rate of DNA synthesis and is linear up to 30 min. Requirements for synthesis demonstrated a dependence upon added
deoxynucleoside triphosphates. ATP and PEP are also necessary as breakdown of triphosphates was rapid in their
absence. Synthesis was inhibited by N-ethylmaleimide and
ara-CTP, the nucleoside of which is a potent inhibitor of
cellular synthesis (19). The apparent Km for dTTP here
was $4-6 \times 10^{-6}$M. Synthesis occurs in the nucleus, as the
incorporated counts are recovered in the nucleus when separated from the cytoplasm.

The size of the products synthesized in the lysate
has been examined in two ways. First, the continuity of
newly synthesized material has been demonstrated by measuring the extent of bromodeoxyuridine triphosphate (BdUTP)
incorporation. The products synthesized in the presence
of BdUTP were analyzed by equilibrium sedimentation in
CsCl after sonication to an average molecular weight of
300,000 daltons. The density of nearly all fragments of
this size was hybrid indicating the original material was
BdUTP substituted at several times this size. The density
of the denatured product (before sonication) was fully
substituted, and $1-2 \times 10^{6}$ daltons in size.

A second characteristic of the synthesis is the discontinuity of the early product. Short term labeling was
found preferentially in a 4S sedimentation region on alkaline sucrose gradients. In a 10 sec pulse period, over
60% of the incorporated counts were found in this region.
With addition of excess unlabeled nucleotide, most of the
discontinuous fragments were "chased" (within 5 min) into
DNA several million daltons in size. These analyses demonstrate the ability of the lysate system to synthesize
large regions of DNA by a discontinuous mechanism.

A portion of rapidly labeled (nascent) DNA in microorganisms and animal cells is single-stranded, presumably
released from the replication fork upon stopping synthesis (20-22). A simple method for detection of the single-
stranded nascent DNA is ultracentrifugation of the Sarkosyl lysate, which pellets bulk DNA leaving the released
single-strands in the supernatant (23). The in vitro product with short labeling times also showed this tendency to
dissociate from the template (Table 1). The fraction of
total counts in the supernatant decreased from 25% with a
10 sec label to 6% with a 2 min label. Further decrease
was not observed possibly due to a fraction of contaminating degraded DNA. Dilution of a 10 sec pulse label results
in "chase" of the majority of counts into the pellet or
bulk DNA fraction (Table 1). This property of the in
vitro product is consistent with results previously reported for nascent DNA (20-23).

Nuclear System: Nuclei, obtained by several washes, are
free of almost all cytoplasmic tags by inspection in the
phase microscope. In comparison with synthesis in the
lysate, the nuclei are about one-half to one-quarter as
active on a DNA weight basis. Synthesis also continues
for a shorter duration (10 min) than in whole lysates (up
to 30 min). The product analyzed by alkaline sucrose
gradient grew from the initial 4S size to approximately
8S with little evidence for synthesis of higher molecular
weight DNA, in contrast to the lysate, in which high molecular weight DNA is the primary product with the longer
time of synthesis. Supplementation of nuclei with the
cytosol fraction allowed the maturation of the 8S fragments to high molecular DNA. Preliminary characterization
of the activity(s) has indicated a macromolecular nature
(non-dialyzable), and sensitivity to trypsin, heat (60° C,
10'), and dilution; further investigations are in progress.
The activity in the cytosol caused relatively little enhancement of net synthesis in contrast to its pronounced
effect on the maturation of discontinuous nascent intermediates into high molecular weight DNA. The relationship
of this activity to one that primarily stimulates DNA synthesis (15,24) is unknown.

DNA Chain Growth: Further analysis of the steps involved
in chain growth has been made with both lysate and nuclei

TABLE 1

Fractionation of Nascent DNA from the Lysate
by Sedimentation of Unsheared Bulk DNA

Incubation Time	Supernatant fraction (cpm)	Pellet fraction (cpm)	Supernatant total (%)
10 sec	1,592	4,770	25
30 sec	1,421	12,220	10
60 sec	1,929	20,900	8
120 sec	2,910	40,800	6
300 sec	5,810	84,900	6
10 sec + chase (10 min)	509	7,900	6

Incubation mixture (0.3ml) contained 80mM sucrose, 20mM KCl, 0.5mM CaCl$_2$, 9mM MgCl$_2$, 1.5mM KPO$_4$, 20mM glucose, 45mM MOPS pH 7.6, 20mM trisodium phosphoenol-pyruvate, 0.1mM DTT, 2.5mM ATP; 0.2mM each of GTP, CTP, UTP, dCTP, dGTP, dATP; 6µM [3H]dTTP (15 Ci/mmole),lysate of cultured lymphocytes (prepared with 0.25% NP-40) equivalent to 4x10^7 cells/ml. Incubations were at 37° C.

systems. To examine growth of the small intermediates, the products were electrophoresed in 6% polyacrylamide gels in urea-acetate buffer calibrated with internal markers of tRNA and a 140-nucleotide RNA fragment from T4-infected cells (25). The 4S intermediate is a broad peak (10 sec pulse, Fig. 1) centered about 120 nucleotides which, during a 1 min chase with unlabeled nucleotide, grows to a size too large to enter the gel (exclusion limit approximately 400 nucleotides). Further examination of the growth was carried out at 26°C, at which temperature the incorporation was 30% of the rate at 37°C. In pulse-chase experiments at the lower temperature, it was possible to observe continuous elongation of the nascent 120-nucleotide material until a size was achieved (2 min) which was excluded from the gel. Thus, growth of the 120-nucleotide intermediate appears to occur to several hundred nucleotides before joining to high molecular weight daughter DNA strands. The mechanism of this growth is unclear at present but two possibilities are suggested. Chain growth of the 120-nucleotide fragments can occur by ligating several such fragments to each other to form a larger fragment before joining to the large molecular weight strand; or, the fragment can be extended by nucleotide addition to form a larger discontinuous fragment before joining to high molecular weight DNA.

Recently, the role of RNA in initiating DNA synthesis has been demonstrated in several organisms (26-30, this volume). The in vitro product of the lysate and the nuclei systems have been analyzed for the existence of primer RNA. An absolute dependence of DNA synthesis upon added ribonucleoside triphosphates has not been observed with the in vitro systems described here. However, a small (20%) but reproducable increase in the initial velocity of synthesis has been observed with the nuclei system when all ribonucleotides are added, even though added at one-tenth the level of ATP.

To demonstrate a physical RNA-DNA covalent bond, a nearest neighbor transfer experiment has been employed (31,32). After incorporation with one α-^{32}P-labeled deoxynucleoside triphosphate as the source of radioactivity, the products were treated with alkali and then analyzed for the presence of ^{32}P-labeled ribonucleoside monophosphates. Transfer of ^{32}P-label occurs only when the 5'-P of a labeled deoxynucleotide is attached to the 3'-OH of

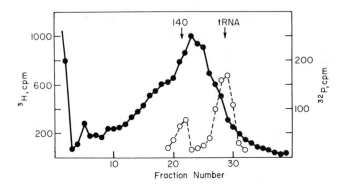

Fig. 1. Polyacrylamide gel electrophoretic analysis
of pulse-labeled nascent DNA in lysates. Sample, labeled
with a 10 sec pulse as described in Table 1, was electrophoresed on a 6% polyacrylamide gel (150/1, acrylamide/bis)
in 6M urea, 0.02M sodium acetate pH 5, 2mM EDTA. Internal
markers of ^{32}P-labeled E. coli tRNA and ^{32}P-labeled 140-
nucleotide T4 RNA (25) were included. The counts in the
first fraction (not shown) were 40% of the total incorporated.

a ribonucleotide. The experiment, carried out separately
for the four deoxynucleotides, showed that all ribonucleotides were adjacent to all four deoxynucleotides in roughly
similar proportions. Similar results have been obtained
with the DEAE-Dextran system and with the nuclear system.
This result is similar to the random junctions observed
for polyoma (28).

Two experiments implicated RNA-DNA junctions in the
synthesis of DNA. After fractionating the product of synthesis by agarose column chromatography, the separated
sizes were analyzed for the RNA-DNA linkage. About 80% of
the ^{32}P-label transferred to ribonucleotides was found in
sizes centered about 120 nucleotides, the length of nascent fragments. The ratio between total counts and
transferred counts corresponds to chain length of the
fractions within 10-15%. This indicates that one RNA-DNA
bond exists in each DNA chain, and the observation from
other experiments, that fragment length on polyacrylamide
gels is not affected by alkali shows that the RNA-DNA
junction is at or near fragment termini.

An examination of the lability of the RNA-DNA bond
has been carried out using both lysate and nuclei systems.
A 3-min incorporation of α-^{32}P-labeled dTTP in nuclei was
followed by excess unlabeled dTTP, with samples taken 3
and 5 min after starting the chase (Table 2). ^{32}P counts
in ribonucleotides decreased to less than half during the
15 min chase period. In contrast, when ara-CTP was present, which resulted in inhibition of DNA synthesis by 90%,
the counts in the rNMP's remained. A similar experiment
with the lysate system showed a similar lability of the
RNA-DNA bond but with a half-time for disappearance of 2
min rather than 14 min as observed in the nuclei system.
The half-life of the RNA-DNA bond in the two systems
correlates with the faster rate (5-10x) for growth of the
nascent fragments in the lysate compared to the nuclear
system, measured on polyacrylamide gels. This correlation
along with stabilization of RNA-DNA bonds by ara-CTP,
indicates a close coupling between DNA synthesis and removal of the RNA junction. This is interpreted as the
removal of RNA primer fragments from nascent DNA as the
DNA is extended and joined.

To examine the complexity of RNA nucleotides at the
RNA-DNA junction the synthetic product was examined again
by ^{32}P-transfer from incorporated α-^{32}P-labeled deoxynu-

TABLE 2

Lability of the RNA-DNA Bond and the Effect of
ara-CTP in the Nuclear System

Total Incubation Time (min)	Chase (min)	ara-CTP	% Label transferred from incorporated [^{32}P]dTMP
3	0	−	0.17
6	3	−	0.15
18	15	−	0.06
6	3	+	0.20
18	15	+	0.18

Incubation mixture as described in Table 1 except lysate
was replaced by washed nuclei (8×10^6 cell equivalents/ml),
the temperature lowered to 26° C, unlabeled deoxynucleotides at 25µM, and [α-^{32}P]dTTP (35 Ci/mmole; 10^{-5}M). For
the chase, a 100-fold excess of unlabeled dTTP was added.
In cases where ara-CTP was present, it was added (10^{-4}M)
at the same time as the unlabeled dTTP. Transfer counts
were determined after alkali digestion (0.3M KOH, 16 hr,
37° C) by electrophoresis at pH 5.5 (0.6% pyridine acetate). The transferred counts were confirmed to be 2'(3')
ribomonophosphates by further electrophoresis (0.03M
ammonium formate, pH 3.5) and chromatography (67 vol isopropanol, 16 vol HCl, 17 vol H$_2$0). The total incorporated counts in all cases were 7×10^5 cpm.

cleotides, but after digestion with RNase A instead of alkali. In this case only the nearest-pyrimidine neighbors are analyzed. A ^{3}H-labeled internal control for the digestion products of complex RNA was synthesized with E. coli RNA polymerase on denatured lymphocyte DNA. After RNase A digestion, the oligonucleotides were separated according to chain length by DEAE-Dextran chromatography in the presence of urea (33). Various nucleotide lengths were analyzed by chromatography and electrophoresis for base composition. The ^{32}P nucleotides were established to be ribonucleotides by alkaline hydrolysis and chromatography. Two comparisons can be made which indicate a diverse rather than unique set of ribonucleotide sequences are joined to the DNA. The ratio of counts in mononucleotides to dinucleotides is similar for ^{32}P transferred counts and for ^{3}H counts from complex RNA. Second, among the dinucleotide sequences, the ^{3}H and ^{32}P ratios are similar for ApCp to GpCp and for ApUp to GpUp. This suggests that the RNA nucleotides attached to DNA are not a specific set of sequences, but reflect the base composition of the genome dinucleotide population. Low ^{32}P transfer counts precluded analysis of oligonucleotides larger than dinucleotides. Alkaline digestion of the nucleic acids bound to the DEAE column showed ^{32}P-transfer counts exclusively in GMP and AMP, indicating complete digestion (by the RNase A) of the junction with pyrimidine ribonucleotides.

Additional support for the attachment of RNA to nascent DNA comes from preliminary experiments using polynucleotide kinase, which indicate that nascent DNA isolated from growing cells also has RNA attached at its 5'-terminus. A previous study, which employed equilibrium density gradient analysis, suggested that RNA is associated with nascent DNA in cultured human lymphocytes (23), but subsequent studies have shown that most or all of that association is noncovalent (34). Nevertheless, the results cited here do indicate that RNA, probably serving a primer function, is covalently associated with nascent DNA. Our inability to detect an increased density of nascent DNA suggests that the RNA segment is quite short, i.e., less than 10% of the total 120 nucleotide nascent chain length. From the work described above it may be inferred that the RNA pieces do not have a simple sequence. Consistent with this, the recent work from Reichard and associates (30) indicates that the RNA primer of polyoma DNA is a deca-

nucleotide with a non-unique sequence at the 5' end of
the RNA.

ACHNOWLEDGEMENTS

The results described here will be detailed in later
publications.

Supported by Grant CA11705 from the National Institutes of Health, and Grant NP-102C from the American Cancer
Society.

REFERENCES

1. Kornberg, A. DNA Synthesis, W.H. Freeman and Co.,
   San Francisco, 1974.
2. Huberman, J.A. and Riggs, A.D. J. Mol. Biol. 32,
   327 (1968).
3. Hand, R. and Tamm, I. J. Mol. Biol. 82, 175 (1974).
4. Blumenthal, A.B., Kriegstein, H.J. and Hogness, D.S.
   Cold Spring Harbor Symp. Quant. Biol. 38, 205 (1973).
5. Schandl, E.K. and Taylor, J.H. Biochem. Biophys.
   Res. Comm. 34, 291 (1969).
6. Nuzzo, F., Brega, A. and Falaschi, A. Proc. Nat.
   Acad. Sci. (U.S.A.) 65, 1017 (1970).
7. Schandl, E.K. and Taylor, J.H. Biochim. Biophys.
   Acta 228, 595 (1971).
8. Sato, S., Ariake, S., Saito, M. and Sugimura, T.
   Biochem. Biophys. Res. Comm. 49, 270 (1972).
9. Waqar, M.A. and Huberman, J.A. Biochem. Biophys. Res.
   Comm. 51, 174 (1973).
10. Berger, Jr., H. and Huang, R.C.C. Cell 2, 23 (1974).
11. Fareed, G.C. and Salzman, N.P. Nature New Biology
    238, 274 (1972).
12. Magnusson, G., Pigiet, V., Winnacker, E.L., Abrams,
    R. and Reichard, P. Proc. Nat. Acad. Sci. (U.S.A.)
    70, 412 (1973).
13. Francke, B. and Hunter, T. J. Mol. Biol. 83, 99.
    (1974).
14. Friedman, D.L. and Mueller, G.C. Biochim. Biophys.
    Acta 161, 455 (1968).
15. Kidwell, W.R. and Mueller, G.C. Biochem. Biophys.
    Res. Comm. 36, 756 (1969).
16. Lynch, W.E., Umeda, T., Uyeda, M. and Lieberman, I.
    Biochim. Biophys. Acta 287, 28 (1972).

17. Winnacker, E.L., Magnusson, G. and Reichard, P. J. Mol. Biol. 72, 523 (1972).
18. Elford, H.L. Biochem. Biophys. Res. Comm. 33, 129 (1968).
19. Roy-Burnam, P. Recent Results in Cancer Res. 25, 66 (1970).
20. Oishi, M. Proc. Nat. Acad. Sci. (U.S.A.) 60, 691 (1968).
21. Okazaki, R., Okazaki, T., Sakabe, K., Sugimoto, K. and Sugino, A. Proc. Nat. Acad. Sci. (U.S.A.) 59, 598 (1968).
22. Habener, J.F., Bynum, B.S. and Shack, J. J. Mol. Biol. 49, 157 (1970).
23. Fox, R.M., Mendelsohn, J., Barbosa, E. and Goulian, M. Nature New Biology 245, 234 (1973).
24. Hershey, H., Steiber, J. and Mueller, G.C. Biochim. Biophys. Acta 312, 509 (1973).
25. Paddock, G. and Abelson, J. Nature New Biology 246, 2 (1973).
26. Brutlag, D., Schekman, R. and Kornberg, A. Proc. Nat. Acad. Sci. (U.S.A.) 68, 2826 (1971).
27. Sugino, A., Hirose, S. and Okazaki, R. Proc. Nat. Acad. Sci. (U.S.A.) 69, 1863 (1972).
28. Pigiet, V., Eliasson, R. and Reichard, P. J. Mol. Biol. 84, 197 (1974).
29. Hunter, T. and Francke, B. J. Mol. Biol. 83, 123 (1974).
30. Reichard, P., Eliasson, R. and Söderman, G. Proc. Nat. Acad. Sci. (U.S.A.) 71, 4901 (1974).
31. Flügel, R.M. and Wells, R.D. Virology 48, 394 (1972).
32. Sugino, A. and Okazaki, R. Proc. Nat. Acad. Sci. (U.S.A.) 70, 88 (1973).
33. Tener, G.M. in Methods in Enzymology XII(A), Grossman, L. and Moldave, K. (eds.), Academic Press, N.Y., 1967, p. 398.
34. Mendelsohn, J.M. and Goulian, M. Unpublished results.

EVIDENCE FOR RNA LINKED TO NASCENT DNA
IN EUKARYOTIC ORGANISMS

M. Anwar Waqar, Robert Minkoff, Alice Tsai
and Joel A. Huberman
Department of Biology
Massachusetts Institute of Technology
Cambridge, Massachusetts 02139

ABSTRACT. In this paper we report the results of our
efforts to detect RNA linked to nascent DNA in two kinds
of eukaryotic organisms: the slime mold Physarum polycephalum and cultured mammalian cells (in most experiments, CHO cells). We have tried to detect such linkages
because of the possibility that RNA may serve as a primer
for the synthesis of DNA.

In Physarum nascent DNA chains appear to have a
density greater than that of single-stranded, mature
Physarum DNA. If the nascent DNA chains are treated
with KOH or RNase, then less density difference is found,
suggesting that the density difference is due to RNA attached to the nascent DNA. When Physarum DNA is
labeled for 2 min or 5 min with α -^{32}P-deoxyribonucleo-
side triphosphates which have been injected into the plasmodium, then alkaline hydrolysis of the labeled DNA
reveals the presence of a small proportion of ribonucleotides on the 5' side of the labeled deoxyribonucleotides,
consistent with the covalent attachment of a stretch of
RNA to the 5' end of the nascent DNA chain.

Although nascent DNA in mammalian cells does not
differ in density from mature DNA, alkaline hydrolysis
of mammalian DNA labeled in cell-free lysates with α -
^{32}P-deoxyribonucleoside triphosphates does reveal occasional RNA-DNA junctions within the nascent DNA.
Thus, although evidence for covalent linkage of RNA to
nascent DNA is much harder to find in mammalian cells
than in Physarum, it is likely that such junctions are
present in mammalian DNA. Further experiments will be
needed to determine whether the differences between
Physarum and mammalian cells in ease of detection of

334

RNA-DNA junctions can be explained by more rapid processing of the putative RNA primer in mammalian cells.

INTRODUCTION

The possibility that RNA may play a role in DNA synthesis by serving as a primer for the initiation of new Okazaki fragments has already received a great deal of attention at this conference. Work from the Okazaki group (1-3, this conference) and the Reichard group (4-7, this conference) and others (8) has been instrumental in demonstrating that stretches of RNA can be detected covalently linked to the 5' ends of Okazaki fragments in E. coli and in polyoma virus. The paper just presented by Goulian (9, this conference) supports similar linkages between RNA and nascent DNA in human lymphocytes, and we, as will be described shortly, have also obtained evidence for such linkage both in the slime mold Physarum polycephalum and in mammalian cells.

But demonstration of these links between RNA and nascent DNA has not been simple. If one RNA-DNA link exists per Okazaki fragment, then, in an exponentially growing population of mammalian cells, there is probably less than one RNA-DNA link for every 10^6 base-pairs of DNA. The methods needed to detect such rare links must be extremely sensitive and, as previous speakers have noted, the possibilities for artifacts are numerous.

In this talk we present evidence for RNA-DNA links in the slime mold, Physarum polycephalum, and in cultured mammalian cells, and we discuss additional experiments which need to be done to verify and extend our findings.

RESULTS

Physarum polycephalum - density shift experiments: The slime mold Physarum polycephalum offers several advantages for studies of DNA synthesis and function. In its plasmodial form, Physarum grows as a single, flat syncytial cell which may contain over 10^8 nuclei. Because these nuclei share common cytoplasmic signals, they enter mitosis and start DNA synthesis in nearly perfect natural synchrony. In plasmodia of Physarum the mitotic cycle lasts about 12 hours. S phase begins immediately after mitosis (there is no G1) and lasts 3-4 hours. The location of a plasmodium in the mitotic cycle can be determined by microscopic inspection of nuclei from a small sample of the plasmodium.

Two years ago (10) we found that when Physarum plasmodia, in S phase, were pulse-labeled briefly at room temperature with ^{3}H-thymidine, the pulse-labeled DNA would band in Cs_2SO_4 gradients at densities somewhat greater than that of single-stranded Physarum DNA (Figure 1). If the pulse-labeled DNA was first treated with alkali or RNase (in order to hydrolyze RNA) then its density was normal (in the case of alkali) or nearly normal (in the case of RNase) (see Figure 1). As previous speakers have noted, although this kind of result does suggest association between pulse-labeled DNA and RNA, the association could easily be non-covalent. In fact, the partial resistance to RNase could be due to the presence of RNA-DNA hybrid, non-covalent duplexes (or to the presence of polyribopurine stretches). In the case of Physarum, two kinds of evidence argue for either covalent association or a strong and specific aggregation. First, notice that in Figure 1 only the pulse-labeled DNA and not the marker DNA is shifted to higher density. In previous experiments (10) we have shown that DNA pulse-labeled for 20 min is shifted in density to a much smaller extent than DNA pulse-labeled for 1 min. Thus the association between RNA and DNA is specific for DNA labeled in a very brief ($\sim$ 1 min) pulse.

In addition, we have subjected the RNA-DNA complex to a second cycle of heat denaturation. The graphs in Figure 2 show that when pulse-labeled DNA from the high density region of a first Cs_2SO_4 gradient is pooled, dialyzed, redenatured (100° for 10 min) and then centrifuged again in a second Cs_2SO_4 gradient, it still bands at the same high density. Thus the bonds between RNA and pulse-labeled DNA appear to be stable or rapidly-reforming ones.

Physarum polycephalum - nearest neighbor experiments:
Better evidence for the existence of covalent bonds between RNA and DNA can be obtained by a variation of the nearest neighbor analysis technique (11). The rationale for this technique is illustrated by the example in Figure 3. The hypothetical stretch of ribonucleotides linked to

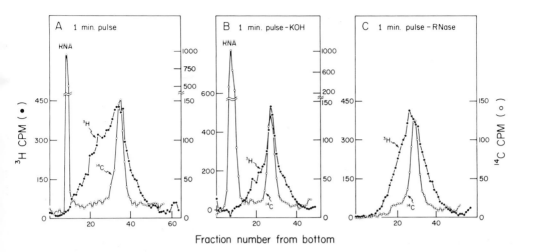

Fig. 1.   Density of nascent <u>Physarum polycephalum</u>
DNA in Cs$_2$SO$_4$ gradients.

Growth and labeling of <u>Physarum</u> plasmodia were
as described earlier (10).   The plasmodia were labeled
with ^{3}H-thymidine for 1 min in early S phase.   Short,
pulse-labeled strands were isolated on neutral sucrose
and formamide gradients, denatured in formamide,
treated as indicated below, mixed where indicated with
marker ^{14}C-HeLa cell ribosomal RNA and single-
stranded <u>Physarum</u> DNA, and centrifuged to equilibrium
in Cs$_2$SO$\overline{4}$ gradients as described earlier (10).

(A)   No treatment to remove RNA.

(B)   Nascent DNA was hydrolyzed with 0.3 M KOH
(18 hrs at 37º) (10).

(C)   Nascent DNA was hydrolyzed with pancreatic
RNase A (1 mg/ml at 37º for 3 hrs) (10).

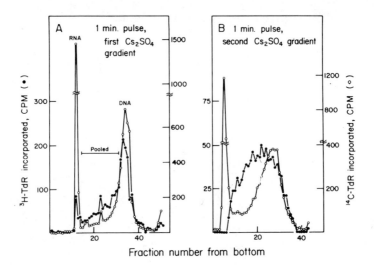

Fig. 2. Resistance of density shift in nascent Physarum DNA to redenaturation and rebanding in a Cs$_2$SO$_4$ gradient.

(A) Physarum plasmodia were pulse-labeled for 1 min with ^{3}H-thymidine, then lysed, and short strands were isolated, denatured, and banded in Cs$_2$SO$_4$ as in Fig. 1A.

(B) Indicated fractions from (A) were pooled, dialyzed against 10 mM Tris, 10 mM EDTA, 0.1% Sarkosyl, pH 7.4, denatured by heating (100° for 10 min), then mixed with Cs$_2$SO$_4$ and ^{14}C-RNA and single-stranded DNA markers, and centrifuged to equilibrium.

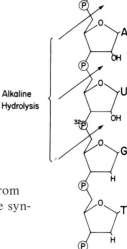

Fig. 3. Diagram illustrating the transfer of ^{32}P from dGMP to rUMP in a hypothetical RNA-DNA sequence synthesized with α-^{32}P-dGTP.

deoxyribonucleotides shown in Figure 3 is presumed to
have been synthesized with α-^{32}P-dGTP   Clearly, if
this stretch is treated with alkali, the labeled phosphate
group linking rUMP with dGMP will be transferred to released rUMP. In actual practice, nascent DNA is labeled
with any one of the 4 α-^{32}P-deoxyribonucleoside triphosphates (dNTP's) and the labeled DNA is purified,
hydrolyzed with alkali, and then acid-precipitated. The
supernatant left after precipitation of the DNA is examined by chromatographic or electrophoretic techniques
for ^{32}P-labeled ribonucleoside monophosphates (rNMP's).
We have been able to detect as little as one RNA-DNA
junction per 10^5 deoxynucleotides incorporated, when
care was taken to ensure that the α-^{32}P-dNTP's used
for synthesis were not contaminated by other ribo- or
deoxyribonucleotides.

Ordinarily α-^{32}P-dNTP's, like other charged
molecules, are not incorporated by living cells. Permeabilized cell preparations (2, 12, 13) or more disrupted in
vitro systems (14, 15) must be used for incorporation of
α-^{32}P-dNTP's. Fortunately, plasmodia of Physarum
are sufficiently large that microinjection of α-^{32}P-
dNTP's directly into the plasmodial cytoplasm is relatively easy to accomplish. When α-^{32}P-dNTP's in
quantities of about 10-25 μCi are injected into a single
plasmodium, about 1% of the injected label is incorporated into DNA within 5 min, primarily in an area of the
plasmodium less than 1/2 cm from the site of injection.
We (16) have partially characterized the DNA synthesized
with injected α-^{32}P-dNTP's. As shown in Figure 4a,
after short injection periods (up to 5 min) the incorporated
label is primarily in short DNA strands (less than 300
nucleotides). After 60 min of injection (Figure 4c) the
label is found primarily in longer DNA strands, consistent with a discontinuous mechanism of DNA synthesis.
Similar results are obtained when plasmodia are labeled
in vivo with ^{3}H-thymidine (Figure 4b, d). Such minor
differences must be borne in mind when interpreting results from injection experiments.

When Physarum DNA pulse-labeled by injection of
α-^{32}P-dNTP's for 2 or 5 min is analyzed for RNA-DNA
junctions, the results shown in Table 1 are obtained.
Transfer of ^{32}P label occurs from all 4 dNTP's to all 4
rNMP's, but the extent of transfer from dGTP is much

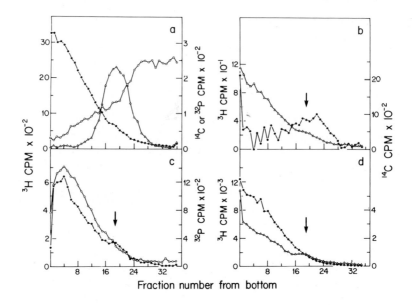

Fraction number from bottom

Fig. 4. Comparison of the sedimentation proper-
ties of DNA strands made either with injected α -^{32}P-
dCTP or with absorbed ^{3}H-thymidine. Pulse-labeling,
injection, lysis, partial purification of DNA, and sedimentation through alkaline sucrose gradients were as
previously described (16).
    (a) α -^{32}P-dCTP was injected 2 min before lysis
during early S phase. The plasmodium had previously
been labeled with ^{3}H-thymidine for 32-38 hrs. o———o,
^{32}P; •———•, ^{3}H; $\triangle$———$\triangle$, ^{14}C marker DNA, about 300
nucleotides long, prepared by sonication of uniformly
labeled Chinese hamster cell DNA (prepared and characterized by H. Horwitz and given to us).
    (b) A plasmodium was pulse-labeled with ^{3}H-thy-
midine for 2 min during early S phase. The plasmodium
had previously been labeled with ^{14}C-thymidine for 32-
38 hrs. •———•, ^{3}H; $\triangle$———$\triangle$, ^{14}C. The arrow shows
the position to which ^{14}C-labeled marker DNA of 300
nucleotides sedimented in a parallel tube, see (a).
    (c) As in (a), but plasmodium was lysed 60 min
after injection.
    (d) As in (b), but labeling lasted for 60 min.

340

greater than from the other 3 dNTP's. In every case, transfer occurs to all 4 rNMP's in approximately equal proportions.

Because the frequency of phosphate transfer from dGTP is so high, it is very easy to demonstrate that all of the observed transfer is from incorporated dGMP to ribonucleotides. When DNA labeled by injection of α-^{32}P-dGTP for 5 min is isolated and hydrolyzed to completion by pancreatic DNase I and snake venom phosphodiesterase, $>$ 99.99% of the radioactivity is recovered as 5'-dGMP (16). Thus the 5% of radioactivity which is recovered in ribonucleotides after alkaline hydrolysis must be transferred from adjacent 5'-dGMP residues; there is no other possible source of this radioactivity.

From the data in Table 1, and from the base composition of Physarum DNA, we have calculated (16) that there is one RNA-DNA junction for every 70 nucleotides in the DNA labeled by injection. This value is consistent with the average size of the pulse-labeled DNA (Figure 4) suggesting that there may be just one RNA-DNA junction per pulse-labeled DNA strand (this calculation is rough; accurate determination of the size of Okazaki fragments in Physarum has not yet been accomplished). Thus the evidence provided by density determinations in Cs_2SO_4 gradients and by nearest neighbor analyses is consistent with the possibility that RNA serves as a primer for DNA synthesis in Physarum polycephalum. Before we discuss the additional kinds of experiments needed to test this possibility more completely, we wish to compare the experiments on Physarum which we have just presented with similar experiments on mammalian cells.

Mammalian cells - density shift experiments: When we first observed the abnormally high density of pulse-labeled Physarum DNA (Figures 1, 2), we guessed that the same method should reveal an abnormally high density for pulse-labeled mammalian DNA. In our attempts to detect density-shifted mammalian DNA, we used a variety of methods for nucleic acid isolation and several methods for preservation of RNA, and we tested a number of different mammalian cell lines (CHO, HeLa, 3T6, and 8866). We also used both Cs_2SO_4 gradients and CsCl gradients, with and without formaldehyde. We tried heat denaturation and formamide denaturation. Nevertheless, in over

TABLE 1

Distribution of radioactivity in 2'(3') rNMP's after alkaline hydrolysis of products formed with injected α-32P-dNTP's. Injections were for periods of 2 min. or (in parentheses) for 5 min.

α-32P-dNTP	% of incorporated 32P transferred to ribonucleotides	Distribution of 32P in ribonucleotides (%)			
		UMP	AMP	CMP	GMP
dATP	0.37 (0.41)	22 (24)	27 (20)	22 (20)	29 (37)
dTTP	0.19 (0.15)	26 (29)	26 (27)	26 (27)	21 (20)
dGTP	5.70 (4.30)	20 (24)	27 (24)	26 (24)	27 (28)
dCTP	0.70 (1.60)	18 (26)	33 (19)	24 (28)	24 (27)

Growth of Physarum plasmodia, injection of label, purification of nucleic acids, and nearest neighbor analysis were as described earlier (16). Data are from Waqar and Huberman (16), but are presented in different form here.

a year of experimentation, we never detected any difference in density between Okazaki fragments and non-
replicating, single-stranded DNA (example in Figure 6a).
Our findings in this respect are identical to those of
Berger and Huang (17), and of Gautschi and Clarkson who
have calculated from the absence of detectable density
shift that fewer than 10% of the Okazaki fragments tested
in these experiments could contain more than 10 nucleotides of RNA (18).

We would like to point out, however, that a variation of this density shift technique did give false positive
results with mammalian cells. The variation involves
pulse-labeling RNA with ^{3}H-uridine and then looking to
see if any of the pulse-labeled RNA has the density of
DNA in a CsCl or Cs_2SO_4 gradient. Although Mendelsohn,
Castagnola, and Goulian (19; this conference) have also
demonstrated recently that artifactual co-banding of RNA
with DNA can readily occur in cesium salt gradients, we
describe some of our experiments in order to demonstrate how convincing this artifactual co-banding can
seem. We also present the results of a parallel experiment in which the density of ^{3}H-thymidine-labeled
Okazaki fragments was tested, and found it to be identical
to that of DNA.

We started by pulse-labeling CHO cells at 26° for
1 or 2 minutes with ^{3}H-uridine or ^{3}H-thymidine (the cells
pulse-labeled with ^{3}H-thymidine were pre-labeled overnight with ^{14}C-thymidine). The cells were then lysed
with Sarkosyl and sedimented through a neutral sucrose
gradient. In the conditions employed, nearly all of the
bulk DNA formed a rapidly-sedimenting aggregate. When
cells were pulse-labeled with ^{3}H-thymidine, nearly all of
the pulse-label was associated with this aggregate (Figure
5a). When cells were pulse-labeled with ^{3}H-uridine, a
portion of the pulse-labeled material also associated with
the aggregate (Figure 5b). The fractions containing DNA
and associated uridine-labeled material were pooled as
indicated, denatured, and sedimented through formamide-
sucrose gradients. These are denaturing gradients which
separate single nucleic acid strands according to size.
The ^{3}H-uridine labeled strands (Figure 5d) were shorter
than bulk DNA (Figure 5c) and about the same size as
^{3}H-thymidine labeled Okazaki fragments (Figure 5c).

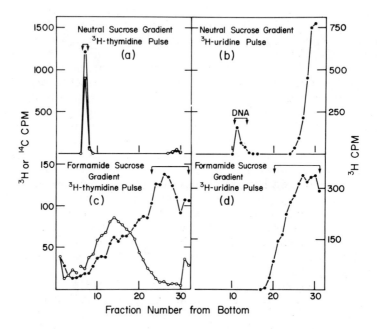

Figure 5

Fig. 5. Sedimentation characterization of nucleic
acids pulse-labeled with ^{3}H-thymidine or ^{3}H-uridine.
(a) ^{3}H-thymidine pulse, neutral sucrose gradient.
CHO cells were grown and radioactively labeled as
described earlier (24). Pulse-labeling with ^{3}H-thymidine
was for 1 min (26°). The pulse was terminated with cold
EtOH, the plates were washed with cold isotonic saline,
and suspended with a rubber policeman in 1.5 ml of isotonic saline. The cell suspension was layered on top of
1.5 ml of 2% Sarkosyl which was on top of a 35 ml
sucrose gradient (5-20% sucrose in 50 mM potassium
phosphate, pH 7.4, 2 mM EDTA, 0.5% Sarkosyl) in a
polyallomer tube. The Sarkosyl and cell suspension
were mixed gently and allowed to stand for 30 minutes
in the cold. The gradient was centrifuged in the Spinco
SW 27 rotor at 20,000 rpm for 2 hrs at 8°. •——•, ^{3}H
cpm; o——o, ^{14}C cpm.
(b) ^{3}H-uridine pulse, neutral sucrose gradient.
CHO cell were pulse-labeled for 2 min at 26° with
^{3}H-uridine (40 Ci/mmole; 400 μCi/ml; New England
Nuclear). The pulse was terminated and the cells were
lysed and centrifuged as in (a). The position of DNA was
determined by viscosity measurements.
(c) ^{3}H-thymidine pulse, formamide-sucrose gradient. Fraction 7 from (a) was pooled with material
from similar neutral sucrose gradients. An equal
volume of formamide was added and mixed. 4.5 ml of
sample were layered on top of 33 ml of a formamide-
sucrose gradient (0-15% sucrose in 99% formamide) in
a polyallomer tube, and centrifuged in the Spinco SW 27
rotor at 25,000 rpm for 22 hrs at 23° C. Aliquots of
each fraction were mixed with Triton-toluene scintillation fluid and counted in a liquid scintillation counter.
•——•, ^{3}H cpm; o——o, ^{14}C cpm.
(d) ^{3}H-uridine pulse, formamide-sucrose gradient.
The indicated fractions from (b) (those having high viscosity) were pooled and denatured by heating to 100° for
15 min followed by quick cooling. The cooled material
(5 ml) was layered onto a 33 ml formamide-sucrose
gradient, centrifuged, and counted as in (c).

Short strands were pooled as indicated, dialyzed to remove formamide and then centrifuged in cesium gradients. The ^{3}H-thymidine labeled Okazaki fragments did not differ significantly in density from bulk DNA (Figure 6a), but a considerable proportion of the ^{3}H-uridine labeled material banded at densities lighter than the marker RNA (Figure 6b). After treatment with alkali (or RNase) nearly all of the ^{3}H-uridine labeled material became acid-soluble (Figure 6b), indicating that most of the ^{3}H-radioactivity was indeed in ribonucleotides.

We checked the possibility that this density-shifted RNA might be involved in DNA replication in three ways. The first check also gave false positive results. When CHO cells were synchronized and pulse-labeled with ^{3}H-uridine in G1 or S phase, then little or no density-shifted RNA could be detected in the G1 cells, but a large amount of density-shifted RNA could be detected in the S phase cells (Figure 6c, d). If we had stopped characterization of the density-shifted RNA at this point, we might well have concluded that it is involved in replication. However, we also tried to measure the kinetics of synthesis of the density-shifted RNA. To our surprise we found that the same proportion ($\sim$ 5%) of total labeled RNA was shifted to light densities whether we pulse-labeled for 2 min at 26° or for 35 min at 37°. Thus the density-shifted RNA was (a) not turned over as rapidly as would be expected for a primer for Okazaki fragments and (b) present in quantities too great to be associated only with replicating DNA. The third test we performed was to treat the ^{3}H-uridine labeled material with DNase I (50 μg/ml, 1 hr, 37°). Although the DNA marker was digested, the amount of ^{3}H-uridine labeled material banding at low density was unaffected (data not shown). The reason for the density shift of ^{3}H-uridine labeled material thus does not appear to be attachment to DNA. The reason for its low density was not investigated further.

Although we decided after our kinetic and DNase experiments that the density-shifted RNA was probably not related to replication and therefore not interesting to us at the time, we note the fact that the density-shifted RNA does seem to be synthesized at particular times during the cell cycle and thus may represent a particular class or classes of RNA molecules.

Mammalian cells: nearest neighbor experiments: Because the nearest neighbor technique is both more sensitive and less ambiguous than the density-shift technique for detection of RNA-DNA joint molecules, we decided that, despite our negative experience with the density shift technique when applied to mammalian cells, it would be worth the effort to use the nearest neighbor technique as well.

In order to incorporate α-^{32}P-dNTP's into the DNA of mammalian cells we have used primarily a lysate of CHO cells prepared by the method of Winnacker, Magnusson, and Reichard (15).

With this system we routinely observe a low level of transfer of ^{32}P from dNTP's to ribonucleotides. For example, an electrophoretic separation of the ^{32}P-ribonucleotides made acid-soluble when DNA (labeled at 25° for 5 min with α-^{32}P-dCTP) was hydrolyzed with alkali is shown in Figure 7. As in our previous experiments with Physarum polycephalum (16), when the four peaks of radioactivity are eluted from electrophorograms like those of Figure 7 and rerun in chromatographic systems, all radioactivity migrates with the appropriate ribonucleotide, not deoxyribonucleotide, markers.

A summary of the results we have obtained to date with lysates of CHO cells and the Reichard in vitro system (15) is shown in Table 2. For comparison purposes we also show the corresponding nearest neighbor frequencies as measured in Syrian hamster (BHK21/C13) cells (20). BHK21/C13 cells are, so far as we know, the closest relatives to CHO cells for which nearest neighbor frequencies have been determined. As with Physarum, transfer is observed from all four dNTP's to all four rNMP's. However, there are several differences between the transfer patterns for Physarum and for CHO cells. First, the transfer from all four dNTP's is considerably lower for CHO cells than for Physarum. Second, the transfer frequency from the different dNTP's varies less than three-fold for CHO cells, while it varies about 30-fold for Physarum. And third, there is more variation in the distribution among the four ribonucleotides in CHO cells than in Physarum.

This variation in distribution is most striking in the very low frequency of transfer from dGTP to rCMP. Notice, however, that the frequency of the dinucleotide

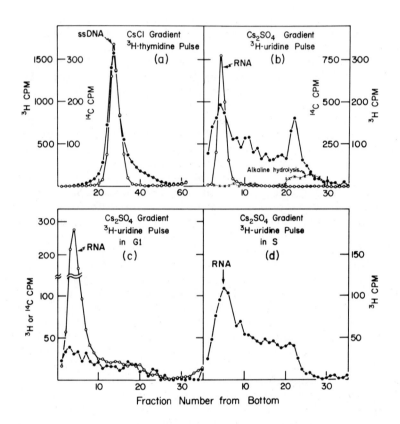

Figure 6.

Fig. 6. Density characterization of nucleic acids pulse-labeled with ^{3}H-thymidine or ^{3}H-uridine.

(a) ^{3}H-thymidine pulse, CsCl gradient. The indicated fractions from Fig. 5c were pooled and dialyzed overnight against buffer (0.1 M Na Acetate, 5-10 μg/ml dextran sulfate, 0.1% Sarkosyl, pH 7.0), and then reduced in volume by rotary evaporation in siliconized glassware. A CsCl step gradient was made by the method of Brunk and Leick (25). To the sample (3 ml) was added 2.5 g CsCl. This was layered in a polyallomer centrifuge tube over a heavy step of 3 ml of buffer and 5 g of CsCl. The gradient was centrifuged in a Spinco Type 65 rotor at 44,000 rpm for 18 hrs at 24°. ●——●, ^{3}H cpm; o——o, added ^{14}C-labeled single-stranded CHO cell DNA.

(b) ^{3}H-uridine pulse, unsynchronized cells, Cs$_2$SO$_4$ gradient. The indicated fractions from Fig. 5d were pooled and dialyzed overnight against 10 mM Tris, 1 mM EDTA, pH 7.4. A small amount of ^{14}C ribosomal RNA was added to the sample. The volume was reduced as in (a), then 2.7 g of Cs$_2$SO$_4$ were dissolved in the sample and the total volume brought to 4.5 ml. The sample was centrifuged in a Spinco SW 50L rotor at 36,000 rpm for 48 hrs at 15° C. For alkaline hydrolysis, half of the above dialyzed sample was made 0.3 N with KOH and incubated at 37° for 20 hrs. After hydrolysis it was centrifuged as above. ●——●, ^{3}H cpm before hydrolysis; X····X, ^{3}H cpm after alkaline hydrolysis; o——o, added ^{14}C-labeled HeLa cell ribosomal RNA.

(c) ^{3}H-uridine pulse, cells in G1 phase, Cs$_2$SO$_4$ gradient. CHO cells were synchronized by Colcemid reversal (26). At 2 hrs after release from mitosis, 2×10^6 cells were pulse-labeled at 26° for 2 min with ^{3}H-uridine (as in Fig. 5b). The cells were lysed and run on a neutral sucrose gradient as in Fig. 5a. Viscous fractions were pooled and denatured by heating to 100° for 15 min followed by quick cooling. Centrifugation in Cs$_2$SO$_4$ was as in (b). ●——●, ^{3}H cpm; o——o, added ^{14}C-labeled HeLa cell ribosomal RNA.

(d) ^{3}H-uridine pulse, cells in S phase, Cs$_2$SO$_4$ gradient. As in (c), except that cells were pulse-labeled 6 hrs after release from mitosis. Arrow shows position of ribosomal RNA marker.

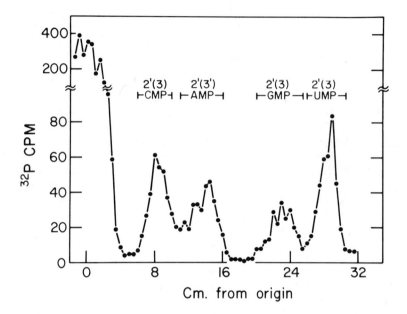

Fig. 7. Paper electrophoresis of 2'(3') ribonucleotides released by alkaline hydrolysis of CHO cell DNA
labeled in vitro with α-^{32}P-dCTP.

CHO cells were grown in suspension culture at 37°
in previously described medium (24, 26). About 1. 7 x 10^{8}
cells were centrifuged, washed twice with the isotonic
Hepes buffer of Winnacker et al (15), then resuspended in
3. 5 ml of the same buffer with 0. 5% NP-40 (5) and lysed
with a Dounce homogenizer. Four ml of cell lysate plus
one ml of assay mixture (15) containing 125 μCi of α-
^{32}P-dCTP (12 Ci/mmole; New England Nuclear) were
mixed and incubated for 5 min at 25° C. The reaction was
stopped by chilling, the nuclei were centrifuged and
washed once in isotonic saline. Finally the nuclear pellet
was suspended in 4 ml of isotonic saline and an equal
volume of 0. 1 M EDTA, 0. 1 M NaCl, 4. 5% Sarkosyl, pH
8. 1, was added to lyse the nuclei. Proteinase K was
added to 150 μg/ml, and the lysate was incubated at 37°
overnight. Chloroform extraction, ethanol precipitation,
chromatography through Sephadex G50, nearest neighbor
analysis, and paper electrophoresis were carried out as
described earlier (16).

sequence (dC)p(dG) is correspondingly low in Syrian hamster cell DNA (this sequence is strikingly low in all vertebrate DNA's examined (11)). While the low frequency of (dC)p(dG) in hamster cell DNA might explain the low transfer frequency generally observed from dGTP to rCMP, the transfer frequencies in general do not exactly mirror their corresponding nearest neighbor frequencies (Table 2). The transfer frequencies and patterns shown in Table 2 are very similar to those obtained by Tseng and Goulian (9, this conference) for DNA made in similar lysates of 8866 (human lymphocyte) cells.

In the limited number of experiments we have performed so far we have found the transfer frequencies to be fairly reproducible even when the cell type or in vitro system is varied. In Table 3 we show that the extent of ^{32}P transfer from dGTP to rNMP's is about 0.05% in CHO or HeLa cell lysates, or in permeabilized 8866 (human lymphocyte) cells, regardless of whether the Winnacker, Magnusson, and Reichard (15), Friedman and Mueller (14), or Benz (12) in vitro systems are employed.

However, when the extent of transfer is examined in purified CHO cell nuclei, it is found to be about fourfold higher than in total cell lysates (Table 3). The reason for this increase is not yet clear, but is consistent with the fact that purified nuclei are deficient in joining of Okazaki fragments into longer DNA strands (21-23, Waqar, unpublished). Perhaps inability to remove the RNA "primer" is one reason why purified nuclei are defective in such joining.

In order to prove that the ^{32}P-labeled ribonucleotides we observe are due to transfer of ^{32}P from adjacent deoxyribonucleotides, we have digested DNA synthesized in isolated nuclei using α-^{32}P-dGTP with pancreatic DNase I and venom phosphodiesterase until the ^{32}P was completely degraded to acid-soluble form. The released ^{32}P was characterized by paper electrophoresis, and more than 99.99% of it was recovered as 5'-dGMP. Thus, the 0.2% of ^{32}P found in ribonucleotides when the same DNA is hydrolyzed with alkali (Table 3) is probably transferred from adjacent 5'-dGMP.

TABLE 2

Distribution of radioactivity in 2'(3')rNMP's after alkaline hydrolysis of CHO cell DNA labeled in vitro with α-^{32}P-dNTP's; comparison with nearest neighbor frequencies in BHK21/C13 cell DNA (in parentheses).

α-^{32}P-dNTP	% of incorporated ^{32}P transferred to ribonucleotides	Distribution of ^{32}P in ribonucleotides (%)			
		UMP(TMP)	AMP	CMP	GMP
dATP	0.09	31 (26)	48 (32)	9 (22)	13 (20)
dTTP	0.12	39 (33)	34 (25)	10 (19)	17 (23)
dGTP	0.05	30 (44)	40 (33)	4 (4)	27 (20)
dCTP	0.13	30 (27)	26 (29)	27 (19)	18 (25)

Nearest neighbor analysis experiments were carried out for each of the four dNTP's as described in the legend to Fig. 7 for dCTP.
Data on nearest neighbor frequencies in BHK21/C13 cell DNA are taken from Suback-Sharpe et al (20).

TABLE 3

Comparison of radioactivity distributions in 2'(3') rNMP's after alkaline hydrolysis of mammalian cell DNA labeled in vitro with α-^{32}P-dGTP.

Cell Type	In Vitro System*	% of incorporated ^{32}P transferred to ribonucleotides	Distribution of ^{32}P in ribonucleotides (%)			
			UMP	AMP	CMP	GMP
CHO	Winnacker, Magnusson, and Reichard (15): total cell lysate	0.05	30	40	4	27
CHO	Winnacker, Magnusson, and Reichard (15): purified nuclei	0.20	24	29	19	28
CHO	Friedman and Mueller (14): total cell lysate	0.05	40	44	3	13
8866	Benz and Strominger (12): permeabilized cells	0.04	28	46	2	23
HeLa	Winnacker, Magnusson, and Reichard (15): total cell lysate	0.05	36	35	6	23
HeLa	Friedman and Mueller (14): total cell lysate	0.05	33	37	6	24

*All pulses were at 25° for 5 min.

DISCUSSION

Both density shift and ^{32}P transfer experiments in
Physarum polycephalum, and ^{32}P transfer experiments
with mammalian cells, appear consistent with the idea
that a short stretch of RNA serves as primer for Okazaki
fragments. But why isn't a density shift observed in
Okazaki fragments from mammalian cells, and why are
the transfer frequencies so much lower in mammalian
cells than in Physarum polycephalum? One possibility
is that Okazaki fragments have a much longer lifetime
before joining in Physarum polycephalum than in mammalian cells, and another possibility is that Okazaki fragments have a relatively longer RNA "primer" in Physarum than in mammalian cells. We have not yet ade-
quately characterized the size or kinetics of joining of
Okazaki fragments in Physarum or the size of the RNA
stretch in Physarum or mammalian cells, but we hope to
do so in the near future.

Additional questions which remain to be answered
before the case for RNA priming of Okazaki fragments
in eukaryotic cells can be considered a strong one are:
(a) what are the kinetics of "primer" synthesis and
degradation?; (b) what is at the 5' end of the RNA "primer"?;
(c) are the sequences of the RNA "primers" on different
Okazaki fragments restricted or random?; and (d) can a
requirement for ribonucleoside triphosphates be demonstrated for DNA synthesis in purified nuclei which have
been depleted of endogenous ribonucleoside triphosphates?
We hope that these questions will soon be answered.

ACKNOWLEDGMENTS

We thank our colleagues Howard Edenberg, Janis
Fraser, and Henry Horwitz for helpful advice and discussions. This research was supported by grants from
the National Science Foundation and the National Institutes
of Health. R.M. was a Special Fellow of the National
Institutes of Health, and J.A.H. is a recipient of a Career
Development Award from the National Institutes of Health.

REFERENCES

1. Sugino, A. , Hirose, S. , and Okazaki, R. Proc. Nat. Acad. Sci. (USA) 69, 1863 (1972).
2. Sugino, A. and Okazaki, R. Proc. Nat. Acad. Sci. (USA) 70, 88 (1973).
3. Hirose, S. , Okazaki, R. , and Tamanoi, F. J. Mol. Biol. 77, 501 (1973).
4. Magnusson, G. , Pigiet, V. , Winnacker, E. L. , Abrams, R. , and Reichard, P. Proc. Nat. Acad. Sci. (USA) 70, 412 (1973).
5. Pigiet, V. , Eliasson, R. , and Reichard, P. J. Mol. Biol. 84, 197 (1974).
6. Eliasson, R. , Martin, R. , and Reichard, P. Biochem. Biophys. Res. Commun. 59, 307 (1974).
7. Reichard, P. , Eliasson, R. , and Söderman, G. Proc. Nat. Acad. Sci. (USA) 71, 4901 (1974).
8. Hunter, T. and Francke, B. J. Mol. Biol. 83, 123 (1974).
9. Tseng, B. Y. and Goulian, M. (Manuscript submitted for publication).
10. Waqar, M. A. and Huberman, J. A. Biochem. Biophys. Res. Commun. 51, 174 (1973).
11. Josse, J. , Kaiser, A. D. , and Kornberg, A. J. Biol. Chem. 236, 864 (1961).
12. Benz, W. and Strominger, J. (Personal communication).
13. Seki, S. , Lemahieu, M. , and Mueller, G. C. Biochim. Biophys. Acta 378, 333 (1975).
14. Friedman, D. L. and Mueller, G. C. Biochim. Biophys. Acta 161, 455 (1968).
15. Winnacker, E. L. , Magnusson, G. , and Reichard, P. J. Mol. Biol. 72, 523 (1972).
16. Waqar, M. A. and Huberman, J. A. Biochim. Biophys. Acta in press (1975).
17. Berger, H. Jr. and Huang, R. C. C. Cell 2, 29 (1974).
18. Gautschi, J. R. and Clarkson, J. M. Eur. J. Biochem. 50, 403 (1975).
19. Mendelsohn, J. , Castagnola, J. M. , and Goulian, M. (Manuscript submitted for publication).
20. Suback-Sharpe, H. , Bürk, R. R. , Crawford, L. V. , Morrison, J. M. , Hay, J. , and Keir, H. M. Cold Spring Harbor Symp. Quant. Biol. 31, 737 (1966).

21.   Kidwell, W. R. and Mueller, G. C.   Biochem. Bio-
      phys. Res. Commun. 36, 756 (1969).
22.   Francke, B. and Hunter, T.   J. Virol. 15, 97
      (1975).
23.   Otto, B. and Reichard, P.   J. Virol. in press
      (1975).
24.   Huberman, J. A. and Horwitz, H.   Cold Spring
      Harbor Symp. Quant. Biol. 38, 233 (1973).
25.   Brunk, C. F. and Leick, V.   Biochim. Biophys.
      Acta 179, 136 (1969).
26.   Huberman, J. A., Tsai, A., and Deich, R. A.
      Nature 241, 32 (1973).

INITIATOR RNA IN DISCONTINUOUS SYNTHESIS OF POLYOMA DNA

Peter Reichard

Medical Nobel Institute, Department of Biochemistry,
Karolinska Institute, Stockholm, Sweden

Isolated nuclei from polyoma virus infected 3T6 cells continue under the proper in vitro conditions to elongate progeny strands of polyoma DNA replicative intermediates formed in the intact cells prior to isolation of nuclei (1). This process involves semiconservative replication of viral DNA (2) by host cell enzymes. Infected nuclei are therefore a useful system for the study of the mechanism involved in mammalian DNA replication.

Previous work has shown that elongation of polyoma DNA progeny strands occurs by discontinuous synthesis (3) as described by Okazaki for microbial systems (4). However, the size of the polyoma DNA fragments (approximately 150 nucleotides, sedimentation constant about 4 s) is considerably smaller than that of the Okazaki fragments of E. coli (approximately 1000 nucleotides). It was first shown by Kornberg's group that RNA is involved as a primer for the initiation of DNA strands of the small phage M 13 (5) and subsequent work of Okazaki and collaborators presented evidence for RNA involvement during initiation of Okazaki pieces in E. coli (6). We reported the participation of RNA during discontinuous synthesis of polyoma DNA (3) and proposed the name initiator RNA (iRNA) for this type of RNA (7).

I should like to stress at the outset that our work concerns initiation of Okazaki fragments and not the first initiation of a new DNA strand at the origin of replication. RNA involved in the priming at the origin would of course also be an initiating RNA. However, in order to avoid confusion, I would like to propose that the term initiator RNA be reserved for RNA priming Okazaki fragments while RNA at the origin might be given a separate name, e. g. origin RNA.

Here I only give a brief description of the characterization of polyoma iRNA, since a more detailed presentation of our work, including experimental data, already has been published. We first obtained suggestive evidence for the presence of RNA at the 5' end of growing DNA strands (3) and then concentrated our efforts on the characterization of nucleotides present at the RNA-DNA link and at the 5' end of the RNA chain.

The first purpose was achieved by studying the transfer of isotope from α-^{32}P labeled deoxynucleoside triphosphates to ribonucleotides after alkaline hydrolysis of the RNA-DNA hybrids and by labeling the 5'-OH ends of DNA strands exposed after alkaline hydrolysis with γ-^{32}P-ATP and polynucleotide kinase (8). The experiments demonstrated that (a) essentially all growing DNA strands contained at least one ribonucleotide at their 5' end, (b) all four deoxynucleotides and ribonucleotides were present at the RNA-DNA link, (c) the doublets at the link did not strictly follow the rules given for nearest neighbour frequencies but the doublet rCpdG was found with low frequency in accordance with the reported low occurrence of CpG in polyoma DNA (9).

The 5' end of iRNA was characterized by quantitation of labeled ribonucleoside tetraphosphates after alkaline hydrolysis of iRNA labeled from ^{3}H and/or β-^{32}P-labeled ribonucleoside triphosphates. For this purpose iRNA was purified as part of polyoma DNA replicative intermediates (RI) and in this way freed from most contaminating RNA. Final purification of iRNA was achieved by gel electrophoresis on polyacrylamid after digestion of RI with pancreatic DNAse. This removed remaining contaminating RNA, tightly bound to RI, since it was much larger in size than iRNA. The fact that iRNA - and not the large sized RNA - was connected to growing DNA chains was demonstrated by the finding that isotope from α-^{32}P-dTTP was exclusively transferred to iRNA after digestion of RI with pancreatic DNAse.

From gel electrophoresis we estimate the size of iRNA to be close to that of a decanucleotide. In spite of the homogeneity in size, iRNA has no unique base sequence. It contains either ATP or GTP (but not UTP or CTP) at the 5' end. Furthermore, digestion with pancreatic RNAse of iRNA labeled at the 5' end from either β-^{32}P-labeled ATP or GTP released a series of different oligonucleotides, implying that transcription of iRNA may start at many points around the circular DNA molecule. The results do, however, not necessarily indicate complete random starts.

Our data suggest that the synthesis of Okazaki fragments of polyoma DNA is initiated by a new class of low molecular weight RNA for which we propose the name initiator RNA. We find no evidence for a specific base sequence but the molecules show a remarkable homogeneity in size, corresponding to that of a decanucleotide. We propose that the size of iRNA may play a role as a signal for the switch from RNA to DNA synthesis. This raises the further more speculative possibility that the size of complete Okazaki fragments may determine new points of initiation of iRNA and thereby of new Okazaki fragments. Thus chain size, rather than base sequence, may play a role as a "signal" during discontinuous DNA synthesis.

There is good reason to believe that the elongation of polyoma DNA is catalyzed by the replication machinery of the host cell and that the results obtained with a viral DNA thus may be of a more general significance for our understanding of how mammalian DNA is replicated.

This work was supported by grants from the Swedish Cancer Society and Magnus Bergvalls Stiftelse.

1. Winnacker, E.-L., Magnusson, G. and Reichard, P., J. Mol. Biol. 72, 523-537, 1972.
2. Magnusson, G., Winnacker, E.-L., Eliasson, R. and Reichard, P., J. Mol. Biol. 72, 539-552, 1972.
3. Magnusson, G., Pigiet, V., Winnacker, E. L., Abrams, R. and Reichard, P., Proc. Nat. Acad. Sci. U.S.A. 70, 412-415, 1973.
4. Okazaki, R., Okazaki, T., Sakabe, K., Sugimoto, K., Kainuma, R., Sugino, A. and Iwatsuki, N., Cold Spring Harbor Symp. Quant. Biol. 33, 129-144, 1968.
5. Brutlag, D., Schekman, R. and Kornberg, A., Proc. Nat. Acad. Sci. U.S.A. 68, 2826-2829, 1971.
6. Sugino, A., Hirose, S. and Okazaki, R., Proc. Nat. Acad. Sci. U.S.A. 69, 1863-1867, 1972.
7. Reichard, P., Eliasson, R. and Söderman, G., Proc. Nat. Acad. Sci. U.S.A. 71, 4901-4905, 1974.
8. Pigiet, V., Eliasson, R. and Reichard, P., J. Mol. Biol. 84, 197-216, 1974.

9. Subak-Sharpe, H. , Bürk, R. R. , Crawford, L. V. ,
Morrison, J. M. , Hay, J. and Keir, H. M. , Cold
Spring Harbor Symp. Quant. Biol. <u>31</u>, 737-748, 1966.

DISCONTINUOUS SV40 DNA SYNTHESIS AND THE DETECTION OF GAP CIRCLE INTERMEDIATES

Philip J. Laipis, Arup Sen, Arnold J. Levine and
Carel Mulder

Department of Biochemistry
Princeton University
Princeton, New Jersey 08540
and
Cold Spring Harbor Laboratory
Cold Spring Harbor, New York 11724

ABSTRACT. When SV40 DNA synthesis is inhibited in infected cells by the drug hydroxyurea, replicative intermediates accumulate, with single-stranded gaps separating short, 4S, Okazaki-like fragments of DNA from the longer progeny strands. A similar gapped structure exists in SV40 replicative intermediates isolated after a brief pulse-label with (^{3}H)-thymidine.

In addition to the gapped replicative intermediate molecules, a class of 16S terminating intermediate molecules has been shown to exist in both drug inhibited, infected cells and normal infected cells. These molecules contain a 3'-hydroxy, 5'-phosphate bounded gap at the terminus of DNA replication. Our results suggest that separation of interlocked daughter SV40 molecules can occur before DNA strand elongation has finished.

INTRODUCTION

The events occurring during SV40 DNA replication may be conveniently divided into four stages: 1) initiation of DNA synthesis, 2) polynucleotide chain elongation, 3) segregation of the two interlocked newly synthesized daughter molecules, and 4) maturation of the progeny daughter molecules. The experiments presented here will deal with two aspects of SV40 DNA replication, the polynucleotide chain elongation step and the maturation or processing of the post-segregational 16S intermediates to form the mature closed-circular viral DNA.

We have previously shown that replicating molecules of SV40 DNA isolated from lytically infected cells treated with either hydroxyurea (HU) or cytosine arabinoside contain short, 4S, Okazaki-like fragments of DNA hydrogen bonded to the circular parental strands (1). These 4S fragments of DNA are separated from each other and from longer progeny strands by single-stranded gaps. The gaps could be repaired and the 4S fragments joined together to form longer polynucleotide chains in an in vitro DNA repair system consisting of T4 DNA polymerase and E. coli polynucleotide ligase. It is possible, however, that the inhibitors we used may have induced some aberrant form of SV40 DNA replication, even though the 4S oligonucleotide fragments were present during all stages of replication and could be chased into mature, viral, closed-circular DNA. In order to rule out this possibility, we have examined the structure of replicative intermediates isolated from SV40 infected cells pulse-labeled for brief periods of time, at room temperature, with (^{3}H)-thymidine. Fareed and Salzman have used this method to isolate replicating molecules labeled preferentially in short, 4S fragments (2).

RESULTS

SV40 infected cells were pulse-labeled at room temperature for 20 seconds with high specific activity thymidine, replicating DNA was extracted by the procedure of Hirt (3) and late replicative intermediates were purified in an ethidium bromide-cesium chloride (EtBr-CsCl) gradient (1). These late replicating forms were sedimented in alkaline sucrose gradients, Figure 1A. Thirty-seven percent of the (^{3}H)-labeled DNA sediments in a position characteristic of short, 4S fragments. These late replicative intermediates were then repaired in an in vitro system containing E. coli ligase alone, T4 DNA polymerase alone, and both enzymes together (1). Incubation with both ligase and T4 DNA polymerase resulted in the joining of the 4S fragments into longer progeny strands, indicating the presence of 3'-OH, 5'-PO$_4$ bounded gaps separating the fragments and progeny strands.

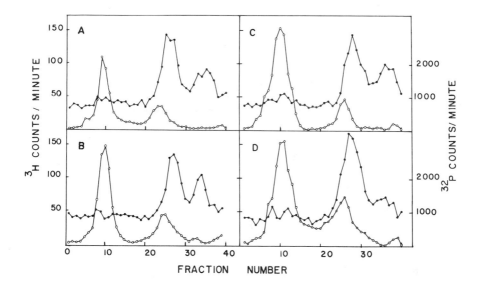

Figure 1.   Alkaline sucrose gradients of late replicating
SV40 DNA molecules isolated from infected cells pulse-
labeled for 20 seconds with 3H-thymidine.
A.   Control gradient.   No in vitro repair
B. ^{3}H-DNA repaired in vitro with E. coli ligase
C. ^{3}H-DNA repaired in vitro with T_4 DNA polymerase
D. ^{3}H-DNA repaired in vitro with both ligase and polymerase
          o-o-o ^{32}P SV40 marker, 55S and 16-18S
          ●-●-● ^{3}H pulse-labeled DNA

Table 1

DNA	Before Digestion		After Digestion		
Normal 16S	16.3S		15.4S	8.6S	
	1.68×10^6		1.45×10^6	3.13×10^5	
HU 16S	15.6S	4.4S	15.8S	8.8S	3.9S
	1.50×10^6	5.4×10^4	1.55×10^6	3.32×10^5	3.9×10^4

Sedimentation coefficients and molecular weights of SV40
16S gap circle molecules after digestion by E. coli R1
restriction endonuclease.   Molecular weights were calculated assuming the 16S SV40 single strand marker has a
molecular weight of 1.6×10^6 daltons.   Data are from the
alkaline sucrose gradients of Figure 4.   Error is approximately ±0.5S, about one fraction.

Fareed et al. (4) have suggested that a likely final intermediate in SV40 DNA replication is a 16S molecule with a progeny strand containing a polynucleotide chain interruption at or near the terminus of DNA synthesis. During the preparation of the late replicative intermediates described above, we detected such gap circle molecules in both normal and HU-inhibited, infected cells. SV40 late replicative intermediate molecules were isolated from normal and HU-inhibited infected cells, and purified by EtBr-CsCl density gradients. These late replicative molecules were sedimented in neutral sucrose gradients to separate the 25S replicating form from 16S gap circle intermediate molecules. The 16S molecules were then repaired and sedimented in alkaline sucrose gradients; the results are presented in Figures 2 and 3. Incubation of the 16S molecules with DNA polymerase plus ligase converted between 35% and 50% of the 16S molecules to the closed-circular form, demonstrating that both the normal 16S intermediate and the HU-inhibited 16S intermediate contain a gap of one or more nucleotides in the newly synthesized progeny strand. To demonstrate that the gap seen in the HU-inhibited 16S intermediate was indeed at the terminus of the DNA synthesis, these molecules and the normal 16S intermediate isolated from uninhibited cells were analyzed with the restriction enzyme E. coli R1 (5). The results we obtained with normal 16S intermediate (Table 1) are in agreement with those of Fareed et al. (4), i.e., the 16S molecules contain an interruption at the approximate terminus of DNA replication, 17% from the R1 scission site. The 16S intermediate isolated from HU-inhibited infected cells also contains a gap at the terminus of DNA synthesis (Figure 4, Table 1).

DISCUSSION

SV40 DNA synthesis proceeds by a discontinuous mechanism which generates short, 4S, Okazaki-like fragments of DNA in both normal infected cells (2), drug-inhibited infected cells (1), and several in vitro nuclei systems extracted from infected cells (5,7). We have shown that the 4S fragments formed during a very short pulse with (^{3}H)-thymidine in SV40 infected cells are separated by 3'-OH, 5'-PO$_4$ bounded gaps from themselves and from longer progeny molecules. Complete repair of these gaps has not

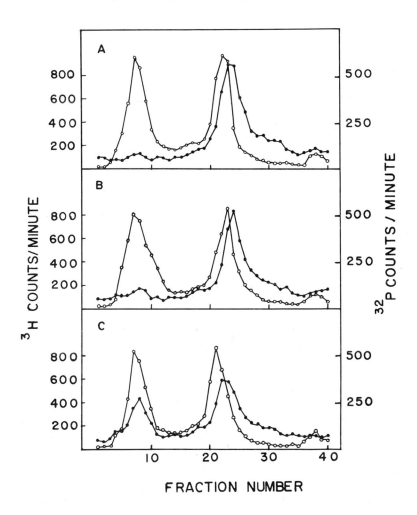

Figure 2. Alkaline sucrose gradients of 16S DNA isolated
from the late replicating fraction of SV40 DNA purified
from infected cells labeled for 10 minutes with ^{3}H thymidine
A. Control gradient. No in vitro repair.
B. ^{3}H-DNA repaired in vitro with E. coli ligase.
C. ^{3}H-DNA repaired in vitro with E. coli ligase and T$_4$ DNA
   polymerase
       o-o-o ^{32}P SV40 marker, 55S and 16-18S
       ●-●-● ^{3}H pulse-labeled DNA

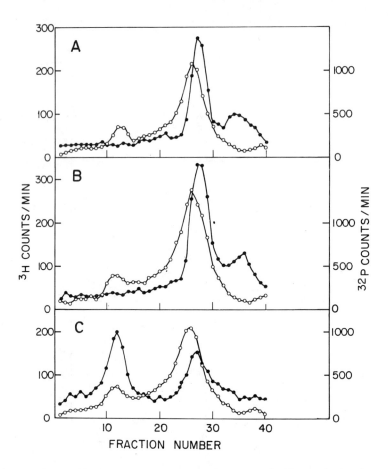

Figure 3. Alkaline sucrose gradients of 16S DNA isolated from the late replicating fraction of SV40 DNA purified from infected cells labeled for 15 minutes with ^{3}H-thymidine in the presence of 10mM HU.

A. Control gradient. No in vitro repair.
B. ^{3}H-DNA repaired in vitro with E. coli ligase
C. ^{3}H-DNA repaired in vitro with E. coli ligase and T$_4$ DNA polymerase

    o-o-o ^{32}P SV40 marker, 55S and 16-18S
    •-•-• ^{3}H pulse-labeled DNA

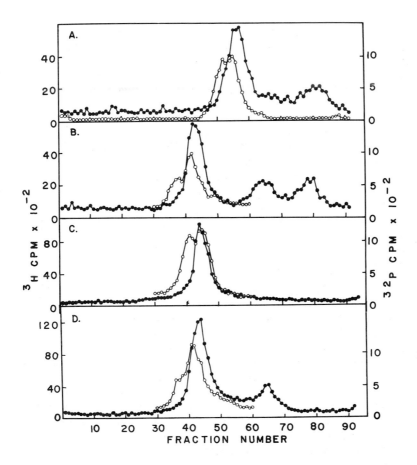

Figure 4. Alkaline sucrose gradients of the 16S DNA's of Figures 2 and 3 digested with E. coli R1 restriction enzyme
A. Control gradient. HU inhibited 16S DNA, before digestion.
B. HU inhibited 16S DNA after digestion with E. coli R1.
C. Control gradient. Normal 16S DNA, before digestion.
D. Normal 16S DNA after digestion with E. coli R1.
    o-o-o ^{32}P SV40 single-stranded DNA markers, 16-18S
    •-•-• ^{3}H pulse-labeled DNA

been accomplished. One possibility is that the 5' end of
the gaps may be blocked by ribonucleotides. In isolated
nuclei systems, polyoma viral DNA is synthesized in short
fragments initiated by a short sequence of ribonucleotides
on their 5' end (5,7). The ribonucleotides may block gap
closure by E. coli ligase.

Our results also suggest that two DNA polymerases are
involved in SV40 DNA replication. One of these DNA polymerases is involved in the synthesis of the short, 4S fragment and the other DNA polymerase is responsible for
filling in the gaps left between fragments and allowing
ligation to occur. This discontinuous synthesis occurs in
both normal and drug inhibited cells, suggesting it is the
common mechanism of DNA synthesis, and not the result of
aberrations in DNA metabolism induced by the short pulse
or drug inhibition.

In addition to late replicating forms of SV40 DNA, we
have examined the gap circle final intermediate in SV40 DNA
replication. This 16S molecule from both normal and HU-
inhibited, infected cells contains a 3'-OH, 5'-PO$_4$ bounded
gap at or very near the terminus of DNA replication. This
gapped circle molecule presumably results from the segregation of the two interlocked daughter molecules of a
single late replicating SV40 intermediate. Our results
indicate that segregation of the daughter molecules occurs
before DNA replication is complete. Attempts to determine
the size of this terminal gap and the sequence of the
deoxyribonucleotides in the gap are currently in progress.

REFERENCES

1. Laipis, P.J. and Levine, A.J., Virology 56, 580 (1973).
2. Fareed, G.C. and Salzman, N.P., Nature (New Biol.) 238, 274 (1972).
3. Hirt, B., J. Mol. Biol. 26, 365 (1967).
4. Fareed, G.C., McKerlie, M.L. and Salzman, N.P., J. Mol. Biol. 74, 95 (1973).
5. Mulder, C. and Delius, H., Proc. Nat. Acad. Sci. 69, 3215 (1972).
6. Magnusson, G., Pigiet, V., Winnacker, E.L., Abrams, R., and Reichard, P., Proc. Nat. Acad. Sci. 70, 412 (1973).
7. Hunter, T. and Francke, B., J. Mol. Biol. 83, 123 (1974).

Part V
Replication of Small
Bacterial Viruses

STRUCTURE AND REPLICATION OF REPLICATIVE FORMS
OF THE ØX-RELATED BACTERIOPHAGE G4

Dan S. Ray and Jeanene Dueber
Molecular Biology Institute and
Department of Biology
University of California
Los Angeles, California 90024

ABSTRACT. The DNA of the ØX174-related bacteriophage G4 is clearly distinguished from that of ØX174 by alkaline equilibrium centrifugation. G4 viral DNA bands at a density of 1.753 g/cc, while ØX174 DNA bands at 1.765 g/cc. Similarly, the complementary strand of G4 replicative form bands at 1.767 g/cc, while that of ØX174 bands at 1.756 g/cc.

Both the viral and the complementary strands of G4 replicative forms are synthesized early in the infectious process. However, only the viral strand is replicated later in infection. Chloramphenicol at 30 µg/ml prevents this transition to an asymmetric mode of replication.

Replication of G4 replicative form DNA is inhibited in E. coli rep⁻ cells and the parental replicative form accumulates in an open circular form having a closed circular complementary strand and a unit-length linear viral strand. The single discontinuity in the viral strand is located at 50% of the genome's length from G4's single Eco RI site. G4 open circular replicative DNA formed during single strand synthesis also has a unique discontinuity at 50% of the genome's length from the Eco RI site. We infer that nicking of G4 parental replicative form DNA prior to its replication occurs at the origin of the viral strand and that the viral strand origin is at the same site for both replicative form replication and single strand synthesis.

INTRODUCTION

Bacteriophage G4 is a highly virulent, rapidly replicating phage related to ØX174 (1). It is morphologically

similar to ØX174 and synthesizes the same number of viral
proteins in infected cells. The sizes of all of the G4
proteins are similar but not identical to those of ØX174.
G4 differs in host range from ØX174 and is not inactivated
by antiserum to ØX174. G4 DNA is circular, single stranded
and of the same molecular weight as ØX174 DNA. Heteroduplexes between G4 and ØX174 show a distinctive pattern of
unpaired single strand regions separated by regions of
duplex DNA.

G4 DNA is a superior template in the formation of the
parental replicative form (RF) in extracts of uninfected
cells (2). The complementary strand is initiated at a
single unique site on the viral DNA by the action of host
proteins. Initiation involves the synthesis of an RNA
primer by the dna G protein, a rifampicin-resistant RNA
polymerase, in the presence of the E. coli unwinding
protein (3). This mechanism of initiation is similar to
that of the filamentous phage M13 which initiates complementary strand synthesis in vitro at a unique site and
utilizes the RNA primer formed by RNA polymerase (4). In
contrast, initiation of ØX174 complementary strands in
vitro involves four additional proteins as well as the dna
G protein and the unwinding protein but does not occur at a
single unique site as for G4 (5).

The rapid rate of replication of G4 DNA, both in vivo
and in vitro, the distinctive requirements for host proteins in the SS (single strand)→ RF conversion in vitro
and the existence of a single Eco RI restriction site in
the RF (6), a valuable reference point in physical studies
of the DNA, prompted us to use G4 for some of our studies
of the mechanism of small phage replication. The work presented here describes the results of investigations of the
structure and mechanism of replication of G4 DNA.

METHODS

Materials and methods have all been described in
detail elsewhere and include the following: media, bacterial strains, growth conditions and labeling of DNA (7,8);
neutral and alkaline ultracentrifugal techniques (8,9);
preparation of ^{32}P-labeled G4 phage, DNA isolation and
sedimentation through neutral sucrose gradients and Eco RI
treatment of DNA (10).

RESULTS

Properties of the Viral and Complementary Strands of G4
DNA: Preparatory to an investigation of G4 DNA replication,
we have determined some of the properties of G4 DNA. To
our surprise, we found that G4 viral DNA differs considerably in buoyant desnity in alkaline CsCl from that of
either ØX174 or M13 (Fig. 1(a)). Since the buoyant density
of DNA in alkaline solution depends on the G + T content of
the individual polynucleotide strands (11), the wide separation of G4 DNA from ØX174 DNA indicates a lower G + T
content for G4 DNA. Thymine accounts for approximately
one-third of the bases in both ØX174 and M13 DNA (12).
Though the base composition of G4 DNA has yet to be determined, we expect, on the basis of its low buoyant density
in alkaline CsCl, that it will differ considerably from
that of ØX174.

The ability to separate the two strands of RF DNA's of
ØX174 and M13 by alkaline equilibrium centrifugation has
been a valuable probe of these DNA species. Fortunately,
the two strands of G4 RF are also separable in alkaline
CsCl. This could not have been predicted a priori, particularly in view of the large difference in the alkaline
buoyant density of ØX174 and G4 viral DNA's.

To prepare labeled G4 complementary strand DNA, we
have made use of the E. coli rep mutation (7,13). Mutant
cells infected with either ØX174 or M13 carry out parental
RF formation (SS→RF) but not RF replication (RF→RF).
Thus, only the complementary strand of RF formed in rep
cells is newly synthesized. ^{3}H-thymidine incorporated into
the RF in this case is confined exclusively to the complementary strand. We have prepared G4 RF labeled with ^{32}P in
the viral strand and ^{3}H in the complementary strand by
infection of E. coli rep$_3$ with ^{32}P-labeled G4 in the presence of ^{3}H-thymidine. Like ØX174 and M13, G4 replication
appears to be blocked in this host prior to RF→RF replication. None of the ^{3}H label in the RF formed in rep$_3$
cells bands at the density of viral strands in alkaline
CsCl. Fig. 1(b) shows the buoyant density distribution of
the DNA strands derived from the RF II (replicative forms
having one or more discontinuities in either or both
strands) formed in rep$_3$ cells. All of the ^{32}P label bands
at the density (1.753 g/cc) of G4 viral strands, as

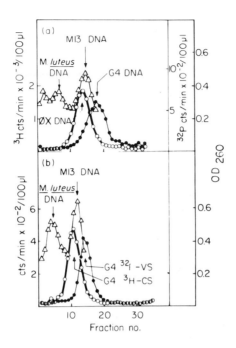

Fig. 1.   Alkaline CsCl equilibrium centrifugation of
G4 viral and complementary strands.   (a) ^{32}P-G4 viral DNA
centrifuged to equilibrium along with ^{3}H-ØX174 viral DNA,
M13 DNA and M. luteus DNA.   Only the latter two DNA's are
present in sufficient quantity to be detected by U.V.
absorption.   Blue dextran, added as a visual marker interferes with the absorption profile in fractions 1-5.
(b) G4 RF II, formed in rep$^-$ cells and containing ^{32}P in
the viral strand and ^{3}H in the complementary strand, centrifuged to equilibrium along with M. luteus and M13 DNA's
as density markers.   Density increases from right to left.
VS, viral strand; CS, complementary strand; o——o, ^{3}H;
●——●, ^{32}P; △——△, O.D.$_{260}$.

373

TABLE 1

Alkaline Buoyant Densities of the Two Strands of
Replicative Form DNA's of M13, ØX174 and G4

DNA Source	Viral Strand Density (g/cc)	Complementary Strand Density (g/cc)	Reference
M13	1.763	1.748	(9)
ØX174	1.765	1.756	(11,14)
G4	1.753	1.767	this work

expected. However, the ^{3}H label bands entirely at a density of 1.767 g/cc, widely separated from the position of
viral single strands. On the basis of these results and
others presented later, we conclude that the component
banding at 1.767 g/cc is the G4 complementary strand. A
comparison of the buoyant densities of the viral and complementary strands of some single stranded DNA phages is
presented in Table 1.

Analysis of G4 RF's Formed at Early and Late Times in the
Infectious Process: To examine the regulation of G4 DNA
replication, we have characterized replicative intermediates pulse labeled at different times after infection.
Fig. 2 shows neutral sucrose gradient sedimentation of
lysates of E. coli 4704 pulse labeled at 0-4, 8-10 and 9-11
minutes after infection by G4. At each of these times, the
labeled material sediments largely as RF II and RF I
(closed circular replicative form). In Fig. 2(c) a small
amount of progeny single strands can be seen sedimenting at
fractions 32 to 39. However, progeny single strands
already encapsulated into phage particles would likely not
be released efficiently under our lysis conditions and
would therefore not be expected to be observed in large
amounts in these gradients.

The distribution of label between the two strands of
the RF's serves as an indicator of the stage of replication
occurring in the cell at the time of the pulse labeling.
Molecules involved solely in asymmetric single strand synthesis would be labeled only in the viral strand while
those involved in RF→RF replication would be labeled in
both strands. To examine the distribution of label between
the strands of the RF, the RF II molecules from each gradient were centrifuged to equilibrium in alkaline CsCl gradients. Since the RF II molecules have one or more single
strand discontinuities, the two strands readily separate
upon alkaline denaturation and band at their respective
equilibrium densities.

The density distributions of each of the pulse-
labeled RF II species are shown in Fig. 3. In each case
^{32}P G4 DNA has been added as a viral strand density marker.
The RF II labeled from 0-4 min after infection shows
labeled material at both viral strand and complementary
strand densities. Since ^{3}H-thymidine was present at the
time of infection, there are likely to be at least two

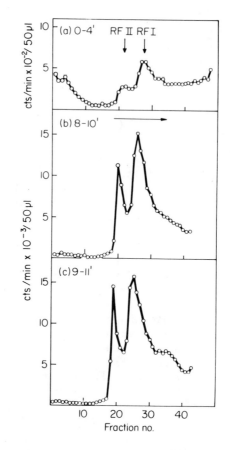

Fig. 2. Neutral sucrose gradient sedimentation of lysates of G4-infected cells pulse-labeled at different times after infection. 30 ml cultures of E. coli 4704 growing in TPA(P/3) medium at 2 x 10^8 cells/ml were infected with G4 at an moi of 1-2 and pulse labeled with ^{3}H-thymidine. (a) 0-4 min pulse (b) 8-10 min pulse (c) 9-11 min pulse. The horizontal arrow indicates the direction of sedimentation; the vertical arrows indicate the sedimentation positions of RF I and RF II.

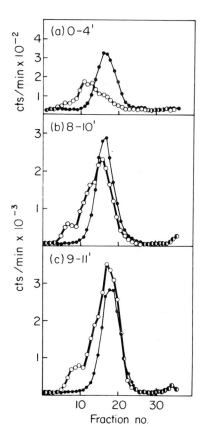

Fig. 3. Alkaline CsCl equilibrium centrifugation of pulse-labeled G4 RF II's. Aliquots of RF II fractions from Fig. 2 were centrifuged to equilibrium in alkaline CsCl along with ^{32}P-G4 DNA as a density marker. (a) 0-4 min pulse (b) 8-10 min pulse (c) 9-11 min pulse. Density increases from right to left. o——o, ^{3}H; ●——●, ^{32}P.

kinds of RF II present here, parental RF molecules formed
during SS→RF conversion and containing label in only the
complementary strand and RF molecules produced by RF→RF
replication and therefore containing label in both strands
of the RF. The presence of the former class of RF would
account for much of the excess label observed at the
density of complementary strands. (Calculation of the
relative amount of labeled viral and complementary strands
is impossible in the absence of knowledge of the relative
thymine content of the two strands.) Most of the pulse
label in the RF II species labeled at 8-10 or 9-11 minutes
after infection is contained in material of viral strand
density. In each case there is a small amount of label
contained in complementary strands and in contaminating
supercoils (fraction 5-8 in Fig. 3(b) and 7-10 in Fig.
3(c)). This strong bias towards viral strand labeling
indicates that the RF molecules are primarily involved in
RF→SS synthesis at these later times.

Analysis of G4 RF's Labeled in the Presence of Chloramphenicol: Due to the relative resistance of ØX174 gene A
expression to chloramphenicol (15), it is possible to
achieve a concentration of the drug sufficient to prevent
RF→SS synthesis but not RF→RF replication. Sufficient
gene A protein is synthesized at a chloramphenicol concentration of 30 µg/ml to allow RF replication to occur even
though single strand synthesis is inhibited. This appears
to also be the case for phage G4. Fig. 4 shows the alkaline equilibrium density distribution of G4 RF II labeled
from 6 to 16 minutes after infection of cells which were
treated with chloramphenicol at 30 µg/ml at 5 minutes after
infection. Unlike the RF II labeled late in infection in
the absence of chloramphenicol (Fig. 3), there is a very
substantial labeling of the G4 complementary strand. Thus,
like ØX174, the transition to asymmetric viral strand replication appears to be prevented by chloramphenicol.

A unique Discontinuity in the Viral Strand of G4 RF: Since
we have previously observed a gene A-specific discontinuity
in the viral strand of ØX174 RF (7,16), it was of interest
to ask whether we could detect and localize such a discontinuity in G4 RF. For ØX174, the site of action of the
gene A endonuclease (17) has not yet been determined. In
the case of G4, it should be possible to localize a site-

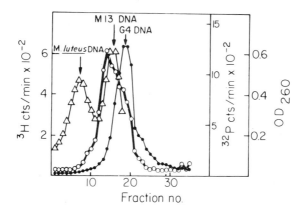

Fig. 4. Alkaline CsCl equilibrium centrifugation of G4 RF II formed in the presence of chloramphenicol at 30 μg/ml. E. coli 4704 cells growing in TPA(P/3) at 2 x 10⁸ cells/ml were infected with G4 at an moi of 1-2. At 5 min after infection, chloramphenicol was added at 30 μg/ml and the cells pulse labeled with ^{3}H-thymidine (10 μCi/ml) from 6 to 16 min. The RF II was isolated by neutral sucrose gradient centrifugation as in Fig. 2. Density increases from right to left. ^{32}P-G4, M. luteus and M13 DNA's were added as density markers. Only the latter two DNA's were present in sufficient quantity to be detected by U.V. absorption. o——o, ^{3}H; ●——●, ^{32}P; △——△, O.D.$_{260}$.

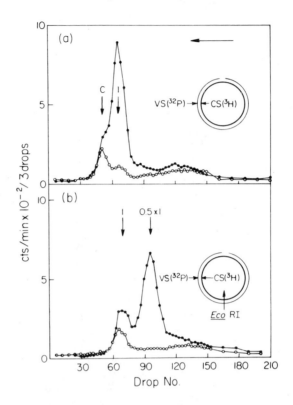

Fig. 5. Sedimentation of untreated and Eco RI-
treated G4 parental RF II under denaturing conditions.
Double-labeled RF II molecules formed upon infection of
E. coli rep⁻ in the presence of ³H-thymidine by ³²P-
labeled G4 was isolated by neutral sucrose gradient centrifugation as in Fig. 2. Portions of the RF II fractions
were concentrated by alcohol precipitation, incubated with
(b) or without (a) Eco RI and sedimented through alkaline
sucrose gradients. The direction of sedimentation is indicated by the horizontal arrow, the sedimentation positions
of circular (c) and unit-length linear (1) strands by
vertical arrows. The inserts show the structure of the
specifically nicked form of RF II and the location of the
discontinuity relative to the Eco RI site. ●——●, ³²P
cts/min (viral strand label); o——o, ³H cts/min (complementary strand label). From Ray and Dueber (1975).

specific discontinuity relative to the single Eco RI restriction site.

We therefore have analyzed G4 RF II formed upon infection of E. coli rep$^-$ cells with ^{32}P-labeled G4 in the presence of ^{3}H-thymidine. All of the ^{3}H label was contained in the complementary strand of the RF II. (Fig. 1(b) shows the alkaline buoyant density distribution of the ^{32}P and ^{3}H labels from such an RF II preparation.) Alkaline velocity sedimentation of RF II molecules formed in rep$^-$ cells shows that the ^{3}H-labeled complementary strands are largely circular while most of the ^{32}P-labeled viral strands are linear (Fig. 5(a)). The RF II molecules giving rise to these alkaline denaturation products must therefore be predominantly of the kind shown in the inset to Fig. 5(a).

Upon cleavage of these RF II molecules with Eco RI, the unit length viral strand is cleaved into two fragments of equal size (Fig. 5(b)), indicating that the discontinuity in the viral strand is opposite the Eco RI site. The circular complementary strand is cleaved to give unit-length linear strands. The unit-length viral strands seen in Fig. 5(b) arise largely from the cleavage of RF II molecules containing circular, rather than linear, viral strands. Such contaminating RF II molecules probably arise from random nicking of RF I during isolation.

In view of the uncertain relationship between the origin of viral strand synthesis and this site-specific discontinuity in the viral strand of pre-replicative RF's, it was of interest to determine the origin of the viral strand. This is most readily accomplished during RF→SS synthesis when large amounts of RF II accumulate due to the massive synthesis of viral strands. This, of course, assumes a single common origin of viral strand synthesis for both RF→RF and RF→SS synthesis. However, such RF II molecules isolated from ØX174-infected cells contain a single unique gap in the viral strand located in the vicinity of gene A (18), a region containing the origin for both RF→RF and RF→SS synthesis (19,20).

To isolate RF molecules containing a unique discontinuity at the viral strand origin, we have pulse-labeled G4-infected cells with ^{3}H-thymidine from 9 to 11 minutes after infection. Fig. 2(c) shows the neutral sucrose gradient sedimentation of the pulse-labeled DNA used for this experiment. The RF II molecules from this gradient were pooled,

381

ethanol precipitated and analyzed further. As shown in
Fig. 3(c), the pulse label was contained almost entirely in
the viral strand of the RF II. Alkaline velocity sedimentation of this RF II (Fig. 6(a)) shows that the ^{3}H-labeled
viral strand sediments at the rate of a unit-length linear
strand, as expected for a viral strand containing only a
single discontinuity. ^{32}P-labeled G4 DNA containing circular and linear unit-length single strands in a ratio of
approximately 2 to 1 was added as a sedimentation marker.

Cleavage of this pulse-labeled RF II with Eco RI
(Fig. 6(b)) yields viral strand fragments of half unit-
length indicating the presence in the pulse-labeled sample
of RF II molecules containing a single discontinuity at the
same unique site as for parental RF II formed in the rep
host. The unit-length viral strands seen in Fig. 6(b)
after Eco RI treatment arise from the cleavage of both contaminating RF I and RF II molecules having an intact viral
strand. The presence of a small amount of RF I in the
pooled RF II fractions can be seen in the alkaline equilibrium gradient in Fig. 3(c). They are not seen in the
alkaline velocity gradient in Fig. 6(a) due to their higher
sedimentation rate. The peak of circular ^{3}H-labeled viral
strands in Fig. 6(a) reflects the presence of the contaminating RF II molecules having circular viral strands.
Nonetheless, these results establish that most of the RF II
molecules pulse labeled during single strand synthesis contain a single unique discontinuity in the viral strand.

DISCUSSION

Even though the gel pattern of proteins from G4-
infected cells is remarkably similar to that of ØX174-
infected cells, the viral DNA's show several large regions
of non-homology by heteroduplex analysis (1). Our results
show another large difference between these DNA species.
G4 and ØX174 DNA's are widely separated in alkaline buoyant
density centrifugation. Furthermore, their respective complementary strands differ greatly in buoyant density and
the density of the G4 complementary strand is greater than
that of the viral strand while the opposite is true for
ØX174.

DNA synthesis early in G4 infection involves the synthesis of both viral and complementary strands. This
likely reflects a period of RF→RF synthesis prior to

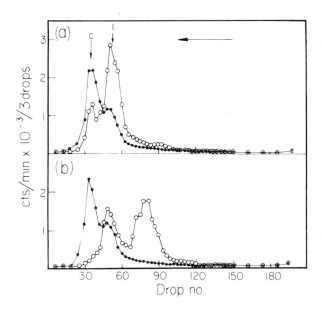

Fig. 6. Sedimentation of untreated and <u>Eco</u> RI-treated
G4 RF II pulse-labeled during single strand synthesis.
Portions of the RF II from Fig. 2(c) were concentrated by
alcohol precipitation, incubated with (b) or without (a)
<u>Eco</u> RI and sedimented through alkaline sucrose gradients.
The direction of sedimentation is indicated by the horizon-
tal arrow, the sedimentation positions of circular (c) and
unit-length linear (l) single strands by vertical arrows.
●———●, ^{32}P cts/min (G4 viral DNA sedimentation marker);
○———○, ^{3}H cts/min (pulse label). From Ray and Dueber
(1975).

RF→SS synthesis as has been observed for both ØX- and
M13-type bacteriophages (12). Later in the infectious
process, at a time when large amounts of phage are being
formed, labeling of the RF is restricted to the viral
strand, another feature common to other single-stranded DNA
phage. The transition to this asymmetric mode of synthesis
can be prevented by chloramphenicol at 30 μg/ml.

G4 replication is inhibited in rep⁻ strains of E. coli
at a point prior to RF→RF synthesis. Under these conditions, much of the parental RF accumulates as an RF II having a single unique discontinuity in the viral strand.
Nicking of the viral strand prior to RF→RF replication is
presumably the result of the G4-equivalent of the ØX174
gene A endonuclease (17). The site of nicking of the G4 RF
is at 50% of the viral genome's length from the single Eco
RI restriction site.

A single unique discontinuity has also been found in
G4 RF II molecules pulse labeled during RF→SS synthesis.
From this observation, we infer that the origin of viral
strand synthesis is also located at this site at 50% of the
genome's length from the ECO RI site. It therefore appears
that the nicking of the G4 parental RF in rep⁻ cells prior
to RF replication is confined to the region of the origin
of viral strand synthesis. This would be necessary, of
course, if the purpose of the nick is to introduce a terminus for the priming of viral strand synthesis. However,
alternatively, the nick might serve to alter the DNA structure to allow propagation of a DNA chain initiated nearby.

It is of interest to note that the origin of the G4
complementary strand in a system using purified proteins
(2) is located within only 5% of the Eco RI site. This
wide separation of the complementary strand origin identified in vitro and the viral strand origin located in vivo
in this work raises the question of the physiological significance of the origin selected by the in vitro system.
The promoter selected by the rifampicin-resistant RNA polymerase specified by the dna G gene (3) could be non-
biological since the template, purified G4 DNA coated with
unwinding protein, is probably not the natural substrate.
For ØX174, the free viral DNA is not a normal intermediate
in parental RF formation. Rather, decapsidation of the
viral DNA is coupled with parental RF formation (8 and
Jazwinski and Kornberg, unpublished).

ACKNOWLEDGEMENT

This work was supported by a grant from the National Institutes of Health (AI 10752).

REFERENCES

1. Godson, G.N. Virol. 58, 272-289 (1974).
2. Zechel, K., Bouche, J.P. and Kornberg, A. J. Biol. Chem. In press.
3. Bouche, J.P., Zechel, K. and Kornberg, A. J. Biol. Chem. In press.
4. Tabak, H.F., Griffith, J., Geider, K., Schaller, H. and Kornberg, A. J. Biol. Chem. 249, 3049-3054 (1974).
5. Schekman, R., Weiner, A. and Kornberg, A. Science 186, 987-993 (1974).
6. Godson, G.N. and Boyer, H. Virol. 62, 270-275 (1974).
7. Francke, B. and Ray, D.S. J. Mol. Biol. 61, 565-586 (1971).
8. Francke, B. and Ray, D.S. Virol. 44, 168-187 (1971).
9. Ray, D.S. J. Mol. Biol. 43, 631-643 (1969).
10. Ray, D.S. and Dueber, J. Proc. Nat. Acad. Sci. USA. In press.
11. Vinograd, J., Morris, J., Davidson, N. and Dove, W.F. Proc. Nat. Acad. Sci. USA 49, 12-17 (1963).
12. Ray, D.S. In "Molecular Basis of Virology." H. Fraenkel-Conrat (ed.), 222-254. Reinhold, New York.
13. Denhardt, D.T., Dressler, D.H. and Hathaway, A. Proc. Nat. Acad. Sci. USA 57, 813-820 (1967).
14. Siegel, J.E.D. and Hayashi, M. J. Mol. Biol. 27, 443-451 (1967).
15. Tessman, E.S. J. Mol. Biol. 17, 218-236 (1966).
16. Francke, B. and Ray, D.S. Proc. Nat. Acad. Sci. USA 69, 475-479 (1972).
17. Henry, T.J. and Knippers, R. Proc. Nat. Acad. Sci. USA 71, 1549-1553 (1974).
18. Johnson, P.H. and Sinsheimer, R.L. J. Mol. Biol. 83, 47-61 (1974).
19. Baas, P.D. and Jansz, H.S. J. Mol. Biol. 63, 569-576 (1972).
20. Godson, G.N. J. Mol. Biol. 90, 127-141 (1974).

A PHYSICAL MAP OF G4 AND THE ORIGINS OF
G4 DOUBLE AND SINGLE STRANDED DNA REPLICATION

G. Nigel Godson
Department of Radiobiology
Yale University School of Medicine
New Haven, Connecticut 06510

SUMMARY. A co-linear Hind/HpaII/Hae restriction enzyme map of G4 is presented together with the locations of the single EcoRI cleavage site and the three RNA polymerase binding sites (promoters). The location of the breaks in early and late G4 RFII DNA (i.e. the origins of double and single stranded DNA synthesis) and the polarity of the G4 cleavage map has been measured by using Pol I in an extended repair synthesis of these RFII DNA molecules. Both are located in the HindA, HpaIIB and Z5 fragments at 5% from the EcoRI cleavage site.

INTRODUCTION

G4 is one of the four ϕX-like G-phages that were isolated by the author in 1974 (1). These phages, together with ϕX174, SI3, STI (2), phage α (3) phage ϕR (4), phage U3 (5) constitute a group of icosahedral coli phages that contain single stranded DNA of approximately 5500 nucleotides as their genome. Many other small single stranded DNA containing, but further uncharacterized phages (6) probably also belong to this group. All of the icosahedral phages that have been studied so far (ϕX174, SI3, G4, G6, G13, G14 and STI) code for the same 8-9 viral proteins which when examined on acrylamide gels vary relatively little in size ($\pm$ 8%, 1, 7, 8,). The nucleotide base sequence of the DNA of these phages, however, varies considerably and by a mismatch that is far greater than can be accounted for by the degeneracy of the genetic code (1, 7). Even the base sequence of SI3 differs considerably from that of ϕX174 (7) despite the fairly high complementation (8) and low recombination (9) that can take place between these phag-

es. This base sequence mismatch between the phages is further exemplified by the very different restriction enzyme cuts that are made in their DNA's (10, 11).

Until recently the only icosahedral phages whose replication has been studied in detail are φX174 and SI3. Now that these studies are being extended into other icosahedral phages surprising differences in their mode of DNA replication is becoming apparent. STI can replicate both its single stranded and double stranded DNA in-vivo without the use of the E. coli DNA proteins dnaB, C and G (12) whereas φX174 (13, 14, 15) and G4 (10) require these proteins for different aspects of their DNA replication. In-vitro, it has been observed that the conversion of the φX174, SI3, G6 G13, G14 and φXh viral DNA strand to a double stranded RF requires E. coli dnaB, C, E and G (plus other proteins & factors), whereas conversion of G4 STI and φXtb single stranded viral DNA to RF requires only dnaE and G of the dna proteins (16, 17). This paper reports the construction of a physical restriction enzyme cleavage map of G4 together with the location of the promoter sites and the origins of both double stranded and single stranded DNA synthesis of G4.

METHODS

Growth and experimentation with G4. In general, all of the methods that are used for φX174 can be used for G4, except that their cell receptor sites are different and some φX174 sensitive E. coli strains are resistant to G4 (i.e. K5252, the ligts 7 strain). Also, G4 is much more virulent than φX174 and replicates and lyses the host cells in half the time required by φX174 (1). Experimental times should therefore be adjusted accordingly. All of the experiments described in this paper were carried out at 30° to slow down the replication.

Repair syntheis of RFII DNA. 1μg of ^{3}H-G4 RFII DNA was incubated with 2 units of E. coli polymerase I (Klenow fragment) in the presence of 670mM K phosphate pH7.6, 6.7m M MgCl$_2$, 1mM βmercaptoethanol, 0.05mM dGTP, dTTP and dATP and ^{32}P dCTP (New England Nuclear) at 37°C for 5 or 15 minutes. The reaction was stopped by adding a final concentration of 10mMEDTA and 0.2M NaCl and the unincorporated counts were removed by passing the reaction mixture over a 23 x 0.8 cm Sephadex G.50 column equilibrated in 0.05M tris pH8, 0.010 M EDTA and 0.2 NaCl. The excluded counts were precipitated with 3 volumes of ethanol and dissolved in 100μl of 6.6mMtris pH7.5, 6.6mM NaCl. 30μl portions were digested with Hind, HpaII and Hae respectively and the restriction enzyme digestion products separated on a 4% polyacrylamide slab gel. Uniformly ^{32}P-labelled G4 RFI restriction enzyme digests were run in adjacent wells as a control. The amount of ^{32}P-dCTP incorporated into the DNA fragments of the repaired RFII was compared with those incorporated into the uniformly labelled RFI DNA fragments by using a Joyce-Loebl microdensitometer to scan an autoradiograph of the dried acrylamide gel.

RESULTS

Restriction enzyme cleavage map of G4. G4 double stranded DNA is cleaved by the Haemophilus influenzae restriction enzyme Hind into 6 fragments (Hind A to E with one doublet), by the Haemophilus parainfluenzae restriction enzyme HpaII into 7 fragments (HpaIIA to F with one doublet) and by the Haemophilus aegyptius restriction enzyme Hae into 12 fragments (Z1 to Z9, with three doublets). This is shown in Table I and described in detail in (11). EcoRI cleaves to G4 double stranded DNA at one unique site which is located on fragments HindB, HpaIIB and Z4 (18). The Hind, HpaII and Hae restriction enzyme fragments have been ordered into a co-linear restriction map (II), which is shown in figure 4.

Location of the RNA polymerase binding sites. G4 DNA has three RNA polymerase binding sites (presumptive promoter sites), one strong one and two weak ones (19). This has been measured by binding E. coli RNA polymerase to G4 double stranded DNA restriction enzyme fragments at various molar ratios (0, 10, 25, 50, and 100 RNA polymerase molecules per DNA molecule) and measuring   which DNA

Table I: Size of G4 Haemophilus influenzae (Hind)
Haemophilus parainfluenzae (HpaII) and Haemophilus
aegyptius (Hae) restriction enzyme fragment of G4 DNA.

Hind		HpaII		Hae	
A	2495	A	2035	Z1	1020
B	950	B	1180	$Z2_1$	750
C	710	C	720	$Z2_2$	750
D	530	C	720	Z3	680
D	530	D	500	Z4	475
E	285	E	285	$Z5_1$	460
	5500	F	60	$Z5_2$	445
			5500	Z6	285
				$Z7_1$	260
				$Z7_2$	260
				Z8	80
				Z9	60
					5525

The sizes are given in nucleotide length and
are taken from Godson (11).

fragments bound the protein using the millipore filter technique of Chen et al (20). The strong RNA polymerase binding site is located on HindA, HpaIIA and Z1; the weaker binding sites are on HindC and Z3 and HindB, HpaIIB and Z5 respectively. These are placed on the map shown in figure 4.

Origin of G4 Single Stranded DNA Synthesis. This was determined by measuring the location of the break in G4 late RFII DNA molecules that had been purified from cells harvested late in infection during the phase of single strand ed DNA synthesis.

In the experiment described in figures 1, 2, and 3, RFII DNA was prepared from HF4704 cells that had been infected with G4 at 30° and pulse labelled with ^{3}H–thymidine at 24 – 25 minutes after infection. Under these conditions, the cells normally lyse after 28 – 30 minutes. The RFII DNA was isolated by an SDS lysis proceedure (18). The ^{3}H–pulse label was exclusively in the open viral DNA strand of the duplex (figure 3). This RFII DNA was then used as a substrate for E. coli polymerase I, under conditions that allow the enzyme to gap fill and then continue to displace the viral DNA strand by extended repair synthesis. This reaction has been described for ϕX174 late RFII DNA molecules (21) and by extended repair synthesis the order of the ϕX174 restriction enzyme fragments and the polarity of the ϕX174 restriction enzyme map were confirmed. When the late G4RFII DNA was incubated with Pol I and ^{32}P–dCTP under these conditions and the distribution of the label measured by digesting the DNA with Hind, HpaII and Hae restriction enzymes, a gradient of label was found. This is illustrated in figure I, which is a comparison of the distribution of ^{32}P– in Hae restriction enzyme fragments from uniformly ^{32}P–labelled G4RFI DNA with that of ^{32}P repaired late G4RFII DNA. When the specific activities of the Hind, HpaII and Hae fragments from ^{32}P–repaired late RFII DNA (i.e. the ^{32}P counts of the fragments from in–vitro repaired RFII versus the ^{32}P counts in the Hae fragments from uniformly labelled RFI DNA), are plotted against their position

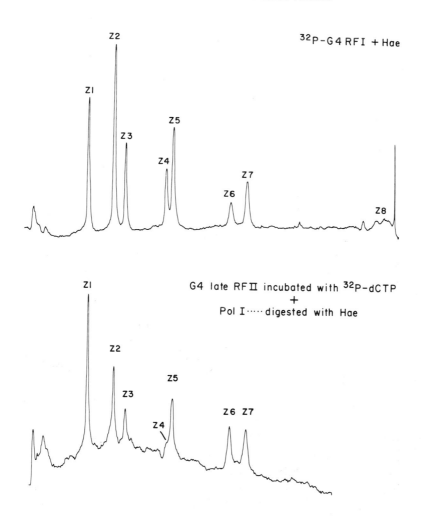

Fig. 1. Distribution of ^{32}P in <u>Haemophilus aegyptius</u> (<u>Hae</u>) restriction enzyme fragments after repair of G4 late RFII DNA with polI and ^{32}P-dCTP.

The digestion is described in the Methods and the fragments were separated on a 4% polyacrylamide gel. The control <u>Hae</u> digest of uniformly labelled ^{32}P-RFI DNA (top scan) was run in a well adjacent to the <u>Hae</u> fragments of the incorporated RFII DNA (bottom).

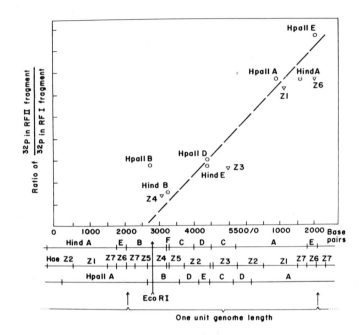

Fig. 2. Relative specific activities of the Hind, Hpa
II and Hae fragments of repaired early G4RFII DNA.

The ratio of the ^{32}P-counts incorporated into the restriction
enzyme fragments of the repaired G4RFII DNA and that incorporated into uniformly ^{32}P-labelled G4RFII DNA fragments
is plotted against their position on the G4 cleavage map
(figure 4). The Hind, HpaII and Hae fragments are all
from the same sample of repaired G4 RFII DNA. As G4 DNA
is circular, the position of the 0-5500 nucleotide scale is
arbitrary.

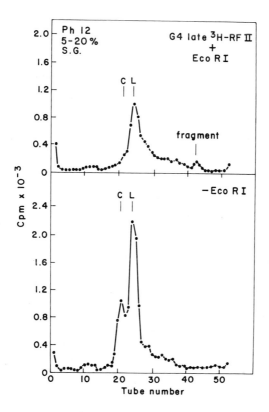

Fig. 3. Relationship of the break in late G4RFII to the EcoRI cleavage site.

[3]H-pulse labelled G4RFII DNA (the same material that was used to for the polI repair reaction described in figures I and 2) was digested with EcoRI at 37° for 2 hours. The reaction was stopped with 10m M EDTA, the products denatured with 0.3N NaOH and the size of the single strand pieces analysed by cent i uging on a 5 - 20 alkaline sucrose density gradient (0.25 m NaOH, 0.3 N NaCl, 0.005 M EDTA) in an SW41 rotor at 33,000 rev/min for 15 hours at 5°.a) ^{3}H-G4RFII digested with EcoRI and b) undigested control.

on the cleavage map (fig. 4), a gradient of label is observed which tends to zero near the EcoRI site (fig 3). In
most experiments, HindB and Z4 contain almost no ^{32}P
counts and therefore must be near the terminus of this reaction. The fragments with the highest specific activity are
HindA, HpaIIE and Z5 or Z7 (Z7 is a doublet and confuses
the analysis). A closer approximation of the location of the
break in the late RFII DNA molecules was obtained by digesting the ^{3}H-pulse labelled RFII (without repairing with
PolI), with EcoRI, then denaturing and examining the size
of the viral DNA strand fragments on an alkaline sucrose
density gradient (figure 3). The EcoRI digestion resulted
in two fragments, a 5% and a 95% piece being released from
the open ^{3}H-labelled viral DNA strand of the RFII duplex.
This places the break in the late RFII at 5% from the EcoRI
site and calculating from the data given in Table I and (II),
the break would be in HindA, I3I, nucleotides (5%) from
the HindB/A junction; in HpaIIB, 755 nucleotides (64%)
from the HpaIIF/B junction and in Z5, I85 nucleotides (40%)
from the Z4/Z5 junction. The location of the origin of the
extended repair reaction halfway along the DNA fragments
accounts for the apparent discrepancies in the labelling of
the fragments containing the origins.

The polarity of the G4 cleavage map can be deduced from
the direction of the synthesis of the polymerase. The chemical 3' 5' direction of the G4 viral DNA strand is anticlock-
wise on the map shown in figure 4.

Origin of G4 double stranded DNA synthesis. Similar experiments with the early RFII DNA prepared from cells pulse
labelled with ^{3}H-thymidine at 6 - 7 minutes after infection
during the period of double stranded DNA synthesis has
located the break in the early G4 RFII DNA in the same restriction enzyme fragments as the break in the late RFII and
at 5% from the EcoRI site. This suggests that a common
origin is used for both double and single stranded DNA replication by G4.

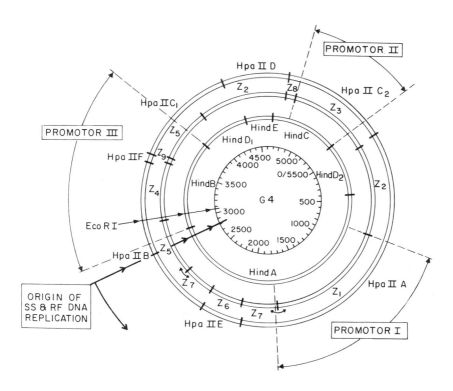

Fig. 4. Physical map of G4.

The order of the Hind, HpaII and Hae restriction enzymes
fragments, the alignment of the three maps and the location of the EcoRI cleavage site is taken from (11). The
location of the RNA polymerase binding sites (promoters)
and the origin of double and single stranded G4 DNA replication is described in this paper. The arrow indicated the
direction that E. coli pol I synthesizes the G4 viral DNA
strand on the complementary DNA strand and represents the
chemical 3' – 5' polarity of the G4 viral DNA strand.

DISCUSSION AND CONCLUSIONS

a)   G4 DNA contains three RNA polymerase binding sites, one strong one and two weaker ones.

b)   RFII DNA that is synthesized in the infected cell in both the phase of single stranded DNA synthesis and the phase of double stranded DNA synthesis contains a break (nick or gap) in the <u>Hind</u>A, <u>Hpa</u>IIB and Z5 restriction enzyme fragments, 275 nucleotides away from the <u>Eco</u>RI cleavage site. The same location has been reported for the unique origin (terminus) of the synthesis of the complementary DNA strand on a G4 circular viral DNA strand template using an in-vitro lysed cell synthetic system (22). This implies that there is a single origin in the G4 genome that is used for all of the phases of G4 DNA replication, i.e., the conversion of the incoming viral single stranded DNA to a double stranded parental RF (SS→ RF), replication of the double stranded DNA (RF→ RF) and synthesis of single stranded progany DNA (SS→RF).

ACKNOWLEDGMENTS

The author acknowledges the skilled technical assistance of Miss P. Svec. This work was supported by USPHS   CA 06519 – 12

REFERENCES

1.   Godson, G. N. Virology 58, 272-289 (1974).
2.   Bradley, D. E. J. Microbiol 16, 965-971 (1970).
3.   Taketo, A. & Kuno, S. J. Biochem. 65, 361-368 (1969).
4.   Burton, A. & Yagi, S. J. Mol. Biol. 34, 481-496 (1968).
5.   Watson, G. and Paigen, K. J. Virol. 8, 669-674 (1971).
6.   Bradley, D. E. Nature. 195, 622-623 (1962).
7.   Godson, G. N. J. Mol. Biol. 77, 467-477 (1973).

8.   Jeng, Y. et al. J. Mol. Biol. <u>49</u>, 521–526 (1970).
9.   Tessman, E. Virology <u>19</u>, 239–240 (1963).
10. Godson, G. N. unpublished.
11. Godson, G. N. Virology <u>63</u>, 320–335 (1975).
12. Bowes, M. J. J. Virol. <u>13</u>, 1400–1403 (1974).
13. Kranias, E. and Dumas, L. J. Virol. <u>13</u>, 146–154 (1974).
14. Greenleigh, L. P. N. A. S. <u>70</u>, 1757–1760 (1973).
15. McFadden, G. and Denhardt, J. Virol. <u>14</u>, 1070–1075 (1974).
16. Schekman, R., Weiner A. and Kornberg, A. Science. <u>186</u>, 987–993.
17. Wickner, S. and Hurwitz, J. This Pub. (1975).
18. Godson, G. N. and Boyer, H. Virology 62, (1974).
19. Godson, G. N. and Ludwig, B. Manuscript in preparation (1975).
20. Chen, C., Hutchison, C. and Edgell, M. Nature N. B. <u>243</u>, 233–236 (1973).
21. Johnson, P. and Sinsheimer, R. J. Mol. Biol. <u>83</u>, 47–61 (1974).
22. Zechel, K., Bouche, J. and Kornberg, A. J. Biol. Chem. in press. (1975).

REPLICATION OF φX174 IN ESCHERICHIA COLI: STRUCTURE OF THE REPLICATING INTERMEDIATE AND THE EFFECT OF MUTATIONS IN THE HOST lig AND rep GENES

David T. Denhardt, Shlomo Eisenberg, Barbara Harbers,
H.E. David Lane and Grant McFadden

Department of Biochemistry
McGill University
3655 Drummond Street
Montreal, Quebec, Canada H3G 1Y6

ABSTRACT. Evidence is presented that the viral and complementary strands of φX174 are synthesized by different
processes in E. coli. The viral strand is synthesized on a
double-stranded replicative form (RF) template by a continuous process starting in a specific region of the genome.
A single gap, with an average size of 13-15 nucleotides,
separates the 5' end of the completed viral strand from the
3' end in the nascent RF molecule containing the newly
synthesized viral strand. Although the gap is located in a
specific region of the genome, the gene A region, the complexity of the pyrimidine tracts characteristic of it
indicates that the gap does not comprise a unique 15-nucleo-
tide sequence. In contrast, the complementary strand is
synthesized in a discontinuous fashion on a single-
stranded DNA template. The gaps separating the nascent
fragments can be found in many regions of the genome.
However, the base composition and the distribution of
radioactivity among the pyrimidine tracts suggest that
these gaps in the complementary strand are not randomly
located. Polynucleotide ligase is required in vivo to join
together the nascent complementary strand fragments when
the gaps are filled in. The E. coli rep gene product, like
the gene A product of φX, is required for RF replication
but not for synthesis of the complementary strand during
parental RF formation; its function is unknown and
although not essential for E. coli DNA replication its
absence impairs the functioning of the cells replication

398

forks, causing them to progress more slowly.

INTRODUCTION

Several different sorts of studies suggest that during
ϕX174 replicative form (RF) replication the replicating
intermediate has a structure like that illustrated in Fig.1.
Salient points include: 1). initiation of synthesis of the
progeny viral strand at a special location in one direction
on the double-stranded genome; 2). subsequent synthesis of
the complementary progeny strand on the now displaced and
single-stranded parental viral strand; 3). joining together
of the ends of the newly synthesized progeny strands to
form the complete progeny strand by a reaction requiring
a DNA polymerase, polynucleotide ligase, and probably a
nuclease. Other systems exhibit variations on this theme
such as: 1). whether the initial initiation event results
in synthesis in two directions (bidirectional, e.g. SV40,
λ) or in one direction, (unidirectional, e.g. ϕX, P2);
2). the timing of synthesis on the displaced single-
stranded parental molecule relative to the time of its
displacement; 3). the relative sizes of the nascent fragments synthesized on the two parental strands. Aspects of
this scheme that are poorly understood include: 1). the
nature of the event that serves as a start signal for
synthesis of the progeny strands (this includes both initiation of a round of replication and initiation of nascent
fragments); 2). the mechanism used to prime the polymerization catalyzed by DNA polymerase; 3). the functions of
the several proteins that have been identified as part of
the replication machinery. Each of these aspects will be
discussed at the appropriate place.

One of the germinal observations in this field was
reported in 1966 in a short note in Biochim. Biophys. Acta
by Sakabe and Okazaki (1); that is that E. coli DNA pulse-
labeled with [3H]thymidine existed as fragments very much
smaller than expected if the new genome were being
synthesized in a continuous fashion. These observations
have been confirmed in many laboratories. The discovery
of polynucleotide ligase provided the mechanism for joining
the fragments (see the 1968 Cold Spring Harbor Symposium
Vol. 33). Concurrently, research by A. Kornberg (2) and
many others has established that DNA polymerases synthesize

REPLICATING φX 174 REPLICATIVE FORM

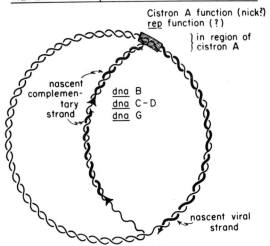

pol C (DNA polymerase III) elongates both chains
pol A (DNA polymerase I) fills in gaps

Fig. 1.   Structure of the intermediate in φX174 RF replication. Synthesis of the viral strand is initiated in the
region of gene A by a mechanism requiring both the cistron
A function and the rep function; a nick concomitantly
appears in the parental viral strand (23, 24), perhaps to
permit unwinding of the parental duplex.   The viral strand
is synthesized by a continuous process. Synthesis of the
complementary strand succeeds in time synthesis of the
viral strand and requires multiple initiations on the
single-stranded parental viral strand template.   The
initiation mechanism is different from that used by the
viral strand, and occurs at many, but not random,
locations.   The growing fork (at 5 o'clock in the diagram)
proceeds clockwise around the genome in a unidirectional
fashion (26, 27); growing fork progression in the
counterclockwise direction does not seem to occur.   DNA
polymerase III appears responsible for chain elongation,
whereas DNA polymerase I seems to have as its major role
the filling in of the gaps left between nascent strands
(6, 13).

DNA only in a 5' to 3' direction by adding deoxynucleotides onto the 3' end of a pre-existing primer chain in association with a template. Visions of new kinds of "replicases" arising from J. Cairns' discovery of the polA mutant (3) have been dissipated by the finding of new DNA polymerases in E. coli and the demonstration that small amounts of DNA polymerase I were present in viable polA strains (4).

Prior to the advent of the polA mutant most of the advances in our understanding of DNA replication had been made as the result of in vivo studies. In vitro studies prior to the discovery of the polA mutant were fraught with difficulty (5), in hindsight because of the presence of the viciously active DNA polymerase I. Recent work in in vitro systems using extracts derived from the polA strains have been tremendously revealing and are described elsewhere in this volume.

Research in our laboratory for the past few years has focussed primarily on aspects of the in vivo replication of φX174 DNA. Because of the single-stranded nature of the φX genome and its small size (1.7 megadaltons, not much larger than the nascent E. coli DNA segments) we have been able to dissect some of the events involved in φX174 replication. In 1971 we reported (6) the unexpected finding that φX RF II had gaps in the newly-synthesized strands. Although Sadowski et al. (7) had reported earlier that the nascent fragments of E. coli DNA were separated by gaps their observation has gone unremarked. The discovery (6) that in a polA mutant the gaps persisted in the RF II for a longer time was consistent with the observation (8) that the nascent DNA fragments were joined together more slowly in polA mutants.

It was suggested by Denhardt (9) that the gaps occurred in specific regions in the genome where a hypothetical protein (π protein) interacted with a DNA polymerase and a deoxyribonucleotide in such a way as to permit DNA polymerase to initiate synthesis of a DNA molecule; an additional hypothesis was that an ATP-dependent "replication-nuclease" was required to nick and later repair the duplex to permit local unwinding. Specific predictions of this model were that ATP should not be required for the conversion of single-stranded φX DNA to RF and that a small fraction of the E. coli genome should be repeated roughly 1000x. The

former prediction has been disproved (10). Kato <u>et al.</u> (11)
were unable to obtain any evidence in support of the latter
prediction (but see (12) for a positive result). Meanwhile,
evidence has accrued from a variety of systems (discussed
elsewhere in this volume) consistent with the idea that a
short sequence of ribonucleotides is synthesized by an RNA
polymerase and that this is used to provide a primer for
DNA polymerase. In a formal sense the RNA polymerase is
equivalent to the π protein, but the consequence of the
former mechanism is that the 5' end of nascent DNA chains
will, at least transiently, be joined to RNA. Although
the evidence is good that in some systems RNA may serve as
a primer it is not clear how universal a mechanism this is.

In this paper we shall emphasize S. Eisenberg's
meticulous studies of the gaps in nascent φX RF II. These
gaps have now been characterized in considerable detail,
but we do not know if any of them represent the location of
RNA primers. We shall also briefly consider the role of
various proteins involved in the replication of φX and
E. coli DNA and the question of the continuous nature of
the synthesis of φX DNA.

RESULTS

<u>Characteristics of the Gaps in Nascent RF II</u>. Gaps can be
identified in RF molecules by making use of the ability of
T4 DNA polymerase to fill them in. T4 DNA polymerase is
used because it has no 5' to 3' nuclease activity and is
incapable of strand displacement, thus any radioactivity
incorporated into an ostensibly double-stranded, circular
φX RF molecule is incorporated into gaps in the duplex.
Rather rigorous controls are necessary of course to insure
that one is not misled by incorporation into contaminating
E. coli DNA fragments. Fig. 2 shows neutral and alkaline
sucrose gradients of RF II molecules that had been labeled
during <u>in vivo</u> replication with [^{3}H]thymidine, purified,
and then labeled <u>in vitro</u> with [^{32}P]deoxynucleotides in a
gap-filling reaction with T4 DNA polymerase and α[^{32}P]
deoxynucleoside-5'-triphosphates. Some of the strands that
are labeled are shorter than unit-length while others
are unit-length; the shorter fragments can be joined
together by polynucleotide ligase. In published studies
we have shown that there is only one gap in the viral

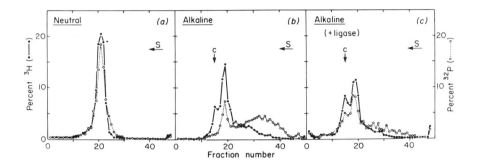

Fig. 2.   Sedimentation velocity analysis of the RF
molecules labeled with ^{32}P in the gaps. φX RF was purified
from cells infected with φX and labeled with [^{3}H]thymidine
for 3 min during the period of RF replication; these
molecules were then labeled in the gaps with α[^{32}P]dCTP
using T4 DNA polymerase (6, 13, 14).
(a)   Neutral sucrose gradient of the [^{3}H,^{32}P]RF molecules.
Centrifuged in a high salt, 5-20% sucrose gradient for 3h,
50 krpm, 15°C in the SW 50.1 rotor in a Beckman L2-65B
centrifuge.
(b)   Alkaline sucrose gradient of denatured [^{3}H,^{32}P]RF
DNA. Centrifuged in a high salt, 0.2 M NaOH, 5-20% sucrose
gradient for 4 hours, 50 krpm, 15°C in the SW 50.1 rotor
in a Beckman L2-65B.
(c)   Alkaline sucrose gradient of denatured [^{3}H,^{32}P]RF
DNA.   The DNA centrifuged here was the same DNA as
centrifuged in (a) and (b) except that in the gap-filling
reaction T4 polynucleotide ligase and ATP were included
also.   Conditions of sedimentation as in (b).   Sedimentation is from right to left.   The arrows "c", in (b)
and (c) indicate the position of 21 S circular single-
stranded viral DNA.   All data are plotted as the percent
of the total recovered radioactivity found in each
fraction. Solid circles (●——●) represent the ^{3}H incorporated in vivo; open circles (o---o) represent the ^{32}P
incorporated into gaps.

strand of the newly synthesized RF II, whereas there are
several gaps at many different locations in the complementary strand; if polynucleotide ligase is present during
the gap-filling reaction then the shorter strands are
joined together to form linear or circular unit-length
strands (13,14). DNA-DNA hybridization studies and
analyses with restriction enzymes were used to establish
the viral nature of the labeled DNA.

The size and base composition of the gaps in the viral
and complementary strands were determined separately from
the average number of each of the four deoxynucleotides
incorporated into the gaps. In order to make this calculation it was necessary to know the actual number of gaps
that were labeled; this was accomplished by determining
the number of 5' ends present in the circular duplexes by
labeling them with $\gamma[^{32}P]$ATP and polynucleotide kinase.
In order to be sure that the same population of molecules
was used in the calculation , the $\gamma[^{32}P]$ATP and the
$\alpha[^{32}P]$dNTP labeled molecules, separately incorporated in
identical reactions, were converted to RF I with ligase
(and T4 DNA polymerase in the case of the 5' terminal-
labeled molecules) and the ^{32}P normalized to the ^{3}H
uniformly present in RF I molecules only. The results of
these experiments are given in Table 1. It can be seen
that the gaps in the viral and complementary strands were
about the same size, but that the viral strand gaps were
enriched for C whereas the complementary strand gaps were
enriched for T relative to intact strands; the complementary purine was correspondingly reduced. The fact that
the average base composition of the gaps in the complementary strand was not the same as the base composition of
the complementary strand suggested that the gaps in the
complementary strand were not randomly located.

The suggested non-randomness of the location of the
gaps in the complementary strand was unexpected so
S. Eisenberg and B. Harbers sought to confirm it by studying the pyrimidine tract distribution of these gaps. Thus
molecules were labeled with $\alpha[^{32}P]$dTTP and $\alpha[^{32}P]$dCTP and
the pyrimidine tracts separated by two-dimensional iono-
phoresis/homochromatography after digestion with formic
acid/diphenylamine. Plate 1 shows autoradiograms of the
tracts present in the complete complementary strand (a)

404

TABLE 1

Average Size and Base Composition of the Gaps In Nascent φX RF II

	Molecules dNMP Incorporated per Gap		Base Composition of the Intact Strands	
	Viral strand	Complementary strand	Viral strand	Complementary strand
C	3.7 (26%)	3.6 (27%)	18.5%	24.1%
T	4.8 (34%)	5.0 (37%)	32.7%	24.6%
G	2.0 (14%)	2.2 (16%)	24.1%	18.5%
A	3.6 (25%)	2.7 (20%)	24.6%	32.7%

The number of gaps in a population of RF II molecules was determined by labeling with $\gamma[^{32}P]ATP$ and polynucleotide kinase, after removal of 5' terminal phosphates from the RF with alkaline phosphatase. The amount of each of the four dNMPs incorporated (using T4 DNA polymerase and $\alpha[^{32}P]dNTP$ substrates) into each strand of the RF II was determined after separating the individual strands by reannealing in the presence of excess unlabeled viral strands; separation of the single-stranded labeled viral strands from the double-stranded, labeled complementary strands was accomplished by chromatography on hydroxyapatite. In separate experiments the numbers were reproducible to within 20%. Further details will be described elsewhere (Eisenberg et al., to be published). The base composition of the complete strands is taken from Sinsheimer (34).

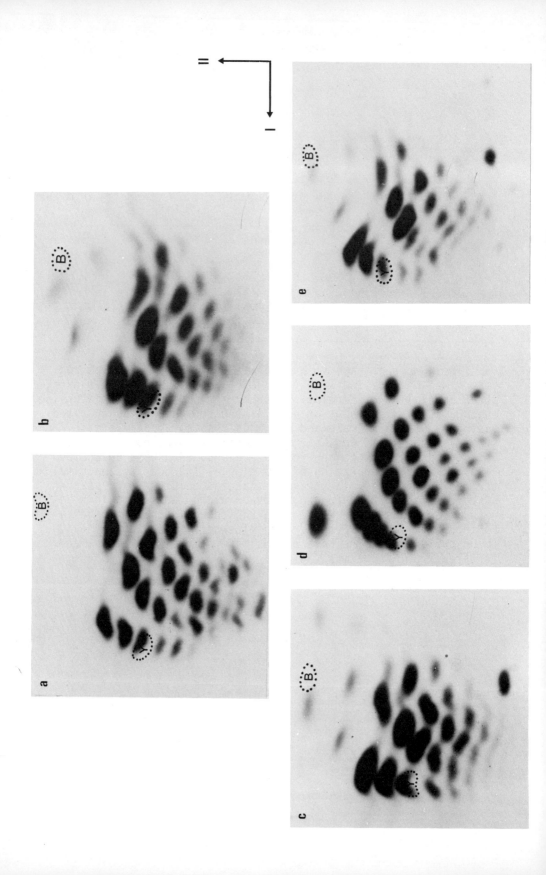

Legend to Plate 1. Pyrimidine tracts present in ^{32}P-labeled φX DNA. The purified DNA was depurinated with 2% diphenylamine in 67% formic acid at 30° for 18 h, extracted three times with ether, dried and dissolved in a small volume of water (36). The depurination products were fractionated by ionophoresis-homochromatography (37). Electrophoresis in the first dimension (I) was performed on a cellulose acetate strip (3 x 50 cm) at pH 3.5 for 20-30 min at 4.5 kV in a buffer containing 4.5% acetic acid, 0.5% formic acid, 7 M urea and 5 mM EDTA. The separated oligonucleotides were transferred to DEAE cellulose thin layer plates (20 x 20 cm) and chromatographed in the second dimension (II) at 60° in a 2% solution of partially hydrolyzed yeast RNA. Autoradiography of the plates was then performed with Kodak X-ray film. Further details will appear elsewhere (Eisenberg et al., submitted for publication).

(a) Pyrimidine tracts characteristic of the intact complementary strand of φX DNA. This strand was synthesized in vitro on a viral template using DNA polymerase I and α[^{32}P]-labeled pyrimidine deoxynucleoside triphosphates(6).

(b) Complementary strand DNA containing ^{32}P only in the gaps. This DNA came from RF that was purified from cells during the period of RF replication. The gaps were filled in with T4 DNA polymerase, α[^{32}P]dCTP, α[^{32}P]dTTP, dATP and dGTP. The viral and complementary strands were separated by competition-reannealing (13) and chromatography on hydroxyapatite.

(c) Viral strand DNA containing label only in the gap. This DNA is the viral strand DNA isolated in the same experiment as the complementary strand DNA in (b).

(d) The intact viral strand. In this case the viral DNA was isolated from phage uniformly labeled with ^{32}P. Thus there is a strong spot of inorganic phosphate in the upper left hand corner.

(e) Viral strand DNA containing label only in the gaps. In this experiment the RF had been isolated from E. coli C at 40 minutes after infection with φXam3 at 37°. By this time single-stranded DNA synthesis had been in progress for at least 25 minutes and all of the RF II molecules with gaps in them should have been exclusively the product of single-stranded DNA synthesis.

and in the gap-filled complementary strands containing ^{32}P only in the gaps (b). The distribution of labeled pyrimidine tracts is clearly different; if the gaps were randomly located in the complementary strands then one would expect to observe similar pyrimidine tract distributions in (a) and (b). However, these data show that the nucleotide sequences labeled in the total population of gaps in the complementary strand are not equivalent to the complete complementary strand. Plate 1 also shows the autoradiogram of pyrimidine tracts obtained from RF molecules labeled only in the gap in the viral strand, (c),and again it is compared with the complete viral strand (d). The pattern is obviously too complex to result from molecules with a single, unique gap of some 14 nucleotides in length. We defer until later discussion of this observation.

Assuming that the gaps in the RF II molecules reflect the history of their origins during replication we deduce that there is an asymmetry in the way the viral and complementary strands are replicated, perhaps at the stage of initiation of synthesis. The presence of one gap in the viral strand and many gaps in the complementary strand suggests that the former is synthesized by a continuous process and the latter by a discontinuous one.

The autoradiogram in Plate 1 (e) shows the pyrimidine tract distribution characteristic of the gap found in nascent RF II present in the cell at late times during progeny viral DNA synthesis. RF II in the cell at this time is known to contain one gap in the viral strand in the region of gene A and the similarity between the gap in these RF II molecules and the gap in the viral strand during RF replication (c) is apparent.

Continuous and Discontinuous Synthesis of φX DNA. As mentioned in the Introduction there is a requirement for ligase action during discontinuous, but not during continuous, DNA replication. Consequently a study of the sizes of nascent strands synthesized in the presence of reduced ligase activity (caused by a mutation in the host,ligts7, resulting in a thermosensitive ligase, 16) should permit one to distinguish these two types of synthesis. DNA synthesized by a continuous process will be unaffected in size whereas discontinuously synthesized DNA will remain

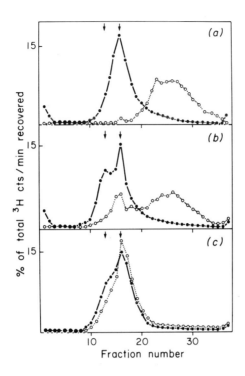

Fig. 3. Sedimentation velocity analysis of the φX DNA synthesized under conditions of ligase sufficiency or deficiency. The ligase deficient strain was E. coli KS 252 (ligts7) generously provided by Dr. B. Konrad (16). These cells were shifted from the permissive to the restrictive temperature (41°) 30 min before infection with φX. Cultures of the lig⁻ and a lig⁺ control were pulse-labeled with [³H]thymidine for 30 sec during the periods of parental RF formation, (a); RF replication, (b); and single-stranded DNA synthesis, (c). The labeled φX DNA was extracted, purified (6) and centrifuged on alkaline sucrose gradients together with ³²P circular and linear φX DNA derived from RF II; the positions of these markers are indicated by the left and right arrows respectively. Sedimentation is from right to left, and the data are plotted as the percentage of the total recovered radioactivity found in each fraction. Solid symbols (●——●) represent the lig⁺ infection; open symbols (o——o) represent the lig⁻ infection. Circular DNA from the lig⁺ infection in (a) is not seen because it is part of an RF I molecule and has pelleted.

D.T. DENHARDT *et al.*

TABLE 2

Competition annealing of φX174 DNA labeled during RF replication

Fraction	Single strand specific nuclease	^{3}H cpm x 10^{-3} in φX DNA competed	^{3}H cpm x 10^{-3} in φX DNA uncompeted	Competed/ Uncompeted ratio
A	-	43.0	67.2	0.64
"	+	42.9	70.3	0.61
B	-	53.3	88.8	0.60
"	+	47.1	81.2	0.58
C	-	20.5	31.1	0.66
"	+	19.1	23.5	0.81

A culture (30 ml) of E. coli KS252 was grown to 4×10^8 cells/ml in TKCaB at 30°. Immediately after infection with φXam3 at a multiplicity of infection of 3-5, chloramphenicol was added to 30 µg/ml. [Me-3H]thymidine was added at 3 min intervals to a final concentration of 10 µCi/ml. At 22 min after infection, the labelling was terminated in an equal volume of ethanol-phenol, the viral DNA extracted, purified by isopycnic centrifugation in neutral CsCl and then velocity sedimentation in neutral sucrose. Fractions sedimenting with 16S RF II, 21 S RF I and the more-quickly-sedimenting intermediates (designated A, B, and C, respectively) were pooled, precipitated in isopropanol and resuspended in 10 mM NaCl, 10 mM MgCl$_2$ and 10 mM Tris, pH 8.1. One-half of each sample was supplemented with N. crassa single-strand-specific nuclease at 1-2 units/ml and incubated at 36°C for 3 hr. Digestions were terminated by addition of EDTA to 20 mM followed by heating at 100°C for 5 min. The six fractions were precipitated and analyzed by reannealing the denatured DNA, after sonication, in the presence ("competed") and absence ("uncompeted") of a 25-fold excess of unlabeled φX viral strand DNA (13). [32P]RF was added in all annealing preparations and 3H cpm normalized to 50% competition of the [32P] control.

410

as small fragments. Fig. 3 shows the effect of the temperature-sensitive lesion on the synthesis of the φX DNA
during the three stages of φX DNA replication (open circles,
mutant ligts; closed circles, lig$^+$ control). In panel (a)
it is clear that the complementary strand synthesized on
the parental viral single-stranded DNA is produced by a
discontinuous process since only short fragments are
obtained. Evidence that this is also the case during RF
replication is shown in panel (b). The short fragments
were shown (McFadden, to be published) to be complementary
strand DNA by hybridization techniques; the unit-length
strands in contrast have viral strand polarity and appear
to be synthesized by a continuous process.

During single-stranded progeny viral DNA synthesis
still a third situation obtains. In this case, as illustrated in panel (c), viral strands appear to be synthesized
by a continuous process, but now a significant proportion
of the incorporated radioactivity can be found in strands
which are longer than unit-length if the pulse is short
enough (17,18). For reasons that are not evident it is
much more difficult to demonstrate the extra-long strands
during RF replication. Together these results suggest that
in vivo the viral strand is always synthesized by a continuous process whereas the complementary strand is always
synthesized by a discontinuous process.

Clearly there is a physical asymmetry in the way the
viral and complementary strands are synthesized. Is there
also a temporal asymmetry? In other words is one strand
synthesized before the other? G. McFadden asked this
question by isolating replicating intermediates that were
uniformly labeled with [^{3}H]thymidine and subjecting them
to treatment with a single-strand specific nuclease. The
relative amounts of viral and complementary strand DNA in
the replicating intermediates before and after digestion
with the single-strand specific nuclease were determined
by a competition-hybridization procedure. The data are
presented in Table 2. The enrichment for complementary strand
sequences in the replicating intermediates (C) after exposure to the nuclease shows that all of the complementary
sequences are in duplex DNA, whereas a portion of the viral
sequences are not in a base-paired structure. This supports
the model shown in Fig. 1 in so far as the viral strand is

411

TABLE 3

Requirements for Various Gene Products for φX DNA Replication

Gene	Mode I	Mode II	Mode III
φX cistron A	no[a]	yes[a]	?
φX cistrons, B,D,F,G	no[b]	no[b]	yes[b]
φX cistron C	no[c]	no[c]	yes[c] (leaky)
polA	yes[d] ?	yes[d,e] ?	yes[d] ?
polB	no[f,g]	no[f,g]	no[f,g]
polC (dnaE)	no, in vivo[g,h,i] yes, in vitro[j]	yes[g,h,i]	yes[g,h,i]
dnaA	no[k,l,m,n]	no[k,l]	no[k,l]
dnaB	yes[o,s]	yes[o,s]	no[s], yes[o]
dnaC-D	yes, in vitro[j] no, in vivo[p]	yes[p]	yes[p]
dnaG	yes[j,q]	yes[q]	yes[q]
rep	no[r]	yes[r]	?

a.   Hutchison, C.A. III, Sinsheimer, R.L. J. Mol. Biol. 18, 429 (1966).
b.   Iwaya, M. and Denhardt, D.T. J. Mol. Biol. 57, 159 (1971).
c.   Funk, F.D. and Sinsheimer, R.L. J. Virol. 6, 12 (1970).
d.   Schekman, R.W., Iwaya, M., Bromstrup, K. and Denhardt, D.T. J. Mol. Biol. 57, 177 (1971).
e.   Eisenberg, S. and Denhardt, D.T. Proc. Nat. Acad. Sci. U.S.A. 71, 984 (1974).
f.   Campbell, J.L., Soll, L. and Richardson, C.C. Proc. Nat. Acad. Sci. U.S.A. 69, 2090 (1972).
g.   Denhardt, D.T., Iwaya, M., McFadden, G. and Schochetman, G. Can. J. Biochem. 51, 1588 (1973).
h.   Dumas, L.B. and Miller, C.A. J. Virol. 11, 849 (1973).
i.   Greenlee, L.L. Proc. Nat. Acad. Sci. U.S.A. 70, 1757 (1973).
j.   Wickner, R.B., Wright, M., Wickner, S. and Hurwitz, J. Proc. Nat. Acad. Sci. U.S.A. 69, 3233 (1972).
k.   Manheimer, J. and Truffaut, N. Mol. Gen. Genetics 130, 21 (1974).
l.   Takedo, A. Mol. Gen. Genetics 122, 15 (1973).
m.   Wright, M., Wickner, S. and Hurwitz, J. Proc. Nat. Acad. Sci. U.S.A. 70, 3120 (1973).
n.   Schekman, R., Weiner, A. and Kornberg, A. Sci. 186, 987 (1974).
o.   Steinberg, R.A. and Denhardt, D.T. J. Mol. Biol. 87, 525 (1968).
p.   Kranias, E.G. and Dumas, C.B. J. Virol. 13, 146 (1974).
q.   McFadden, G. and Denhardt, D.T. J. Virol. 14, 1070 (1974).
r.   Denhardt, D.T., Iwaya, M. and Larison, L.L. Virology 49, 486 (1972).
s.   Dumas, L.B. and Miller, C.A. J. Virol. 14, 1369 (1974).

synthesized by a continuous process and only subsequently
is the complementary strand synthesized. Thus there is
an asymmetry in the time of synthesis of the two strands.

Differences in Gene Product Requirements. Table 3 summarizes research from many laboratories, including our own,
into the role of viral and bacterial genes in the three
stages of φX174 DNA replication. An attempt has been
made to indicate whether the gene product is necessary for
complementary strand synthesis (mode I), viral strand
synthesis (mode III), or both (mode II). A question mark
indicates that there are no data available.

There are several puzzles. These include the apparent
necessity for the dnaE and dnaC gene products in vitro and
their equally apparent nonessentiality in vivo for the
formation of the parental RF. Further work is needed to
resolve this. The fact that some mutants behave differ-
ently from others complicates the picture. It seems
difficult at this time to understand the requirement (19)
for the dnaG gene product for progeny single-stranded DNA
synthesis if the dnaG gene product is indeed an RNA
polymerase (20). If progeny viral DNA is made by a
'rolling-circle' mechanism (18) then the need for an RNA
primer is not apparent.

The role of the rep protein is especially interesting
because it, in contrast to the other bacterial dna gene
products listed, is essential for φX RF replication but
not for E. coli DNA replication. D. Lane has made the
intriguing discovery that the rep mutation causes the
replicating E. coli chromosome to adopt an altered configuration (21). This is illustrated in Fig. 4 where a
steeper gradient of marker frequencies is shown to exist
in the mutant. The data described in Table 4 show that
in the absence of the rep function the rate of travel of
the replication fork is reduced, either because the rate
of elongation of the nascent strands is lowered or because
the frequency with which synthesis of the intermediate
fragments is initiated is reduced. Since both mutant and
wild-type have approximately the same growth rate, and
hence rate of initiation of rounds of replication, the
mutant chromosomes must contain more growing forks.

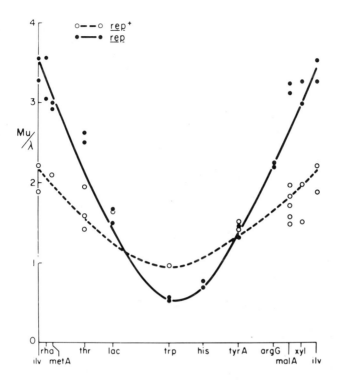

Fig. 4.   Relative frequencies of the genetic markers in
rep and rep⁺ E. coli.   DNA was extracted from isogenic,
but doubly lysogenic for μ and λ , strains of rep and
rep⁺ cells growing in L broth under steady-state conditions and analyzed for the relative amounts of μ and λ
DNA present by DNA-DNA hybridization using the techniques
of Bird et al. (35).   The strains differed depending on
the location of μ in the chromosome.   From the amount of
μ DNA present relative to the amount of lambda DNA
(always at its normal att site) gradients of marker
frequencies found under Rep⁻and Rep⁺ conditions were
determined (21).

TABLE 4

Reduced velocity of replication fork progression in
the rep mutant

Experiment	Replication rate (μm/min)	
	rep[+]	rep
I	16.6	10.1
II	15.2	8.3

Experiment I: Chromosome replication in cells aligned by
successive amino acid and thymine starvations was re-initi-
ated by adding [^{3}H]thymine of low specific activity. The
cells were then pulse-labeled with [^{3}H]thymidine of high
specific activity and killed. The DNA was extracted,
spread on slides, and exposed to a photographic emulsion.
The lengths of high grain density regions, in tracks containing two such regions separated by a low density region,
were measured.

Experiment II: Cells aligned as above but in heavy (^{15}N,
^{2}H) medium were transferred to light (^{14}N, ^{1}H) medium
containing [^{3}H]thymine. At intervals the DNA was extracted
and centrifuged to equilibrium in CsCl buoyant density
gradients. The rate of transfer of DNA from the heavy to
the hybrid position was determined. The replication rate
was calculated by assuming a total length of 1300 μ for
the chromosome and a bidirectional mode of replication.

DISCUSSION

The asymmetries we have reviewed above are most easily discussed by considering the different modes of φX DNA replication.

Mode I Synthesis. Formation of the parental RF. This is so far the only one to have been successfully characterized in vitro (2, 20, 22). On the basis of our in vivo data we believe that this type of DNA synthesis occurs both during the formation of the parental RF and during the formation of the complement to the single-stranded viral strand displaced during RF replication. Synthesis of the complementary strand is discontinuous in vivo and apparently can be initiated at many different sequences on the viral strand. If the same events occur here as occur during the complementary strand synthesis phase of RF replication then these sequences are not randomly located.

The evidence is good that the dnaB and dnaG gene products are required for this reaction, and, in vitro at least, the dnaE gene product is also necessary. From the in vitro work (20) and the requirement for the dnaG gene product (19) it is reasonable to believe there is an RNA primer. Equating the size of the gaps in nascent RF II to the size of the RNA primer leads to the conclusion that it is some 13-15 nucleotides long, and that there are many locations in the genome where the RNA can be synthesized. However, it should be noted that Wickner and Hurwitz (22) have not observed a requirement for RNA synthesis in the in vitro conversion of φX174 viral DNA to RF.

Mode II Synthesis. RF replication. To obtain RF replication it is necessary to add at least two functions to those observed to be required for mode I synthesis. These are controlled by the A gene of φX and the bacterial rep gene. Regarding the nature of the gene A function there is evidence that when, and only when, the gene A product is active a "nick" is found in a location, probably specific, in the viral strand of the RF (23, 24). There is no clue to the function of this nick. It could be required to permit unwinding of the DNA duplex, or to provide a 3' OH primer for synthesis of the new viral strand. Alternatively the nick could serve simply as a signal to allow the rep-controlled DNA replication apparatus to recognize the DNA; there is evidence that the nicks in T4 DNA may serve as a signal for late transcription (25).

Once synthesis of a new viral strand is initiated it is completed as a growing fork moves in a continuous, uni-directional (26,27) fashion around the circular genome. This is in contrast to the situation during phage P2 repli-cation where both strands of the unidirectionally-synthe-sized P2 DNA are synthesized discontinuously (28). Synthesis of the viral strand must be completed quickly because within the limits of detection of our experiments we detect few shorter than unit-length viral strands. Schröder and Kaerner (29) have detected the presence of viral strands longer than unit-length and have presented evidence that synthesis of the viral strand precedes synthesis of the complementary strand.

One puzzling observation is the contradiction between the size of the gap measured by quantitation of the nucleo-tides incorporated (13-15) and the size estimated from the complexity of the pyrimidine tracts (at least 100 nucleo-tides). One interpretation of this is that initiation of synthesis occurs in the region of the nick (which we assume to be uniquely located in the gene A region), but that there is some "play" with regard to its precise location. Alternatively, the gap could vary considerably in size depending on how long it had existed after separation of the two progeny duplexes. On this model the 5' end of all the gaps would be identical, but the 3' end would vary depending on how large the gap was.

Mode III Synthesis. Progeny viral single-stranded DNA synthesis requires several new gene products in addition to those required during RF replication - specifically the products of viral genes B, C, D, F, and G (30, 31). Synthesis in this mode resembles mode II synthesis in so far as a duplex DNA molecule is the substrate and a new RF II molecule is produced with a gap in the viral strand (15, 27) that appears identical to the gap in the viral strand observed in RF II during RF replication (cf. the pyrimidine tract distributions in Plate 1 (c) and (e)). In this mode, however, synthesis of a complementary strand is suppressed so that only a circular single-stranded viral

DNA molecule is produced. It has been argued (30) that the function of the additional gene products required for mode III synthesis is not solely the negative one of inhibiting complementary strand synthesis, but the nature of any positive role is unknown. One approach to elucidating their function could utilize in vitro systems since mode III synthesis has been observed in vitro (5,32,33). Also, using in vitro systems capable of mode I synthesis, it would be interesting to search for component(s) that inhibited mode I synthesis.

Concluding Speculations: Of obvious interest are questions concerning the signals perceived by the enzymes involved in DNA replication. One class of signals includes those that involve recognition of specific nucleotide sequences in the DNA. That proteins can recognize specific nucleotide sequences is particularly evident from the variety of restriction enzymes that have been characterized. If our present ideas about the action of the gene A protein of φX are confirmed, then it seems plausible that at least in this system a signal that is required for the initiation of a "round" of replication of the φX genome is the insertion of a nick into a specific nucleotide sequence in the chromosome. Extrapolating to higher organisms, it is possible that the initiation of replication of eucaryote replication units is controlled by specific "replication nickases" that act on replicons, making them available to the replication machinery.

Another class of signals includes those that are not dependent on a particular nucleotide sequence. It is reasonable to believe that when a polymerase synthesizes a polynucleotide strand on a template it uses precursors in a particular conformation (e.g. C (3')-endo or C (3')-exo, 38) and that the resulting duplex is left in a particular conformation (e.g. the A form or the B form respectively, 39). Thus, for example,the "classical" RNA polymerases that transcribe mRNA from a DNA duplex could be imagined to use C (3')-exo substrates and to produce an unstable "B-like" RNA-DNA duplex. As a result the RNA would be readily displaced from its template and the intact DNA-DNA duplex reformed. In contrast, the RNA polymerase that synthesizes the RNA required as a primer for DNA synthesis would use the C (3')-endo substrate and synthesize an RNA-DNA duplex in the stable A conformation. As soon as a sufficiently long (10-20 nucleotides) RNA primer was synthesized so that the DNA-RNA hybrid was locked into the A conformation then the RNA polymerase would automatically be ejected; this would happen because to adopt the stable A conformation the sugars in the complementary strand DNA would all have to cooperatively undergo a C (3')'exo to a C (3')-endo conformation and we are suggesting that this is the signal that causes the RNA polymerase to be released. According to this model there would then be no special sequence at the 3' end of the RNA primer, though there could be one at the 5' end where synthesis of the RNA primer was initiated.

In order to utilize the RNA primer efficiently it would be advantageous to use a DNA polymerase that had a high affinity for a primer-template in the A conformation and would therefore prefer to synthesize DNA in the A conformation and to use substrates in the C (3')-endo conformation. However, since the B form of the DNA duplex is the more stable, at some point during the elongation of the nascent DNA chain the entire newly synthesized duplex would revert from the A to the B conformation. This sudden cooperative A to B transition would have catastrophic consequences at both ends of the duplex segment. At the 5' end the RNA would be forced out of the duplex, or at least those ribonucleotides adjacent to the DNA would be forced to adopt a distorted conformation that would make them susceptible, for example, to "repair" nucleases. At the 3' end the DNA polymerase would no longer find as favorable a primer-template conformation, and for efficient DNA synthesis a second DNA polymerase adapted to interact efficiently with a template-primer in the B conformation and using precursors in the C (3')-exo conformation would be favored. This could explain the common observation that cells have more than one DNA polymerase involved in replicating their DNA. The variable length of the nascent DNA fragments observed during DNA replication may then reflect a competition between elongation of the nascent DNA in the A conformation and its tendency to revert to the B form.

ACKNOWLEDGEMENTS

We thank John Spencer for the generous use of his laboratory facilities (supported by MRC grant MT 1453), Angus Graham for making much of our work possible, and Linda Pallett for her usual superb typing. The research reported in this paper was supported by the Medical Research Council, the National Cancer Institute of Canada, and the Conseil de la recherche medicale du Québec.

REFERENCES

1. Sakabe, K. and Okazaki, R. Biochim. Biophys. Acta 129, 651 (1966).
2. Kornberg, A. DNA Synthesis. W.H. Freeman and Co., San Francisco (1974).
3. DeLucia, P. and Cairns, J. Nature 224, 1164 (1969).
4. Lehman, I.R. and Chien, J.R. J. Biol. Chem. 248, 7717 (1973).
5. Denhardt, D.T. and Burgess, A.B. Cold Spring Harbor Symp. Quant. Biol. 33, 449 (1968).
6. Schekman, R.W., Iwaya, M., Bromstrup, K. and Denhardt, D.T. J. Mol. Biol. 57, 177 (1971).
7. Sadowski, P., Ginsberg, B., Yudelevich, A., Feiner, L. and Hurwitz, J. Cold Spring Harbor Symp. Quant. Biol. 33, 165 (1968).
8. Kuempel, P.L. and Veomett, G.E. Biochem. Biophys. Res. Commun. 41, 973 (1970).
9. Denhardt, D.T. J. Theor. Biol. 34, 487 (1972).
10. Denhardt, D.T., Iwaya, M., McFadden, G. and Schochetman, G. Can. J. Biochem. 51, 1588 (1973).
11. Kato, A.C., Borstad, L., Fraser, M.J. and Denhardt,D.T. Nucleic Acids Res. 1, 1539 (1974).
12. Lin, H.J. Biochim. Biophys. Acta 349, 13 (1974).
13. Eisenberg, S. and Denhardt, D.T. Proc. Natl. Acad. Sci. U.S.A. 71, 984 (1974).
14. Eisenberg, S. and Denhardt, D.T. Biochem. Biophys. Res. Commun. 61, 532 (1974).
15. Johnson, P.H. and Sinsheimer, R.L. J. Mol. Biol. 83, 47 (1974).
16. Konrad, E.B., Modrich, P. and Lehman, I.R. J. Mol. Biol. 77, 519 (1973).
17. Sinsheimer, R.L., Knippers, R. and Komano, T. Cold Spring Harbor Symp. Quant. Biol. 33, 443 (1968).

18. Gilbert, W. and Dressler, D.   Cold Spring Harbor Symp. Quant. Biol. 33, 473 (1968).
19. McFadden, G. and Denhardt, D.T.   J. Virol. 14, 1070 (1974).
20. Schekman, R., Weiner, A. and Kornberg, A.   Science 186, 987 (1974).
21. Lane, H.E.D. and Denhardt, D.T.   J. Bacteriol. 120, 805 (1974).
22. Wickner, S. and Hurwitz, J.   Proc. Natl. Acad. Sci. U.S.A. 71, 4120 (1974).
23. Francke, B. and Ray, D.S.   J. Mol. Biol. 61, 565 (1971).
24. Henry, T.J. and Knippers, R.   Proc. Nat. Acad. Sci. U.S.A. 71, 1549 (1974).
25. Riva, S., Cascino, A. and Geiduschek, E.P.   J. Mol. Biol. 54, 103 (1970).
26. Baas, P.D. and Jansz, H.S.   J. Mol. Biol. 63, 569 (1972).
27. Godson, G.N.   J. Mol. Biol. 90, 127 (1974).
28. Konrad, E.B., Modrich, P. and Lehman, I.R.   J. Mol. Biol. 90, 115 (1974).
29. Schröder, C.H. and Kaerner, H.-C.   J. Mol. Biol. 71, 351 (1972).
30. Iwaya, M. and Denhardt, D.T.   J. Mol. Biol. 57, 159 (1971).
31. Funk, F.D. and Sinsheimer, R.L.   J. Virol. 6, 12 (1970).
32. Denhardt, D.T.   Nature 220, 131 (1968).
33. Linney, E.A. and Hayashi, M.   Biochem. Biophys. Res. Commun. 41, 669 (1970).
34. Sinsheimer, R.L.   Proc. Natl. Acad. Sci. 8, 115 (1968). ed. J.N. Davidson and W.E. Cohn, Academic Press, N.Y.
35. Bird, R.E., Louarn, J., Martuscelli, J., and Caro, L. J. Mol. Biol. 70, 549 (1972).
36. Burton, K. and Peterson, G.B. Biochem. J. 75, 17 (1960).
37. Brownlee, G.G. and Sanger, F. Europ. J. Biochem. 11, 395 (1969).
38. Altona, C. and Sundaralingam, M.   J. Am. Chem. Soc. 95, 2333 (1973).
39. Arnott, S., Hukins, D.W.L., Dover, S.D., Fuller, W. and Hodgson, A.R.   J. Mol. Biol. 81, 107 (1973).

BACTERIOPHAGE M13 REPLICATION IN E. COLI dnaB STRAINS

J. Michael Bowes
Department of Bacteriology, University of
California, Davis, California   95616

ABSTRACT. By means of a single step growth procedure it
was shown that all the dnaB strains tested (dnaB-70, -107,
-279, -43, -6, -8,454, and -313) except dnaB-266 (see
below) produced 10-100 phages per cell after infection at
the restrictive temperature. Using this procedure it was
possible to show that (i) λ phage production (which has
been shown previously to require the host cell dnaB function) still specifically required the dnaB function and (ii)
the production of plaque forming units of M13 specifically
required the dnaA, dnaC(D), dnaE and dnaG gene products.
After infection of dnaB mutants at 42 C with M13 there was
a marked stimulation of DNA synthesis as measured by incorporation of 3H-thymidine. All three major forms of
phage specific DNA (RFI, RFII, and SS-DNA) as well as 3H-
thymidine labeled phage were made in amounts approximating
the amounts made in parallel infections of dnaB+ trans-
ductants or revertants at the non-permissive temperature.
     One dnaB mutant (dnaB-266) was found not to produce
M13 at restrictive temperature. However, no phage were
produced at 42 C in a 42 C resistant transductant of
dnaB-266 either.   In both these strains, RF replication
continued in the functional absence of dnaB and the defect
in phage production seems to be due to some as yet
undefined defect in SS-DNA synthesis which is present in
both strains.
     The results presented in this paper, in contrast to
previous reports, clearly establish that M13 does not
require the dnaB function for progeny RF replication and
in fact no step in M13 replication specifically requires a
functional dnaB product.

J.M. BOWES

TABLE 1

Bacteria

Strain Designation	Strain No. in JMB Collection	Characteristics and Comments	Reference	Source
dnaB-70	JMB-35	HfrH thy pro str^r dnaB-70	6	C.E. Dowell
JMB-108	JMB-108	Spontaneous, 42 C resistant revertant of dnaB-70	*	
CRT 43	JMB-36	F$^-$ thr leu thy lac T4^r str^r dnaB-43	6	C.E. Dowell
W4580/CR34	JMB-44	F'thy+ arg+/thr leu lac Su+ thy	*	D. Pratt
dnaB-43	JMB-129	F'thy+ (W4580/CR34) derivative of CRT43	*	
JMB-130	JMB-130	Spontaneous, 42 C resistant revertant of dnaB-43	*	
E279	JMB-16	F$^-$ thr leu thy lac B$_1$ tonA str^r suII dnaB-279	7	J.A. Wechsler
dnaB-279	JMB-142	F' thy+ (W4580/CR34) derivative of E279	*	
CRT266	JMB-2	F$^-$ thr leu met B$_1$ thy dnaB-266	7	Y. Hirota
AB1157	JMB-163	F$^-$arg his leu pro B$_1$ thy lac ara str^r T6^r xyl mann	8	C.I. Davern
JMB-198	JMB-198	F' thy+ (W4580/CR34) derivative of AB1157	*	
dnaB-266	JMB-250	F' thy+ (JMB-198) derivative of AB1157	*	
K38	JMB-242	HfrC (λ)	9	D. Pratt
JMB-257	JMB-257	42 C resistant transductant of dnaB-266 by P1kc (K38)	*	
DM48	JMB-240	HfrC phoA	10	D.A. Marvin

*This paper.

424

INTRODUCTION

It has been reported that the filamentous, male specific bacteriophage M13 (1) requires the E. coli dnaB (2) function for replication. No phage and little or no progeny double-stranded replicative form (RF) DNA or single-stranded (SS-) DNA were found to be made at the restrictive temperature (3,4). However once an M13 infection has been established at the permissive temperature in an E. coli dnaB-70 strain, both phage production and SS-DNA synthesis occur at essentially wild type levels after shift to the non-permissive temperature (3,4). Based on these results it has been proposed that only M13 progeny RF replication has a specific requirement for the dnaB function. Furthermore, since M13 SS-DNA synthesis apparently did not require the dnaB function whereas RF replication apparently did, it has been proposed that these two processes in M13 DNA replication occur by separate mechanisms (5).

The results presented in this paper, contrary to these previous reports, establish that the dnaB function is not required for any step in the replication of M13 DNA and the production of infective phage. Since neither SS-DNA synthesis nor progeny RF replication require the dnaB function, there seems to be no a priori reason for hypothesizing separate mechanisms of replication of M13 RF and SS-DNA.

MATERIALS & METHODS

Bacteria: All the bacteria used in this paper are described in Table 1. Each strain was checked for its auxotrophic characteristics prior to use.

Bacteriophage: M13 was received from David Pratt. One high titer M13 preparation was used for all the experiments described in this paper. This preparation was made by infecting 800 ml of log phase E. coli K-38 at about 10^8 cells per ml in H-broth (9) at 37 C with M13 at a multiplicity of about 10 phage per cell. After about 6 hours the cells were removed by centrifugation and the phage were concentrated and purified from the supernatant by two precipitations with polyethylene glycol (PEG, 4% final) and NaCl (0.5 M final) as previously described (11). The PEG precipitated phage was further purified by equilibrium CsCl centrifugation (12). M13 was assayed for plaque

forming units (PFU) per ml as previously described (9).
For ether treatment, anhydrous ether was added to the
sample to at least 20% (v/v). The sample was then vortexed,
allowed to sit at room temperature for 1-15 minutes, and
then diluted and assayed for PFU/ml.

Bacteriophage λ was also received from David Pratt. A
clear plaque derivative was used which was assayed in the
standard fashion using DM48 as the host bacteria.

Bacteriophage P1kc was obtained from B. Pugashetti,
who in turn received it from Stanley Streicher (Massachusetts Institute of Technology, Cambridge, Massachusetts).
Transductions were performed as described (13).

Bacterial Growth: Single colony isolates of particular
bacterial strains were grown overnight in H-broth at the
appropriate temperature (usually 33 C) in test tubes without any attempt at aeration. These single colony isolates
were used (i) to test the strains for their specific auxotrophic requirements and (ii) to generate fresh overnight
cultures by 1/100 dilution into fresh H-broth. These overnight cultures were then used to inoculate the experimental
cultures by 1/100 dilution into M9 medium (14). For thymine
requiring strains, thymine was added to M9 to a final
concentration of from 2-40 micrograms/ml. If the cultures
growing in M9 were diluted or resuspended in fresh M9
medium, it contained 1/100 volume of H-broth. Growth was
monitored periodically by reading the optical density (OD)
at 450 nm using a Zeiss PM QII spectrophotometer. An
OD 450nm of 0.5 corresponded to 0.5 to 2 x 10^8 cells/ml
depending upon the bacterial strain.

Single Step Growth Procedure with M13: Cells were grown
in M9 medium at the appropriate temperature (usually 33 C)
to an OD 450nm of about 0.5. They were then concentrated
10-fold by centrifugation (at room temperature) and
resuspension of the pelleted cells in 1/10 volume M9
medium prewarmed at the non-permissive temperature (usually
42 C). From this step on all additions were prewarmed at
42 C prior to addition. After 15 minutes incubation at
42 C, 0.5_2ml of cells were infected with 0.02 ml M13 (at
1-2 x 10^{12} PFU/ml). After 6 minutes of infection 0.05 ml
of anti-serum (K approximately 1,000; 9) was added, and at
9 minutes of infection (3 minutes of anti-M13 serum treatment) the cells were diluted 1/1000 into fresh M9 medium

prewarmed at either 42 C or 33 C. Samples were then taken at timed intervals and assayed for PFU/ml of M13.

Measurements of DNA Synthesis: The ability to synthesize DNA at different temperatures was measured by determining the incorporation of 3H-thymidine (in thy+ cultures) or 3H-thymine (in thy⁻ cultures) into acid insoluble material. Cells were grown in M9 medium at the permissive temperature (usually 33 C) to OD 450nm of about 0.5. For the incorporation of 3H-thymidine, the culture was then diluted (see each experiment for the dilution) into fresh M9 medium containing uridine at 250 micrograms/ml to inhibit thymidine phosphorylase. In some cases deoxyadenosine was used instead of uridine. However deoxadenosine was found to be inhibitory to thy⁻ strains which were also dra⁻ (15) while uridine was not, so the use of deoxyadenosine was later discarded. The diluted culture (1-3 ml) was then added to a 50 ml flask containing 3H-thymidine, (specific activity of 20 Ci/mmole from New England Nuclear) at a final concentration of 5 to 10 microcuries/ml. After 60 minutes incorporation at permissive temperature, a portion of the culture (usually 1 ml) was shifted to another 50 ml flask prewarmed to the non-permissive temperature. Samples of 50 to 100 microliters were taken at timed intervals and placed directly onto NaOH soaked and dried GFA filters. The filters were then washed and the radioactivity in each determined by liquid scintillation counting as previously described (16). For the incorporation of 3H-thymine the procedure was essentially the same except that (i) neither uridine nor deoxyadenosine were added, (ii) non-radioactive thymine was present (at 4 micrograms per ml), and (iii) 3H-thymine (specific activity of 10 Ci/mmole from New England Nuclear) was added to 10-20 microcuries/ml.

Preparation of Lysates for Sedimentation Analysis: This was done essentially as previously described (17), except that (i) 5 ml samples were quickly added to 15 ml ice cold NaCl-EDTA-Tris-NaCN (17), (ii) sarkosyl was added right after lysozyme and the mixture stored on ice for 5-10 minutes, and (iii) the lysates were incubated at 45 C for another 5 minutes and then stored frozen no longer than 36 hours before sedimentation analysis.

<u>Sedimentation Analysis</u>: Sedimentation in preformed, neutral CsCl gradients and the collection procedure have been described (18). 32-P labeled M13 phage and SS-DNA marker preparations were done as described for bacteriophage fd (12).

RESULTS

<u>Strain dnaB-70</u>: A single step growth procedure was used to study M-13 phage production in various strains without the complications due to the presence of a large excess of free phage particles. These particles persist due to the slow rate of adsorption characteristic of this filamentous male specific phage (19). The procedure uses a high concentration of both M13 and cells to promote rapid and synchronous infection, followed by treatment with anti-M13 serum to inactivate the large excess of free phage particles which have not infected a cell. In most infections of this type more than 95% of the PFU (plaque forming units) remaining right after treatment with anti-M13 serum are due to infected cells and not free phage particles. This distinction can be made based on the sensitivity to ether of plaque formation by infected cells but not free phage particles. Since most of the PFU's after treatment with anti-M13 serum are ether sensitive, the production of just a few phage per cell can be seen as a large increase in either resistant PFU's (see Figure 1A).

Previous experiments indicating that participation of the <u>E. coli dnaB</u> function was required for M13 RF replication used strains which contained the <u>dnaB-70</u> mutation (2, 3,4,5). In this paper the strain designated <u>dnaB-70</u> is the same strain used by Primrose et al. (3) which they called Bonhoeffer strain 5. It is a pro⁻, strR derivative of the strain (HfrH 165/70 or strain 2 in reference 6) used by Hofschnieder's group (4,5). When strain <u>dnaB-70</u> was subjected to this single step growth procedure, it was found that about 10 phage were produced per infected cell at 42 C even though host cell DNA synthesis was shut off (Figure 1). Since phage can be produced it is not likely that RF replication is blocked by the <u>dnaB-70</u> mutation.

In Figure 2 it is shown that λ but not M13 specifically requires the <u>dnaB-70</u> function for phage production at 42 C. Both strain <u>dnaB-70</u> and its 42 C resistant revertant (JMB-108) produce equivalent amounts of M13 at 42 C

428

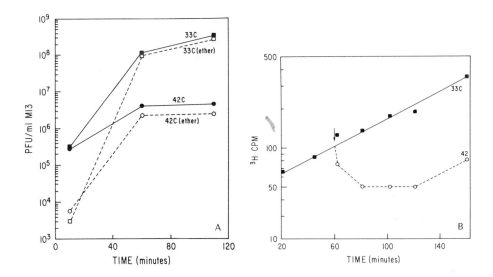

Fig. 1. M13 phage production and DNA synthesis in dnaB-70. Strain dnaB-70 was grown in M9 at 33 C to an OD 450nm of about 0.5 and then treated in two ways: (A) A portion of the culture was concentrated 10-fold into 42 C prewarmed M9, infected with M13, and then treated with anti-M13 serum. The infected cells were then diluted to either 42 C or 33 C and samples taken and assayed for PFU/ml of M13 with or without ether treatment (see Methods). (B) Another portion of the culture was diluted 1/8 into fresh M9 containing 4 microgram/ml of thymine, and DNA synthesis was measured by the incorporation of 3H-thymine (10 microcuries/ml) as described in the Methods section.

429

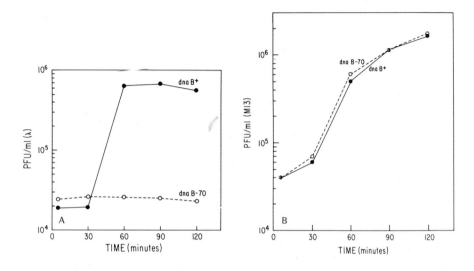

Fig. 2. M13 and λ phage production at 42 C by dnaB-70
and dnaB+. Strain dnaB-70 and its 42 C resistant revertant
(JMB-108) were grown in TKPT [TK broth (31) containing
40 microgram/ml each of proline and thymine] to an OD 650nm
of about 0.35. Both cultures were washed once by centrifugation and resuspension, concentrated 10-fold into starvation media (0.01 M $MgSO_4$ containing proline and thymine at
40 microgram/ml each) and incubated at 33 C for 15 minutes.
Each culture was split into two halves and all 4 portions
were shifted to 42 C for another 15 minutes incubation.
Each culture was then infected with either phage λ or M13
(both prewarmed at 42 C). For phage λ infections (A) λ
was added to 2×10^8 PFU/ml. After 12 minutes infection
anti-M13 serum was added as described for M13 infections.
After a total of 15 minutes infection, the infected
cultures were diluted 1/10,000 into 42 C prewarmed TKPT
broth. Samples were taken and assayed for PFU/ml of λ. M13
infections (B) were done in the same way except that a
higher multiplicty of infection was used (see Methods).

(Figure 2A) while only the dnaB+ revertant can produce λ
at 42 C (Figure 2B). The inability to produce λ at 42 C by
strain dnaB-70 is reversible since diluting the 42 C
infected cells to 33 C instead of 42 C will allow a normal
burst of λ (results not shown). Since λ has previously
been shown to require the dnaB function (20), these
experiments, which were done in an identical manner in the
presence of anti-M13 serum, show that the dnaB-70 mutation
was not phenotypically reverted by the single step growth
procedure using anti-M13 serum.

Using this same single step growth procedure M13
replication has been examined in other E. coli mutants
defective in DNA synthesis. It was found that M13 required
dnaA, dnaC(D), dnaE, and dnaG as defined by the inability to
produce M13 phage in this single step growth procedure at
the restrictive temperature. In contrast when this
procedure was used on various dnaB mutants they were all
(dnaB-107, -279, -500, -313, -454, -6, -8, and -43, see
reference 2) capable of producing 10-100 PFU's per infected
cell except dnaB-266 (see below). These results indicate
that the dnaB function is not required for M13 replication.

The earlier experiments showing a block to M13 phage
production at 42 C in dnaB-70 mutants (3,4) were done
using a low multiplicity of infection (about one). It was
thought that there might be some problems in interpreting
M13 infections of dnaB strains at 42 C when a low multiplicity of infection was used. Strain dnaB-70 and its
42 C resistant revertant (JMB-108) were infected with M13
at low MOI in the presence or the absence of chloramphenicol at 42 C. If the dnaB-70 mutation were specifically
blocking RF replication then both it and chloramphenicol
should have similar results since RF replication is the
first step in M13 DNA blocked by chloramphenicol (14). It
can be seen (Figure 3) that in M13 infections of chloramphenicol treated cultures of dnaB-70 and dnaB+ at 42 C,
the PFU's gradually decrease with time. A similar
decrease is seen during the initial stage of the M13
infected dnaB-70 culture which was not treated with
chloramphenicol. However later on the PFUs in the
untreated culture rose to a level about 10 times greater
than in the chloramphenicol treated culture even though
there was no overall increase in PFU's over the initial
input level. This indicates that the dnaB-70 cells
infected at a low multiplicity are producing phage but
that phage production is probably being masked by (i) the

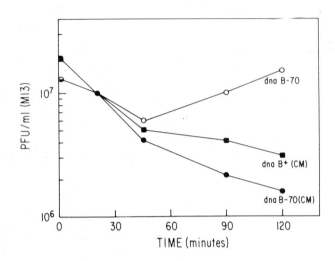

Fig. 3. Low multiplicity M13 infection of dnaB-70
and dnaB+ at 42 C with and without chloramphenicol. Strain
dnaB-70 and its dnaB+ revertant (JMB-108) were grown in
M9 to an OD 450nm of about 0.5, diluted to an OD 450nm of
about 0.2, and shifted to 42 C. After 5 minutes at 42 C
chloramphenicol was added to 180 microgram/ml to a portion
of the dnaB-70 culture and to the dnaB+ culture. Chloramphenicol was omitted from another portion of the dnaB-70
culture. After a total of 15 minutes at 42 C all three
cultures were infected with M13 at a multiplicity of about
1. Samples were taken and assayed for PFU/ml of M13 at the
times indicated.

slow rate of infection (19), (ii) the decreased amount of phage produced at 42 C (unpublished results), and (iii) the loss of the ability to support M13 infection with continued incubation at 42 C.

Another possibly contributing problem to the analysis of M13 replication in strains temperature sensitive in dnaB was encountered. "Good" male strains (as determined by M13 sensitivity at 33 C) containing dnaB mutations could not be maintained unless the maleness could be directly selected. Different single colony isolates of dnaB-70 (an HfrH strain) gave variable results in terms of sensitivity to M13 even at the permissive temperature. A similar situation was observed for F'lac derivatives of some other dnaB mutants. This somewhat anomalous behaviour may be a partial explanation for some results of Olsen et al. (4) which showed almost no incorporation of 3H-thymidine into viral specific DNA in M13 infected HfrH 165/70 relative to a 42 C resistant revertant.

Strains dnaB-43 and dnaB-279: Because of the problem of maintaining good males in dnaB mutants, F'thy+ male derivatives of dnaB strains were constructed as they could be continually and easily selected by merely omitting thymine from the medium. Figure 4 shows phage production after M13 infection (at 42 C) of the F'thy+ containing strain dnaB-43 and its 42 C resistant revertant (JMB-130). Strain dnaB-43 produced about 20-30 PFU per infected cell at 42 C even though host cell DNA synthesis was shut off upon shift from 33 C to 40 C (Figure 4B). That M13 replication does not require the dnaB function is illustrated further by the stimulation in the incorporation of 3H-thymidine into DNA which occurs upon infection of dnaB-43 with M13 at 42 C (Figure 5). This 3H-thymidine is being incorporated into viral DNA as most of it cosediments with added 32-P SS-DNA in a neutral CsCl velocity gradient. The amount of viral DNA made in the dnaB-43 mutants and its 42 C resistant revertant are comparable (Figures 6A and 6B respectively). While these sedimentations were not run for a long enough time for good separation of SS-DNA from RFI and RFII, it is clear that a considerable amount of viral specific DNA is made and that there is no obvious block to replication. To get another indication that phage particles are being produced at 42 C, the cell free supernatants from the same samples used to analyze for viral specific DNA by sedimentation

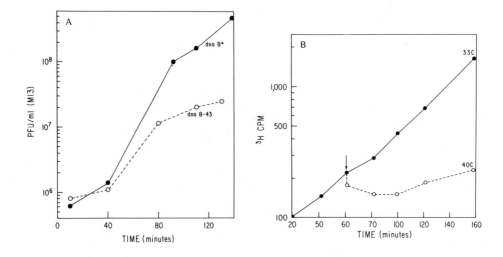

Fig. 4. M13 phage production and DNA synthesis in dnaB-43 and a 42 C resistant revertant. Strain dnaB-43 and its 42 C resistant revertant (JMB-130) were grown in M9 to an OD 450nm of about 0.1 and then treated in two ways. (A) A portion was concentrated 10-fold into 42 C M9, infected with M13, and then treated with anti-M13 serum. The infected cells were then diluted 1/1,000 into fresh media prewarmed at 42 C. Samples were taken for assay of PFU/ml. (B) Another portion of the dnaB-43 culture was diluted 1/20 into M9 medium at 33 C containing 250 microgram/ml deoxyadenosine and DNA synthesis was measured by examining the incorporation of 3H-thymidine (10 microcuries/ml) into DNA at 33 C and after shift to 40 C. Samples of 100 microliter were placed onto NaOH soaked filters at the times indicated and assayed for radioactivity (see Methods).

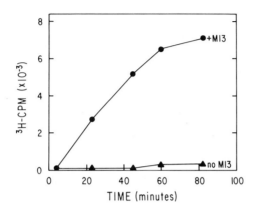

Fig. 5.   Incorporation of 3H-thymidine into dnaB-43
with and without M13 infection.   Strain dnaB-43 (JMB-129)
was grown in M9 at 33 C to an OD 450nm of about 0.5
divided into two portions and shifted to 42 C.   After 10
minutes at 42 C, uridine (to 250 micrograms/ml) and 3H-
thymidine (to 5 microcuries/ml) were added to each culture.
Five minutes later (15 minutes at 42 C) one culture was
infected with M13 (0.01 volume at about 2 x 10^{13} PFU/ml).
Samples (50 microliters) were withdrawn at the times
indicated to NaOH treated GF/A filters which were then
washed, dried, and placed in scintillation vials for
determination of radioactivity (see Methods).

were treated with polyethylene glycol (PEG) and NaCl to
precipitate phage particles (11). The resuspended, PEG
precipitated material was then sedimented in the same type
of preformed neutral CsCl velocity gradients used to
analyze the viral DNA species. In these gradients the
viral particles sediment to a position near their equilibrium density where they then stop. 3H-thymidine labeled
phage particles are made in comparable amounts at 42 C in
both dnaB-43 and its 42 C resistant revertant JMB-130
(Figures 6C and 6D).

Another F'thy+ containing strain dnaB-279 was tested
in ways identical to those already shown for dnaB-43 in
Figures 4 and 5. Strains dnaB-279 (i) produced about 100
PFU per infected cell at 42 C, (ii) immediately shut off
DNA synthesis after shift to 42 C, and (iii) was stimulated
by M13 infection to synthesize DNA at 42 C (as measured by
the incorporation of 3H-thymidine into acid insoluble
material). Neutral CsCl sedimentation of the lysates of
dnaB-279 infected with M13 at 42 C show that with
increasing times of infection more viral specific DNA is
made (Figure 7A). The amount of RFI (which sediments 5-6
fractions behind the arrow which indicated the position of
the added 32-P SS DNA marker) increased approximately 20-30
times from 15 to 90 minutes after infection. The phage in
the cell free supernatant of the 90 minute sample were
precipitated with PEG and sedimented to a "quasi-
equilibrium" position as already described for dnaB-43. As
expected 3H-thymidine labeled material, which was made and
released into the surrounding medium by the 42 C infected
cells, sedimented to the same position as the added 32-P
phage marker (Figure 7B).

Strain dnaB-266: Of all of the dnaB strains tested only
dnaB-266 did not produce phage at 42 C (Figure 8A). However this strain, which immediate shuts off DNA synthesis
upon shift to 42 C (Figure 8B), was stimulated by M13
infection at 42 C to incorporate 3H-thymidine into DNA
(Figure 8C). A 42 C resistant transductant of dnaB-266
(JMB-257) was also defective in phage production at 42 C
(Figure 9A) even though the rate of DNA synthesis by
uninfected cells considerably increased upon shift from
33 to 42 C (Figure 9B). The viral specific DNA made in
these strains after infection at 42 C was examined by
sedimentation in neutral CsCl gradients. Even after 90
minutes of infection very little SS-DNA had accumulated

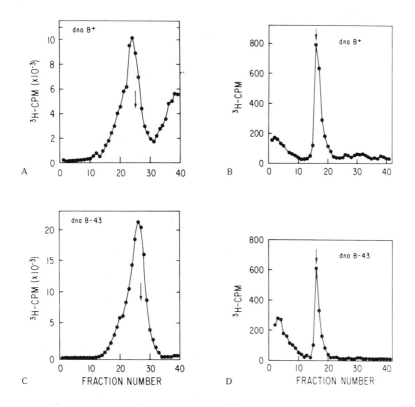

Fig. 6. Synthesis of 3H-thymidine labeled M13 DNA and phage particles at 42 C by dnaB-43 and dnaB+. After 45 minutes of infection, 5 ml samples were removed from the M13 infected dnaB-43 culture described in Figure 5 and from a similarly treated dnaB+ revertant (JMB-130) culture. Each sample was prepared for sedimentation in neutral CsCl to detect either intracellular phage DNA (A,B) or extracellular phage particles (C,D) as described in the Methods section. The sedimentation in neutral CsCl was reduced to 2 hours due to a power failure. In A and B the arrow indicates the position of 32-P labeled M13 SS-DNA. In C and D the arrow indicates the position in the gradient of 32-P labeled M15 phage particles.

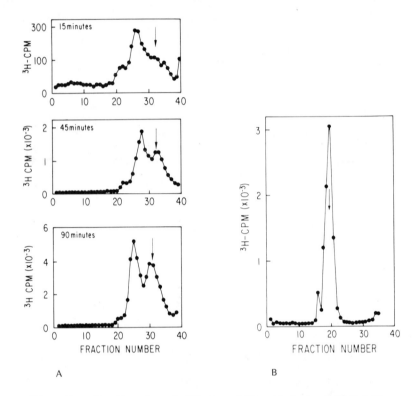

A

B

Fig. 7. Synthesis of 3H-thymidine labeled M13 DNA and phage particles at 42 C by dnaB-279. Strain dnaB-279 was grown and treated as described in Figure 6A for dnaB-43 except three samples were removed at 15, 45, and 90 minutes after infection. The samples were then analyzed for intracellular 3H-thymidine labeled M13 DNA (A) by sedimentation in neutral CsCl gradients. (B) The 90 minute sample was also analyzed for the presence of 3H-thymidine labeled phage particles as described for dnaB-43 in Figure 6C. In (A) the arrows indicated the position of the added 32-P labeled SS-DNA marker while in (B) the arrow indicates the position of the added 32-P labeled M13 phage particles.

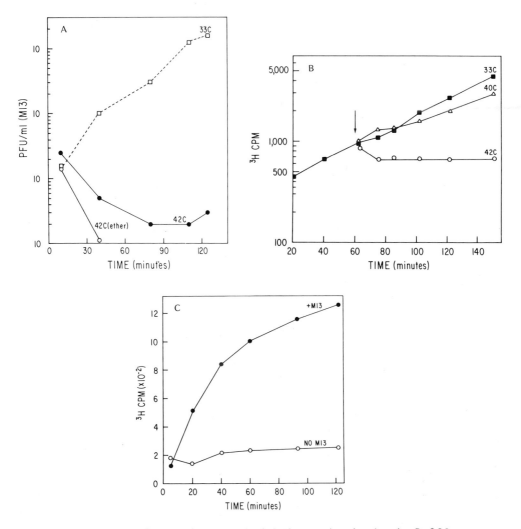

Fig. 8. M13 production and DNA synthesis in dnaB-266.
Strain dnaB-266 (JMB-250) was grown at 33 C in M9 and
treated as described for dnaB-43 in Figure 4. (A) M13
phage production at 33 C and 43 C after 42 C infection. (B)
The culture was diluted 1/4 into fresh M9 at 33 C containing
uridine. 3H-thymidine was added (to 5 microcuries/ml) and
DNA synthesis was measured at 33 C and after shift to 40
and 42 C. Samples of 100 microliter were taken at the times
indicated and treated as described (see Methods). (C) The
effect of M13 infection on 3H-thymidine incorporation at
42 C was done as described for dnaB-43 in Figure 5 except
that 100 microliter samples were taken.

439

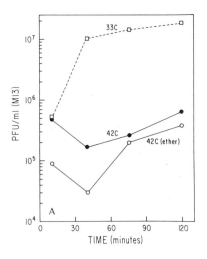
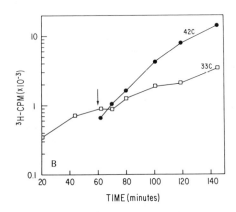

Fig. 9. M13 phage production and DNA synthesis in a
dnaB+ transductant of dnaB-266. JMB-257 was grown and
treated as described in Figure 8 for dnaB-266. (A) M13
phage production (B) DNA synthesis (3H-thymidine incorporation at 33 C and after shift to 42 C).

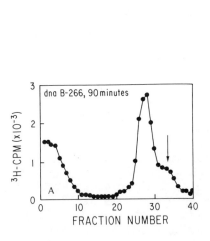
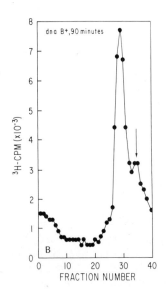

Fig. 10. Analysis of 3H-thymidine labeled M13
specific DNA made at 42 C by dnaB-266 and its dnaB+
transductant by sedimentation in neutral CsCl. Strains
dnaB-266 and JMB-257 were grown in M9 at 33 C and treated
as described for dnaB-43 and JMB-130 in Figures 6A and
6B respectively except that the samples were removed at 90
minutes instead of 45 minutes after infection.

in the dnaB-266 strain (Figure 10A). The 42 C resistant transductant made only slightly more SS-DNA (Figure 10B). The slightly increased amount of SS-DNA made by the 42 C resistant transductant may account for the slight recovery in phage production (about one per cell) observed in the dnaB+ strain relative to dnaB-266 (compare Figures 8A and 9A). Neither strain made 3H-thymidine labeled phage which could be detected as PEG precipitable material sedimenting to the same position as 32-P labeled marker phage added as a marker (results now shown). While it is not certain why dnaB-266 and its dnaB+ transductant are defective in phage production and SS-DNA sythesis it is certain that there is no defect in RF replication as this is the predominant viral DNA species made at 42 C in both strains (Figure 10).

DISCUSSION

The results presented in this paper have shown that (i) infective phage are made at 42 C in the single step growth procedure in all dnaB strains examined except dnaB-266; (ii) there is a stimulation in DNA synthesis at 42 C upon infection with M13 in all the dnaB strains examined including dnaB-266; (iii) in the two strains examined (dnaB-279 and dnaB-43) intracellular viral specific DNA replication was reasonably normal and 3H-thymidine labeled phage were made at the restrictive temperature; and (iv) even in dnaB-266 which could not make phage at 42 C, RF replication was certainly not defective. These results have clearly established that the dnaB function is not required for any specific step in the replication of M13.

Based on the presumption that the dnaB gene product was specifically required for M13 progeny RF replication and not for replication of SS-DNA, Staudenbauer et al. (5) concluded that these two processes occurred via separate mechanisms. One corollary to the conclusion that M13 replication does not require the dnaB function is that there is no a priori reason to hypothesize separate mechanisms for the replication of RF and SS-DNA in M13, and both processes could use the same general mechanism.

Other small, SS-DNA containing bacteriophages have been examined to determine whether the dnaB function was required. Bacteriophage ØX174 (21) has been shown to require the dnaB function (22, 23) as well as dnaC(D), dnaE, and dnaG (24,25,26). There is still some confusion as to

whether dnaA is required by ØX174 (27,28). The requirement for dnaB in ØX174 seems to reflect the status of the
proteins found in the viral particle since certain host
range mutants of ØX174 do not require the dnaB function
(29,30). On the other hand, bacteriophage ST-1 (31) does
not require the dnaB function and only seems to require
dnaE [DNA polymerase III, (32)] and not dnaA, dnaC(D), or
dnaG for phage production (33). Clearly the various SS-DNA
phages differ in their dependence upon the various host
cell products required for E. coli DNA replication. These
differences might reflect basic differences in the specific
mechanism used for DNA replication in the viruses despite
apparent similarities in the overall replication process.
More extensive studies are required to elucidate the nature
and extent of these differences.

At present the defect to M13 phage production observed
in dnaB-266 is somewhat perplexing. Presumably this strain
cannot make phage at the restrictive temperature because it
cannot make SS-DNA. A wild type transductant of dnaB-266
was also defective in phage production and SS-DNA synthesis
even though it was normal for host cell DNA synthesis. These
results seem to indicate that a second, unlinked mutation is
affecting the ability to make SS-DNA. However when the
dnaB-266 mutation was transduced into another strain by
cotransduction with malB and the resulting strain then
made F'thy+, this strain was also defective in phage production. A 42 C resistant transductant of this strain was
able to produce M13 at 42 C. These results seem to imply
the existence of a second mutation which is close enough to
dnaB (malB) for occasional cotransduction. Further experiments are necessary to determine the actual nature and
genetic locus of this defect in SS-DNA synthesis and phage
production.

ACKNOWLEDGEMENT

I would like to extend my deepest appreciation to
David Pratt for allowing this work to be completed in his
laboratory and for critically reading the manuscript.
This work was supported by Public Health Service grant
AI-10201 from the National Institute of Allergy and
Infectious Diseases and in part by Public Health Service
postdoctoral research fellowship GM-52104 from the National
Institute of General Medical Sciences.

REFERENCES

1. Marvin, D. A. and Hahn, B. Bacteriol. Rev. 33, 172 (1969).
2. Gross, J. D. Curr. Top. Microbiol. Immunol. 57, 39 (1971).
3. Primrose, S. B., Brown, L. R., and Dowell, C. E. J. Virol. 2, 1308 (1969).
4. Olsen, W. L., Staudenbauer, W. L., and Hofschneider, P. H. Proc. Nat. Acad. Sci. U.S.A. 69, 2570 (1972).
5. STaudenbauer, W. L., Olsen, W. L., and Hofschneider, P. H. Eur. J. Biochem. 32, 247 (1973).
6. Bonhoeffer, F. Z. Verebungslehre 98, 141 (1966).
7. Wechsler, J. A., and Gross, J. D. Mol. Gen. Genet. 113, 273 (1971).
8. Shubach, W. H., Whitmer, J. D., and Davern, C. I. J. Mol. Biol. 74, 205 (1973).
9. Salivar, W. O., Tzagoloff, H., and Pratt, D. Virol. 24, 359 (1964).
10. Hohn, B., H. Lechner, and Marbin, D. A. J. Mol. Biol. 56, 143 (1971).
11. Yamamoto, K. R., Alberts, B. M., Benzinger, R., Lawhome, L., and Treiber, G. Virol. 40, 734 (1970).
12. Tseng, B. Y., Hohn, B., and Marvin, D. A. J. Virol. 10, 362 (1972).
13. Miller, J. H. Experiments in Molecular Genetics (Cold Spring Harbor Laboratory, Cold Spring Harbor, N. Y. 11724, 1972).
14. Pratt, D. and Erdahl, W. S. J. Mol. Biol. 37, 181 (1968).
15. O'Donovan, G. A., and Neuhard, J. Bact. Rev. 34, 278 (1970).
16. Tseng, B. Y. and Marvin, D. A. J. Virol. 10, 384 (1972).
17. Ray, D. S., and Schekman, R. W. Biochim. Biophys. Acta 179, 389 (1969).
18. Lin, N. S. C. and Pratt, D. J. Mol. Biol. 72, 37 (1972).
19. Tzagoloff, H., and Pratt, D. Virol. 24, 372 (1964).
20. Fangman, W. L. and Novick, A. Genetics 60, 1 (1968).
21. Sinsheimer, R. L. Progress in nucleic acid research and molecular biology. Davidson, J. N. and Cohn, W. E. (ed.), Academic Press Inc., New York (1968).
22. Dumas, L. B. and Miller, C. A. J. Virol. 14, 1369 (1974).

23. Steinberg, R. A. and Denhardt, D. T. J. Mol. Biol. 37, 525 (1968).
24. Dumas, L. B. and Miller, C. A. J. Virol. 11, 848 (1973).
25. Kranias, E. G. and Dumas, L. B. J. Virol. 13, 146 (1974).
26. McFadden, G. and Denhardt, D. T. J. Virol. 14, 1070 (1974).
27. Schekman, R., Wickner, W., Westergaard, O., Brutlage, D., Geider, K., Gertsch, L. L., and Kornberg, A. Proc. Nat. Acad. Sci. U.S.A. 69, 2691 (1972).
28. Taketo, A. Mol. Gen. Genet. 122, 15 (1973).
29. Vito, C. C., Primrose, S. B., and Dowell, C. E. J. Virol. 15, 281 (1975).
30. Bone, D. R. and Dowell, C. E. Virol. 52, 330 (1973).
31. Bowes, J. M., and Dowell, C. E. J. Virol. 13, 53 (1974).
32. Gefter, M. L., Hirota, Y., Kornberg, T., Wechsler, J. A., and Barnoux, C. Proc. Nat. Acad. Sci. U.S.A. 68, 3150 (1971).
33. Bowes, J. M. J. Virol. 13, 1400 (1974).

REPLICATION OF MYCOPLASMAVIRUSES

Jack Maniloff and Jyotirmoy Das
Departments of Microbiology and of
Radiation Biology and Biophysics
The University of Rochester
Rochester, New York 14642

ABSTRACT

The MVL51 virion, a mycoplasmavirus with single stranded circular DNA, has been found to have four protein species. Two of these are found in the SSII replicative intermediate. Antibiotic studies indicate a coupling between penetration and DNA replication. These data are discussed, together with previous reports, to describe the molecular details of MVL51 replication. Host modification and restriction of an enveloped mycoplasmavirus has also been observed.

INTRODUCTION

The mycoplasmas are the smallest procaryotes, with only a limiting cell membrane, and carry at least three morphologically distinct viruses (1,2). Group L1 viruses are naked bullet-shaped particles, about 14 x 70 nm; Group L2 are enveloped viruses, 50-125 nm in diameter; and Group L3 are 60 nm icosahedral particles with short tails. Group L1 and L2 viruses are not lytic and infected mycoplasmas continue to grow while producing progeny virus, but Group L3 viruses are most probably lytic. All three groups are DNA viruses.

MVL51, a Group L1 virus, has been studied most extensively. The viral chromosome is a single stranded circular DNA of molecular weight 2×10^6 daltons (3). When used in a transfection system, this DNA has been shown to infect both host cells and a mycoplasma species that cannot be infected by virus particles (4).

The intracellular replication of MVL51 begins with the conversion of the parental chromosome to a double stranded DNA replicative intermediate (5). This can exist in two forms: covalently closed circles (RFI) or relaxed ("nicked") circles (RFII). Single stranded progeny chromosomes (SSI) are produced from the RF and become associated with viral proteins to form a protein-DNA complex (SSII). SSII is the final viral intermediate which has been isolated and is the form of the progeny DNA which is extruded through the cell membrane without lysing or killing the cell.

We report here on further studies of the molecular details of the structure and replication of MVL51 and on observations of host modification.

RESULTS

MVL51 Proteins: For these studies, viruses were washed from lawns in phosphate buffered saline (5), pH 7.4, filtered through a 0.22 μm filter, and given two cycles of differential centrifugation. Each cycle consisted of a low speed centrifugation (at 10,000 rpm for 10 min at 4°C), removal of the low speed supernatent, a high speed centrifugation (at 40,000 rpm for 3 hrs at 4°C in a Beckman 42.1 rotor) of the low speed supernatent, and resuspension of the high speed pellet in 0.1 M phosphate buffer. After the second differential centrifugation cycle, the concentrated virus suspension was layered on a CsCl step gradient in a Beckman SW50.1 rotor: each step was 0.9 ml of a CsCl solution (in 0.01 M Tris-HCl buffer, pH 7.4, containing 2 mM MgSO$_4$) of densities 1.1, 1.2, 1.3, and 1.4 g/ml. The gradient was centrifuged for 1-2 hrs at 30,000 rpm at 4°C and the viral band was collected and dialyzed against Tris buffer. The viruses were then put on a 10-40 percent (w/v) sucrose gradient and centrifuged at 24,000 rpm for 30 min at 4°C in a Beckman SW27 rotor. The virus band was collected, dialyzed, and the proteins of this purified virus analyzed in 10 percent SDS polyacrylamide gels, as described by Laemmli (6).

The MVL51 virion was found to contain four proteins (Table 1), of 70,000, 53,000, 30,000 and 19,000 daltons. From the data in Table 1, the approximate stoichiometric ratios of these proteins are 6.7 : 1.0 : 1.8 : 8.3. These four proteins could account for about 75 percent of the viral chromosome's coding capacity.

SSII, the replicative intermediate complex of protein and progeny chromosomes was collected from sucrose gradients of artificial (SDS) cell lysates (5), dialyzed, and the proteins were analyzed on gels as described above. SSII was found to have two of the virion proteins, at 70,000 and 53,000 daltons, with stoichiometric ratios about 1.4 : 1.0.

Effect of rifampicin (RIF) and chloramphenicol (CAM):
Previous studies (7) have examined the effects of RIF and CAM when these antibiotics were added after the start of MVL51 infection. In this report, the studies were done using cells pretreated with the antibiotic and infection was carried out in the presence of the antibiotic. The cells used were RIF-resistant host cells (7).

RIF (100 µg/ml) was added to the cells for 10 to 30 min, virus was then added at an MOI of 3-5, virus specific DNA components were labeled, and sucrose gradient analysis of cell lysates was performed, as described previously (5). No viral DNA intermediates were observed. In order to investigate whether the parental viral DNA was able to get into a RIF treated cell, the experiment was repeated with ^{32}P-labeled virus. No incorporation of parental label was found, indicating that MVL51 DNA penetration requires an RNA polymerase. Similar observations have been reported for the filamentous phage M13 (8) suggesting that penetration is coupled to replication of the parental single stranded DNA to form a double stranded RF.

When CAM (100 µg/ml) was used instead of RIF, it was found that CAM treated cells were able to synthesize about five percent as much viral specific DNA as untreated controls. This must represent synthesis by cell proteins made before CAM treatment and MVL51 infection. Gradient analysis showed that almost all the nascent viral DNA was in SSI; no SSII was formed, confirming the absence of protein synthesis in CAM treated cells. Hence endogenous host cell proteins can carry out the replicative steps from parental chromosomes to the production of progeny chromosomes.

Host modification and restriction: These studies used MVL51 (a Group L1 virus), MVL2 (a Group L2 virus), and *Acholeplasma laidlawii* strains JA1 and 1305 (Table 2).

447

TABLE 1

PROTEINS OF MVL51 VIRION AND SSII REPLICATIVE INTERMEDIATE

Molecular weight[a]	Relative amount (percent) in[b]	
	Virion	SSII
70,000	64	65
53,000	7	35
30,000	7	-
19,000	22	-

[a]The electrophoretic mobility as a function of molecular weight of the gels was calibrated using five proteins of molecular weight 23,500 to 135,000 daltons.

[b]Estimated from staining density of polyacrylamide gels stained with Coomassie brilliant blue.

TABLE 2

RELATIVE NUMBER OF VIRUS PLAQUES
ON *ACHOLEPLASMA LAIDLAWII* LAWNS

Virus[a]	Relative PFU on A. *laidlawii*	
	Strain JA1	Strain 1305
MVL51 · JA1	1	0.2
MVL51 · 1305	1	0.6
MVL2 · JA1	1	10^{-3}
MVL2 · 1305	10^{-3}	1

[a]The symbols after the dot indicate the last host on which the virus was grown.

MVL51 shows slight modification by 1305, but no restriction. However, MVL2 shows host modification and restriction. Although no data are available yet on the modification mechanism, it is interesting to speculate that, since MVL51 and MVL2 are both DNA viruses but only the enveloped MVL2 is modified and restricted, perhaps it is the viral membrane, rather than the viral DNA, that is being affected.

DISCUSSION

There are now sufficient data to describe the replication of MVL51. Penetration is coupled to the replication of the circular single stranded parental DNA. It is possible that a virion protein may modify a cellular RNA polymerase for this step, since penetration is RIF-sensitive in RIF-resistant cells (described above). Endogenous cell proteins are also sufficient for synthesis of double strand DNA replicative forms (RFI and RFII) and synthesis of circular single stranded progeny DNA (SSI), since these can occur in CAM treated cells (described above). The sensitivity of RF replication to RIF (7) indicates the involvement of RNA polymerase for this process; this must be a modified polymerase since it is RIF-sensitive in RIF-resistant cells. Preliminary experiments show that the parental RF and most progeny RFI and RFII are membrane-associated (9). SSI interacts with two of the four virion proteins (described above) to form SSII (5), this process is sensitive to RIF and CAM (7, also described above). Both SSI and SSII are free cytoplasmic intermediates (9). Final assembly and maturation must take place as SSII is extruded through the cell membrane without killing the cell.

The replicative details of MVL51 resemble those of filamentous phage (8), although the viral structures are significantly different. MVL51 seems to have a more complex virion morphology and to utilize few viral gene products for intracellular replication.

ACKNOWLEDGEMENTS

We thank Mr. David Gerling for his assistance. These studies were supported by USPHS Grant AI-10605. J.M. is the recipient of a USPHS Research Career Development Award (AI-17480).

REFERENCES

1. Maniloff, J. and Liss, A. In "Viruses, Evolution and Cancer", E. Kurstak and K. Maramorosch, eds., Academic Press, New York, p. 583 (1974).
2. Gourlay, R. N. CRC Crit. Rev. Microbiol. $\underline{3}$, 315 (1974).
3. Liss, A. and Maniloff, J. Biochem. Biophys. Res. Commun. $\underline{51}$, 214 (1973).
4. Maniloff, J. and Liss, A. Ann. N. Y. Acad. Sci. $\underline{225}$, 149 (1973).
5. Das, J. and Maniloff, J. (Manuscript submitted for publication).
6. Laemmli, U. K. Nature (Lond.) $\underline{227}$, 680 (1970).
7. Das, J. and Maniloff, J. (Manuscript submitted for publication).
8. Kornberg, A. DNA Synthesis, W. H. Freeman & Co., San Francisco, p. 247 (1974).
9. Das, J. and Maniloff, J. (Manuscript in preparation).

Part VI
Phage λ Replication

STRUCTURAL ANALYSIS OF INTRACELLULAR λ DNA

Manuel S. Valenzuela,* and Ross B. Inman
Biophysics Laboratory and
Department of Biochemistry
University of Wisconsin
Madison, Wisconsin 53706

RESULTS AND DISCUSSION

Does λ DNA terminate at a unique site at the completion of the first round of replication?

When double branched and single branched circular replicative intermediates from the first round of λ DNA replication are analyzed by electron microscopic denaturation mapping, the position of growing points can be determined relative to the left or right end of the mature λ DNA molecule. Plots of growing point position vs. % replication then yield an estimate of the origin and termination of the first round of replication.

Previous data obtained in studies of λcIIcIII replication show that the origin of replication (as determined from the above type of plot) is situated at 13.40 μ and the round terminates at an average position of 6.49 μ for a mature λ DNA length of 17.06 μ (Fig. 1). Our previous estimate for the origin (13.94 μ) was based, in part, on the average mid-point of each pair of bidirectional growing points (1). Because replication towards the right proceeds a little further than to the left, it appears more reasonable to use plots of growing point position vs. % replication to estimate the origin and termination positions.

Does the termination position (6.49 μ) represent a unique site, where the round must end, or does the round

---

*On leave from the Graduate Department of Biochemistry, Brandeis University, Waltham, Massachusetts 02154

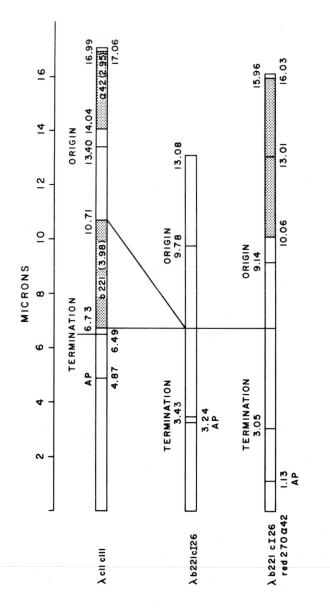

Fig. 1. Maps showing experimentally determined average position of origin and termination of the first round of replication. AP indicates the position of the antipode to the origin. The dotted areas show the position of the b221 deletion and the a42 tandem duplication.

stop wherever growing points happen to collide? In principle this question can be answered by analysis of replicative intermediates in which DNA has been deleted, just to the left, and/or added, just to the right, of the origin. Phage λb221cI26 and λb221cI26red270a42 have a deleted segment (b221) between 6.73 and 10.71 μ, as determined by heteroduplexy, and the latter also contains a tandem duplication (a42) of the segment between 14.04 and 16.99 μ (2). The b221 deletion and a42 duplication are therefore suitable for the above mentioned test (see Fig. 1).

Results of this experiment show that the origin of replication, for λb221cI26 and λb221cI26red270a42, is situated at 9.78 and 9.14 μ, respectively; this is in reasonable agreement with the λcIIcIII origin (13.40 μ) when allowance is made for the b221 deletion (3.98 μ). The experimentally determined termination positions were found to be 3.46 and 3.05 μ, respectively (Fig. 1) and therefore these data indicate that termination probably occurs wherever growing points happen to collide.

Plots of bidirectional growing point positions are shown in Fig. 2(a,b,c). In each case a large amount of scatter can be observed and this detracts from the above experiment. In any event it can be seen that a significant number of growing points move to the left across the putative λcIIcIII termination site at 6.49 μ (see Fig. 2b and c).

The large amount of scatter observed in bidirectional plots (Fig. 2a,b and c) may be due to gross temporal interruptions in growing point propagation. Fig. 3(a and b) shows examples of unidirectional replication taken from studies of P2 and 186 phage (3,4). In these cases very little scatter is observed. Gross interruptions in growing point propagation will introduce scatter into bidirectional but not in unidirectional plots.

Evidence for cross-strand exchange in intracellular λ DNA.

When λb221cI26red270a42 infects a rec A strain of E. coli about 3% of all first round circles and circular intermediates possess a novel junction. Of fourteen molecules exhibiting junctions, six were single branched circles (Fig. 4a) and eight involved a junction between a circle and a linear duplex segment (Fig. 4b).

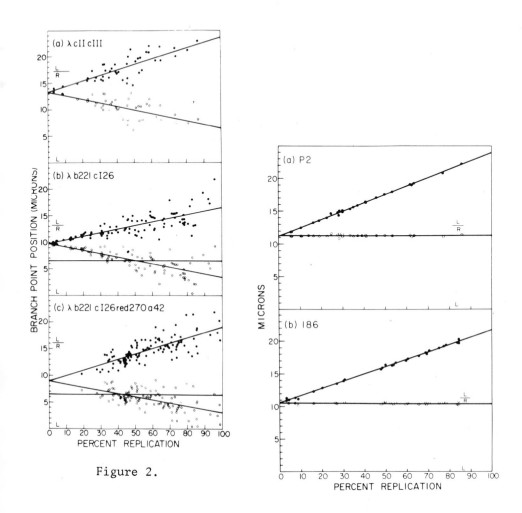

Figure 2.

Figure 3.

Fig. 2. Plots of bidirectional growing points vs. % replication. The positions of the left and right ends of the mature molecules are indicated by L and R, respectively. The horizontal line at 6.49 μ shows the position of the putative λcIIcIII termination site.

Fig. 3. Plots of unidirectional growing points vs. % replication.

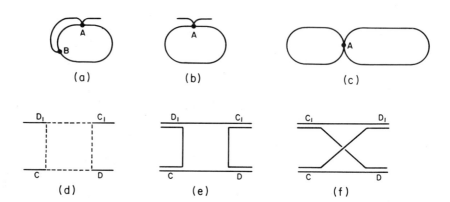

Fig. 4(a,b,c)
Drawings of molecules involving junctions (A). B indicates
a branch point. In the case of (c), one circle is monomeric
while the other can be either monomeric, dimeric or trimeric.

(d)
Fine structure often observed at junctions. Solid and dotted
lines indicate double and single stranded DNA, respectively.

(e,f)
Molecular interpretation of junctions, (e) corresponds to the
observed structure (denatured sites in trans) while (f) shows
the structure redrawn so that denatured sites are in cis.

CD and C_1D_1 show DNA polarity as deduced from denaturation
mapping.

In a study of late rounds of λcIIcIII replication about 0.3% of all circular types also involved junctions. Of twelve molecules exhibiting junctions, four were of the type shown in Fig. 4(b) while eight had junctions between two circles (Fig. 4c), where one circle of each pair was monomeric and the other either monomeric, dimeric or trimeric length.

Partial denaturation mapping showed that in every case so far examined (26 molecules), junctions occur at homologous positions on each partner. However they can occur at different positions from molecule to molecule.

Junctions often exhibit single stranded fine structure of the type shown in Fig. 5 and drawn in Fig. 4(d). Out of 26 junctions, 14 were judged to have this structure while the rest were either not well resolved or ambiguous in this respect. The molecular interpretation of the junction fine structure is drawn in Fig. 4(e).

Denaturation mapping shows that all molecules (26) exhibit matching denatured sites in trans on either side of the junction. The DNA polarities can therefore be deduced and are shown as CD and C_1D_1 in Fig. 4(d and e). When this structure is redrawn so that the base sequences are congruent (denatured sites in cis) the resulting structure (Fig. 4f) appears to involve a cross-strand exchange.

Two general ways to explain the presence of junctions involve either recombinational events or replication growing points. Junctions have the properties expected of recombinational intermediates. Popular models for recombination can be resolved into a cross-strand exchange of the specific types discussed by Holliday (5), Broker and Lehman (6), Stahl et al. (7), Meselson and Radding (8) and Sobell (9) or the general types listed by Sigal and Alberts (10). Furthermore according to model building tests by Sobell (11) the transition between the two cross-strand isomers discussed by Sigal and Alberts (10) proceeds via an intermediate structure similar to that proposed by Broker and Lehman (6), and this structure corresponds to the open junctions observed in the present investigation.

On the other hand junctions could also arise by branch migration of a replication growing point as shown in Fig. 6. This type of migration might account for the structures drawn in Fig. 4(a and b) but perhaps not so

457

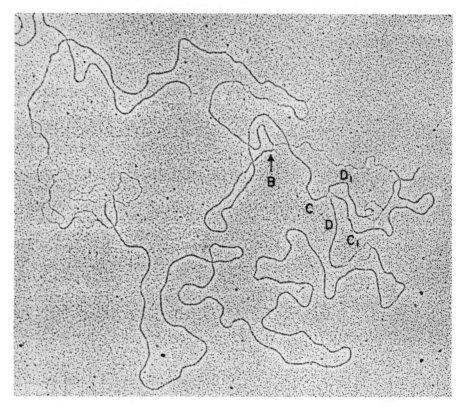

Fig. 5. Electron micrograph of a single branched
circle involving a junction. The branch point is indicated
at B and the DNA polarities at the junction are shown
(CD and C_1D_1).

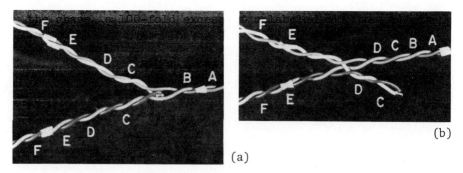

(a)

(b)

Fig. 6. Branch migration of a replicating growing
point can in principle give rise to a junction.

easily explain structures of the type shown in Fig. 4(c). Similar junctions have been observed in studies of replicating mitochondrial DNA from sea urchin oocytes and it was concluded that they resulted from branch migration of replicating growing points (12).

Junctions were found to be ten-fold more frequent in the λb221cI26red270a42 studies (where maturation of recombinational intermediates was blocked by the phage red and host rec A mutations) than in the λcIIcIII experiment. Although this difference lends support to the notion that the junctions do in fact arise from recombinational events, it is obviously necessary to test this conclusion more rigorously.

ACKNOWLEDGMENT

This work was supported by grants from the National Institutes of Health and the American Cancer Society.

REFERENCES

1. Schnös, M. and Inman, R. B. J. Mol. Biol. 51, 61 (1970).
2. Emmons, S. W. and Thomas, J. O. J. Mol. Biol. (in press).
3. Schnös, M. and Inman, R. B. J. Mol. Biol. 55, 31 (1971).
4. Chattoraj, D. K. and Inman, R. B. Proc. Nat. Acad. Sci. 70, 1768 (1973).
5. Holliday, R. Genetic Res. 5, 283 (1964).
6. Broker, T. R. and Lehman, I. R. J. Mol. Biol. 60, 131 (1971).
7. Stahl, F. W., Chung, S., Crasemann, J., Faulds, D., Haemer, J., Lam, S., Malone, R., McMilin, K., Nozu, Y., Siegel, J., Strathern, J. and Stahl, M. In Fox, E. and Robinson, W. (ed.) Virus Research, Academic Press, p. 487 (1973).
8. Meselson, M. and Radding, C. Proc. Nat. Acad. Sci. 72, 358 (1975).
9. Sobell, H. M. Proc. Nat. Acad. Sci. 69, 2483 (1972).
10. Sigal, N. and Alberts, B. J. Mol. Biol. 71, 789 (1972).
11. Sobell, H. M. In Grell, R. (ed.) Mechanisms in Recombination, Plenum Press, p. 433 (1974).
12. Matsumoto, L., Kasamatsu, H., Piko, L. and Vinograd, J. J. Cell Biol. 63, 146 (1974).

MOLECULAR MECHANISMS IN THE CONTROL OF λ DNA REPLICATION: INTERACTION BETWEEN PHAGE AND HOST FUNCTIONS

A. Skalka, M. Greenstein and R. Reuben
Roche Institute of Molecular Biology, Nutley, N. J. 07110

ABSTRACT: Our studies of the molecular events involved in the switch from early (circle) to late (rolling circle) replication of phage DNA have revealed several ways in which host and phage recombination and repair systems can exert both negative and positive control. Under one set of conditions (fec⁻) a phage mutation (gam), which results in a failure to inhibit the host's recBC function, restricts replication to the early mode. Under a second set of conditions (feb⁻), combined lesions in the phage general recombination (red) system and the host's repair (PolI or ligase) system produce a similar result.

Our most recent experiments with the fec⁻ system, in which the BC nuclease activity was manipulated by temperature shifts after infection with a temperature sensitive gam mutant, suggest that initiation of rolling circle replication is inhibited because the BC nuclease attacks some transient intermediate present during the switch from early to late replication and/or that it attacks at some transient stage in which the rate of DNA synthesis is limited. Structural analysis on a relaxed circular molecule which accumulates in this infection has revealed a uniquely positioned interruption which may be the origin of rolling circle replication. Analysis of feb⁻ infections indicate that the red recombination system and a polI + ligase repair system may provide alternate routes for the transition from circle to rolling circle replication. Intermediates along these routes may contain the BC nuclease-sensitive structure alluded to above.

A dynamic model emerges from all of this in which circle replication provides a driving force that can be controlled, either to remain as such or to be shifted into rolling circle replication, depending on the disposition of recombination, repair, and possibly morphogenetic functions.

460

INTERNAL CONTROL OF GENE EXPRESSION AND MODULATION BY ENVIRONMENTAL FACTORS

1. Transcription

Development of bacteriophage λ is under the direction of both positive and negative control elements known to operate at the level of gene transcription. Examples of such elements include proteins specified by λ genes N, Q, cI, II & III and cro (1 - 6). These controlling proteins operate "internally": their synthesis is governed by the phage, and they affect expression of phage genes. The ways in which they operate to switch certain pathways on or off are generally understood. Such understanding reveals much about viral control of its own growth and suggests general mechanisms which might be important in the regulation of transcription (7). However some aspects concerning the function of these control elements are not well understood. One aspect involves the modulation of control by environmental factors. Examples of such modulation are: 1) the effects of the physiological state or genotype of the host bacterium on the choice between lytic or lysogenic growth, and (2) induction of a repressed prophage by relatively nonspecific agents (mitomycin-C, UV light) which seem to affect host DNA synthesis. Response to these modulating influences indicate that the phage can adjust to a changing variety of molecules in its environment in the cell. The ability to adapt its development to such changes is important to the survival of lambda. Understanding of interactions of this kind can provide us with insight concerning bacterial as well as viral control mechanisms (8).

2. Replication

The replication of lambda DNA also appears to be subject to internal control. The initiation of DNA synthesis depends on phage DNA transcription, both directly (as summarized by Hayes in this volume) and indirectly. Lambda DNA is not replicated in the absence of phage O and P proteins which are presumably part of a replication machine that includes bacterial proteins. Thus, lambda replication can also be modulated by intracellular environmental factors represented by bacterial replication

proteins (9). This report will summarize results from our
analysis of the interaction between phage and bacterial
modulating elements which influence the program of DNA
replication in lambda.

TWO STAGES FOR REPLICATION OF λ

During the lytic phase, replication of λ DNA occurs
in two stages (Fig. 1). In the early stage, the chromosome is replicated as a circular molecule from a fixed
origin and often in both directions like its host's DNA.
In a normal infection, replication via this early mode
occurs only once or twice, after which rolling circle
(late) replication predominates. The two modes of replication can be most easily distinguished by the structure
of accumulated products. Early replication produces
monomeric circular molecules; late replication produces
concatemers--presumably from the tails of rolling circles.
The concatemers (but **not** circular monomers) are the forms
which become packaged into phage heads (10). The modulating influences that we will discuss affect the switch
from the early to late mode of replication.

MODULATING EFFECTS OF THE BACTERIAL REC BC NUCLEASE

1. The Role of the Viral Gamma Protein

Earlier studies have shown (10) that the transition
from the early to late mode of replication depends on the
presence of the viral protein, gamma (gam). In gam⁻
infection, replication is confined to the early mode and,
provided that recombination is inhibited by additional
mutations, covalently closed and relaxed circular monomers
accumulate. Infection (or induction) with gam⁻ phage
under recombination defective (recA⁻, red⁻) conditions has
been defined as fec⁻ (11, 12). Under fec⁻ conditions
there is an abnormal accumulation of early replication
products (an average of 30-40 phage equivalents/infected
cell) at late times. This fact suggests that a direct,
internal control gene for the turn-off of early replication, analagous to the cro gene which turns off early
transcription, does not exist. If it does exist, then
it must not be expressed in the absence of late replication. We describe below a model which relates the turn-
off of early replication to the onset of late replication.

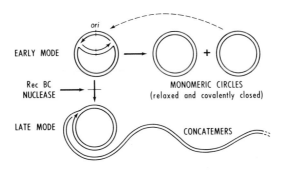

Fig. 1.   Two stages for replication of lambda DNA.   Schematic indicating stage at which modulation by BC nuclease
can be detected (10).

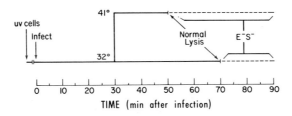

Fig. 2.   Scheme for activation of BC nuclease by a temperature shift.   UV-irradiated recA⁻ cells (E. coli 204, thy⁻
su⁻) were infected with Eam4 red (bet)113gamts23 Sam7
phage at 32°C.   At 30 min. postinfection, the BC nuclease
was activated by shifting the temperature to 41°C.
Methods for growth, irradiation, and infection were as
described previously (10).

The function of gam in replication is related to modulation: the gam protein does not affect phage gene expression directly, and the gam mutation has no detectable
effect on phage replication in recBC nuclease deficient
hosts. These and other observations (10, 13) indicate
that the critical function missing in gam⁻ infection is
inhibition of this bacterial nuclease.

2.   Mechanisms of Modulation by RecBC Nuclease

We have considered two possible ways in which the
recBC nuclease might inhibit late replication: (1) by
destruction of some transient structure, present only at
the time of transition; (2) by destruction of the late
replicating fork itself.   In an attempt to distinguish
between these two, we looked for conditions under which we
could activate the BC nuclease after the switch to late
replication had taken place.   We believed that if
mechanism (1) were correct, such activation should be
without consequence.   Preliminary tests with several of
the gam mutants isolated by Zissler (12) indicated that
with one of these, ts23, the inhibitory effect could be
reversed in vivo after incubation at high temperature.
(Karu et al.   (14) have recently demonstrated in vitro
reversal for Zissler's ts37).
The basic design of our temperature-shift experiments
is shown in Fig. 2.   Cells were irradiated with UV light
to inhibit host DNA synthesis.   They were then infected
and maintained at 32° until about halfway through the
lytic cycle, at which time late replication was well
underway.   A portion was shifted to 41° to inactivate the
gamts protein and thereby activate the BC nuclease.

DNA Concatemers are Degraded by the BC Nuclease

The first experiments were designed to test the
stability of replication products labeled just before the
temperature shift.   Accordingly, DNA was labeled from
26 to 30 min. with ^{3}H thymine.   The isotope was then
chased with an excess of unlabeled thymidine at both 32°C
and 41°C. In order to accumulate all replication products
and to follow the fate of DNA beyond the normal time of
lysis, DNA maturation and cell lysis were blocked while
the DNA synthesis period was extended by inclusion of

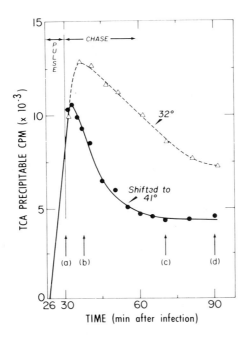

Fig. 3.   Degradation of late phage DNA after activation
of the BC nuclease.   The protocol is outlined in Fig. 2.
Phage DNA was labeled at 32° by addition of ^{3}H-thymine
at 26 min. postinfection as described in Table 1.   At
30 min. isotope was chased with excess unlabeled thymidine
at 32°C and 41°C.   The breakdown of labeled phage DNA was
measured by assaying the amount of radioactivity remaining
in trichloroacetic acid–insoluble material during the
chase (10).

● Shifted to 41°C   Δ Maintained at 32°C

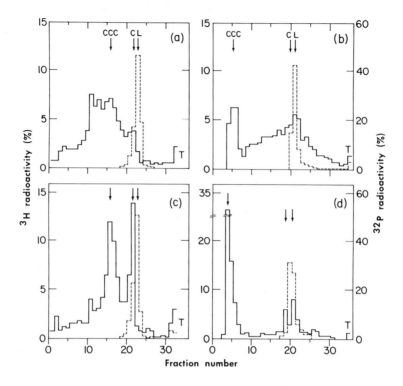

Fig. 4. Sedimentation analysis of late intracellular
phage DNA before and after a temperature shift. (a) and
(c), neutral sucrose; (b) and (d), alkaline sucrose. The
protocol outlined in Fig. 2 was employed. Methods for
labeling, extraction of the DNA and sucrose analysis have
been described previously (10). Infected cells were
pulsed with ^{3}H-thymidine from 26 to 30 min (see Table 2)
and chased with excess unlabeled thymidine at the elevated
temperature. DNA was isolated immediately after the pulse
and during the chase at 40, 70 and 90 min after infection.
DNA was analyzed by neutral and alkaline sucrose gradient
sedimentation. Each gradient contained at least 10^3 [3]H
CPM. The results presented here represent DNA samples
from the pulse at 32°C ((a) and (b)) and the 41°C chase at
90 min ((c) and (d)). ^{32}P-labeled linear monomer λ DNA,
extracted from phage particles, was employed as a marker
(broken lines). In the neutral gradients, the position
of the linear marker (L) and positions calculated for relaxed circular monomers (C) and covalently closed monomeric circles (CCC) are indicated. In the alkaline gradients, the position of the single-stranded linear marker
(L) and positions calculated for single-stranded circular
monomers (C) and covalently closed circular molecules
(CCC) are shown. Sedimentation is from right to left.

additional mutations in the phage genes E (capsid morphogenesis) and S (lysis and DNA stop) (15, 16, 17).   The
results from such an experiment (Fig. 3) show that late
DNA labeled at 32° was rapidly degraded at 41° to a
plateau where about 30% of the DNA remained intact.   Thus,
we conclude that about 70% of the labeled species were
sensitive to the activated nuclease and about 30% were
resistant.   In this infection, we detected significant
degradation even at 32° suggesting that the gamma protein
was partially defective even at the "permissive" temperature.   These effects are independent of E and S function
because infection with otherwise isogenic E^+S^+ phage gave
similar results (data not shown).   As can be predicted
from comparison of the data of Fig. 3 and the time scale
in Fig. 2, degradation in E^+S^+ infection was not completed
at the time of lysis.   Infection with isogenic gam^+ phage
(data not shown), resulted in no degradation at either
32° or 41° verifying that we had observed a gam-specific
effect.

       To identify the labeled species which were sensitive
and resistant to degradation at 41°, DNA was extracted
(see Fig. 3) (a) right after the pulse (to identify all
labeled species), and at times during the chase when (b)
the maximum rate of degradation and (c) and (d) completed
degradation were observed.   The structure of the labeled
species present in the extract was inferred from their
characteristic sedimentation in neutral and alkaline
sucrose (Fig. 4).   Fig. 5 shows the fate of the various
pulse-labeled species as determined in these analyses.
Immediately after the pulse, approximately 70% of the
label was in concatemers and 30% in monomeric circles,
both relaxed (about 13%) and covalently closed (about 21%)
forms.   The synthesis of these circular molecules at late
times suggests that in this infection both early and late
replication were occurring simultaneously at 32°.   This may
also reflect the partial defectiveness of the temperature
sensitive gamma protein.   During the chase, label in
concatemers disappeared almost completely, labeled relaxed
circles appeared to be completely stable, and the proportion of labeled covalently closed circles appeared to
increase slightly.   The degradation of the concatemer
species alone seems to account for the bulk of the breakdown observed in Fig. 3.

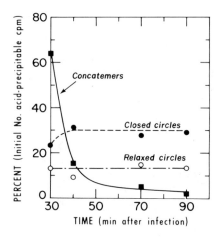

Fig. 5. The fate of pulse-labeled molecules after
activation of BC nuclease as determined by sucrose sedimentation analysis. These data were derived from the
experiment described in Figure 4. Percentages of the
total amount of radioactivity incorporated into relaxed
circles, covalently closed circles, and concatemers
during the pulse are plotted as a function of time during
the chase at 41°C. o=relaxed circles - neutral gradient,
fractions 21-23; ●=closed circles - alkaline gradient,
fractions 4-6; □=concatemers- neutral gradients, fractions
10-18, minus closed circles.

Degradation of Concatemers Does Not Result in Inhibition of Late Replication

a. Degradation is slow as compared to synthesis.

Examination of the data in Fig. 3 shows that the slope for incorporation of ^{3}H thymine into DNA at 32° (the pulse) was at least 3 times greater than that of the maximal slope for degradation at 41°. This difference seemed even more striking when we considered that the rate of DNA synthesis at 41° must have been even higher than that at 32°. Since our bacterial host was a thymine-requirer, we were able to calculate the rates at both temperatures from data shown in Table 1. From 30-38 min after infection, this value for the non-shifted culture was calculated to be about 1.38 phage equivalents/min/cell. For the shifted culture it was 5 phage equivalents/min/cell.

We know from our sedimentation analyses that three different species of molecules were labeled during our pulse at 32°C, while only one species, concatemers, was being degraded (Fig. 4). The average length of these concatemers estimated from the data of Fig. 4 was approximately 7.1 phage equivalents and, although breakdown seemed to start immediately, it would have taken about 13 min. for them to be completely degraded (Fig. 5). Thus, the average rate of degradation is 0.55 phage equivalents/min in the same cells, approximately one-tenth the rate of synthesis at 41°.

b. Concatemer synthesis continues long after a temperature shift.

Rates of synthesis were measured at various intervals before and after a temperature shift in cells infected with the temperature sensitive gam E⁻S⁻ phages and an isogenic gam⁺ phage (Table 2). It can be seen that for most intervals there was only a slight difference between the two values, even after the temperature shift. Sedimentation analysis of DNA pulse-labeled both soon (40 min) and late (80 min) after the temperature shift showed that concatemers were made at both times in relative amounts almost indistinguishable for that detected in DNA extracted before the shift (Compare Fig. 4 and 6). A simple model which explains the simultaneous synthesis and degradation

469

TABLE 1

KINETICS OF ^{3}H-THYMINE INCORPORATION INTO GAMTS DNA IN INFECTED RECA$^-$ CELLS

Sample time (min) After Infection	Time Interval (for Rate)	32°C		41°C	
		Total CPM	Rate CPM/min	Total CPM	Rate CPM/min
31	30 – 31	1,208	1,208	4,065	4,065
32	31 – 32	4,022	2,814	13,040	8,975
33	32 – 33	7,504	3,482	23,686	10,646
34	33 – 34	11,298	3,794	32,907	9,221
35	34 – 35	–	–	43,838	10,931
36	35 – 36	18,865	3,784*	54,464	10,626
37	36 – 37	23,974	5,109	61,478	7,014
38	37 – 38	27,897	3,923	73,300	11,822

*Corrected for 2 min interval, 34 – 36 min.

The protocol was as outlined in Fig. 2 except that the recA strain QR48 (thy$^-$, su$^-$) and E$^+$S$^+$ phage were employed. At 30 min., ^{3}H-thymine at final specific activity of 10 μCi/μg thymine was added to the 32°C and 41°C samples. At one-minute intervals after addition of isotope, samples were removed and the amount of label incorporated into trichloroacetic acid-insoluble material was assayed. The average rate was determined from the slope of a plot of Total CPM vs. Time. The rate of labeling, in terms of λ phage DNA equivalents, was calculated from the specific activity of exogenous thymine (the only source for this thymine-requiring strain) and the thymine content of lambda DNA (estimated at 4.8 x 10^{-12}μg/phage equivalent) (10, 35).

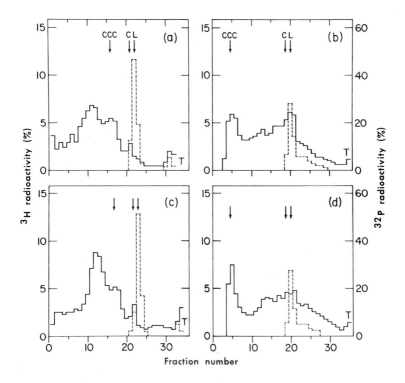

Fig. 6. Sedimentation analysis of phage DNA synthesized after a temperature shift. (a) and (c), neutral sucrose; (b) and (d), alkaline sucrose. The protocol is outlined in Fig. 2. Phage DNA, labeled with ^{3}H-thymidine during the intervals 40-42 min and 80-82 min, and extracted immediately thereafter was analyzed by neutral and alkaline sucrose sedimentation with methods described previously (10). The symbols marking the positions of specific molecules and the direction of sedimentation are the same as in Fig. 4. Each gradient contained at least 4×10^3 ^{3}H CPM.

(a) and (b), 40-42 min; (c) and (d), 80-82 min.

TABLE 2

RELATIVE RATES OF DNA SYNTHESIS; ^{3}H-THYMIDINE (CPM)
INCORPORATED IN A 1 MIN. PULSE

Interval (min)	gamts		gam$^+$	
	32°	41°	32°	41°
8 – 9	3,856	–	5,178	–
14 – 15	5,907	–	8,628	–
24 – 25	9,026	–	14,756	–
30	Temperature Shift			
34 – 35	13,245	55,598	17,682	111,317
44 – 45	14,262	69,754	18,239	95,222
59 – 60	14,685	47,603	17,092	44,498
69 – 70	15,373	54,688	17,522	37,119
79 – 80	16,431	42,325	21,489	32,061

The protocol is described in Fig. 2. An isogenic gam$^+$
phage was studied simultaneously. At the indicated times,
samples were pulsed for 1 min. with ^{3}H-thymidine (final
conc. of 40 μCi/ml), and the amount of label incorporated
into trichloroacetic acid–insoluble material was assayed,
as described previously (10).

is shown in Fig. 7. In this scheme degradation of concatemers proceeds from the ends of the tails of rolling
circles, but at a rate much slower then synthesis. Thus,
all of the pulse label can be lost from these molecules
without any damage to the replicating fork itself. New
label added after the shift enters concatemers at the fork
of this structure in the same way as label before the
shift. If this scheme is correct, the double-strand
specific exonuclease activity of the BC nuclease is the
one which destroys concatemers. Other data support the
view that this is the most significant of the three
nucleolytic activities of the BC nuclease in vivo (18,
19, 20).

Our results seem to rule out mechanism (2) posed at
the outset. Although it is obvious that activation of
the BC nuclease at a late time was not without consequence, the late replicating fork itself did not seem to
be destroyed by the BC nuclease. We conclude, therefore,
that mechanism (1) is probably correct, and that the BC
nuclease blocks the initiation of rolling circle replication by attack on some transient structure present only
at the time of transition. Another possibility is that,
for some unknown reason, during the transition the rate
of DNA synthesis is slower than the rate of degradation.

The following section summarizes results from previous studies which introduced a model that depicts
potentially sensitive transitional intermediates.

A COMPLEMENTARY ROLE FOR THE PHAGE RED RECOMBINATION SYS-TEM AND THE BACTERIAL POL I + LIGASE REPAIR SYSTEM IN THE GENERATION OF ROLLING CIRCLES

The gam⁻ defect discussed above was first detected
because of the inability of gam mutants to grow (i.e. make
plaques) under recombination defective (red⁻, recA⁻) conditions. Similar observations were made concerning gam
or red mutants in a large class of bacterial mutants,
called feb⁻ (12), later found to include polymerase I⁻
(polI⁻) and ligase⁻ (lig⁻) strains (12, 21, 22). Our
analysis of λ replication in such infections (23, 24, and
work in progress) indicated that the growth defect is at
the same stage of development as in fec⁻, namely at the
transition from early to late replication. In feb⁻, as
in fec⁻ conditions, replication does not stop but con-

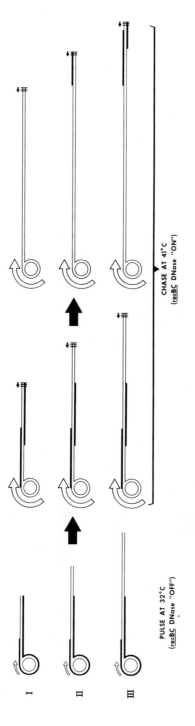

Fig. 7. Model for concurrent replication and recBC DNase degradation of late lambda DNA intermediates. The pulse labeling pattern of lambda rolling circles in various stages of tail elongation and the fates of the labeled segments after BC nuclease activation are illustrated. Darkened areas of the molecules represent the label incorporated during the pulse. The large, open and small, solid arrows indicate the relative directions of tail elongation and recBC DNase digestion of the tail, respectively. Arrow sizes indicate the relative rates of these processes. (Figure adapted from Greenstein and Skalka, 35).

tinues in the early mode throughout infection until some 40-60 phage equivalents of predominately circular monomers accumulate. The results with gam⁻ infection are not too surprising. We already know why the gam protein is required for late replication. Our speculations about how red might be required at first focused on the possibility that λ exo might provide some enzymatic activity missing in the pol I⁻ (Pol A of DeLucia & Cairns, 25) cells. We later rejected this hypothesis for two reasons:

1. Further analysis in two other feb⁻ infections, including lig⁻ (N2668; 21) or pol I⁻ hosts defective in the pol-associated 5'→ 3' exonuclease but not the polymerase (RS5052; 26), gave similar results (23). It was difficult to see how either of the two components of red, λ exonuclease or β protein, might complement all of these enzymatic activities (27).

2. red mutants make DNA at a reduced rate and with qualitative differences as compared to red⁺ phage even in wild type hosts (10, 23). This suggested that pol or lig could not completely compensate for the red defect.

Another hypothesis was then proposed. We supposed that in some way the phage's red recombination pathway and the host's repair pathway (which includes pol I and lig) could provide alternate routes to the same end-namely the generation of rolling circle replicating structures. Figure 8 shows a diagram of how we imagine these pathways might work. This model is incomplete and probably incorrect in some (if not all) of its details. It is intended mainly as a framework for thinking about how these alternate routes might work and how the various points might be tested.

A critical feature of the model is the notion that the initiation of rolling circles depends on the appearance of an interruption in a replicating circle at a position in front of the replication fork. The pathway which is then taken is determined by whether the nick is on the strand which is being copied in the forward direction or in the reverse direction relative to the movement of that fork (Fig. 8 intermediates I & I'). If the interruption

is in the "forward" strand (I), the replication machine
will run off, leaving a broken arm. The resultant
structure (II) can become a rolling circle after polI +
lig repair of a gap opposite the break. If the interruption is in the "reverse" strand (I') the replication
machine may or may not fall off. In our diagram we have
it continuing (III). What is important in this structure
(II') is the production of a 3' single-stranded end which
we perceive as a possible intermediate in red-promoted
recombination. Recombination of this end within the
same circle or, as shown here, with a second circle could
generate a rolling circle. Details of how this recom-
binant structure (IV) might be resolved are unspecified
here, but several schemes have been proposed, some requiring additional enzymatic activities (23, 27, 28).

Possible Sites of Action by the BC Nuclease

The formation of rolling circles from structure II
depends on repair of the gap opposite the broken arm. If
such repair were inhibited, for example as a result of
exonucleolytic degradation at the gap by BC nuclease, then
rolling circles could not be generated. A possible site
for exonucleolytic digestion on structure II' is the 3'
end necessary for formation of a recombinational synapse.
Destruction of this part of the molecule would also block
formation of rolling circles (and perhaps also decrease
the frequency of red⁻ promoted recombination, 12).

Further Comments About the Nick

As stated at the outset of this section, the appearance of a nick in the replicating circle is crucial to the
model. If the nicks are at random location, the probability that one is encountered by a replication fork may
depend only on time. There are, however, many site-
specific enzymes inside the cell that could make such a
nick. Some examples are: the site-specific endonuclease
presumed to be associated with int-xis recombination (29),
the endonuclease associated with O & P proteins (30), and
the DNA maturation system. We are particularly interested
in this aspect of the model because of the observation of
a possible candidate in the fec⁻ system.

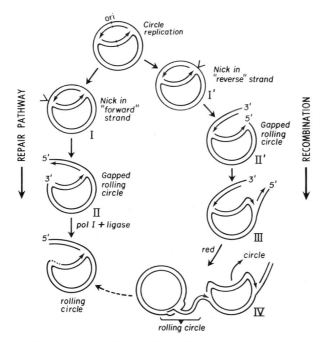

Fig. 8.   Alternate pathway for the generation of rolling
circles from bi-directional circle replication.   Schematic
adapted from Skalka (23).

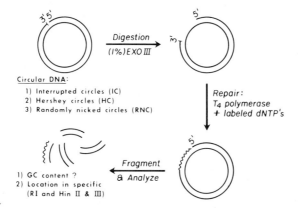

Fig. 9.   Scheme for labeling and analysis.   Uniformly
³H-labeled DNA substrates (interrupted circles, Hershey
circles, and randomly nicked circles) were treated with
E. coli exonuclease III under conditions such that an
average of 1% of the nucleotides were removed (Reuben &
Skalka, manuscript in preparation).   The Exo III was inactivated by heating, and the resultant gap was repaired
with T4 polymerase and ³²P-labeled dNTPs (36).   The DNA
was subsequently analyzed as described in the legends
to Figs. 11 and 13.

LOCALIZATION OF THE INTERRUPTION IN RELAXED CIRCLES WHICH ACCUMULATE IN A FEC⁻ INFECTION

Recall (Fig. 1) that the population of monomeric circular products of early DNA replication include both covalently closed and relaxed forms. Previous results have shown that the relaxed circles contain only one interruption per molecule and that this interruption can be on either strand of the DNA. Furthermore, the interruption is probably not a simple nick or gap because it cannot be sealed by T4 polynucleotide ligase (+ T4 polynucleotide kinase) either before or after treatment with T4 DNA polymerase (10). This inability to be sealed by enzymes analogous to those presumed to be present inside the cell might account for the stability of relaxed circles observed in Fig. 5. Recent in vitro experiments have shown that these molecules are resistant to purified BC nuclease (Reuben & McKay, personal communication. In the following section we discuss results which indicate that the interruption in these circles is in a specific location.

Scheme for Identifiction of the Site of Interruption in
Relaxed Circles

This scheme (Fig. 9) is based on the observation that a gap can be created at the site of the interruption in the relaxed circle with the use of E. coli exonuclease III, an enzyme that removes nucleotides from the 3' end of double stranded DNA. In our experiments, we compared results for interrupted circles (IC), purified from a red⁻ gam⁻, recA⁻ induction (31), with randomly nicked circles (RNC) (prepared by mild pancreatic DNAse treatment of covalently closed circles purified from the same induction) and with "Hershey" circles (HC) (prepared by annealing mature linear λ DNA; 32). An average of 1% of the nucleotides were digested from each type of molecule. After inactivation of the exo III, the resulting gap was repaired with labeled nucleotides employing T4 DNA polymearse. These molecules, specifically labeled in regions flanking the sites of interruption, were then fragmented either enzymatically (with Eco RI or Hin II + III restriction enzymes) or physically, and the fragments analyzed.

The basis for our analyses is shown in Fig. 10, which contains a physical map of lambda DNA indicating both the heterogeneous distribution of nucleotides and the specific location of Eco RI restriction sites. In our first exper-

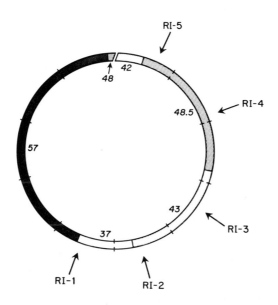

Fig. 10. Physical map of circular lambda DNA indicating
G + C content and Eco RI cleavage sites. G + C content
of the lambda chromosome is indicated by shading with
percentages specified by numbers (37). Eco RI restriction
sites are indicated by arrows (38).

479

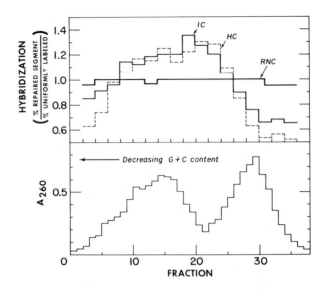

Fig. 11. Analysis of G + C content in repaired segments.
Lambda DNA was hydrodynamically sheared to fragments
approximately 5% of the genome length. The fragments
were separated on the basis of G + C content by equilibrium centrifugation in Cs_2SO_4-Hg^{++} density gradients
(lower panel, 37). Fractions containing DNA of varying
G + C content were used to prepare DNA filters for
hybridization (10).

The three DNA substrates prepared as in Fig. 9 were
denatured, sheared and hybridized to the λ DNA of varying
G + C content. Each analysis included at least 2 x 10^4
^{32}P CPM and 2x10^4 ^3H CPM. The percentage of specific
label hybridized (repaired section) was normalized to the
percentage of uniform label hybridized to each filter.
This ratio ($^{32}P/^3H$) is plotted as a function of G + C
content. IC = Interrupted circles. HC = Hershey circles;
RNC = randomly nicked circles.

iment we determined the G + C content of the repaired segment. The lower panel in Fig. 11 shows sheared fragments
consisting of about 5% of the genome length separated
according to their G + C content. The upper panel shows
results from hybridization of denatured fragments of
specifically labeled DNA to the λ fragments of varying
G + C content. As expected, repaired segments from randomly nicked circles·hybridized with equal efficiency to
all fragments. The repaired interrupted circles, however,
hybridized best to fragments from the 43 and 48% G + C
segments of λ DNA and very poorly to fragments from the
37 and 57% G + C segments. The hybridization pattern of
repaired interrupted circles (IC) was almost identical
to that for the repaired Hershey circles (HC). This result indicated that the interruption in relaxed circles
was not in a random location. The G + C content of the
repaired segment was consistent with its being located
either at the site of mature ends of linear DNA or some-
where in the right half of the molecule (see Fig. 10).

Figure 12 shows results from analysis of RI restriction fragments. Fragments 1 and 6 are not found in
digests of IC or RNC because these molecules do not contain mature ends; instead we find a fragment consisting
of 1 + 6. Repaired randomly nicked circles contained the
expected amount of isotope in all of the 5 fragments.
When we compared the ratio of label in each fragment from
a digest of repaired interrupted circles to the randomly
nicked molecules (IC/RNC) we found that fragments near
the molecular ends were preferentially labeled. Furthermore, the relative amount of label seemed to decrease in
proportion to the fragment's distance from the mature
end. The content of isotope in repaired interrupted
circles as compared to Hershey circles (IC/HC) was identical for all fragments. This result indicates that the
location of the interruption is probably near or at the
molecular ends. Results with Hin II + III digests, in
which many (approximately 100) fragments are formed (33),
also showed a non-random pattern of labeling. As with
shear and RI fragments, interrupted and Hershey circles
gave identical results (these data are not shown).

These results suggest that the relaxed circles which
accumulate inside the cell consist of a mixed population
of molecules containing one interruption/molecule at one
or the other of the sites of the staggered nicks which

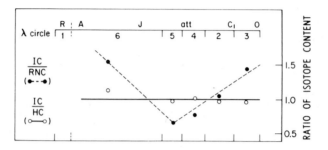

Fig. 12. Location of repaired section among Eco RI
restriction fragments. The 3 specific DNA substrates
were prepared as described in Fig. 9. The DNA was
cleaved by restriction endonuclease RI yielding the 5 (or
6) fragments indicated on the λ map. Each analysis included at least $1x10^4$ ^{32}P CPM. Fragments were separated
by Agarose gel electrophoresis (39) and the radioactive
content of each fragment determined for the 3 cleaved
forms of DNA. Under the appropriate fragment is plotted
the ratio of the radioactivity contained in the fragments
from interrupted circles as compared to those from Hershey circles (IC/HC, 0———0), or as compared to those
from randomly nicked circles (IC/RNC, ●———●).

produce the mature linear chromosome. It seems possible
that the phage A protein might be involved since A
function has been implicated in DNA maturation (34). Analysis of relaxed circles from A mutant infections is in
progress. Further study should help us to decide if this
interruption plays any role in the control of late DNA
synthesis.

To the extent that late replication depends on a
nick in the DNA, its generation by a phage protein could
be considered a form of internal control. Action of the
host's BC nuclease and PolI + ligase on intermediates
generated from this nick would then be examples of modu-
lations of this control by intracellular environmental
elements.

Summary

The model that emerges from all of this is a dynamic
one, with circle replication providing a driving force
that can be controlled and modulated by both internal and
environmental elements. The significance of this for
lambda is obvious. Lambda can replicate in one of three
states: (1) as a circle and then a rolling circle in
the lytic state, (2) as an autonomous circle in the plasmid state, and (3) as part of the host's circular chromosome in the prophage state. Its ability to choose the
best state at any time depends on its ability to "sense"
changes in its environment.

The significance of these interactions in terms of
understanding molecular mechanisms of the host is more
subtle, but equally important. (1) We now know more
about the in vivo activity of the BC nuclease. We know
what kinds of molecules are resistant to its action inside the cell, and we have some idea about in vivo rates.
(2) It appears that the enzymatic activity of polymerase I
and ligase can be reduced to a level where the "repair"
function is deficient while the residual activity is
sufficient to support replication. However, for lambda,
and perhaps for the bacterium as well, there are conditions under which a defect in repair (i.e. in the feb
system) can be just as lethal as a defect in replication.

REFERENCES

1. Thomas, R., In The Bacteriophage Lambda, ed. by A. D. Hershey, p. 211-220. Cold Spr. Harb. Labs. Cold Spr. Harb., N. Y.
2. Ptashne, M.,In The Bacteriophage Lambda, op cit p. 221-238.
3. Eisen, H. and Ptashne, M., In The Bacteriophage Lambda, op cit p. 239-246.
4. Echols, H.,In The Bacteriophage Lambda, op cit p. 247-270.
5. Reichardt, L. and Kaiser, A. D.,Proc. Nat. Acad. Sci. 68, 2185 (1971).
6. Echols, H. and Green, L.,Proc. Nat. Acad. Sci. 68, 2190 (1971).
7. Ptashne, M.,The Harvey Lectures, Series, 69, 143 Academic Press, N. Y. (1975)
8. Roberts, J. and Roberts, C.,Proc. Nat. Acad. Sci. 72, 147 (1975).
9. Georgopoulos, C. P. and Herskowitz, I., In The Bacteriophage Lambda, op cit p. 553-564.
10. Enquist, L. W., & Skalka, A.,J. Mol. Biol., 75, 185 (1973).
11. Manly, K. F., Signer, E. R. and Radding, C. M., Virology 37, 177 (1969).
12. Zissler, J., Signer, E. R. & Schaefer, F., In The Bacteriophage Lambda, op cit p. 455-468.
13. Sakaki, Y., Karu, A. E., Linn, S., and Echols, H., Proc. Nat. Acad, Sci., U.S.A. 70, 2215 (1973).
14. Karu, A., Sakaki, Y., Echols, H., and Linn, S.,In Mechanisms in Recombination, ed. by R. J. Grell, Plenum Press, New York, N. Y. p. 95-106 (1974).
15. Skalka, A., Poonian, M. & Bartl, P.,J. Mol. Biol., 64, 541 (1972).
16. Wake, R. G., Kaiser, A. D., and Inman, R. B.,J. Mol. Biol., 64, 519 (1972).
17. Reader, R. W., and Siminovitch, L., Virology, 43, 607. (1971).
18. Benzinger, R., Enquist, L. W., and Skalka, A., J. Virol., in press (1975).
19. Kushner, S.,J. Bacteriol., 120, 1213 (1974).
20. Kushner, S.,J. Bacteriol., 120, 1219 (1974).

22. Konrad, E. B., Modrich, P., Lehman, I. R., J. Mol. Biol. 77, 519 (1973.

23. Skalka, A., In Mechanisms in Recombination, ed. by R. J. Grell, Plenum Press, New York, N. Y. p. 421-432 (1974).

24. Dawson, P., Skalka, A., and Simon, L. D., J. Mol. Biol. in press (1975).

25. DeLucia, P., and Cairns, J. Nature 224, 1164 (1969).

26. Konrad, E. B., Lehman, I. R., Proc. Nat. Acad. Sci. 71, 2048 (1974).

27. Radding, C. M., Ann. Rev. Genet. 7, 87 (1973).

28. Stahl, F. and Stahl, M., In Mechanisms in Recombination R. F. Grell, ed. Plenum Press, N. Y. p. 407-419 (1974).

29. Gottesman, M. E. and Weisberg, R. A., In The Bacteriophage Lambda, op cit, p. 113-138.

30. Shuster, R. and Weissbach, A., Nature 223, 852 (1969).

31. Reuben, R., Gefter, M., Enquist, L. W., and Skalka, A. J. Virol. 14, 1104 (1974).

32. Davidson, N. and Szybalski, W., In The Bacteriophage Lambda, op cit p. 45-82.

33. Smith, H. O. & Wilcox, K., J. Mol. Biol. 51, 379 (1970).

34. Wang, J. C. and A. D. Kaiser, Nature New Biol. 241, 16 (1973).

35. Greenstein, M. and Skalka, A., J. Mol. Biol., submitted (1975).

36. Masamune, Y., Fleischman, R. A., & Richardson, C. C., J. Biol. Chem. 246, 2680 (1971).

37. Skalka, A., In Method in Enzymology (Nucleic Acids), Part D, Academic Press, N. Y., Vol. 21, p. 341 (1971).

38. Allet, B., Jeppesen, P. G. N., Katagiri, K. J., and Delius, H., Nature (London) 241, 120 (1973).

39. Helling, R. B., Goodman, H. M., and Boyer, H. W., J. Virol. 14, 1235 (1974).

ROLE OF oop RNA PRIMER IN INITIATION OF COLIPHAGE
LAMBDA DNA REPLICATION

Sidney Hayes* and Waclaw Szybalski

McArdle Laboratory for Cancer Research
University of Wisconsin, Madison, WI   53706

ABSTRACT: The products of E. coli genes dnaB, dnaG and
dnaE, the β-subunit of RNA polymerase, and lambda genes 0,
P, and the cis-dominant ori site are required for initiation
of DNA synthesis from the ori region of coliphage λ, which
lies between genes cII and 0 and includes the ori site.
The same genes, except for dnaE, are also required for initiation of the synthesis of two short RNAs from the l strand
of λ DNA: a 4S RNA, termed oop, promoted from within
the ori region, and lit RNA (about 600 nucleotides long)
starting from a site 2200 nucleotides farther downstream
(Fig. 1).   Synthesis of lit requires the cis action of
gene 0 and the ori site, and the trans action of gene P
product.   Replication is coupled to transcription through
the requirement of the same factors for initiation of DNA
synthesis and oop and lit transcription.   DNA replication,
per se, is not required for oop and lit transcription, since
their synthesis occurs at the normal rate when λ DNA synthesis is blocked in a dnaE⁻ host.   Thus, the initiation
of λ DNA synthesis proceeds in several stages, including:
(1) interaction of the products of genes dnaB, dnaG, 0 and
P, the β-subunit of RNA polymerase and the ori site, which
results in synthesis of oop (and lit) RNA; (2) formation of
a replication complex by interaction of DNA polymerase(s)
with the oop primer RNA at the ori site, and (3) creation

---

*Present address: Department of Molecular Biology
and Biochemistry, University of California-Irvine, Irvine,
California 92664

of a DNA replication fork in the λ ori region, leading to
initiation of the first round of DNA synthesis. λ DNA
synthesis, as measured by phage burst from an induced lysogen and DNA titration, is inhibited at high temperature in
conditional-lethal polA hosts, suggesting that polymerase I
or its 5'-3' exonucleolytic activity is also required at
some undetermined stage. Covalent linkage between oop RNA
and λ DNA, formed by the extension of the 3'-end of oop RNA
into DNA of ori region, was detected by employing CH_2O-
Cs_2SO_4 gradient centrifugation and various hybridization
procedures.

Dependence of in vitro λ DNA synthesis on oop RNA
was studied in a system containing a λ DNA template, DNA-
free extracts from induced λ lysogens and all four dNTP's
and rNTP's. DNA synthesis depends on exogenous oop RNA but
does not require active O or P products. Neither tRNA nor
rightward or leftward λ transcripts substitute for oop RNA.
The ability of oop to prime in vitro DNA synthesis depends
on a dominant component in the cell extract we have termed
OSA (for oop specifying activity) which is present only in
extracts from induced lysogens. Several mutations abolish
the oop-dependent DNA synthesis in the extracts, including
the conditional lethal host mutation polAts580 and the N x
double mutation, which blocks both the rightward and leftward transcription in the induced prophage.

These results suggest that λ DNA replication depends
on oop RNA, OSA, and DNA polymerases I and III. One might
postulate that the λ gene O product acts in cis, parti-
cipating in the promotion of oop RNA synthesis, and that
oop in turn, acts in trans to initiate λ DNA replication
at or near the ori site (and possibly at other secondary
sites "opened" by transcription or movement of a replication fork).

INTRODUCTION

The replicon model of Jacob et al. (1) suggests that
initiation of a round of DNA synthesis is triggered by the
interaction of a diffusible or trans-acting protein(s)
termed the "initiator" with a specific cis-dominant target
site on the replicon termed the "replicator". Much of
what is known about the initiation of coliphage λ DNA
replication seems to approximate this general model. Two
λ genes, O and P, could be considered to code for the

"initiator" proteins which act at the ori site, or "replicator" (Fig. 1). Mutations in 0 and P prevent phage replication (2,3,4 and Table 1), probably on the level of
initiation of λ DNA synthesis (5,6). The position of
the ori site was determined by a combination of genetic,
transcriptional and electron micrographic mapping (4,7,8)
and found to be located within a stretch of about 500 base
pairs between gene cII and the middle of gene 0 (Fig. 1),
which stretch we term here the "ori region", and which
contains the origin for bidirectional λ DNA replication
(4,7). Twelve cis-dominant ori mutants were described
and all map within the ori region (8-11). The λori⁻
prophage has to be maintained as a lysogen since it is
unable to replicate on its own, even when supplied with
all the wild-type λ gene products. It is not clear whether
all ori⁻ mutations represent the same function, since more
than one cis-dominant replication function could be located
in the ori region, including the origin of replication and
the p_o promoter for the oop RNA. Indeed, two classes of
ori⁻ mutants were observed, with respect to their rate of
reversion and ability to form small clear-plaque recombinants in crosses with λimm21 (9).

In addition to the ori site and the 0 and P genes,
several other factors were found which control λ DNA
replication. These include the cis-dominant "transcriptional activation", which depends on the rightward transcription of the ori region (8,11), the oop transcript from
the l strand in the ori region (12,13), dependence on host
functions, including genes dnaB and dnaG and DNA and RNA
polymerases (3,13), and the regulatory effects of the
repressor and tof product (14-19).

The concept of "transcriptional activation", i.e. the
cis-dominant dependence of λ DNA replication on the rightward transcription of the ori region in addition to all
other replication factors (11), was based on the original
experiments of Thomas and Bertani (14), who showed that the
repressed λimmλ phage cannot replicate even if supplied
with all the replication products by the replicating
heteroimmune phage λimm434. They tentatively concluded
that the λ repressor interferes directly with the λimmλ
replication. Dove et al. (8,11) and others isolated λ mutants which were insensitive to such immunity-specific
"replication inhibition" by the repressor, and it was shown
that these mutants acquired new repressor-insensitive

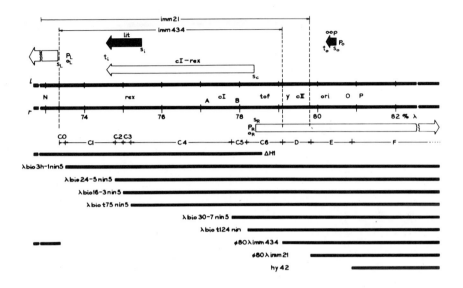

Fig. 1. Expanded physical and genetic maps of the N through P region of phage λ and of several deletion and substitution mutants. For details see reference 12. The symbols p, s and t indicate promoters, startpoints for RNA synthesis and terminators for transcription, respectively. Reprinted with permission from (12).

Table 1. Effect of Phage and Host Mutations on λ DNA Replication and oop and lit RNA Synthesis

Phage mutant	λ DNA	oop(or lit)	Host mutant	λ DNA	oop(or lit)
+	13	50	dnaA	17	72
O⁻	1	1	dnaF	15	75
P⁻	1	1	dnaB	1	3
ori⁻	1	1-3	dnaG	1.5	1
+ (noninduced)	1	1-2	dnaE	1	70
			polAts (5'-3' exots)	1-4	

Cultures of λcI857 lysogens were grown at 30°C to A_{575} = 0.35 and thermally induced by transferring to 42.5°C. The DNA assays were patterned after those of Stevens et al. (4). The results are the average hybridization values for the att-N and x-0-P tester ^{3}H-RNAs and indicate the increase in copies of λ DNA 20 minutes after thermal induction (level after induction/level of noninduced control). The dnats strains used will be specified elsewhere. Replication of λ in induced lysogens with conditional lethal mutations in polymerase I (polAts580, see Konrad and Lehman, this volume) was compared with λ replication in the induced parent KS463(λcI857). To measure RNA synthesis, the lysogens were pulsed with ^{3}H-uridine for 1 minute either at 30°C (0 time) or 20 minutes after transferring culture to 42.5°C. The ^{3}H-labeled oop and lit RNAs were assayed by liquid hybridization as previously described (12). Columns 3 and 6 show relative levels of lit and oop synthesis. The data for oop RNA are the actual hybridization values, expressed as a percentage (x1000) of the total input of 3H-RNA (12,13). The relative transcription rates, calculated as hybridization values per 1%λ length, are 77 for lit, 260 for oop, 21 for cI-rex and 650 for the major leftward transcription (12). Table modified and reprinted with permission from (13).

490

promoters for the rightward transcription of the ori region
(8,11,20). Not only the new promoter ri^c mutants of Dove
et al. (11) or the analogous c17 mutant (14,21), but also
the o$_R$-constitutive mutants, in which rightward transcription is insensitive to the λ repressor, escape the replication inhibition (8,22,23). Conversely, mutations in the
rightward promoter (e.g., x13) block λ replication even in
the absence of repressor when other products are supplied
by the complementing phage. The experiments on "transcriptional activation" provided the first evidence for involvement of RNA synthesis in the initiation of DNA replication,
but the mechanism of this cis-dominant phenomenon is still
far from clear (8). This is certainly not a requirement
for the 0 or P proteins coded by the rightward λ messenger,
since some ri^c promoters are within the 0 gene and, moreover, the 0 and P products are being supplied by the heteroimmune helper. A nondiffusible protein seems to be an
adequate explanation for the cis-action of gene A in P2
phage (24), but this notion was not considered plausible in
the case of λ by Inokuchi et al. (25), who argued that
transcriptional activation was required for replicative
cutting of DNA at the ori region. This and other (26)
studies, however, do not discriminate between template cutting, which might be required for the initiation of replication, and nicking which occurs after initiation. Dove
et al. (8) discuss the hypothetical mechanisms of the
transcriptional activation of λ replication.
     The possible cis-dominant action of gene 0 and the
role of the primer RNA are the subject of this paper. Our
in vivo and in vitro studies on the role of these factors
and the effect of several host products will be discussed.

RESULTS

Leftward Transcription of the ori Region: In the uninduced λ prophage about 80% of the transcription from the
l strand originates within the immunity region (cI-rex
mRNA, Fig. 1), 2% is from the ori region (oop RNA representing 0.001% of the pulse-labeled input RNA), and the remainder is distributed between the int and b2 regions (12).
The rate of independent lit synthesis from the uninduced
prophage is difficult to measure since it overlaps with
the cI-rex transcript, but could be estimated to be 0.0025%
of the pulse-labeled input RNA. Upon induction of λ tof$^+$
lysogens accomplished by thermal inactivation of the ther-

491

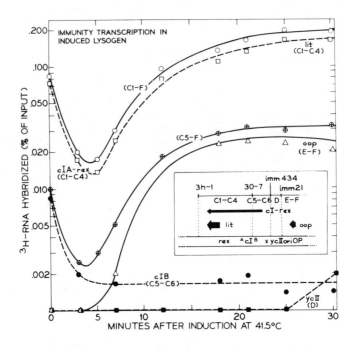

Fig. 2.   Leftward transcription in the immunity-ori
region after prophage induction. The RNA of the SA431
lysogen (Fig. 5) was labeled with ^{3}H-uridine for 1 minute
at 30°C (zero time) and at the indicated times after induction at 41.5°C. The extracted ^{3}H-RNA was hybridized and
the percent of ^{3}H-RNA hybridized per interval calculated as
in ref. 12. The cI-rex transcription (see insert) is drasti-
cally reduced during the first few minutes after induction,
evidenced by the drop in transcription in intervals cIA-rex
(C1-C4) and cIB (C5-C6). Subsequently, the coordinate
synthesis of two new RNAs, lit in (C1-C4) interval and oop
in (E-F) interval, become predominant, whereas there is no
significant transcription in the intervening (C5-C6) and (D)
intervals (see insert). Similar results were obtained for
W3350(λcI857). Figure reprinted with permission from (12).

mosensitive cI857-coded repressor, the cI-rex transcription is rapidly turned off (Fig. 2). After a brief
lag of 5-7 minutes a 600 nucleotide-long transcript,
denoted lit, appears in the left part of gene rex and
terminates at the same site as the cI-rex mRNA (Fig. 2
and ref. 12). The manifold increase in lit transcription
parallels the 30-100 fold increase in synthesis of oop
RNA, as if both transcripts originated from a common
promoter or were positively regulated by a common factor,
although the oop startpoint, s_o, is located at least 2200
nucleotides upstream from the lit startpoint, s_i (12).

Control of oop and lit Transcription is Coupled to
Initiation of Replication: Does the repressor directly repress the transcription of oop and lit, or is their synthesis "positively" stimulated by the product of genes
whose transcription is blocked by the repressor? The
effect of the repressor on oop and lit synthesis, which is
already in progress, was investigated in transiently
derepressed λ lysogens. Repression was lifted by heating
a lysogen with a reversibly thermosensitive cI857-coded
repressor to 42°C and then quickly restored by cooling the
culture to 30°C to permit renaturation of the repressor.
Renatured repressor blocked the major, o_L-controlled, leftward att-N transcription but did not repress the oop
or lit RNA synthesis (12,13), indicating that lit and oop
are regulated by λ genes which are under control of the
repressor. A search for such λ genes revealed that the
coordinate stimulation of oop and lit transcription
observed after prophage induction or phage infection is
blocked by phage mutations p_R^- (x13 mutation), 0^-, P^- and
ori⁻. Moreover, host genes dnaB and dnaG, and the β-subunit of RNA polymerase are also required. Synthesis of oop
and lit is sensitive to rifampicin but this block is
overcome by the β-mutation to rifampicin resistance (13 and
unpublished).
    The host and phage functions required for oop and lit
synthesis were also found to be required for initiation of
λ DNA replication (Table 1). The host dnaA, dnaC-D
(strain NY170, restriction⁻; ref. 27) and dnaF genes were
not required for oop, lit or λ DNA synthesis. Thus, lambda
replication is coupled to transcription through the requirement of the same factors (ori, 0, P, dnaB, dnaG) both for
initiation of DNA synthesis and for oop and lit transcript-

ion (Table 1). Conditional lethal mutations in polA reduce
or prevent λ DNA synthesis. In addition, a defect in
dnaE, which codes for polymerase III (28), abolishes λ
replication but does not prevent oop or lit synthesis.
This result indicates that multiple lambda DNA templates
(gene dosage effect) are not required for the observed
30-100 fold increase in oop and lit synthesis. Maximal
oop and lit synthesis can occur from a single prophage
DNA copy when λ DNA replication is blocked either by a
dnaE⁻ mutation or by nalidixic acid (13 and Table 1). We
asked if the level of oop and lit synthesis is stimulated
at very late times (30-90 minutes) after induction when
as many as 100 to 1000 phage DNA copies per cell can be
made. The level of oop and lit synthesis was compared
after induction in two strains which do not lyse the
cell: SA431, a defective lysogen which can initiate DNA
synthesis but cannot synthesize phage structural genes
(see Fig. 5), and an S⁻ strain which both replicates and
packages its DNA (29). It was found (Fig. 3) that the
oop and lit synthesis remained high through 90 minutes
after induction of SA431, but turned off by 60-80 minutes
for the S⁻ strain. These results could be interpreted to
mean that packaging of the S⁻ DNA prevents oop and lit
synthesis, or that some late gene deleted in SA431 acts
to limit their synthesis. Clearly, however, oop and lit
synthesis is not gene dosage dependent.

   Is oop RNA a Primer for Initiation of DNA Synthesis?
Synthesis of oop can proceed in the absence of measurable
DNA synthesis. Thus, oop transcription, though linked to
the control of initiation of replication, would precede
formation of the replicaton forks. Initiation of λ
replication may include the following steps: (a) The
products of genes O, P, dnaB and dnaG, the β-subunit of
RNA polymerase and the ori site promote vigorous synthesis
of oop (and lit). (b) A replication fork is created by
initiation of the first round of DNA synthesis.
   Does oop prime the DNA synthesis by serving as a
substrate (primer) for the addition of deoxynucleotides
to its 3'-OH end by one of the DNA polymerases, as shown
in Figure 4? We have attempted to isolate DNA covalently
linked to oop RNA. ³H-oop RNA and newly synthesized
lambda DNA from the l strand of the ori region were isolated from SA431 (refer to Fig. 5 and 8). The induced

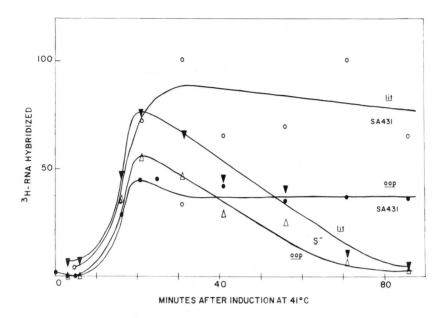

Fig. 3. Kinetics of oop and lit RNA synthesis at late times after induction of lysogens. SA431 (circles) and W3101(λcI857S7) (triangles) lysogens contain defective prophages able to replicate but not to lyse the host through 90 minutes after induction. Only the S⁻ strain is able to package its DNA. The RNA was pulse labeled for one minute with ³H-uridine at the indicated times, extracted and hybridized as in (12). The hybridization values for lit and oop ³H-RNA were divided by the corresponding cpm's per 5 µl of the total extracted RNA collected only in the 0 to 30 minute interval, and plotted on the relative 0-100 scale.

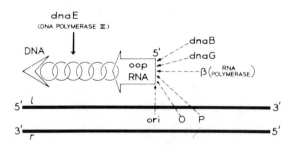

Fig. 4. A model for the role of oop RNA in the initiation of leftward DNA synthesis. The oop RNA synthesis
starts at the ori site and requires the products of λ genes
O and P and host genes dnaB and dnaG, in addition to the
host RNA polymerase for its full expression. The 3'OH terminus of the oop RNA primer is extended by the host DNA
polymerase III, the product of gene dnaE. This diagram
does not imply that the oop RNA cannot be utilized as a
primer at a site other than the site at which it is synthesized. Reprinted with permission from (13).

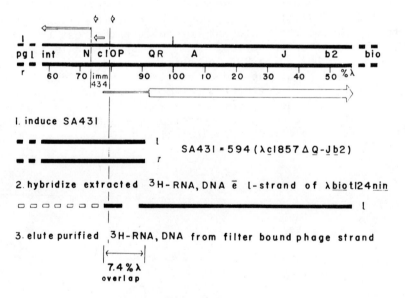

Figure 5.

Fig. 5 Preparation of ^{3}H-oop RNA-DNA. Cultures of
SA431 were grown at 30°C to A_{575} = 0.35, thermally induced
at 41-42°C for 5,8,11,14,20 and 25 minutes and pulse labeled with ^{3}H-uridine during the last minute of induction.
Cells were pelleted, converted to spheroplasts, centrifuged again and taken up in one ml of a solution of 0.175M
NaCl, 0.021M EDTA, 0.5% sodium dodecyl sulfate (SDS) and
0.011M Tris·HCl (pH 7.4). The lysis of spheroplasts was
aided by heating at 60°C for 3 minutes. The very viscous
lysates were sonicated in an ice bath at 40 watts for 2
minutes, then frozen, thawed, and extracted three times
with phenol at 4°C. The aqueous layer was adjusted to pH
7.4, heated for 3 minutes at 100°C, rapidly cooled and
hybridized at 67°C for 5 hours with the l strand of
λbiot124nin. The hybrid, consisting of the l strand of
λbiot124nin hydrogen-bonded to ^{3}H-oop RNA and newly synthesized DNA from the ori region, was adsorbed to a nitrocellulose filter, treated with pRNase (which was subsequently inactivated with iodoacetate as in ref. 12), and
the oop RNA-DNA was eluted from the washed filter at 93°C
(10 min) in 1 ml of sterile 1/100xSSC. The eluates, were
pooled and then divided into two equal portions, one of
which was treated with pancreatic DNase (100 μg/ml, free
of RNase, at 27°C for 30 min). Each portion was then
heated at 100°C for 10 min, quickly cooled and run at
35,000 rpm for 60 hrs in a Cs_2SO_4 equilibrium gradient.
The results are shown in Fig. 6. -- In several experiments a second hybridization to the l strand of ø80λimm21
(see ref. 12) was employed to eliminate traces of ^{3}H-RNA
from the λ immunity region (the bio genes of the host are
removed by the chlA deletion in SA431) and to narrow the
region of purified newly synthesized DNA from 7.4% (see
Fig. 5) to about 5.6%. The additional step proved unnec-
essary since the only detectable ^{3}H-RNA, purified by
hybridization to λbiot124nin, was oop RNA.

cultures of SA431 were pulsed for one minute with ^{3}H-uridine, lysates were sonicated, and the ^{3}H-RNA and DNA extracted and denatured as in Fig. 5. The ^{3}H-oop RNA and r DNA made in the tof-ori-P region was purified from other λ and E. coli ^{3}H-RNA and DNA by hybridization to the l strand of λbiotl24nin DNA as indicated in Fig. 5. This technique (12) permits purification of (a) an r-strand DNA synthesized from (and adjacent to) the ori region and (b) the oop RNA, which, after induction, is the only detectable transcript from this region. The purified filter-bound ^{3}H-oop RNA and r-strand DNA from the ori region were eluted from the filter. The eluate was split, and half was treated with RNase-free pDNase (30). Both the DNase-treated and untreated fractions were run to equilibrium in Cs_2SO_4 gradients (Fig. 6). The DNased sample ran as a single broad peak with maximum density at 1.58 g/cm^3, whereas the untreated sample had three peaks at densities 1.59, 1.56 and 1.43. The latter ^{3}H-RNA peak, which bands near the density of λ DNA, was pooled, treated briefly with RNase-free pDNase and then split into aliquots which were treated either by alkali, pRNase in high or low salt, or S1 nuclease. Each product was then heated to 100°C for 2.5 minutes in 1% CH_2O, quickly cooled and rerun in a Cs_2SO_4 gradient containing 1% CH_2O, (Fig. 7). After partial DNase treatment, the ^{3}H-RNA-DNA peak (which banded previously at 1.43 g/cm^3; Fig. 6) ran as a broad peak throughout the Cs_2SO_4-1% CH_2O gradient. Treatment with either alkali or pRNase in 1/200xSSC eliminated all precipitable counts, indicating that all the ^{3}H label was in RNA. Treatment with pRNase in 2xSSC digests 92% of the precipitable ^{3}H-oop RNA. The difference between the two RNase curves (high and low salt) indicates that only 8% of the ^{3}H-oop RNA in the gradient was either double stranded or hydrogen bonded to DNA. The remaining label must represent ^{3}H-oop RNA which is covalently linked to varying lengths of single stranded DNA. The incomplete digestion with the single-strand-specific S1 nuclease reflects either the formation of a limit digest or resistance of the self-complementary double-hairpin loops of oop (31) to S1 but not pRNase. From these experiments we have tentatively concluded that oop RNA can be covalently joined to DNA.

The Effect of ori Mutants on lit Transcription: The coordinate synthesis of oop and lit suggests that either:

(a) One promoter located within the ori region serves both transcripts, i. e., the effective RNA polymerase enters DNA at the oop promoter, p_0, synthesizes oop RNA, terminates transcription at t_0, but instead of being released from the template, drifts (movement of RNA polymerase without transcription) along 2200 nucleotides from t_0, the terminator for oop transcription, to startpoint s_i for lit transcription, and then synthesizes lit RNA, terminating at t_i, or (b) lit and oop each have independent promoters. The possibility that the products of genes dnaB, dnaG, O, P and the β-subunit of RNA polymerase form a new polymerase which also recognizes a promoter for lit is rather unlikely, since ori$^-$ mutants impose a cis dominant block on lit transcription. Could the initiation of lit synthesis require the oop RNA or its translation product as a cofactor? The absence of lit synthesis in ori$^-$ mutants (Table 1) might be explained by either hypothesis if the ori$^-$ mutations inactivate the oop promoter, p_0. We have tested whether oop (or its product) is required as a cofactor for lit synthesis, by preparing double lysogens of ϕ80λimm434cIts (which does not code for lit RNA) and immλ phages defective in the ori site (Fig. 8). The imm434 phage is ori$^+$ and produces normal levels of O$^+$ and P$^+$ products and oop RNA; however, it cannot complement the ori$^-$ (Fig. 8 line 3) or delori strain (line 2) for lit synthesis (compare with the wild type level, line 1). Thus, (a) oop or the ori product does not act as a cofactor, and (b) the ori site appears to be directly required for lit transcription.

Gene O Product is Required in cis for Synthesis of lit: Are the O and P products required for lit synthesis? Fig. 8 (line 4) shows that the immλori$^+$O$^-$P$^-$ prophage is not complemented for lit synthesis by phage imm434, suggesting that either O or P are required in cis for lit synthesis. Experiments with O$^-$ or P$^-$ defective λimmλ (lines 5 and 6) show that it is the O$^+$ product of ϕ80λimm434O$^+$P$^+$, which does not complement in trans, whereas the P$^+$ product can do so. These results suggest that (a) the cis-dominant lambda ori site and the products of genes O and P are directly required for lit as well as oop synthesis, and the initiation of λ replication; (b) the O product acts in cis with regard to lit synthesis. The O product could act in cis by binding to the ori site, since the

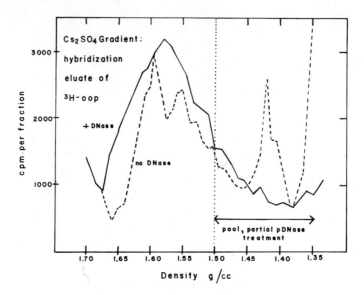

Fig. 6. Cs$_2$SO$_4$ equilibrium gradient of ^{3}H-oop RNA and newly synthesized DNA from, and adjacent to, the ori region of SA431. Two preparations (untreated and DNase treated) were purified as described in Fig. 5. All label is in ^{3}H-oop RNA. The density of each fraction was determined using a refractometer. Aliquots of each fraction were precipitated with 5% trichloroacetic acid.

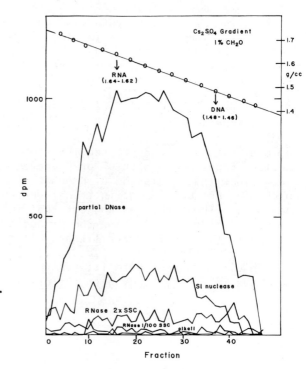

Figure 7.

Fig. 7. Cs_2SO_4-1% formaldehyde(CH_2O) gradient of pooled fractions (see Fig. 6., no DNase, dashed line) which contained ^{3}H-oop RNA banding with λ DNA. The pooled material was dialyzed against sterile 1/100xSSC, treated with RNase-free pDNase for 5 minutes at 27°C (30 μl 1M magnesium chloride, 10 μl 1M calcium chloride, 20 μl 1M Tris.HCl, pH 7.4, 20 μg DNase per 2 ml solution), heated at 100°C for 9 minutes and quick cooled; RNase-free carrier tRNA (20 μg/ml), ammonium acetate (to 0.1 M) and ethanol (2 volumes) were added and the mixture held overnight at -60°C. The pellet (12,000 rpm for 30 minutes) was preci- pitated twice in ethanol and then taken up in sterile 1/100 xSSC. Aliquots of this sample were (a) adjusted to 2xSSC with 20xSSC and treated with 40 μg/ml of pRNase (free of DNase) for one hour at 22°C; (b) adjusted to 0.5 M NaOH and incubated at 80°C for 1 hour; (c) adjusted to 1/200xSSC and treated with RNase as in (a); or (d) treated with S1 nu- clease (a gift from Dr. D. Zarling) in solution containing 3mM sodium acetate, pH 4.5, 0.18 mM zinc chloride, 0.3 M sodium chloride, and 12.5 μg denatured calf thymus DNA, for one hour at 37°C; or (e) untreated. Each treated aliquot was precipated in ethanol after addition of tRNA carrier and ammonium acetate, as described above. The pellets were taken up with 0.5 ml 1/100xSSC and 0.10 ml distilled water; 0.030 ml CH_2O was added, the samples were heated at 100°C for 2.5 minutes, and quick cooled; 0.70 ml 1/100xSSC, 0.12 ml CH_2O and 1.75 ml of Cs_2SO_4 (refraction 10°17') were added and the mixtures centrifuged for 60 hours at 35,000 rpm. The fractions were collected, and the density (refraction) and 5% TCA-precipitated cpm's deter- mined. Reagent-grade 37% CH_2O (containing 12% methanol) was used in these experiments and was neutralized with NaOH and phosphate buffer, pH 7.8, as in (39). The pro- cedure for denaturation and centrifugation in 1% CH_2O was according to Thomas and Berns (40).

ori⁻ mutation blocks the activity of $\underline{O}$ with regard to the requirement for lit synthesis, as well as preventing the initiation of replication.

The products of genes $\underline{O}$ and P of λ have both been considered to be diffusible because one can complement $\underline{O}^-$ and P⁻ mutants (3,4,11,25). Superinfection of a cell by phages λimm434O⁺P⁺ and λ immλO⁻P⁺ results in approximately equal numbers of each phage. The O⁺ and P⁺ products of the imm434 phage could function in trans to initiate replication of the immλO⁻P⁻ phage. It might also be argued, however, that either or both the $\underline{O}$ or P products of the imm434 phage act in cis on the imm434 template resulting in a trans acting product(s) which participates in replication of both templates. Rambach (9) has shown that his ori⁻ phage cannot be helped to replicate and, conversely, that certain ori⁻O⁺ phage weakly complement ori⁺O⁻ phage. We found his strains r95 and r96 to be defective in oop and lit synthesis (Table 1). The weak complementation of ori⁺O⁻ phage by his other mutants could be explained by hypothesis 1 (above) if these strains are leaky for the synthesis of oop and lit, as in mutant til2 (unpublished). Kleckner (32 and personal communication) has also suggested that the lambda $\underline{O}$ product might act in cis but only under N⁻ conditions.

In vitro Priming Activity of Isolated oop RNA: We have developed a cell-free system in which in vitro λ DNA synthesis depends on the oop RNA, a λ DNA template and an extract from the induced λ lysogen (33). The extract was prepared as described by Wickner et al. (34) and other details are listed in Table 2. We used preparative hybridization to isolate oop RNA (which was labeled with ³H to permit an easy quantitative assay) according to a procedure similar to that outlined in Fig. 5.

As shown in Table 2, the DNA synthesis in this system fully depends on the amount of λ DNA template, on oop RNA, on the cell extract and on the four deoxynucleoside triphosphates. There seems to be only a partial requirement for the ribonucleoside triphosphates and for Mg²⁺. E. coli tRNA will not substitute for oop RNA. Similarly, concentrated preparations of other λ RNAs, including leftward att-N RNA and rightward tof-ori-P RNA, do not substitute for oop RNA. Since the latter RNA contains the rightward ori transcript, it might be concluded that there is no

A | lit | oop
imm λ | ori O P

B | oop
imm 434 | ori O P

lit synthesis (A+B)		oop (A)	(A+B)		
68	+ + +	24			
3	− − −	2	16	+	+ +
4	− + +	2	39	+	+ +
7	+ − −	3	20	+	+ +
43	+ + −	1	20	+	+ +
6	+ − +	2	23	+	+ +

Fig. 8. Effects of mutations in the ori site and complementation by λ O and P products on lit RNA synthesis. The following single lysogens containing various replication defective prophages (see A) were employed: SA297 = 594(λcI857dely-Jb2) = ori⁻O⁻P⁻; SA439 = 594(λcI857delP-Jb2) = ori⁺O⁺P⁻ (data not shown but similar to that for the other ori⁺O⁺P⁻ strain); M72su⁺(λN7N53cI857ori7) and M72su⁺ (λN7N53cI857ori8) (respectively, su⁺ derivatives of strains r95 and r96 from Rambach (9) via W. Dove) = ori⁻O⁺P⁺; 594(λcI85708P3) = ori⁺O⁻P⁻; 594(λcI857P3) = ori⁺O⁺P⁻; 594(λcI85708) = ori⁺O⁻P⁺; and the control strain SA431 = 594(λcI857delQ-Jb2) = ori⁺O⁺P⁺. All these prophages designated (A) contain the lit-coding immλ DNA and are inserted at the regular λ attachment site between E. coli genes gal and bio. Double lysogens of these strains were prepared by inserting at the att80 site (between the tdk and trp genes of E. coli) a phage ø80att80exo-Nλimm434y-Pλ (ø80λimm434cIts), which carries a cIts mutation in its 434 repressor, and cannot code for lit RNA (its structure shown in Fig. 1 was confirmed by heteroduplex mapping by Dr. E. Szybalski). Both the single lysogens and their double lysogens carrying also the ø80λimm434cIts helper (prophage B) were grown and pulse labeled with ³H-uridine for one minute at 30°C (zero time) and at 7,11,15,20 and 25 minutes after induction at 42°C. The extracted ³H-RNA was prehybridized to the l strands of ø80λimmλ and λbio30-7nin5 and the eluted ³H-RNA hybridized with the l strand of various λ deletion DNAs to measure immunity-ori transcription in intervals C1-C4 (rex-cIA before induction, lit after induction), C5-C6 (cIB), D(y-cII) and E-F (ori-O-P) as in Fig. 2 and ref. 12. The indicated levels of lit synthesis (lines 2-6) represent the average levels (for 7th-25th minutes after induction) of ³H-RNA hybridizing in the C1-C4 interval for the double lysogens (A+B) from which the analogous control values for the single lysogens (A) were subtracted. The oop columns specify the average levels of oop synthesis after induction of the single immλ lysogen (A) and double immλ + imm434 lysogen (A+B), respectively. The top line lists data for the immλ single lysogen only. All values represent the percent (x10³) of total input ³H-RNA hybridized.

Table 2. In Vitro DNA Synthesis: Reaction Components

Addition or Omission	pmol ^{32}P-dTMP Incorporated (% Activity)
Complete (cell extract, 1 µl oop RNA, λcb2 DNA)[a]	
2.0 µg λDNA[b]	100
0.4 µg λDNA	30
0.2 µg λDNA	12
-λDNA	2
-cell extract	0
-oop	2
-oop + tRNA[c]	2
-dATP, dGTP, dCTP, dTTP	0
-UTP, CTP, GTP	44
-ATP	33
-Mg^{2+}	19

[a] Cell extract was prepared from SA431 lysogen (Fig.5) grown at 30°C in Hershey broth (supplemented with 10 µg/ml of thiamine and 20 µg/ml of thymine) to A_{575} = 0.45 and thermally induced for 15 minutes at 41-43°C. The cells were centrifuged for 5 minutes, resuspended in 0.002 volume of 10% sucrose containing 50 mM Tris·HCl (pH 7.5), and rapidly frozen in liquid nitrogen. The mixture of thawed cells, 4 µg/ml of lysozyme and 10% Brij-58 (10:1:5, v/v) was incubated for 30 min in ice and centrifuged at 50,000 g for 20 min at 5°C. The clear supernatant was used for in vitro DNA synthesis. Preparations yielding viscous or even slightly turbid supernatants contained DNA and were therefore discarded. The oop RNA was prepared as in Table 3. The 75 µl assay mixtures contained, in addition to 30 µl cell extract, 1 µl oop RNA and λcb2 DNA, 0.05 mM UTP, CTP, GTP, 1.5 mM ATP, 0.04 mM dATP, dGTP, dTTP, plus 300 pmol ^{32}P-dTTP to bring assay to 500 cpm/pmol dTTP, 2 mM spermidine, 4 mM dithiothreitol, 20 mM Tris·HCl, pH 7.6, and 10 mM MgCl$_2$. The reaction mixtures were incubated for 22 min at 36°C, then quickly cooled in ice; 0.3 ml solution of denatured calf thymus DNA (1 mg/ml), BSA (4 mg/ml) and sodium pyrophosphate (0.1 M), and 5 ml 5% TCA were added. The precipitate was collected on GF/C fiber filters, washed with 50 ml 1% TCA and 3 ml ethanol, dried and counted in a scintillation counter. Maximal incorporation of ^{32}P-dTMP (157.3 pmol = 100% activity) into DNA occurred using the complete system with 2 µg λcb2 DNA template (line 1). In all the experiments (except those in lines 2 thru 4) 2 µg/ml of λcb2 DNA was employed.   The ^{32}P counts bound to filters in the absence of cell extract (9.9 pmol) were subtracted from each assay value.

[b] Similar result was obtained when 20 µg/ml of rifampicin was added.

[c] 3 µg E. coli tRNA. Similar result was obtained when leftward att-N or rightward tof-P λ RNA was added.

rightward RNA primer analogous to oop RNA (including an
RNA sequence complementary to oop RNA, which could be
termed "poo"). Moreover, no transcription is required
during the in vitro DNA synthesis, since this reaction is
insensitive to rifampicin. Table 3 shows the stimulating
effect of oop RNA for various DNA templates. The highest
stimulation is observed with a double-stranded λ DNA
template, but not with the single strand of λ DNA. Significant stimulation by oop RNA was observed also with λ-
∅80 hybrid DNA which carries gene P of λ but the ori
region of ∅80. This result shows either a partial compat-
ibility between the oop RNA of λ and the ori region of
∅80 or indicates an imperfect fidelity in our in vitro
system. With the coliphage T7, E. coli and calf thymus
DNA templates the background TMP-incorporation values in
the absence of oop are high and oop stimulation is low but
possibly significant. The significance of the high background with heterologous templates in the absence of oop
RNA will be discussed in conjunction with somewhat analogous data shown in Table 4 (lines 1 and 2).

Absence of oop stimulation for the l strand λ DNA template, which could form a perfect hybrid with oop RNA, and
good stimulation for the hy42 DNA template, which does not
hybridize with oop RNA, indicates that the oop priming
does not require extensive complementary pairing between
DNA and oop RNA. At present we favor the hypothesis that
in the first step of the priming the rU_6 sequence near the
3' terminus of oop RNA forms a triple-stranded helix with
the $dA_{>6} \cdot dT_{>6}$ sequence on DNA (35). Such a dPy·dPu·rPy
triple helix could be stabilized by the replication complex
or any of its protein components. It might be significant
that a short ∅80 transcript analogous to the λ oop RNA,
has a similar 3'-terminal $U_5A(A)$-sequence (36). We do
not know whether oop-stimulated initiation of in vitro
λ DNA replication faithfully represents the in vivo
initiation. Experiments using λ DNA fragments cleaved by
EcoRl restriction endonuclease have shown that oop will
stimulate DNA synthesis from all the fragments, although
the ori-carrying fragment is the best template (data not
shown). However, even in in vivo experiments with λ O⁻
or P⁻ mutants, replication-initiation D-loops were shown
to appear throughout λ DNA (37).

The experiments in Tables 2 and 3 employed extracts
from induced λ lysogens. The DNA synthesis catalyzed by

Table 3. In Vitro DNA Synthesis: Template Requirements

DNA Template (2.0 µg)	pmol ^{14}C-dTMP Incorporated per Assay		Stimulation of DNA synthesis by oop RNA +oop/-oop (OSA)
	No oop primer	+oop RNA (1µl)	
none	--	1	--
λcb2	4	63	15.8
λcb2 l-strand (5 µg)	--	1	--
λatt80imm80ori80P-Rλ (hy42)	4	43	10.8
T7	18	43	2.4
E. coli	17	77	4.5
calf thymus	34	92	2.7

Reaction mixtures and assay conditions are described in Table 2 (line 2), but the reaction was carried out at 30°C. DNA was phenol extracted (3 times at 4°C) and exhaustively dialyzed against 10mM Tris·HCl + 10 mM NaCl (pH, 7.9; 4°C). The phage DNA molecules were intact (as determined by Dr. E. Szybalski using electron microscopy) but E. coli and calf thymus DNAs were fragmented. The oop RNA was prepared from induced cultures of SA431 labeled with 3H-uridine for 1 minute between 15-25 minutes after induction. The cells were harvested and the 3H-RNA was extracted (3 times with phenol at 60°C) and purified by hybridization to the l strand of λbiotl24nin in a manner analogous to that outlined in Fig. 5 and reference 12.

extracts from λ-free E. coli, uninduced lysogens and
induced defective lysogens is shown in Table 4. There are
two major features of DNA synthesis in extracts from λ-
free E. coli or noninduced lysogens (lines 1 and 2): the
background activity in the absence of oop RNA is high and
there is no stimulation by the oop RNA. This result suggests that some products of λ induction are present in the
extract and that they (a) reduce the level of the oop
primer-independent (background) DNA synthesis, probably by
reducing the nicking and repair of the λcb2 template, and
(b) permit oop RNA to stimulate the λ DNA synthesis. We
have tentatively called these activities (a) BLA, for
background lowering activity and (b) OSA, for oop specifying activity. We report elsewhere (38) that both BLA and
OSA activities are dominant in mixed extracts from induced and noninduced lysogens. To identify which λ genes
participated in these activities, cell extracts were
prepared from induced defective λ lysogens and tested for
BLA and OSA activity (Table 4,lines 4-11). Deletion of
genes P-Q-A-J (lines 8 and 9), or blocking the expression
of genes att-N or tof-J (Fig. 5) by the N⁻N⁻ or p_R^-(x13)
mutations (lines 5 and 6), or a point mutation in tof gene
(line 11), has no effect on BLA, or OSA. Point mutations
in genes O and P (line 4) or the additional deletion of
genes between y and P (line 10) lowers the level of OSA
about two-fold, but has no effect on BLA. OSA is eliminated
and BLA is reduced in extracts of λN⁻N⁻x⁻ lysogens (line
7) in which both the rightward transcription from p_R and
transcription to the left of N are blocked. Recent experiments indicate that these activities are present in extracts
from inducible lysogens of deltof-A-J (ΔHl, Fig. 1) but
not in NN⁻deltof-A-J (38). These results suggest that no
special λ RNA or proteins correspond to OSA or BLA. They
cannot be a p_R-promoted product since x⁻ and ΔHl strains
produce OSA and BLA; they are not cI-rex since uninduced
lysogens have no activity, and they are not N or an N-
dependent leftward product since N⁻N⁻ strains produce OSA
and BLA. There are no genes left, except for the lit
region, which has not been tested. We tentatively conclude
that OSA and BLA (activities) are brought about by the
induction of lambda transcription, which affects somehow the
E. coli nucleus. It might be significant that chloramphenicol blocks the appearance of both BLA and OSA when added
before induction whereas rifampicin blocks only OSA (38).

Table 4. In Vitro DNA synthesis: Requirement for Cell Extract from Induced Lambda Prophage

Cell extract (cells grown at 30°C to A_{575} = 0.45; heated 15 min at 43°C)	pmol ^{14}C-dTMP Incorporated Per Assay		Stimulation of DNA synthesis by oop RNA +oop/-oop (OSA)
	No RNA primer	+oop RNA	
W3350	12.8	14.3	1.1
W3350(λcI$^+$)	22.5	18.4	.8
W3350(λcI857)	3.0	36.0	12.0
W3350(λcI857) lysogens with following prophage mutations:			
O$^-$P$^-$	3.0	17.0	5.7
N N	0.1	17.0	17.0
x13	1.0	16.0	16.0
N$^-$N$^-$x13	6.7	10.4	1.6
delQ-Jb2	4.2	63.4	15.1
delP-Jb2	4.8	61.6	12.8
dely-Jb2	3.6	24.6	6.8
cro	3.3	50.4	15.3

Extracts of the lysogens and nonlysogens were made as indicated in Table 2. The assays included 2 µg λcb2 DNA, 1µl ^{3}H-oop RNA, (column 3), and were run for 20 minutes at 28°C or 30°C. The optimal stimulatory effect of oop was recently found to occur at 26°C (unpublished). Symbol del denotes deletions.

DISCUSSION

Several lines of evidence suggest that oop RNA primes λ DNA synthesis: (1) oop is the only RNA transcribed from the l strand in the ori region; (2) its synthesis is positively regulated by the products of phage and host replication genes; (3) the stimulation of oop transcription is independent of λ DNA replication, i. e., there is no gene dosage effect; (4) oop RNA synthesis precedes the initiation of replication; (5) ^{3}H-oop RNA covalently linked to λ DNA is found in induced replicating λ lysogens; and (6) addition of oop RNA stimulates in vitro DNA synthesis from double stranded λ DNA templates in E. coli whereas all other λ RNAs, including those from the r-strand cII-ori-0-P region, or tRNAs have no such stimulatory effect.

Since there are no known DNA polymerases which could initiate DNA replication in the middle of double-stranded DNA, it is attractive to postulate that the 3'-terminus of the oop RNA is utilized as a primer for initiation of λ DNA synthesis. Phage λ has developed an elaborate mach-inery, partially utilizing the host systems, to control oop synthesis and other transcriptional events which are coupled to the initiation of DNA replication. However, once the initiation occurred, the actual DNA synthesis may proceed with a less elaborate machinery. It is important to stress that we are studying here the first initiation event on double-stranded DNA, which leads to formation of a unidirectional replication fork and an asymmetric D replication loop in which only one strand will be replicated. To attain bidirectional replication, another initiation event has to occur, but this time on the single-stranded DNA within the replication loop. This secondary initiation event and further initiations leading to synthesis of so-called Okazaki fragments were discussed by several participants of this Symposium. Al-though the detailed mechanism of DNA synthesis initiation described here might be applicable only to phage λ, we suspect that most of the other replicons are similarly primed by special RNA molecules, synthesis of which is controlled by various permutations of the host (and viral) factors, and that the availability of the primer RNA controls the primary initiation event in DNA synthesis.

ACKNOWLEDGMENTS

These studies, which were primarily carried out in McArdle Laboratory, University of Wisconsin, Madison, Wisconsin were supported by grants from the National Cancer Institute (CA-07175) and the National Science Foundation (GB-2096). The work done at U.C.-Irvine was supported by Grant R01 GM21958-01. We thank Tina Chow, R. Jay Rau, Bob Hayes and Connie Hayes for expert technical assistance, and Eric C. Rosenvold for experimental data cited in ref. 38.

REFERENCES

1. Jacob, F., Brenner, S. and Cuzin, G., Cold Spring Harbor Symp. Quant. Biol. $\underline{28}$, 329 (1963).
2. Brooks, K., Virology $\underline{26}$, 489 (1963).
3. Kaiser, D., In: The Bacteriophage Lambda (A.D. Hershey, ed.) p. 195. Cold Spring Harbor Lab., Cold Spring Harbor, N.Y. 1971.
4. Stevens, W., Adhya, S. and Szybalski, W., In: The Bacteriophage Lambda (A.D. Hershey, ed.) p. 515. Cold Spring Harbor Lab., Cold Spring Harbor, N.Y. 1971.
5. Ogawa, T. and Tomizawa, J., J. Mol. Biol. $\underline{38}$, 217 (1968).
6. Tomizawa, J., In: The Bacteriophage Lambda (A.H. Hershey, ed.) p. 549. Cold Spring Harbor Lab., Cold Spring Harbor, N.Y. 1971.
7. Schnös, M. and Inman, R.B., J. Mol. Biol. $\underline{51}$, 61 (1970).
8. Dove, W.F., Inokuchi, H. and Stevens, W.F., In: The Bacteriophage Lambda (A.D. Hershey, Ed.) p. 747. Cold Spring Harbor Lab., Cold Spring Harbor, N.Y. 1971.
9. Rambach, A., Virology $\underline{54}$, 270 (1973).
10. Eisen, H., Fuerst, C., Siminovitch, L., Thomas, R., Lambert, L., Pereira da Silva, L. and Jacob, F., Virology $\underline{30}$, 224 (1966).
11. Dove, W. F., Hargrove, E., Ohashi, M., Haugli, F. and Guha, A., Japan. J. Genet. $\underline{44}$, suppl. 1, 11 (1969).
12. Hayes, S. and Szybalski, W., Molec. Gen. Genet. $\underline{126}$, 275 (1973).
13. Hayes, S. and Szybalski, W., In: Molecular Cytogenetics (B.A. Hamkalo and J. Papaconstantinou, eds.) p. 277, Plenum Press New York, 1973.
14. Thomas, R. and Bertani, L.E., Virology $\underline{24}$, 241 (1964).

15. Heinemann, S.F. and Spiegelman, W.G., Cold Spring Harbor Symp. Quant. Biol. 35, 315 (1970).
16. Brachet, P., Eisen, H. and Rambach, A., Molec. Gen. Genet. 108, 266 (1970).
17. Herskowitz, I. and Signer, E., Cold Spring Harbor Symp. Quant. Biol. 35, 355 (1970).
18. Kumar, S. and Szybalski, W., Virology 41, 665 (1970).
19. Szybalski, W. and Szybalski, E.H., In: Viruses, Evolution and Cancer (E. Kurstak and K. Maramorosch, eds.) p. 563. Academic Press, New York, 1974.
20. Nijkamp, H.J.J., Szybalski, W., Ohashi, M. and Dove, W.F., Molec. Gen. Genet. 114, 80 (1971).
21. Pereira da Silva, L. and Jacob, F., Ann. Inst. Pasteur 115, 145 (1968).
22. Packman, S. and Sly, W.S., Virology 34, 778 (1968).
23. Ptashne, M. and Hopkins, N., Proc. Nat. Acad. Sci. 60, 1282 (1968).
24. Lindahl, G., Virology 42, 522 (1970).
25. Inokuchi, H., Dove, W. and Freifelder, D., J. Mol. Biol. 74, 721 (1973).
26. Shuster, R.C. and Weisbach, A., Nature 223, 852 (1969).
27. Wechsler, J.A., J. Bact 121, 594 (1975).
28. Gefter, M., Hirota, Y., Kornberg, T., Wechsler, J., and Barnoux, C., Proc. Nat. Acad. Sci. 68, 3150 (1971).
29. Adhya, S., Sen, A. and Mitra, S., In: The Bacterio-phage Lambda (A.D. Hershey, ed.) p. 743. Cold Spring Harbor Lab., Cold Spring Harbor, N.Y. 1971.
30. Zimmerman, S.B. and Sandeen, G., Anal. Biochem. 14, 269 (1966).
31. Dahlberg, J.E. and Blattner, F.R., In: Virus Research: Second ICN-UCLA Symposium on Molecular Biology (C. F. Fox and W.S. Robinson, eds.) p. 533. Academic Press. New York. 1973.
32. Kleckner, N.E., Ph.D. Thesis, M.I.T., Cambridge, Mass. 1974.
33. Hayes, S. and Szybalski, W., Feder. Proc. 33, 1492 (1974).
34. Wickner, R.B., Wright, M., Wickner, S. and Hurwitz, J., Proc. Nat. Acad. Sci. 69, 3233 (1972).
35. Szybalski, W., Abstr. of Third Int. Congress for Virology, Madrid, Spain, 1975.

36. Pieczenik, G., Barrell, B.G. and Gefter, M.L., Arch. Biochem. Biophys. 152, 152 (1972).
37. Inman, R.S., and Schnos, M., In: DNA Synthesis in Vitro. (R.D. Wells and Inman, R.S.). University Park Press, Baltimore, Md., 1973.
38. Hayes, S. Rosenvold, E.C., Rau, R.J. and Szybalski, W., Feder. Proc. 34, 639 (1975).
39. Freifelder, D. and Davison, P., Biophys. J. 3, 49 (1963).
40. Thomas, C. and Berns, K.I., J. Mol. Biol. 4, 309 (1962).

Part VII
Plasmid and Plastid Replication

MODES OF PLASMID DNA

REPLICATION IN ESCHERICHIA COLI

D. R. Helinski, M. A. Lovett, P. H. Williams, L. Katz,
J. Collins, Y. Kupersztoch-Portnoy, S. Sato,
R. W. Leavitt, R. Sparks, V. Hershfield,
D. G. Guiney and D. G. Blair
Department of Biology
University of California, San Diego
La Jolla, California 92037

ABSTRACT. Various approaches have been employed to
examine the replication properties of plasmid elements in
Escherichia coli. Results presented were obtained from an
analysis of the effect of known chromosomal DNA replication mutants on plasmid DNA replication, the isolation
and characterization of plasmid DNA replication mutants,
isolation and characterization of plasmid DNA-protein complexes, analysis of plasmid DNA replication intermediates
isolated from intact cells and an analysis of RNA-containing plasmid DNA molecules synthesized in intact cells in
the presence of chloramphenicol.

INTRODUCTION

Plasmids are circular duplex DNA molecules that
exist stably in the extrachromosomal state in a wide variety of bacterial species. Recently, information has been
obtained that provides some insight to the mechanisms
responsible for the maintenance of these elements in the
autonomous state. These studies have received their im-
petus from the increased medical importance of plasmid
elements with respect to the control of infectious diseases and the advantages that plasmids provide as model
systems for the study of the mechanism of replication of
duplex circular DNA molecules.
    While plasmid elements in general exhibit the
characteristic features of covalently closed and duplex

TABLE 1: General properties of plasmids

Plasmid	Molecular Weight	No. Copies Per Chromosome	fi Character	Other Phenotypic Properties	Reference
ColE1	4.2×10^6	10–15	none	Colicin E1	1,2
ColE2	5.0×10^6	10–15	none	Colicin E2	1
ColE3	5.0×10^6	10–15	none	Colicin E3	1
F1	62×10^6	1–2	fi^+	–	3,4
Flac	95×10^6	1–2	fi^+	Lac operon	5
ColV2	94×10^6	1–2	fi^+	Colicin V2	4
R100	70×10^6	1–2	fi^+	$Cm^R Sm^R Su^R Tc^R$	6
ColIb	61×10^6	1–2	fi^-	Colicin Ib	7
R64	78×10^6	limited	fi^-	$Tc^R Sm^R$	8
R6K	26×10^6	13	fi^-	$Am^R Sm^R$	9
R28K	44×10^6	2–3	fi^-	Am^R	9

TABLE 2: Effect of chromosomal DNA replication mutants on ColE1 DNA replication*

Mutant	Type	ColE1 Replication at the Non-Permissive Temperature
dna A (CRT484, CRT83)	Initiation	Normal
dna C/D (CR284, CT284, PC1, PC7)	Initiation	Greatly reduced rate
pol A (TS214)	DNA polymerase I	Stops immediately
pol C (E486)	DNA polymerase III	Initial rate 20% of wild-type
dna F (E101)	Diphosphoribonucleotide reductase	Immediate reduction in rate
dna G (BT308)	Elongation	Stops immediately

*The studies with the pol A mutant were carried out by Kingsbury and Helinski (10,12). All other mutant studies were carried out by Collins et al. (11).

516

circular DNA molecules, they vary considerably in size and in their complement of genes determining a variety of bacterial properties. As shown in Table 1, the plasmids studied in our laboratory can be grouped into three categories on the basis of their molecular weight, the number of copies per chromosome, and their sexual properties. The first group of plasmids includes the non-self-transmissible colicinogenic plasmids of the E type that are of a relatively small size and are present to the extent of 10-15 copies per chromosome. The second group includes the self-transmissible plasmids of the fi$^+$ type (plasmids of this type repress the fertility properties of an F factor present in the same cell) that are present to a limited extent in the cell and are of a relatively large molecular weight. The third group includes the relatively large molecular weight, self-transmissible plasmids of the fi$^-$ type that are present with the exception of the R6K plasmid to the extent of a limited number of copies per host cell.

GENETIC CONTROL OF PLASMID REPLICATION

Various approaches have been used in the study of genetic and biochemical factors that are responsible for the regulated maintenance of plasmid elements in a bacterial cell. In addition to the characterization of plasmid DNA replicative intermediates and plasmid DNA-protein complexes that may be involved in the replication of these elements, the effect of known chromosomal DNA replication mutations on plasmid DNA replication has been examined and plasmid DNA replication mutants have been isolated and characterized. As indicated in Table 2, it has been found that ColE1 replication in Escherichia coli is independent of the dna A gene function and dependent on the dna C, D, F, and G functions of the host cell (11). At least a partial dependency has also been demonstrated for DNA polymerase III since although ColE1 DNA replication continues in a pol C mutant at the non-permissive temperature, it proceeds at a greatly reduced rate (11). Earlier studies have demonstrated a requirement for DNA polymerase I for the replication of plasmid ColE1 in E. coli (10,12). The results with the dna A mutant differ from earlier reports of Goebel (13,14,15) who reported a dependency of ColE1 replication on the dna A function and the independence of ColE1 DNA synthesis on DNA polymerase III. Nevertheless,

it is clear from these mutant studies that not only are
certain chromosomal gene functions required for plasmid
DNA replication, as initially shown in the work of Jacob,
Brenner and Cuzin (16) with the Flac plasmid, but certain
of these chromosomal gene functions play a role in both
chromosomal and plasmid DNA replication. In the case of
the ColEl plasmid, specific mutational events in the plasmid itself also will result in defective plasmid DNA replication, indicating a requirement for certain plasmid genes
in the maintenance of this element in a bacterial cell
(17).

CONTINUED REPLICATION OF ColEl IN THE ABSENCE
OF PROTEIN SYNTHESIS

Although the mutant studies indicate a requirement
for proteins determined by both chromosomal and plasmid
genes for the replication of ColEl DNA, this plasmid element is unusual in its ability to continue to replicate in
the absence of protein synthesis (18,19,20). In the presence of concentrations of the protein inhibitor chloramphenicol (CM) that will prevent the initiation of chromosomal DNA synthesis, ColEl DNA replication continues for
10-15 hours resulting in the accumulation of 1,000-3,000
copies of the supercoiled DNA form of ColEl DNA per cell
(Fig. 1). It is of interest with respect to the replication and molecular vehicle properties of ColEl that the
covalent insertion of an additional segment of DNA into
this plasmid does not appear to alter significantly its
replication properties in the absence or presence of chloramphenicol. It has been found that the insertion of a
segment of DNA from the bacteriophage ϕ80pt190($\underline{trp}^+$) containing the tryptophan operon of $\underline{E.\ coli}$ in one case and a
segment of DNA determining resistance to the antibody
kanamycin ($\underline{kan}$) obtained from the R plasmid pSC105 in the
other case (these plasmids are designated ColEl-$\underline{trp}$ and
ColEl-kan, respectively) did not effect the ability of the
plasmid to continue to replicate in the presence of CM
(21). The insertion of DNA into the ColEl plasmid was
carried out using the restriction endonuclease EcoRl since
ColEl DNA possesses a single site cleaved by this enzyme
(22). The final yield of supercoiled plasmid DNA for normal
ColEl, ColEl-$\underline{trp}$ and ColEl-$\underline{kan}$ after incubation of the
cells in the presence of chloramphenicol, was approximately

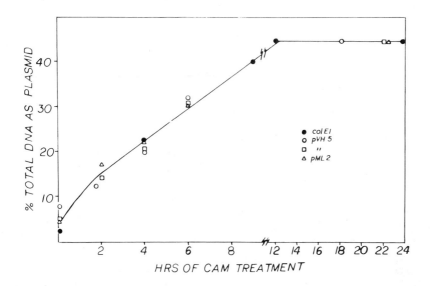

Fig. 1. The extent of replication of plasmids ColEl, ColEl-trp and ColEl-kan in the presence of chloramphenicol. E. coli strains carrying the different plasmids were grown to the logarithmic state, CM added, and samples removed at 0, 2, 3, 6 and 22 hours after the addition of the CM. The level of supercoiled plasmid DNA in the cells from each time point was determined. ColEl, ColEl-trp and ColEl-kan exhibit molecular weights of 4.2×10^6, 14.3×10^6 and 8.7×10^6, respectively and possess one, three and two EcoRl sensitive sites, respectively. ●---●, ColEl; o---o and ▫--▫ ColEl trp (separate experiments); Δ---Δ ColEl-kan. (From Hershfield et al., [21]).

the same in each case, equivalent to 45% of the total DNA
of the cell (Fig. 1). This corresponds to approximately
360 copies of the ColEl-<u>trp</u> plasmid, 560 copies of the
ColEl-kan plasmid, and 1,200 copies of normal ColEl DNA
per cell (21). In addition, both the normal and the hybrid
plasmids are maintained at a level of approximately 10-15
copies per chromosome, or 20-30 copies per cell, in bacteria growing under normal logarithmic growth conditions
(21). Thus, in addition to its relatively small size and
the presence of a single EcoRl sensitive site in the molecule, the presence of the ColEl plasmid as multiple copies
per bacterial cell and its continued replication in the
presence of CM are particularly advantageous properties for
the use of ColEl as a vehicle for the cloning and amplification of a specific segment of DNA. It should also be
noted at this time that while the insertion of DNA at the
single EcoRl site on the molecule does not effect its
replication properties it does result in a loss of the
ability of the plasmid to produce the antibiotically active
protein colicin El (21). This particular property can be
taken advantage of in the use of ColEl as a molecular vehicle in distinguishing between cells that have been transformed with normal ColEl or ColEl hybrid DNA molecules (21).

ColEl REPLICATES AS A COVALENTLY-CLOSED CIRCLE

The replicative intermediates of ColEl DNA obtained
from intact cells of E. <u>coli</u> exhibit similar physical-
chemical properties as the covalently-closed replicative
intermediates that have been described for the circular
DNA of the animal virus SV40 (23,24). Using a procedure
that enriches for replicating molecules involving a brief
period of thymine starvation in the presence of cyclic AMP
followed by pulsing for a short period with radioactive
thymidine (25), the following properties of replicating
ColEl DNA molecules have been observed (26). (a) ColEl
replicating intermediates sediment more rapidly than mature
supercoiled DNA in a neutral sucrose gradient. (b) The
replicating molecules band at a less dense position than
mature supercoiled DNA in an ethidium bromide-cesium chloride equilibrium gradient. (c) The superhelical density
of the replicating molecules decreases as replication becomes more extensive. (d) At least one of the two newly
synthesized strands is synthesized as short discontinuous

fragments. (e) Replicating molecules obtained from an ethidium bromide-cesium chloride gradient appear in the electron microscope as partially supercoiled structures containing two open circular branches of equal size. Nicking of the partially supercoiled replicating molecules with pancreatic DNAase I or exposure to light in the presence of ethidium bromide yields open circular molecules with an internal loop of varying size. While the data with the plasmid R6K is not as extensive as obtained for the ColEl plasmid, the properties of replicated forms of the R6K plasmid are also consistent with replication of this plasmid element as a covalently closed circle (27).

ColEl REPLICATION IS UNIDIRECTIONAL FROM A FIXED ORIGIN

Using the EcoRl site on the ColEl DNA molecule as a reference point on EcoRl cleaved replicating intermediates of this plasmid, it has been demonstrated that ColEl DNA replication proceeds unidirectionally from a fixed origin (28). As shown in Figs. 2 and 3, linear molecules of the contour length of open circular ColEl DNA and with internal loops of varying size possess one unreplicated arm (U1) that remains constant in size and another unreplicated arm (U2) that progressively decreases in size. This is the result expected for a unidirectional mode of replication from a fixed origin that is located in the case of the ColEl plasmid 18.2 ± 0.2% of unit length from one EcoRl end (28). Similar results have been obtained for the mode of ColEl replication in mini-cells (29) and in an in vitro system (30).

R6K REPLICATION IS BIDIRECTIONAL FROM A FIXED ORIGIN

The R6K plasmid is an fi$^-$, self-transmissible element of a molecular weight of 25×10^6 that determines resistance to the antibiotics penicillin and streptomycin (31). Cleavage of R6K DNA with the EcoRl restriction endonuclease generates two fragments of molecular weights of 14.9×10^6 (fragment A) and 9.3×10^6 (fragment B) (32). The majority of pulse labeled replicating molecules of R6K are isolated as covalently closed circular DNA forms using the procedures previously employed for the isolation of replicating forms of the ColEl plasmid (27). Cleavage of the

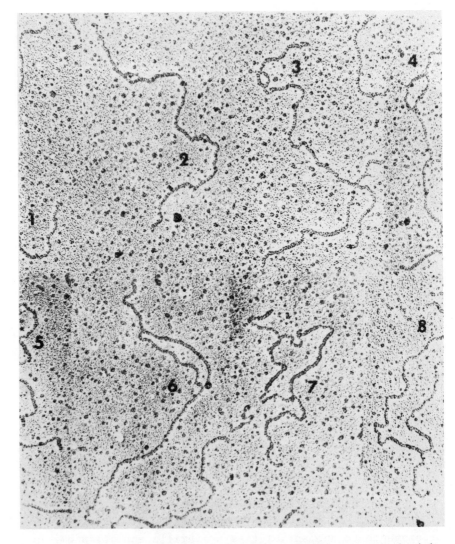

Fig. 2. Replicating ColE1 DNA molecules treated with the EcoR1 restriction endonuclease. Replicating molecules of intact E. coli cells were pulse-labeled and purified by ethidium bromide-CsCl equilibrium centrifugation and neutral sucrose gradient sedimentation. The pool of replicating molecules was treated with EcoR1 endonuclease and spread for electron microscopy. Micrographs of selected molecules are arranged in order of increasing extent of replication. Panel 1 represents an unreplicated linear molecule. (From Lovett et al., [28]).

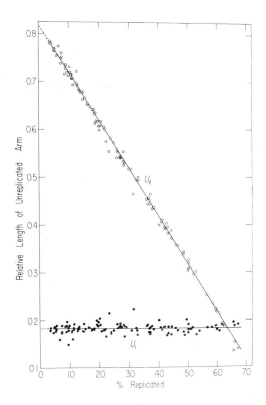

Fig. 3. Analysis of the length of the two unreplicated segments of replicating ColE1 DNA molecules after cleavage with EcoR1 as a function of the extent of replication.
Replicating ColE1 DNA molecules purified and treated with
the EcoR1 restriction endonuclease as described in Fig. 2
were used in this analysis. U1 and U2 are the unreplicated
branches joining an internal loop comprised of R1 and R2.
The relative lengths of U1 and U2 for each molecule was
determined from the formula T = U1 + U2 + 1/2 (R1 + R2),
where T = total length. The % replicated = 1/2 (R1 + R2)/T.
(From Lovett et al., [28]).

population of replicating R6K DNA molecules with the
enzyme EcoR1 generates branched forms of fragments A and B
(27). Only Y structures (molecules with a single unreplicated segment joined by two replicated segments of equal
length) have been observed in the case of fragment A. The
fact that looped structures (molecules with two linear segments joining a symmetrical loop of replicated DNA) were
not observed for fragment A indicates that replication of
fragment A progresses in one direction. Two kinds of
structures, representing replicating DNA, were observed for
fragment B after EcoR1 digestion of the replicating R6K
molecules; looped structures and Y structures. As in the
case of fragment A, molecules with a double Y structure
were not observed. The various types of branched forms of
fragments A and B observed by electron microscopy are
shown in Fig. 4. The analysis of the change of length of
the unreplicated arms of the branched structures (designated U1 and U2, respectively) as a function of extent of
replication for 91 looped structures of fragment B, derived
from replicating R6K molecules, is shown in Fig. 5. As
shown in Fig. 6, these results are consistent with a mode
of replication of R6K that is asymmetric, proceeding from
a fixed origin in fragment B in two directions to a unique
termination point or region of fragment B (27). The distance between the origin and terminus is approximately 20%
of the R6K genome with the terminus located at a position
that is approximately 7.5% of the total distance of fragment B from an EcoR1 end. The data shown in Fig. 5 are
consistent with sequential replication in the two directions in the case of most of the molecules, proceeding
first from the origin to the terminus in one direction, and
then from the origin in the other direction with the consequence that fragment A is replicated unidirectionally (27).
Consistent with this mode of replication of R6K is the
finding of only single Y structures in the case of the
branched molecules of fragment A (27).

PROPERTIES OF RNA-CONTAINING ColE1 DNA MOLECULES

Supercoiled ColE1 DNA synthesized in the presence of
chloramphenicol (CM-ColE1) has been examined in considerable detail in the hope of obtaining some insight to the
role of RNA in the initiation and elongation events of
replication of the ColE1 plasmid. Supercoiled ColE1 DNA

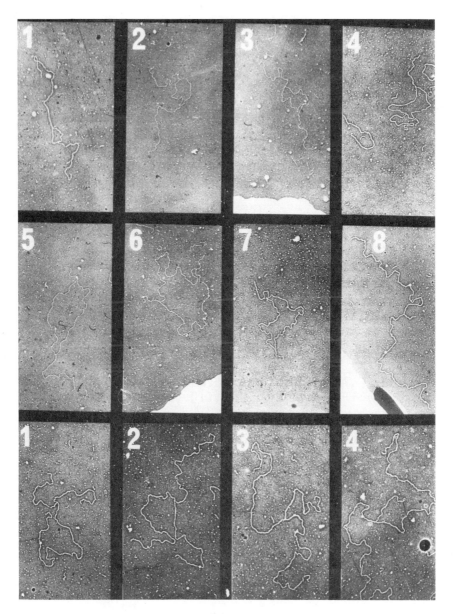

Fig. 4. Replicating R6K molecules treated with the
EcoRl restriction endonuclease. Frames 1 to 8 represent
electron micrographs of molecules of fragment B arranged in
order of increasing extent of replication. Frames 9 to 12
represent molecules of fragment A arranged in order of increasing extent of replication. (From Lovett et al., [27]).

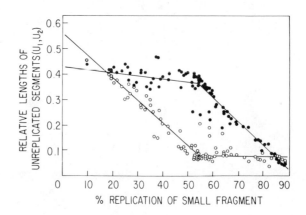

Fig. 5. Analysis of the length of the two unreplicated segments of the looped structures of fragment B of R6K
as a function of the extent of replication. U1 and U2 are
the unreplicated segments joining an internal loop comprised of two segments of replicated DNA. The relative
lengths of U1($\bullet$) and U2 (o) for each molecule and the percent replicated were determined as described in Fig. 3.
(From Lovett et al., [27]).

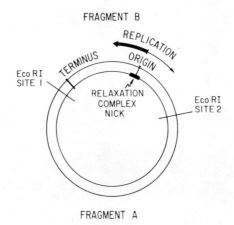

Fig. 6. Model of replication of R6K DNA. The relative positions of the origin, terminus, site of the nick in
the relaxed complex and the two EcoRl cleavage sites are
indicated. The size of fragment A is 14.9 x 10^6 daltons.
The molecular weight of fragment B is 9.3 x 10^6.

molecules synthesized in the presence of chloramphenicol contain ribonucleotides as part of their covalently closed structure (33). The majority of the molecules that are sensitive to ribonuclease A, ribonuclease H, or alkali (pH13) contain ribonucleotides at a single site per molecule with the site occurring with equal probability in either strand (light and heavy) of DNA (33). The RNA segment of the CM-ColE1 DNA contains GMP at the 5' end and comprises on an average for the two strands 25-26 ribonucleotides with a base composition of 10-11G, 3A, 5-6C, and 6-7U (34). The possibility that the ribonucleotide segments represent part of or complete primer RNA is supported by the data of Clewell and co-workers (35,36) that demonstrated a direct dependence of ColE1 DNA replication on RNA synthesis using various inhibitors of RNA polymerase. Recently, it has been found that the composition of ribonucleotides in the RNA segment in the light strand of CM-ColE1 DNA molecules differs from the ribonucleotide composition in the RNA segment of the heavy strand (37). The light strand RNA segment consists of 38 ribonucleotides with a base composition of 17G, 5A, 8C and 8U. The 5'-terminus of the RNA segment exists as a unique sequence (37) that has been determined to be: 5'...p(rG)p(rG)p(rG) p(rG)p(rA)p(rC)...3'. Furthermore, the deoxyribonucleotide-ribonucleotide junction at the 5' end of the RNA segment in the light strand of CM-ColE1 DNA is specifically ...p(dC)p(rG)...(37). The RNA segment in the heavy strand of CM-ColE1 DNA consists of 15 ribonucleotides with a base composition of 5G, 2A, 4C and 4U (37). The partial 5'-terminal sequence of ribonucleotides in the RNA segment of the heavy strand also is unique and determined to be: 5'...p(rG)p(rA)p(rA)p(rU)p(rG)...3' (37). As in the case of the RNA segment in the light strand, the deoxyribonucleotide-ribonucleotide junction at the 5'-end of the RNA segment is unique and in the case of the heavy strand is ...p(dA)p(rG)... (37). The possibility that the RNA segment in the light strand of CM-ColE1 DNA represents initiating primer RNA at the origin is supported by preliminary evidence indicating that this RNA segment is present at the origin of replication in a majority of molecules (38). Unlike the RNA segment in the light strand of CM-ColE1 DNA, the heavy strand RNA segment appears to be distributed randomly or at multiple unique sites on the heavy strand (38); consistent with the possibility that the RNA segment in the heavy strand of CM–ColE1 DNA represents Okazaki priming RNA.

The correspondence of the RNA segments CM–ColE1 DNA with primer RNA is supported by the demonstration of the requirement of ColE1 DNA synthesis for the generation of RNA containing ColE1 DNA molecules and the inhibition of formation of these RNA sensitive molecules by the addition of rifampin to cells actively synthesizing ColE1 in the presence of CM (33). An analysis of the formation of these RNA-containing ColE1 DNA molecules has indicated that the RNA-containing strands of ColE1 DNA are unable to serve as a template for further rounds of replication (39). Evidence has been obtained for a repair mechanism in E. coli cells that can function to remove the RNA segment from the CM–ColE1 DNA (39). The removal of this RNA segment in intact cells is carried out effectively in DNA polymerase I mutants that are defective in polymerizing or 5'-3' exonuclease activity, however, this process is defective in certain recBC and uvrA mutants (29). Fig. 7 illustrates a possible model for the formation of RNA-containing ColE1 DNA and the possible role of these RNA segments as primers for the initiation of DNA synthesis at the origin of replication and the discontinuous synthesis of one of the DNA strands during the elongation steps.

PROPERTIES OF PLASMID RELAXATION COMPLEXES

It is reasonable to propose that at least one break in one of the two polynucleotide strands of covalently closed plasmid DNA must occur at some point during replication to facilitate the unwinding of the parental strands during the replication process and, ultimately, the separation of the two daughter molecules. The relaxation complex of various supercoiled plasmid DNA molecules and protein has been examined in considerable detail since these complexes exhibit unusual biochemical properties and may play a role as a specific nicking enzyme complex required during the replication process. The characteristic feature for the relaxation complex of plasmid DNA isolated from E. coli cells is a conversion of the complexed supercoiled DNA to the open circular DNA form upon treatment with certain proteases or specific agents that denature protein (40). Table 3 summarizes the extent to which various plasmids examined in our laboratory have been found to exist in the

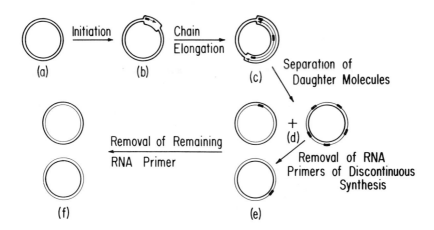

Fig. 7. Model for the replication of ColEl DNA in the presence of chloramphenicol. (a) Closed circular plasmid DNA. (b) Rifampicin sensitive synthesis of an initiation primer RNA molecule complementary to a region of DNA at the origin of replication. (c) Unidirectional replication by continuous synthesis of one strand from the initiation primer, and discontinuous synthesis, also primed by RNA molecules, of the complementary daughter strand. (d) Separation of the daughter molecules containing covalently integrated primer ribonucleotides. In the absence of CM the RNA primers may be removed before closure of the DNA strands. (e) Rapid removal of all but one of the primers in the discontinuously made daughter strand, the remaining primer being distinguished by reason of its position, its size, or its base composition. (f) Removal of the initiation primer RNA and the remaining primer of discontinuous synthesis. This repair is inhibited by mutations in the uvrA and recBC genes. (From Williams et al., [39]).

529

form of a relaxation complex. As indicated in Table 3, in the case of the plasmids ColE1, ColE2, F1 and R6K, the nicking event has been shown to be strand-specific (2,41, 43,32) and site-specific in the case of ColE1, ColE2 and R6K (41,32). Heat treatment (60° C, 10 min.) prevents the induced conversion of the supercoiled DNA in the complex to the open circular DNA form and results in the dissociation of protein from the supercoiled DNA in the case of the ColE2, F1 and R6K complexes (41,43,32). These results strongly suggest that relaxation of complexed supercoiled DNA involves an SDS-activatable endonucleolytic cleavage of a covalently closed circular DNA molecule. A similar event is probably involved in the induced relaxation of the complexed ColE1 DNA since it has been found that exposure of this complex to ethidium bromide plus Mg^{2+} results in a dissociation of the protein from the supercoiled DNA (49).

The analysis of the protein associated with the supercoiled DNA in the relaxation complex has revealed the following unique properties. The induced conversion by SDS or alkali results in a loss of approximately 1/2 of the total protein (^{3}H-leucine labeled) in the case of the ColE1 complex (50). The other 1/2 of protein material is found as a 60,000 molecular weight protein associated at the 5'-terminus of the nicked strand (50,51). This protein remains associated with the nicked strand after treatment with hot SDS (60° C, 60 min.), after incubation with 8M urea, 2M KSCN, or 2M LiCl, or sedimentation in an alkaline (pH12.5) sucrose gradient or centrifugation to equilibrium in a neutral or alkaline (pH 12.5) cesium chloride gradient (50,52). These observations indicate that the induced relaxation of these complexes involves an activation of a strand-specific endonuclease and subsequent formation of a covalent linkage between the 60,000 dalton component of the complex (at least in the case of ColE1 complex) and the 5'-terminus of the nicked strand. The model shown in Fig. 8, based upon these various properties in the relaxation complex, proposes that the relaxation complex provides the transient nicking and closure activity that has been proposed to take place during the process of replication of plasmid DNA. Support for this role of the relaxation complex in ColE1 DNA replication comes from the recent observation that a temperature-sensitive mutant of ColE1 DNA that ceases to replicate at 43° C possesses a relaxation complex that exhibits in vitro an altered response to high

530

TABLE 3: Relaxation complexes of supercoiled plasmid DNA and protein in E. coli

Plasmid	% Complexed Molecules	Strand Specificity of Relaxation Event	Reference
ColE$_1$	20 – 95%	heavy strand (poly UG)	40
ColE$_2$	40 – 80%	heavy strand (poly UG)	41,42
ColE$_3$	70 – 80%	not known	42
F$_1$	~ 50%	heavy strand (poly UG)	43
Flac	~ 50%	not known	44
ColV2	~ 50%	not known	45
R100	> 80%	not known	46
ColIb	> 80%	not known	47
R64	60%	not known	48
R6K	60 – 70%	heavy strand (poly UG)	48,32
R28K	70%	not known	48
pSC101	80%	not known	22

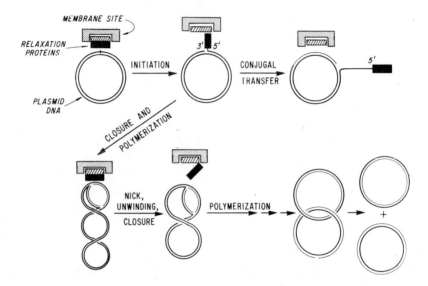

Fig. 8. Possible role of the relaxation complex in
the replication and conjugal transfer of plasmid DNA. In
this model several functions of the relaxation complex are
proposed. (a) Promoting the association of the plasmid
DNA with a membrane replicator site. (b) Catalysis of a
strand and site specific nick in the the plasmid DNA in response to an appropriate "signal" followed by covalent linkage of one of the protein components with the 5'-terminus
of the nicked strand. (c) Closure of the nicked strand
after unwinding of the DNA strands. (d) Facilitation of
transfer of the nicked strand to the recipient cell during
bacterial mating. Light lines represent parental DNA;
heavy lines represent newly synthesized DNA.

temperature incubation (53). Revertants of this mutant are normal in ColEl DNA replication at 42° C and the relaxation complex isolated from these revertants exhibits a normal in vitro response to temperature incubation (53). Support for a role of relaxation complex in plasmid DNA replication also comes from the finding that the origin-terminus for ColEl replication is indistinguishable from the location of the single strand break induced in the ColEl DNA-protein relaxation complex (22,28). Finally, as shown in Fig. 6, the origin of DNA replication and the site of the nick in the open circular DNA form of the relaxation complex also are indistinguishable in the case of the R6K plasmid (27).

POSSIBLE EVENTS IN PLASMID DNA REPLICATION

On the basis of the various observations to date on the replication of ColEl DNA in E. coli, the following events are being considered to be involved in the replication of plasmids. (a) Formation of the relaxation complex between the covalently closed, circular plasmid DNA and protein. (b) Association of the DNA-protein complex with a membrane replicator site. (c) Activation of the relaxation complex by an initiation event followed by a strand- and site-specific nick of the covalently closed plasmid at or very near the origin of replication. At least in the case of ColEl plasmid, the nicking event is associated with the covalent attachment of a specific protein to the 5' terminus of the nicked strand and this covalent linkage is considered to facilitate the unwinding of the plasmid DNA and the closure of the nicked strand after the unwinding event. (d) Primer RNA is synthesized at a fixed site near the replication origin and unidirectional synthesis in the case of ColEl DNA and bidirectional synthesis in the cases of R6K DNA extends from the RNA primer with nicking and sealing of the replicative intermediate, catalyzed by the relaxation complex, taking place during the replication process. It is conceivable that synthesis of the RNA primer at the origin is the event that results in activation of relaxation complex and nicking at a unique site. (e) Primer RNA is removed prior to or immediately after final closure of the newly synthesized plasmid DNA strand. (f) Finally, the completed covalently closed daughter molecules of ColEl are released from the membrane replicator site. Further testing of this molecule presently is being carried out with the ColEl and R6K plasmids and various in vitro generated chimeric plasmid DNA molecules.

ACKNOWLEDGMENT

Studies summarized in this report were supported by U.S. Public Health Service Research Grant AI-07194 and National Science Foundation Grant GB-29492.

REFERENCES

1. Bazaral, M. and Helinski, D. R. J. Mol. Biol. 36, 185 (1968).
2. Clewell, D. B. and Helinski, D. R. Biochemistry 9, 4428 (1970).
3. Kline, B. C. and Helinski, D. R. Biochemistry 10, 4975 (1971).
4. Sharp, P. A., Hsu, M. T., Ohtsubo, E. and Davidson, N. J. Mol. Biol. 71, 471 (1972).
5. Collins, J. and Pritchard, R. H. J. Mol. Biol. 78, 143 (1973).
6. Nisioka, T., Mitani, M. and Clowes, R. C. J. Bacteriol. 103, 166 (1970).
7. Clewell, D. B. and Helinski, D. R. Biochem. Biophys. Res. Commun. 41, 150 (1970).
8. Vapnek, D., Lipman, M. B. and Rupp, W. D. J. Bacteriol. 108, 508 (1971).
9. Kontomichalou, P., Mitani, M. and Clowes, R. C. J. Bacteriol. 104, 34 (1970).
10. Kingsbury, D. T. and Helinski, D. R. J. Bacteriol. 114, 1116 (1973).
11. Collins, J., Williams, P. H. and Helinski, D. R. Molec. Gen. Genetics, in the press (1975).
12. Kingsbury, D. T. and Helinski, D. R. Biochem. and Biophys. Res. Commun. 41, 1538 (1970).
13. Goebel, W. Eur. J. Biochem. 15, 311 (1970).
14. Goebel, W. Nature New Biol. 237, 67 (1972).
15. Goebel, W. Biochem. Biophys. Res. Commun. 51, 1000 (1973).
16. Jacob, F., Brenner, S. and Cuzin, F. Cold Springs Harbor Symp. Quant. Biol. 28, 329 (1963).
17. Kingsbury, D. T. and Helinski, D. R. Genetics 74, 17 (1973).

18. Bazaral, M. and Helinski, D. R.   Biochemistry 9, 399 (1970).
19. Clewell, D. B. and Helinski, D. R.   J. Bacteriol. 110, 1135 (1972).
20. Clewell, D. B.   J. Bacteriol 110, 667 (1972).
21. Hershfield, V. H., Boyer, H. W., Yanofsky, C., Lovett, M. A. and Helinski, D. R.   Proc. Natl. Acad. Sci. U.S.A. 71, 3455 (1974).
22. Lovett, M. A., Guiney, D. and Helinski, D. R.   Proc. Natl. Acad. Sci. U.S.A. 71, 3854 (1974).
23. Jaenisch, R., Mayer, A. and Levine, A.   Nature New Biol. 233, 72 (1971).
24. Sebring, E. D., Kelly, T. J., Jr., Thoren, M. M. and Salzman, N. P.   J. Virol. 8, 478 (1971).
25. Katz, L. and Helinski, D. R.   J. Bacteriol. 119, 450 (1974).
26. Katz, L., Williams, P. H., Sato, S., Leavitt, R. and Helinski, D. R., manuscript in preparation.
27. Lovett, M. A., Sparks, R. and Helinski, D. R., manuscript submitted for publication.
28. Lovett, M. A., Katz, L. and Helinski, D. R.   Nature 251, 337 (1974).
29. Inselburg, J.   Proc. Natl. Acad. Sci. U.S.A. 71, 2256 (1974).
30. Tomizawa, J., Sakakibara, Y. and Kakefuda, T.   Proc. Natl. Acad. Sci. U.S.A. 71, 2260 (1974).
31. Kontomichalou, P., Mitani, M. and Clowes, R. C.   J. Bacteriol. 104, 34 (1970).
32. Kupersztoch-Portnoy, Y. M., Lovett, M. A. and Helinski, D. R.   Biochem. 13, 5484 (1975).
33. Blair, D. G., Sherratt, D. J., Clewell, D. B. and Helinski, D. R.   Proc. Natl. Acad. Sci. U.S.A. 69, 2518 (1972).
34. Williams, P. H., Boyer, H. W. and Helinski, D. R.   Proc. Natl. Acad. Sci. U.S.A. 70, 3744 (1973).
35. Clewell, D. B., Evanchik, B. G. and Cranson, J. W.   Nature New Biol. 237, 29 (1972).
36. Clewell, D. B. and Evanchik, B. G.   J. Mol. Biol. 75, 503 (1973).
37. Williams, P. H., Boyer, H. W., Leavitt, R. W. and Helinski, D. R., manuscript in preparation.
38. Leavitt, R. W. and Helinski, D. R., unpublished results.

39. Williams, P. H., Blair, D. G., Lovett, M. A. and Helinski, D. R., manuscript in preparation.
40. Clewell, D. B. and Helinski, D. R. Proc. Natl. Acad. Sci. U.S.A. 62, 1159 (1969).
41. Blair, D. G., Clewell, D. B., Sherratt, D. J. and Helinski, D. R. Proc. Natl. Acad. Sci. U.S.A. 68, 210 (1971).
42. Clewell, D. B. and Helinski, D. R. Biochem. Biophys. Res. Commun. 40, 608 (1970).
43. Kline, B. C. and Helinski, D. R. Biochemistry 10, 4975 (1971).
44. Kupersztoch-Portnoy, Y. M., unpublished observation.
45. Leavitt, R. W., unpublished observation.
46. Morris, D. F., Hershberger, C. L. and Rownd, R. J. Bacteriol. 114, 300 (1973).
47. Clewell, D. B. and Helinski, D. R. Biochem. Biophys. Res. Commun. 41, 150 (1970).
48. Kupersztoch-Portnoy, Y., Gabor-Miklos, G. L. and Helinski, D. R. J. Bacteriol. 120, 545 (1974).
49. Guiney, D., unpublished observation.
50. Lovett, M. A. and Helinski, D. R., manuscript submitted for publication.
51. Guiney, D. and Helinski, D. R., manuscript submitted for publication.
52. Blair, D. G. and Helinski, D. R., manuscript submitted for publication.
53. Collins, J. and Helinski, D. R., manuscript in preparation.

CONTROL OF RTF AND r-DETERMINANTS REPLICATION IN COMPOSITE R PLASMIDS

Robert H. Rownd, Daniel Perlman*, Nobuichi Goto[†] and
Edward R. Appelbaum
Laboratory of Molecular Biology and
Department of Biochemistry
University of Wisconsin
Madison, Wisconsin 53706

ABSTRACT. When Proteus mirabilis harboring the R plasmid
NR1 is cultured in drug-free medium, the R plasmid DNA
exists in the form of a basic composite structure consisting of a resistance transfer factor (RTF) component and
an r-determinants component. After growth of the cells in
medium containing appropriate drugs, poly-r-determinant
R plasmids consisting of a single copy of the RTF component
and multiple, tandem copies of the r-determinants component
and autonomous poly-r-determinants (not attached to an RTF)
are formed. This mechanism of gene amplification is referred to as a "transition". By reducing the rate of DNA
chain elongation, it is possible to increase substantially
the fraction of R plasmid DNA which is in the process of
replication. The analysis of replicating molecules of R
plasmid NR1 DNA by denaturation mapping has shown that the
basic composite structure has two origins of replication;
one is in the RTF component and the other is in the r-
determinants component. Replicating R plasmid molecules
isolated from the same culture of cells were found to have
initiated replication at either of these two origins of
replication. Possible factors which might influence the
selection of an origin for the initiation of replication
are discussed. Several lines of evidence have suggested

* Present address: Department of Biology,
Massachusetts Institute of Technology, Cambridge,
Massachusetts 02138
† Present address: Department of Microbiology, School
of Medicine, Tokyo Medical and Dental University, Tokyo
113, Japan

that subjecting P. mirabilis cells to conditions of
"physiological stress" in some way stimulates the replication of the r-determinants component in order to make
available the extra copies of r-determinants required for
the amplification of the drug resistance genes during the
transition. The entire r-determinants component appears to
be excised and amplified during the transition of the R
plasmid NR1, whereas it is possible to amplify only a segment of the r-determinants compoment of the R plasmid NR84
during a transition. Since each r-determinants has an
origin of replication, poly-r-determinant structures consisting of multiple, tandem sequences of this component
must have multiple origins of replication on the same
plasmid DNA molecule.

INTRODUCTION

Extrachromosomal genetic elements such as viral DNA,
bacterial plasmid DNA, and the DNA in eucaryotic organelles
offer many advantages for studying the structure and the
control of the replication of DNA. Owing to its relatively
small size, it is usually possible to isolate extrachromosomal DNA in either the non-replicating or the replicating
state without fragmentation of the DNA. This has provided
the opportunity of studying the structure of intact
replicating DNA molecules and to determine the origin and
the direction of replication using several different procedures. The initial studies on these problems were
carried out by Schnös and Inman (1) who showed that the
initiation of the replication of phage λ DNA occurred at a
unique site on the λ chromosome and that the majority of λ
DNA molecules were replicated in a bidirectional manner
during the first round of replication. About one-third of
the replicating λ DNA molecules, however, were replicated
in a unidirectional manner in which replication proceeded
in either direction from the same origin used for bidirectional replication. Subsequent studies on the origin
and the direction of replication of mitochondrial DNA (2),
phages P2 and 186 DNA (3,4), and colicin E1 DNA (5,6) have
also shown that replication is initiated at a unique site
and that replication is unidirectional. A unique origin of
replication has also been demonstrated for SV40 and polyoma
DNA and replication proceeds bidirectionally in both of
these animal viruses (7,8). In all of these cases, the

replicating molecules were double-branched circles [theta
(θ)-type structures], except for replicating P2 and 186 DNA
molecules which were single-branched circles [sigma (σ)-
type structures]. With replicating mitochondrial DNA and
SV40 and polyoma DNA, the parental strands remained as a covalently closed circular duplex and the newly synthesized
(daughter) strands were not covalently attached to the parental DNA (2,9,10,11). Recent evidence from several laboratories also indicates that an undetermined fraction of
the replicating molecules of several bacterial plasmids also
remain covalently closed circular during the process of
replication (5,6,12). Phage T7 DNA replicates bidirectionally from a unique internal origin of replication and replicating molecules are linear structures during the first
round of replication (13). In phages λ and T7 DNA replication, long concatemeric replicating molecules are observed
after the initial rounds of replication (14,15,16,17).

Drug resistance plasmids (R plasmids) in bacteria are of
particular interest in the study of the control of DNA replication since many R plasmids are composite structures consisting of two distinguishable components (18,19,20): a
resistance transfer factor (RTF) which mediates the transfer
of these plasmids during bacterial mating and an r-
determinants component which harbors the drug resistance
genes of the plasmid. Each of these elements is capable of
autonomous replication; that is, each is a "replicon" (18,
21,23,24,25,26). In most genera of the Enterobacteriaceae
these two components are united to form a composite structure (18,21,23,25). This is also true in Proteus mirabilis
when this host strain is cultured in drug-free medium (22,
25,26). However, when P. mirabilis is cultured in medium
containing any of the drugs to which the r-determinants
component confers resistance, R plasmids are formed which
consist of a single copy of the RTF component and multiple,
tandem copies of the r-determinants component (poly-r-
determinant R plasmids). Autonomous poly-r-determinants
consisting of multiple copies of the r-determinants component are also formed (19,22,23,24,25,26). Since both the
RTF component and the r-determinants component must contain
an origin at which replication is initiated (26), the replication of R plasmids in P. mirabilis provides a model
system for studying the replication of chromosomes containing multiple origins of replication.

TRANSITION OF COMPOSITE R PLASMID DNA

The dissociation and the reassociation of the RTF and r-determinants components of composite R plasmids to form poly-r-determinant structures during growth of R^+ P. <u>mirabilis</u> in medium containing appropriate drugs (referred to as a "transition") (22,23,24,25,26) is illustrated in Figure 1. In the case of the composite R plasmid NR1, the TC resistance genes reside on the RTF component (RTF-TC) (26,27,28) which has a density of 1.710 g/ml in a CsCl gradient and a molecular weight of 49 x 10^6. The chloramphenicol (CM), streptomycin/spectinomycin (SM/SP) and sulfonamide (SA) drug resistance genes reside on the r-determinants component (1.718 g/ml; 14 x 10^6). When P. <u>mirabilis</u> host cells are cultured in drug-free medium, the RTF-TC and r-determinants components are united to form a simple composite structure (1.712 g/ml; 63 x 10^6). In this state the R plasmid DNA appears as a single satellite band to the host chromosomal DNA (1.700 g/ml) in a CsCl density gradient (Fig. 2A). After prolonged growth in medium containing appropriate concentrations of any of the drugs to which the r-determinants component confers resistance, poly-r-determinant R plasmids consisting of a single copy of the RTF-TC component and multiple, tandem copies of the r-determinants component and autonomous poly-r-determinants consisting of multiple copies of the r-determinants component are formed (Fig. 1). The extra copies of r-determinants required for this mechanism of gene amplification appear to be provided by the dissociation of the composite R plasmid into the RTF-TC and r-determinants components and the subsequent autonomous replication of r-determinants. In this transitioned state, poly-r-determinant R plasmid DNA and autonomous poly-r-determinants DNA have a high molecular weight and are heterogeneous in size since these species contain multiple, tandem copies of r-determinants. The density of transitioned R plasmid DNA containing a large number of copies of r-determinants approaches 1.718 g/ml as a limiting value since most of the mass of the R plasmid DNA is due to r-determinants (Fig. 2B). The percentage of the R plasmid DNA in the transitioned state is considerably greater than the percentage of the 1.712 g/ml R plasmid DNA present in cells grown in drug-free medium since the incorporation of additional copies of r-determinants into R plasmids

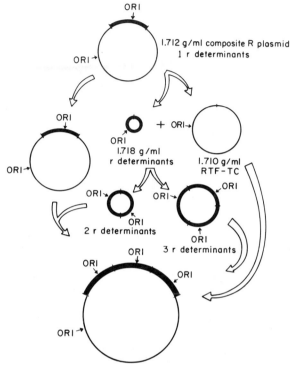

ORI

1.712 g/ml composite R plasmid
1 r determinants

ORI →

ORI

ORI →

ORI

ORI →
ORI
1.718 g/ml
r determinants

+ ORI→

1.710 g/ml
RTF-TC

ORI

ORI →   ORI →
ORI
2 r determinants

ORI

↑
ORI
3 r determinants

ORI   ORI

ORI →

ORI

ORI →

1.714 g/ml poly-r-determinant R plasmid
(3 r deteminants)

Fig. 1. Schematic diagram illustrating the dissociation and the reassociation of the RTF-TC and r-determinants
components of composite R plasmids in P. mirabilis and the
density increase that accompanies the incorporation of
multiple copies of the r-determinants component into
individual R plasmids to form poly-r-determinant R plasmids.
After the association of a large number of copies of the r-
determinants component with a single copy of the RTF-TC
component, the poly-r-determinant R plasmid DNA would have
essentially the same density as r-determinants DNA
(1.718 g/ml), since most of the mass of the DNA is due to
the r-determinants component. The origins of replication
in the RTF-TC component and the r-determinants component
are designated ORI.

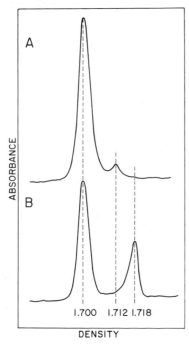

A

B

1.700   1.712 1.718

ABSORBANCE

DENSITY

Figure 2.

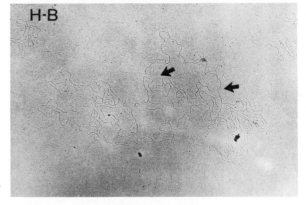

H-A

H-B

Figure 3.

Fig. 2. Density profiles of composite R plasmid <u>NR1</u>
DNA in non-transitioned and transitioned <u>P. mirabilis</u>.
(A) Density profile of DNA isolated from non-transitioned
cells which were cultured in drug-free Penassay broth.
(B) Density profile of DNA isolated from transitioned cells
which had been cultured in Penassay broth containing
100 µg/ml CM for 40 generations. A more detailed description of the methods used for cell culture, DNA isolation,
and centrifugation in an analytical CsCl density gradient
have been described previously (24,29).

Fig. 3. Two examples of replicating molecules of
composite R plasmid <u>NR1</u> DNA. The two arrows indicate the
two branch points (replication forks) of the replicating
molecules. These two replicating molecules were isolated
from R$^+$ <u>P. mirabilis</u> cells after addition of hydroxyurea to
the growth medium as described previously (31). One molecule has initiated replication at the origin in the RTF-TC
component (H-A) and the other at the origin in the r-
determinants component (H-B) of the composite R plasmid DNA.
Composite R plasmid <u>NR1</u> DNA molecules using either origin
of replication are replicated as a θ-type structure in
which both daughter branches are joined to the non-
replicated DNA at the two branch points.

increases their size. The skewing of the 1.718 g/ml R
plasmid DNA band to the less dense side (Fig. 2B) is due to
the presence of R plasmid DNA molecules having an intermediate density between 1.712 and 1.718 g/ml. These intermediate density molecules could correspond to poly-r-
determinant R plasmids harboring only a few tandem copies
of the r-determinants component (Fig. 1) or to fragments of
larger poly-r-determinant R plasmids which were produced by
breakage of the DNA during isolation and handling.

ISOLATION OF REPLICATING R PLASMID DNA

The small size of plasmid DNA is advantageous in
studying the structure of these molecules since it mini-
mizes the artifactual breakage of the DNA during isolation
and handling. However, it also introduces the problem that
only a small fraction of the plasmid DNA will be in the
process of replication at any instant in a random culture
of bacterial cells. This is because it only takes a small
fraction of the cell division cycle to duplicate a replicon
which is only a few percent (or less) of the size of the
bacterial chromosome (29,30). We have recently shown that
there is a substantial increase in the fraction of replicating R plasmid DNA in bacterial cells as a consequence of
decreasing the rate of DNA chain elongation by limiting the
concentration of DNA precursors (31). Substrate limitation
was achieved by growth of a thymine auxotroph of
P. mirabilis in medium containing limiting concentrations
of thymine or by the addition of hydroxyurea (an inhibitor
of ribonucleoside diphosphate reductase) to the growth
medium. Under these conditions, there is an increase in
the time required for a DNA replication fork to traverse
the plasmid DNA from the origin to the terminus of replication (S period). Since it requires a larger fraction of
the bacterial division cycle to replicate the plasmid DNA,
more plasmids will be in the process of replication during
substrate limitation. Bacterial mass synthesis is not
significantly affected during the first few hours
of substrate limitation (31,32). It appears that the rate
of initiation of plasmid replication continues at near the
normal rate during this period.

TWO ORIGINS OF REPLICATION IN COMPOSITE R PLASMID <u>NR1</u> DNA

Using either thymine limitation or hydroxyurea treatment,
we have isolated replicating composite <u>NR1</u> DNA molecules
from non-transitioned cells of <u>P. mirabilis</u> which were cultured in drug-free medium by fractionating the DNA in a CsCl
preparative gradient. The DNA was spread for electron
microscopy at a pH of 10.7 under the conditions described by
Inman and Schnös (33) so that approximately 20 small denaturation sites would be present in the R plasmid DNA. We
could thus compare the distribution of denaturation sites in
the replicating molecules with the denaturation map of composite <u>NR1</u> DNA which we had previously established using
non-replicating composite <u>NR1</u> DNA molecules (26,28) in order
to orient the replicating molecules in a unique manner.

As shown in Figure 3, composite <u>NR1</u> DNA molecules replicate as a θ-type (Cairns) circular structure in which both
daughter branches are joined to the non-replicated duplex
DNA at the two branch points (26,31). The position of
the two branch points on replicating <u>NR1</u> DNA molecules was
determined by comparing the distribution of denaturation
sites on the replicating molecules with the denaturation map
of composite <u>NR1</u> DNA. In Figure 4 the positions of the replicated regions of molecules in various stages of replication are superimposed on the <u>NR1</u> DNA denaturation map. For
convenience of presentation, the circular <u>NR1</u> DNA molecules
are represented by a linear denaturation map. In this presentation the r-determinants component is located at the
left side (0.0-0.23) of the map and the RTF-TC component is
located at the right side (0.23-1.0) of the map. Analysis
of the positions of the two branch points in replicating
molecules has revealed that there are two different origins
at which the replication of composite R plasmid DNA can be
initiated (26,34). One is located in the RTF-TC component
at a position of 0.80 on the denaturation map (Fig. 3A,
Fig. 4) and the other is located in the r-determinants component at a position of 0.14 on the denaturation map (Fig.
3B, Fig. 4). Replicating molecules isolated from the same
culture of cells were found to have initiated replication
at either of the two origins of replication. In the case of
either origin, replication can be either bidirectional or unidirectional in either sense. The replication branch points
were usually not symmetrically arranged about either origin of
replication. Replicating molecules have not been observed

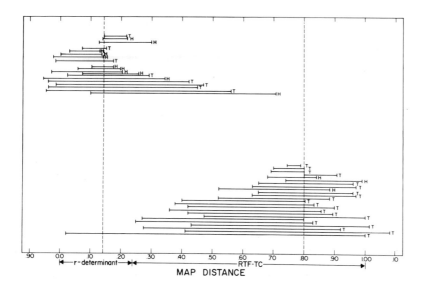

Fig. 4. Positions of the replicated regions of composite R plasmid NR1 DNA molecules in various stages of replication superimposed on the NR1 DNA denaturation map. The T or H at the right terminus of the replicated region of individual molecules designates whether thymine limitation (T) or hydroxyurea treatment (H) was used in the experiments to enrich for replicating molecules. Since the replicating molecules are circular structures, the denaturation map was extended 0.1 map unit to the left (i.e., from 0.0 to 0.90) and to the right [i.e., from 1.00 (same as 0.00) to 0.10] to avoid interruption of the continuity of the replicated regions of the individual molecules.

in which both origins were used simultaneously, i.e. with four branch points.

It is interesting to note that in many replicating molecules the branch point to the left of the r-determinants origin of replication and the branch point to the right of the RTF-TC origin of replication tend to cluster around the position 0.95-1.00 on the denaturation map. This suggests that this position could represent a terminus of replication. Although it is conceivable that this position could be a third origin of replication, this seems unlikely for several reasons. First, the positions of the two branch points of all of the replicating molecules which were examined are consistent with the use of either the r-determinants origin or the RTF-TC origin for the initiation of replication. Second, none of the molecules with a small extent of replication (less than 20 percent) were consistent with an origin of replication at 0.95-1.00. Third, if this position was an origin of replication, replication would have to be exclusively unidirectional for this origin, although unidirectional replication could proceed in either direction.

These investigations are presently being extended to include replicating poly-r-determinant molecules which have more than two origins of replication (Fig. 1).

POSSIBLE FACTORS INFLUENCING THE SELECTION OF AN ORIGIN OF REPLICATION FOR THE INITIATION OF REPLICATION

The presence of multiple origins of replication in a replicon raises a number of questions with regard to the involvement of the different replication systems in the control of its replication. While the factors involved in the control of the initiation of plasmid replication are not known in any detail at the present time, it is clear that there are distinct replication patterns which have been observed for different plasmids. The replication of the large, naturally occurring composite R plasmids is especially interesting since dual replication controls may make possible novel mechanisms of gene amplification. For example, in order to amplify the number of copies of the r-determinants component in P. mirabilis during a transition, the autonomous r-determinants component resulting from dissociation of the composite R plasmid apparently must be able to replicate, at least transiently, at a faster rate than the RTF-TC component. Thus, some mechanism must exist for "turning on" the replication of the r-determinants component during growth in medium containing appropriate drugs, so that extra copies of r-determinants would be available for the formation of poly-r-determinant structures.

The transition, however, is a reversible phenomenon. If transitioned cells are subsequently cultured in drug-free medium for a prolonged period, there is a gradual return to the non-transitioned state in which there is only a single copy of the r-determinants component which is attached to an RTF-TC. During this "back-transition", it appears that r-determinants are dissociated from poly-r-determinant structures and are subsequently diluted from the cells, presumably due to a decrease in the rate of their replication in the autonomous state (23,24,25,26,35). Thus, depending on the conditions used to culture the host cells, the autonomous r-determinants component appears to be able to replicate at a faster rate (growth in medium containing appropriate drugs) or a slower rate (growth in drug-free medium) than the RTF-TC component. Our available evidence suggests that there is no difference in the number of copies of the RTF-TC component per cell when the cells are cultured in either drug-free or drug-containing medium. The difference in the control of the replication of these two bacterial replicons which are normally united in the form of a composite structure in most enteric host strains (18,21,23,25) makes possible a novel mechanism of gene regulation at the level of DNA replication. Transitioned P. mirabilis cells have considerably higher levels of R plasmid gene products and of drug resistance than non-transitioned cells due to gene dosage effects (19, 23,35).

The factors involved in the control of the replication of the RTF-TC and r-determinants components are not known in any detail. In the majority of our experiments on the transition, the P. mirabilis cells were cultured in medium containing 100 µg/ml CM. Under these conditions non-transitioned cells can grow, albeit at a slower rate than in drug-free medium (35). In these experiments it was not possible to monitor the replication and structure of the

R plasmid DNA during the transition since ten to twenty
generations of bacterial growth were required to obtain the
transition and only a small fraction of the original cells
undergo the transition (25,35,36). This minor fraction of
the cells outgrows the others to produce the changes in the
R plasmid DNA density profile shown in Figure 2.

At CM concentrations which are high enough to prevent
the growth of non-transitioned cells (250 μg/ml), the
replication and structural changes in the R plasmid DNA
during a transition occur over a period of several hours
(25,36). Under these conditions the cells undergo a growth
lag of 20-30 hours (Fig. 5). During this period they inactivate the CM in the growth medium at a rate of approximately 50 μg/ml/hr/10^8 cells. When approximately 50 percent of the CM has been inactivated, growth of the culture
resumes and a 1.718 g/ml band is observed in the NR1 DNA
density profile. The proportion of this band increases
dramatically during the first few hours of outgrowth. In
Figure 5 the letters which identify the NR1 DNA density
profiles shown on the right correspond to the time at
which the cell samples were taken for DNA isolation at
various times on the growth curve shown on the left.
Electron microscopy and sedimentation in sucrose gradients
have shown that the 1.718 g/ml NR1 DNA in the profile shown
in Figure 5A consists of monomers, dimers, trimers, tetra-
mers, and higher multimeric forms of r-determinants DNA,
i.e. poly-r-determinant structures. At longer times after
outgrowth, the 1.718 g/ml DNA becomes so large in size that
it is impossible to isolate it without fragmentation.
Eventually, the 1.712 g/ml NR1 DNA band disappears
from the density profile. Experiments in which the original
1.712 g/ml NR1 DNA was radioactively labeled and then was
fractionated in a preparative CsCl density gradient indicate that the 1.712 g/ml NR1 DNA has recombined with the
r-determinants DNA to form poly-r-determinant R plasmids
which harbor repeated tandem sequences of r-determinants
(36).

Other experiments have shown that the presence of CM
in the growth medium at the time of outgrowth following the
20-30 lag period is not necessary to obtain the rapid
amplification of autonomous r-determinants and the formation of poly-r-determinant structures. The CM can be

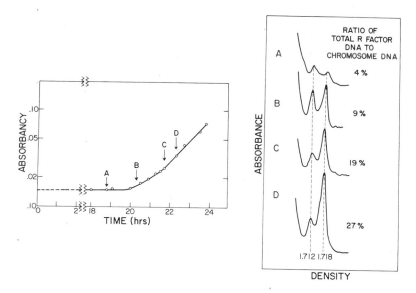

Fig. 5.   Forced transition of the composite R plasmid NR1 in P. mirabilis. A non-transitioned culture of R^+ P. mirabilis in late exponential phase (O.D.$_{650}$=0.8) in Penassay broth was diluted 100-fold into Penassay broth containing 250 µg/ml CM. Growth of the culture was monitored by absorbancy measurements at 650 nm. DNA was prepared from samples of the cells harvested at the times indicated on the growth curve which is shown at the left. The letters identifying the R plasmid DNA density profiles shown on the right correspond to the times at which the cells were harvested during the growth curve shown on the left. The P. mirabilis chromosomal DNA density profiles are not included in this figure in order to emphasize the R plasmid DNA distribution in the density gradient. The ratio of total R plasmid DNA to chromosome DNA for each sample is shown on the right.

removed just prior to the outgrowth period and the rapid amplification of the r-determinants DNA is still observed (36). Thus the cells are in some way preconditioned during the lag period to undergo a very rapid "forced transition" during the subsequent outgrowth period. Under the conditions of these experiments, almost all cells in the population are converted to the transitioned state within a few hours. These and other experiments have suggested that subjecting the cells to conditions of "physiological stress" in some way stimulates the replication of the r-determinants component. The nature of this physiological stress is presently unknown.

AMPLIFICATION OF ONLY A SEGMENT OF THE r-DETERMINANTS COMPONENT DURING THE TRANSITION OF THE R PLASMID NR84

In our experiments on the transition of the R plasmid NR1 in P. mirabilis, the density of the transitioned NR1 DNA has always been found to increase to the limiting value of 1.718 g/ml (the density of the r-determinants component) and transitioned cells have always been found to have an increased level of resistance to both CM and SM (resistance conferred by the r-determinants component) but not to TC (resistance conferred to the RTF-TC component). SA resistance was not examined in these experiments. These findings indicate that the entire r-determinants component is excised and amplified during the transition of the R plasmid NR1. Our recent experiments on the transition of the composite R plasmid NR84, on the other hand, have shown that it is possible to amplify only a segment of the r-determinants component of this R plasmid (37,38). NR84 is an F-like R plasmid which confers resistance to ampicillin (AP), in addition to CM, SM/SP, SA, and TC (39).

When P. mirabilis harboring NR84 is cultured in medium containing either CM, SM or SP, or AP there is an increase in the percentage of the R plasmid DNA relative to the host chromosome DNA, i.e. there is a transition. However, unlike the situation with NR1, the density of the transitioned NR84 DNA does not always increase to a value of 1.718 g/ml. In different experiments the density of the transitioned NR84 DNA band which has been observed has varied between values of 1.709 to 1.718 g/ml. Moreover, the transitioned cells do not always have an increased

550

level of resistance to all of the drugs to which the r-determinants component of NR84 confers resistance (CM, SM/SP, or AP; SA resistance was not examined in these experiments), but rather they have an increased resistance to only one drug or to restricted combinations of these drugs. In Figure 6A is shown the density profile of the DNA prepared from P. mirabilis cells harboring NR84 which were cultured in drug-free medium. The single satellite band of density 1.713 g/ml corresponds to the non-transitioned form of NR84 DNA which consists of a single copy of the r-determinants component joined to the RTF-TC component. In Figure 6B is shown the density profile of DNA prepared from P. mirabilis cells harboring only the RTF-TC component (1.710 g/ml) of NR84. The density profiles shown in Figure 6C-6J are the density profiles of the DNA prepared from transitioned cells of P. mirabilis harboring NR84. In Figure 6C-6J, the single drug listed on the top line of the two lines of drugs immediately to the right of the density profiles designates the single drug which was included in the culture medium in the various transition experiments. On the second line immediately below the single drug is listed the drug or combination of drugs to which the transitioned cells were found to have an increased level of resistance relative to the level of resistance of non-transitioned P. mirabilis harboring NR84[1]. For example, in the experiment shown in Figure 6C, the cells were cultured in medium containing CM and the transitioned cells were found to have an increased level of resistance to only CM. In the case of Figure 6D, the cells were cultured in medium containing AP and the transitioned cells had an increased level of resistance to AP, CM, and SM. SA resistance was not examined in these experiments. The transitioned NR84 DNA in the experiments shown in Figure 6E, 6F and 6G was found to be hetero-geneous in size and of molecular weight similar to transi-tioned NR1 DNA which had been isolated under the same

---

[1]The aminoglycoside antibiotics SM and SP are inacti-vated by the same enzyme, streptomycin adenylate synthe-tase, and either one or both of the two drugs have been used in different experiments to assay for an increased level of drug resistance of transitioned cells. When both SM resistance and SP resistance were examined in the same experiment, this is designated SM/SP↑ in Fig. 6.

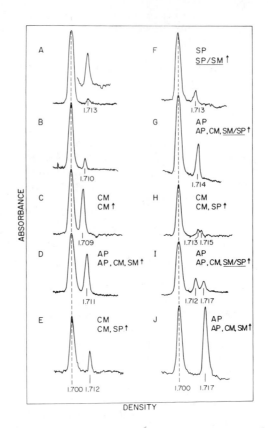

Fig. 6. Transition of the composite R plasmid <u>NR84</u> in P. <u>mirabilis</u>. Non-transitioned cells of <u>P. mirabilis</u> harboring NR84 were cultured in either drug-free Penassay broth (A) or Penassay broth containing the single drug listed on the top line of the two lines of drugs immediately to the right of the density profiles shown in C-J. The cells were usually cultured for approximately 135 generations in Penassay broth containing drugs at the following concentrations: CM (200 µg/ml), SM (200 µg/ml), SP (800 µg/ml), or AP (800 µg/ml). The level of resistance of each of these transitioned cultures to CM, SM, SP, or AP was then determined by plating appropriate dilutions of the cells on agar plates containing various concentrations of the individual drugs. The drug or combination of drugs to which the transitioned cells were found to have an increased level of resistance relative to non-transitioned P. <u>mirabilis</u> harboring <u>NR84</u> is listed on the second line immediately to the right of the density profiles. In B is shown the density profile of the DNA prepared from P. <u>mirabilis</u> harboring an RTF-TC segregant of <u>NR84</u> which had deleted the r-determinants component.

conditions. This indicates that poly-r-determinant structures are also formed in the case of transitioned NR84 DNA (37).

These experiments suggested that only a segment of the r-determinants component of NR84 is amplified during a transition and that the region of the r-determinants component which is amplified in different experiments is not always the same. This interpretation was verified by experiments in which non-transitioned and transitioned NR84 DNA was fractionated and treated with EcoRI restriction endonuclease (40) and the resulting DNA fragments were analyzed by electrophoresis in agarose gels. Amplification of only some of the EcoRI fragments of the r-determinants component DNA was observed in the transitioned NR84 DNA (38). In addition, an unexpected finding was that new EcoRI fragments which were not present in the gel patterns of the DNA from non-transitioned cells were observed in the gel patterns of the DNA prepared from transitioned cells. In different experiments these new fragments were of various sizes, ranging in molecular weight from values of approximately 0.8×10^6 to values exceeding 15×10^6 (the largest fragments observed may be heterogeneous in size).

The origin of these new fragments is not known at the present time, although several possibilities can be considered. For example, if the amplified segments of the r-determinants component do not contain a restriction endonuclease site, then the formation of repeated, tandem sequences of this amplified segment would generate an enlarged region which would also not contain a restriction endonuclease site, as illustrated in Figure 7B. Thus, new large fragments would be observed after treatment of the transitioned NR84 DNA with EcoRI restriction endonuclease. Similarly, if amplification and formation of repeated, tandem sequences brought two of the EcoRI restriction endonuclease sites in the r-determinants component into close proximity, treatment of this type of transitioned NR84 DNA would produce new small fragments as illustrated in Figure 7C. If it is assumed that the segment of the r-determinants component of R plasmid NR84 which is amplified during a transition must be replicated and, hence, contain an origin of replication, the repeated tandem

Fig. 7. Schematic diagrams illustrating the excision of segments of the r-determinants component of the composite R plasmid NR84 and the formation of repeated tandem sequences of these segments.

(A) Diagram of the r-determinants component of non-transitioned NR84 DNA. Only a part of the RTF-TC component on both sides of the r-determinants component is illustrated in order to simplify the diagram. The dashed vertical lines designate EcoRI restriction endonuclease sites in the r-determinants component. The small, numbered circles represent "excision sites" between which the entire r-determinants component (using sites 1 and 5) or only a segment of the r-determinants component (using appropriate internal sites) can be excised. The exact locations of the drug resistance genes (AP, SA, SP, CM) are not known with certainty, so that the designated locations are intended to be only schematic. ORI designates the location of an origin of replication in the r-determinants component of NR84 which is also intended to be schematic.

(B) Diagram of the r-determinants component of transitioned NR84 DNA having a large EcoRI fragment which was not present in the non-transitioned DNA. The r-determinants segment between excision sites 3 and 4 has been excised, amplified by replication, and reassociated in repeated, tandem sequences with the r-determinants component to form the structure illustrated in the diagram. The host cells would have an increased level of resistance (relative to non-transitioned NR84 cells) to only SP since only this drug resistance gene has been amplified in this transition. Since the segment between excision sites 3 and 4 does not contain an EcoRI site, the amplified segment would not be degraded by the nuclease, thus forming a "new large fragment" when the total transitioned NR84 DNA is treated with EcoRI. Assuming that the amplified segment must have an origin of replication (ORI) in order for amplification to occur, the transitioned NR84 DNA would have multiple origins of replication, as illustrated in this diagram.

(C) Diagram of the r-determinants component of transitioned NR84 DNA having an amplification in the relative amounts of three out of the seven r-determinants component EcoRI fragments and having a small EcoRI fragment which was not present in the non-transitioned NR84 DNA. The r-determinants segment between excision sites 2 and 5 would be excised, amplified by replication, and reassociated in repeated, tandem sequences with the r-determinants component to form the structure illustrated in this diagram. The host cells would have an increased level of resistance (relative to non-transitioned cells) to SA,SP and CM since all three of these drug resistance genes have been amplified in this transition. All of the EcoRI fragments between excision sites 2 and 5 would be amplified relative to the other R plasmid EcoRI fragments. In addition, a small fragment which was not present in the non-transitioned NR84 DNA would be present since the EcoRI site immediately to the right of excision site 2 and the Eco RI site immediately to the left of excision site 5 would be brought into close proximity as a result of the formation of repeated, tandem sequences of this segment of the r-determinants component. Assuming that the amplified segments must have an origin of replication (ORI) in order for amplification to occur, the transitioned NR84 DNA would have multiple origins of replication, as illustrated in this diagram.

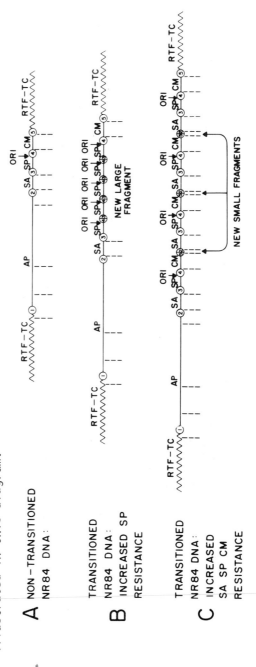

A NON-TRANSITIONED NR84 DNA:

B TRANSITIONED NR84 DNA: INCREASED SP RESISTANCE

C TRANSITIONED NR84 DNA: INCREASED SA SP CM RESISTANCE

NEW LARGE FRAGMENT

NEW SMALL FRAGMENTS

555

sequences of the segment of r-determinants DNA must contain a corresponding number of origins of replication. Thus, as indicated in Figure 7B and 7C, there would be an increase in the number of origins of replication in transitioned R plasmid NR84 DNA.

CONCLUDING REMARKS

These experiments have shown that composite R plasmids have a remarkable plasticity when present in P. mirabilis. These genetic elements are normally present in the form of a basic composite structure consisting of a single copy of the RTF-TC component and a single copy of the r-determinants component when the host cells are cultured in drug-free medium. However, composite R plasmids can dissociate in P. mirabilis and both the RTF-TC and r-determinants components are capable of autonomous replication in this host. During growth of the cells in medium containing appropriate drugs, the number of copies of either the entire r-determinants components or only a segment of this component can be increased and this amplified DNA can form multimeric structures consisting of repeated tandem sequences of the drug resistance genes. This novel mechanism of gene amplification results in a large increase in the level of resistance of the host cells to some or all of the drugs whose resistance is conferred by the r-determinants component.

Poly-r-determinant molecules must contain multiple origins of replication (Figs. 1 and 7), in addition to the two origins of replication normally present in the basic composite R plasmid DNA. Poly-r-determinant structures can thus be considered to be analogues of eucaryotic chromosomes and may serve as a useful model system for the analysis of the replication of chromosomes having multiple origins of replication. Our studies on replicating molecules of the basic composite NR1 DNA structure have shown that either an origin in the RTF-TC component or in the r-determinants component can be used for the initiation of replication of the composite structure. The use of both origins of replication was of course not mandatory since it is possible that the use of one origin for the initiation of replication would be dominant to use of the other origin. At the present time it is not known what

determines the selection of either of the two origins for
the initiation of composite R plasmid replication. It is
of course possible that the use of limiting thymine or
hydroxyurea to increase the fraction of replicating R
plasmid molecules in the cells might have influenced the
selection of one of the two origins for the initiation
of replication. This is a real possibility since
"physiological stress" appears to stimulate the replication
of the r-determinants component. Previous experiments in
this laboratory have indicated that one or more "initiator
protein(s)" are required for the initiation of R plasmid
replication (23,41,42). In future studies it will be of
interest to determine whether there are initiator proteins
which are specific for either of the two origins or whether
the proteins required for the initiation of NR1 replication
can function at either of the two origins. It will also
be of interest to examine whether the presence of additional r-determinants origins of replication in poly-r-
determinant R plasmids will influence the frequency of
selection of the r-determinants origin relative to the
RTF-TC origin for the initiation of replication. Analysis
of replicating poly-r-determinant R plasmids by denaturation mapping should also make it possible to determine
whether there is any specificity involved in the selection
of one of the multiple r-determinants origins for the
initiation of replication.

ACKNOWLEDGEMENTS

This work was supported (in part) by a U.S. Public
Health Service Research Grant (GM 14398) and by a Research
Career Development Award (GM 19206) to R.H.R. N.G. was
supported by a fellowship from the Bureau of Science and
Technology of the Japanese Government during the first year
of his two year stay in Madison. E.A. and D.P. were pre-
doctoral trainees of a U.S. Public Health Service Research
Training Grant.

REFERENCES

1. Schnös, M. and Inman, R.B. J. Mol. Biol. 51, 61
    (1970).
2. Kasamatsu, H. and Vinograd, J. Ann. Rev. Biochem.
    43, 695 (1974).

3. Schnös, M. and Inman, R.B.   J. Mol. Biol. <u>55</u>, 31 (1971).
4. Chattoraj, D.K. and Inman, R.B.   Proc. Nat. Acad. Sci. U.S.A. <u>70</u>, 1768 (1973).
5. Helinski, D., Lovett, M., Williams, P., Katz, L., Kupersztoch-Portnoy, Y., Guiney, D., and Blair, D.   In "Microbiology 1974", Proceedings of the American Society for Microbiology Conference on Extrachromosomal Elements in Bacteria (in press).
6. Lovett, M., Katz, L. and Helinski, D.   Nature <u>251</u>, 337 (1974).
7. Fareed, G.C., Garon, C.F. and Salzman, N.P.   J. Virol. <u>10</u>, 484 (1972).
8. Bourgaux, P. and Bourgaux-Ramoisy, D.   J. Mol. Biol. <u>62</u>, 513 (1971).
9. Robberson, D. and Clayton, D.   Proc. Nat. Acad. Sci. <u>69</u>, 3810 (1972).
10. Jaenisch, R., Mayer, A. and Levine, A.   Nature New Biol. <u>233</u>, 72 (1971).
11. Bourgaux, P. and Bourgaux-Ramoisy, D.   J. Mol. Biol. <u>70</u>, 399 (1972).
12. Scallon, S. and Rownd, R.   Unpublished experiments.
13. Wolfson, J., Dressler, D. and Magazin, M.   Proc. Nat. Acad. Sci. U.S.A. <u>69</u>, 499 (1972).
14. Carlson, K.   J. Virol. <u>2</u>, 1230 (1968).
15. Schlegel, R.A. and Thomas C.A. Jr.   J. Mol. Biol. <u>68</u>, 319 (1972).
16. Wake, R.G., Kaiser, A.D. and Inman, R.B.   J. Mol. Biol. <u>64</u>, 519 (1972).
17. Skalka, A., Poonian, M. and Bartl, P.   J. Mol. Biol. <u>64</u>, 541 (1972).
18. Clowes, R.   Bacteriol. Rev. <u>36</u>, 361 (1972).
19. Davies, J.E. and Rownd, R.   Science <u>176</u>, 758 (1972).
20. Helinski, D.R.   Ann. Rev. Microbiol. <u>27</u>, 437 (1973).
21. Cohen, S.N. and Miller, C.A.   J. Mol. Biol. <u>50</u>, 671 (1970).
22. Perlman, D. and Rownd, R. (Manuscript submitted for publication).
23. Rownd, R., Kasamatsu, H. and Mickel, S.   Ann. N.Y. Acad. Sci. <u>182</u>, 188 (1971).
24. Rownd, R. and Mickel, S.   Nature New Biol. <u>234</u>, 40 (1971).

25. Rownd, R., Perlman, D., Hashimoto, H., Mickel, S., Appelbaum, E. and Taylor, D. In Beers, R.F. Jr., Tilghman, R.C. (ed.) Cellular modification and genetic transformation by exogenous nucleic acids. Proceedings of the 6th Miles International Symposium on Molecular Biology, The Johns Hopkins University Press. p. 115-128 (1973).
26. Rownd, R.H., Perlman, D. and Goto, N. In "Microbiology 1974", Proceedings of the American Society for Microbiology Conference on Extrachromosomal Elements in Bacteria (in press).
27. Sharp. P., Cohen, S. and Davidson, N. J. Mol. Biol. 75, 235 (1973).
28. Perlman, D., Twose, T., Holland, M.J. and Rownd, R. (Manuscript submitted for publication).
29. Rownd, R. J. Mol. Biol. 44, 387 (1969).
30. Bazaral, M. and Helinski, D. Biochemistry 9, 399 (1970).
31. Perlman, D. and Rownd, R. Mol. Gen. Genet. (in press).
32. Barnes, M. and Rownd, R. J. Bacteriol. 111, 750 (1972).
33. Inman, R. and Schnös, M. J. Mol. Biol. 49, 93 (1970).
34. Perlman, D. and Rownd, R. (Manuscript submitted for publication).
35. Hashimoto, H. and Rownd, R. (Manuscript submitted for publication).
36. Perlman, D. and Rownd, R. (Manuscript submitted for publication).
37. Appelbaum, E., Goto, N. and Rownd, R. (Manuscript submitted for publication).
38. Goto, N. and Rownd, R. (Manuscript submitted for publication).
39. Nakaya, R. and Rownd, R. J. Bacteriol. 106, 773 (1971).
40. Hedgepeth, J. Goodman, H.M. and Boyer, H.W. Proc. Nat. Acad. Sci. 69, 3448 (1972).
41. Rownd, R. Fed. Proc. 30, 1313 (1971).
42. Morris, C., Hashimoto, H., Mickel, S. and Rownd, R. J. Bacteriol. 118, 855 (1974).

MECHANISM OF MITOCHONDRIAL DNA REPLICATION
IN MOUSE L-CELLS

Arnold J. Berk and David A. Clayton
Laboratory of Experimental Oncology
Department of Pathology
Stanford University School of Medicine
Stanford, California 94305

ABSTRACT. The replication of L-cell mitochondrial DNA
(mtDNA) proceeds by a mechanism termed displacement
replication in which synthesis of new daughter strands
is highly asynchronous. Approximately 0.6 of the heavy
strand (H-strand) is synthesized unidirectionally from a
unique origin before initiation of light strand (L-strand)
synthesis begins. Early replicative intermediates are θ
forms in which the parental H-strand is single-stranded
and is displaced by the synthesis of daughter H-strand on
the L-strand template. The parental strands of isolated
replicative intermediates are closed circular indicating
that replication proceeds with repeated nicking and
closing of the parental strands. Daughter molecules
segregate as open circular forms as demonstrated by their
buoyant density in ethidium bromide - cesium chloride
gradients. They are then converted to closed circular
molecules with a superhelix density of $\sigma \simeq -0.007$. This
low superhelix density form is first converted to another

closed circular intermediate with a higher absolute
superhelix density, $\sigma \simeq -0.03$, and then to the major
mtDNA form, D-loop mtDNA. The entire replication cycle
is accomplished in approximately 120 minutes. D-loop
mtDNA is the fully mature molecular form which contains
an approximately 450 nucleotide H-strand fragment termed
the 7S initiation sequence. The 7S initiation sequence
is an unstable species and turns over with a half-life
of 7.9 hours.

MtDNA populations isolated from mammalian cells
contain catenated oligomeric forms. It is known that the
formation of catenated dimers of HeLa cell mtDNA involves
breakage and closure of strands. In mouse L-cells,
catenated dimers of mtDNA are converted back to monomers
and the rates of catenane formation and separation are
very rapid.

INTRODUCTION

The replication of L-cell mtDNA proceeds by a
mechanism termed displacement replication (1,2) in which
synthesis of daughter strands is highly asynchronous.
The complementary strands of L-cell mtDNA are resolved in
alkaline CsCl equilibrium -- density gradients into a
heavy (H) strand and a light (L) strand (1,3).
Approximately 0.6 of the H-strand is synthesized unidirectionally before light strand synthesis is initiated
(1,2,4). This displacement synthesis has a unique origin
as demonstrated by EcoRI restriction endonuclease cleavage
of replicating mtDNA molecules (5,6). This origin maps at
the same site as the origin for D-loop synthesis (5,6,
Figure 1). Isolated early replication intermediates are

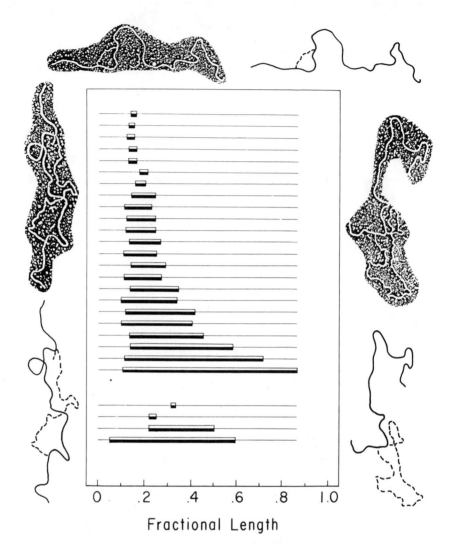

FIG. 1. Array of linear displacement replicative forms
derived by EcoRI cleavage of LMTK⁻mtDNA (5). The extent of
replication increases from top to bottom. Synthesis of a
displacement loop is initiated on the larger EcoRI fragment.
at 11.8 ± 1.2% of the circular genome length from the proximal restriction site (at the left) and proceeds to the right
throughout the length of this EcoRI fragment. The group of
four molecules in the lower part of this array have initiation sites which are greater than ±1 standard deviation for
measurements of the entire population. Micrographs and
accompanying line drawings (with single-strand segments as
dashed lines and duplex segments as solid lines) illustrate
linear replicative forms with synthesis of 16% (top), 74%
(left), and 96% (right) of the circular genome length.
Branch migration is evident at one fork of the linear replicative form shown at the left.

θ forms in which one loop is single-stranded parental H-strand, displaced by the synthesis of daughter H-strand (termed Exp-DmtDNA). A D-mtDNA and Exp-DmtDNA molecule are shown in Figures 2,3. The parental strands of isolated replication intermediates are closed circles indicating that replication proceeds with repeated nicking and closing of the parental strands (4). Daughter molecules segregate as open circles as judged by their bouyant density in the presence of ethidium bromide (EthBr). They are converted to closed circular molecules with a super-helix density of $\sigma \simeq -.007$ and are designated E-mtDNA (Figure 4) (1). E-mtDNA is then converted to the major in vivo form, D-mtDNA (Figure 4) (4). The D-mtDNA molecule has been arrested early in replication when H-strand synthesis has proceeded approximately 450 nucleotides (3). This unique 450 nucleotide H-strand daughter fragment is termed the 7S initiation sequence (1). Recent data indicate that the conversion of E-mtDNA to D-mtDNA proceeds through yet another closed circular intermediate with a superhelix density of $\sigma \simeq -.03$. Closed circular molecules of this topology have been observed previously and have been designated clean closed circles, or C-mtDNA (3), but were not implicated as a replicative intermediate. Flory and Vinograd (7) have shown that the formation of catenated oligomers of mtDNA involves breaking and closing of strands. Results presented here suggest that catenanes are converted back to monomers and that the rates of catenane formation and separation are rapid.

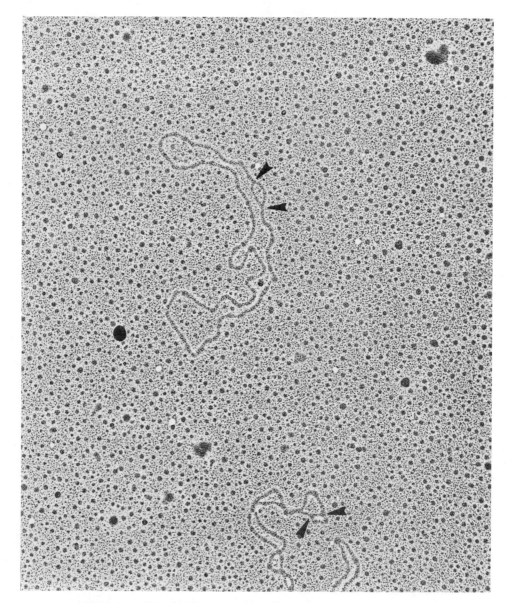

FIG. 2. Electron micrograph of a D-mtDNA molecule
isolated from mouse L cells. The contour length of the
molecule is ∿5 microns and the arrows delimit the D-loop
region.

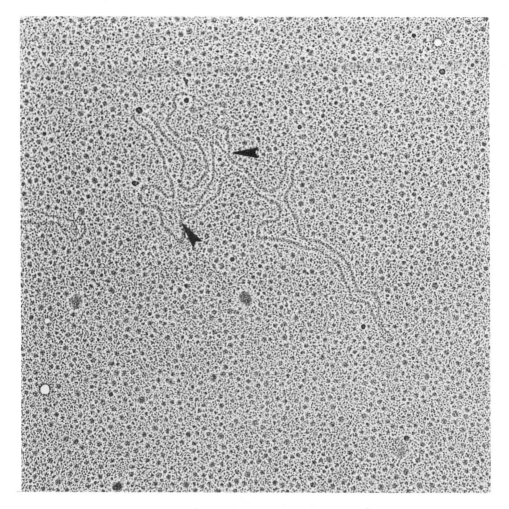

FIG. 3. Electron micrograph of an Exp-D(1)mtDNA
molecule isolated from mouse L cells. The arrows delimit
the replicated portion of the molecule.

METHODS

Growth of cells, in vivo labeling and isolation of mtDNA:

C2-1 cells, a clone of LMTK⁻ cells (1), were used in all
experiments. These cells fail to express the extramitochon-
drial thymidine kinase but do express an intramitochondrial
thymidine kinase and as a result incorporate exogenous
labeled thymidine predominantly into mtDNA (1,8). TMP
produced by endogenous de novo synthesis is also incorporated
into mtDNA in cells grown in media containing thymidine, so
that pretreatment of cells with inhibitors of de novo TMP
synthesis increases the specific activity of mtDNA labeled
with exogenous thymidine (1). Cells were grown in modified
Eagle's MEM containing standard concentrations of all
constituents except Na_2HPO_4 which was reduced from 10^{-2} M
to 10^{-4} M. 0.02 M Tricine was added and the media adjusted
to pH 7.4 with HCl, and made 10% in dialyzed calf serum.
Cells were seeded at 1 x 10^5 per ml in one liter of media
and one mCi of carrier-free ^{32}P phosphate was added. The
doubling time of cells in this media, 24 hours, was the same
in cultures with and without added ^{32}P. When cells reached
a concentration of 5 x 10^5 per ml, 5 µg/ml of 5-fluorodeoxy-
yuridine was added. After one hour of further incubation
cells were harvested at 37°C by centrifugation and resuspended
in 50 ml of the fresh, prewarmed media of the same composition.
Two mCi of [5-methyl-^3H]thymidine, 50-65 Ci/mmol, were added
and cells were incubated with constant stirring at 37°C for

10 minutes, harvested by centrifugation over a period of 3
minutes, and suspended in one liter prewarmed Eagle's MEM
made 10% in calf serum and 50 µM in unlabeled thymidine.
After 60 minutes of incubation in this chase medium, cells
were harvested by centrifugation at 4°C. Cells were
homogenized, mitochondria purified, DNA extracted and
centrifuged to equilibrium in EthBr-CsCl and analyzed by
sedimentation velocity as described previously (1).

RESULTS

Approximately 120 minutes are required for the complete replication of mtDNA in mouse L-cells; i.e., the time
elapsed between the initiation of expansion synthesis on
D-mtDNA molecules and the generation of two daughter molecules (1).

MtDNA was isolated from C2-1 cells pulse-labeled for
10 minutes and chased for 60 minutes in media containing a
great excess of unlabeled thymidine. Pulse-label banding
in the lower band region is present in both the 7S initiation sequence of D-mtDNA synthesized during the pulse
period, and the full length strands of daughter molecules
labeled during the pulse period and closed during the chase
period (1). This DNA exhibits heterogeneous sedimentation
properties with major species sedimenting at ∿27S and ∿37S (1).
Possible forms of mature, closed circular mtDNA found in C2-1
cells expected to band in the lower band of an EthBr-CsCl
equilibrium density gradient and sediment at 27S or 37S are
indicated in Table 1. Candidates for the 27S pulse-labeled
material are E-mtDNA and D-mtDNA, and for the 37S pulse-labeled
material, C-mtDNA and three possible forms of catenated dimer

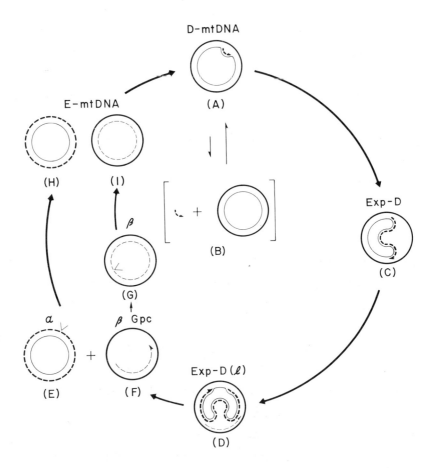

FIG. 4   Replication of mitochondrial DNA in mouse L
cells.   In mouse L cells growing exponentially in culture
the majority of the mtDNA not actively undergoing replication is in the form of a closed circular molecule which
contains a region in which the two circular strands do not
form the Watson-Crick helical structure, but instead one
strand (the "heavy" strand in alkaline cesium chloride,
designated here by thick lines) is displaced as a single
strand and the complementary light strand (symbolized by
thin lines) in this region forms a duplex with a small
linear piece of DNA about 450 nucleotides long.   This small
single-stranded piece of DNA sediments at 7S and is termed

the 7S fragment or 7S initiation sequence. The region of
the molecule containing this unusual structure is termed a
D-loop, and molecules containing D-loops are termed D-mtDNA.
D-mtDNA has a superhelix density in solution which is close
to zero, and thus has the sedimentation properties of nicked
circular mtDNA. It appears that a process by which the 7S
fragment is lost and resynthesized occurs in the mitochondrion.
If the 7S fragment is lost, either in the mitochondrion or
during DNA isolation, the closed circular molecule (B) attains
a significant negative superhelix density and has the altered
hydrodynamic properties characteristic of superhelical DNA.

The replication cycle begins by using the 7S fragment as
a primer for DNA synthesis, displacing the heavy parental
strand as a single strand. (Newly synthesized DNA is represented by dashed lines.) These replicating molecules are
referred to as expanding D-loop molecules (Exp-D) (c), and
can be isolated so that neither parental strand contains a
nick. This, plus the topological requirements of replication,
indicates that the molecule undergoes nicking-ligation events
as replication proceeds. When displacement replication has
proceeded at least 60% of the length of the molecule, synthesis of the other daughter strand, using the displaced heavy
strand as template, begins and proceeds unidirectionally back
toward the origin of replication. Because of the asynchronous
nature of this replication, the heavy daughter strand nears
completion before the light daughter strand. Separation of
the two daughter molecules occurs before completion of the
light strand synthesis, probably at a time near the completion
of heavy strand synthesis, forming the alpha daughter (E), a
circle with a nick or small gap in the heavy strand, and the
beta daughter, (F), a molecule with a large gap in the light
strand. This beta daughter then completes synthesis of the
light strand and, if not complete at separation, the heavy
strand of the alpha daughter is also then completed, leaving
both daughter molecules as simple nicked circles. These
circles are then closed, forming a species (H,I) called
E-mtDNA which concists of non-D-loop-containing circles with
superhelix densities close to zero. 7S fragments are then
synthesized on this E-mtDNA, again with at least one nicking-
ligation event at the end of synthesis, to form D-mtDNA.
The complete cycle is accomplished in about 120 minutes (1).

TABLE 1

SEDIMENTATION COEFFICIENT OF MAJOR STABLE FORMS OF L-CELL CLOSED CIRCULAR MtDNA AND POSSIBLE CATENATED DIMER FORMS

MOLECULE	NATIVE	S^a AFTER DENATURATION & RENATURATION	AFTER NICKING
D-mtDNA	27^b	37^b	26.2 ± 0.2
C-mtDNA	37.1 ± 0.2	$37\pm^c$	26.2 ± 0.2
E-mtDNA	27^b	27^b	26.2 ± 0.2
D-mtDNA-D-mtDNA	$\sim37^d$	51.5 ± 0.5^c	36.5 ± 0.6
D-mtDNA-E-mtDNA	$\sim37^d$	$\sim42^c$	36.5 ± 0.6
E-mtDNA-E-mtDNA	$\sim37^d$	$\sim37^d$	36.5 ± 0.6
E-mtDNA-C-mtDNA	$\sim42^d$	$\sim42^d$	36.5 ± 0.6
C-mtDNA-C-mtDNA	51.5 ± 0.5	51.5 ± 0.5	36.5 ± 0.6

(a) This is $S_{20,\omega}$ as determined by analytical centrifugation or S_{obs} as determined by sedimentation through sucrose gradients as indicated in the following footnotes.

(b) S_{obs} (3).

(c) $S_{20,\omega}$ (9).

(d) Approximate expected S based on conformation of monomer forms and $S_{20,\omega}$ determined for catenated dimers of open circular monomers (9).

TABLE 2

RATIO OF ^3H TO ^{32}P INCORPORATED INTO MtDNA

Molecular Species	Exp. 1	Exp. 2	Exp. 3
D-mtDNA	$0.118\pm.011$	$0.192\pm.004$	$0.132\pm.002$
C-mtDNA	$0.769\pm.019$	$0.719\pm.016$	$0.468\pm.013$
E-mtDNA	$1.952\pm.190$	$1.279\pm.023$	$1.793\pm.029$

molecules: catenated dimers containing two E-mtDNA molecules, two D-mtDNA molecules, one E-mtDNA molecule, and one D-mtDNA molecule.

To determine the molecular form of the pulse-chase label present in the closed circular molecules, the individual 27S and 37S fractions of the velocity sedimentation profile were pooled and thermally denatured under conditions which do not cause nicking of the closed circular strands, and then rapidly reannealed (1,3). After this procedure, E-mtDNA returns to its native conformation and continues to sediment at 27S (3). The 7S initiation sequence of D-mtDNA dissociates from the molecule and sediments at 7S. The full length circular strands of the D-mtDNA molecule are changed from a $\sigma \simeq 0$ to $\sigma \simeq 0.03$ and sediment at 37S (1,3).

An analysis of the total amount of pulse-label relative to mass label in D-mtDNA, C-mtDNA and E-mtDNA shows that the specific activity of these distinct molecular forms after the 60 minute chase is E-mtDNA > C-mtDNA > D-mtDNA (Table 2).

The same experimental protocol can be applied to catenated forms of mtDNA. In this instance the individual submolecules may be regarded as independent. The assignment of form is given by the S value of the catenated species before and after denaturation with subsequent renaturation (Table 1). The data indicate that the majority of catenated dimers exist in the form of D-mtDNA-D-mtDNA, to a lesser extent C-mtDNA-C-mtDNA and that <<5% of catenated dimers have the properties expected of E-mtDNA-E-mtDNA. This suggests that catenated dimers are not formed by aberrant segregation of daughter molecules.

571

To estimate the fraction of pulse-label present in
the closed circular strands of dimer and monomer mtDNA, an
aliquot of unfractionated lower band mtDNA (isolated after
a 10 minute pulse-label and 60 minute chase) was subjected
to sedimentation velocity under alkaline conditions. Under
these conditions catenated dimers sediment at 112S and
monomers at 80S (10). Since mtDNA is sensitive to incubation in alkali (11-13), the DNA was brought to 0.1 N NaOH
and immediately sedimented through alkaline sucrose for the
minimum time required to separate monomers and dimers (10
minutes at 60,000 revs/min in the SW60Ti rotor). The weight
% of dimeric mtDNA determined from this gradient, 22 is in
agreement with the weight % of dimeric mtDNA determined
by electron microscopic scoring of grids of unfractionated
lower band DNA from this experiment (21.7 ± 3.5). The ^{3}H/
^{32}P ratios observed are monomers, 0.445 ± 0.013; dimers,
0.261 ± 0.019. The significance of these ratios is discussed
below.

A mechanism for the generation of catenated oligomers
consistent with the available data involves generation of
linear intermediates of mtDNA at the time of segregation of
newly replicated daughter molecules. Such molecules should
contain a high specific activity of pulse-label after a 10
minute period of pulse and 30 minute chase. After this
period of chase, pulse-label banding in the upper band of
an EthBr-CsCl equilibrium density gradient, where linear
and open circular molecules will be found, reaches maximum
levels (1). Upper band DNA isolated from C2-1 cells pulse
labeled with [^{3}H]thymidine for 10 minutes and chased for 30

minutes was isolated, concentrated and subjected to sedimentation velocity centrifugation. No detectable peak of pulse-label was observed to sediment at 23S, the sedimentation coefficient óf linear DNA with a molecular weight equal to that of mtDNA (14).

DISCUSSION

The ratios of ^{3}H pulse-label to ^{32}P mass label in the major forms of closed circular mtDNA determined in three separate 10 minute pulse — 60 minute chase experiments are presented in Table 2. The relative specific activity of pulse-label after short periods of chase in molecules in which polymerization is complete will be determined by their sequence in the replication process. Assume the sequence of maturation of molecules in which strand elongation has been completed as $A \rightarrow B \rightarrow C$. The first labeled molecules converted to C will be those in which only a small number of nucleotides were added during the pulse period. Molecules in the B form will have been at an earlier point in the replication process during the time of the pulse of label and will therefore have incorporated more labeled nucleotides. Molecules in the A form similarly will contain more labeled nucleotides than those in the B form. A 60 minute chase period was chosen for this work, because after this interval, pulse-label is present in molecules with closed circular strands for no more than 30 minutes (1). In earlier work (1) it was shown that E-mtDNA is a precursor of D-mtDNA both because the specific activity of pulse-label in E-mtDNA was greater than that of D-mtDNA after 60 and 90 minutes of chase, and because chase of pulse-label from E-mtDNA into D-mtDNA

573

was observed. In the three experiments reported here the
relative specific activities of D-mtDNA, C-mtDNA and E-mtDNA
are in the order E-mtDNA > C-mtDNA > D-mtDNA. This indicates
that C-mtDNA is an intermediate in the conversion of E-mtDNA
to E-mtDNA, so that a more complete order of events occurring at the end of replication is: open circular newly
segregated daughter molecules → E-mtDNA → C-mtDNA → D-mtDNA.
C-mtDNA can exist in the classical superhelical conformation
or a conformation in which the Watson-Crick structure is
interrupted and the superhelical turns partially or completely
removed. Such a structure could result from binding
of a single-strand specific binding protein similar to the
T4 gene 32 product (15).

Complex forms of mtDNA including catenated oligomers
were first reported in 1967 (10,16). Two possible mechanisms
have been proposed to account for their formation:
(1) Failure of daughter molecules to properly segregate
at the end of replication. (2) A process involving double-
strand breaks in at least one molecule and reclosing of
strands (see Reference 17 for a review). Flory and Vinograd
(7) have found that 8 hours after initiation of BrdUrd
labeling of HeLa cell mtDNA, quarter-heavy molecular species
are observed in CsCl equilibrium density gradients. The
authors demonstrated that these molecules were predominantly
catenated dimers composed of one half-heavy molecule and one
fully-light molecule. Furthermore, no evidence of half-
heavy catenated dimers was observed. These observations are
compatible with mechanism "2" and incompatible with mechanism
"1". Our data are also inconsistent with mechanism "1". If
catenated dimers were formed by aberrant segregation, one

would expect to find a signficant fraction of pulse-label in catenated E-mtDNA molecules. Such a molecular species is not observed.

After 60 minutes of chase the specific activity of catenated dimer mtDNA is 58.5% of the specific activity of monomer mtDNA. After this period of chase, significant levels of pulse-label have been present in closed circular molecules for no more than 30 minutes (1). Therefore, the rapid mixing of label between monomers and catenated dimers indicates that these molecular forms are rapidly generated from monomers. A specific process for catenane formation is the generation of a linear intermediate at the time of segregation of daughter molecules. It is possible to imagine that a specific percentage of such intermediates would become involved in the formation of a catenane while cyclizing, simply by threading through a neighboring molecule. However, no pulse-label present in the upper band of an EthBr-CsCl equilibrium density gradient was observed sedimenting at 23S, the expected sedimentation coefficient of a linear molecule with the molecular weight of mtDNA. It is therefore unlikely that mtDNA catenanes are generated by such a process.

ACKNOWLEDGEMENT

This research was supported by grant number CA12312 from the National Cancer Institute and grant number NP9 from the American Cancer Society. One of us (A.J.B.) is a Medical Scientist Training Program Fellow supported by grant GM1922 of the National Institute of General Medical Sciences and the other (D.A.C.) is a Senior Dernham Fellow of the American Cancer Society, California Division (D-203).

REFERENCES

1. Berk, A.J. and Clayton, D.A. J. Mol. Biol.
    86, 801 (1974).

2. Robberson, D.L., Kasamatsu, H. and Vinograd, J.
    Proc. Nat. Acad. Sci. 69, 737 (1972).

3. Kasamatsu, H., Robberson, D.L. and Vinograd, J.
    Proc. Nat. Acad. Sci. 68, 2252 (1971).

4. Robberson, D.L. and Clayton, D.A. Proc. Nat. Acad. Sci.
    69, 3810 (1972).

5. Robberson, D.L., Clayton, D.A. and Morrow, J.F.
    Proc. Nat. Acad. Sci. 71, 4447 (1974).

6. Brown, W.M. and Vinograd, J. Proc. Nat. Acad. Sci.
    71, 4617 (1974).

7. Flory, P.J., Jr. and Vinograd, J. J. Mol. Biol. 74,
    81 (1973).

8. Clayton, D.A. and Teplitz, R.L. J. Cell Sci. 10,
    487 (1972).

9. Brown, I.H. and Vinograd, J. Biopolymers 10,
    2015 (1971).

10. Clayton, D.A. and Vinograd, J. Nature 216, 652 (1967).

11. Grossman, L.I., Watson, R. and Vinograd, J. Proc.
    Nat. Acad. Sci. 70, 3339 (1973).

12. Miyaki, M., Koide, K. and Ono, T. Biochem. Biophys.
    Res. Commun. 50, 252 (1973).

13. Wong-Staal, F., Mendelsohn, J. and Goulian, M.
    Biochem. Biophys. Res. Commun. 53, 140 (1973).

14. Studier, F.W. J. Mol. Biol. 11, 373 (1965).

15. Alberts, B.   In Mechanism and Regulation of DNA
    Replication (Kolber, A.R. and Kohiyama, M., eds.),
    pp. 133-148, Plenum Press, New York (1974).
16. Hudson, B. and Vinograd, J.   Nature $\underline{216}$, 647 (1967).
17. Clayton, D.A. and Smith, C.A.   Int. Rev. Exp. Path.
    $\underline{14}$, in press (1974).

REPLICATION OF MITOCHONDRIAL DNA
AND ITS INITIATION SITE

Katsuro Koike, Midori Kobayashi and Shigeaki Tanaka
Cancer Institute (Japanese Foundation for Cancer Research)
Kami-Ikebukuro Toshimaku, Tokyo Japan

ABSTRACT. Closed-circular replicating molecules of mtDNA
were labeled under the conditions of the in vitro system,
and these properties were examined by centrifugal analyses.
From the pulse-chase experiments, the nascent DNA was found
to be of two classes; one class consisted of the fragments
and the other of higher-molecular DNAs. The fragments were
synthesized first and then converted into the higher-molecular DNAs. The appearance of higher-molecular DNAs in
both the heavy and light strands was slightly asymmetrical;
an extension of the heavy strand seems to be preceding the
light one. From DNA-DNA hybridization experiments, the
fragments were found to anneal among themselves. The position of the initiation site was investigated by characterizing the products obtained by cleavage in vivo replicating
molecules with EcoRI restriction endonuclease (EcoRI).
Electron micrographical and agarose gel electrophoretic
analyses revealed that at least five double-strand breaks
were created at specific sites in each molecule. The longest linear structure (2.01 ± 0.07 μ) contained the replicative eye. The specific location for enzyme cleavage occurred at a fixed distance (0.45 ± 0.02 μ) from the initiation
site. Based on in vitro studies and cleavage analysis, a
mode of discontinuous replication of mtDNA is discussed.

INTRODUCTION

It is now well established that the bulk of the DNA in
mitochondria of animal cells is in the form of a closed
circular duplex with a molecular weight of about 10^7 daltons having negative superhelical turns (1,2). Electron
microscopic studies have so far revealed that the replicating structures of mtDNA were characterized as having the
"Cairns" form containing two forks, three branchs, and no
free ends (3). Replication appears to initiate in the
closed circular molecule (4), and is apparently uni-directional (5,6) and discontinuous (7,8). Furthermore, two
alternative modes of mtDNA replication have been reported

(5,7,8).

In order to elucidate the mechanism of mtDNA replication in detail, we have studied an in vitro system of DNA synthesis using isolated newborn rat liver mitochondria containing endogeneous mtDNA templates and enzymes, and also studied the in vivo replicating structure of mtDNA by characterizing the products obtained by EcoRI restriction endonuclease (EcoRI) digestion. In this report we describe our recent results concerning initiation and elongation of the DNA chain, mode of extension of the daughter strands, and position of the initiation site.

METHODS

Incubation of Mitochondria with DNA Precursors: As previously described, mitochondrial fraction was prepared from the new-born rat liver and incubated at $37^{\circ}C$ in the standard reaction mixture (9,10).

Isolation of DNA: Termination of the reaction and isolation of mtDNA have been described previously (9).

Band Velocity Sedimentation and Equilibrium Centrifugation: Band velocity sedimentations in CsCl-ethidium bromide (11) and in alkaline CsCl were described in the previous papers (9,10). Equilibrium centrifugations in CsCl-ethidium bromide and in alkaline CsCl were performed according to the procedures of Radloff et al (12) and Aloni and Attardi (13) respectively.

Electron Microscopy: Specimens were prepared by the formamide technique of Davis et al (14).

DNA-DNA Hybridization: For the detection of DNA-DNA hybrid the batch method of hydroxyapatite chromatography was used (15).

EcoRI Cleavage: Enzymatic restriction of mtDNA was performed by EcoRI under the similar reaction conditions of Yoshimori (16).

Dye-Agarose Gel Electrophoresis: The EcoRI restriction products were examined in the ethidium bromide-1.4% agarose gel according to the procedure described by Sharp et al(17).

RESULTS AND DISCUSSION

Mode of Synthesis of the Nascent mtDNA in the Closed-circle

Fraction: When mitochondria were incubated in the reaction medium, characteristics of the incorporation of [3H]thymidine into acid-insoluble material manifested a specific synthesis of mtDNA (9,10). The radioactive mtDNA labeled for an indicated period was isolated and subjected to the band sedimentation in CsCl-dye. When labeled for 10 min, a major part of the [3H]mtDNA was found in the closed-circle fraction. Continuous labeling for 45 min brought concomitant accumulations of the [3H]mtDNA in both closed and open circle fractions. When the chase incubation was carried out with non-radioactive thymidine added after 10 min labeling, there was a transfer of label from the closed-circle fraction to the open-circle fraction. It was also noted that a shoulder was associated with a dense side of the closed-circle fraction throughout the incubation. By the equilibrium centrifugation in CsCl-dye, majority of the [3H]DNA from the closed-circle fraction was recovered in the lower band and the remaining part was distributed between lower and upper bands. On the other hand, the [3H] DNA from the open-circle fraction was exclusively found in the upper band.

The [3H]mtDNA from the closed-circle fraction was, then, subjected to the band sedimentation in alkaline CsCl. As shown in Fig. 1, nascent DNA so far studied was found to be of two classes; one class consisted of the fragments, and the other of the higher-molecular DNAs (10). After 10 min of labeling (Fig.1a), the major part of the nascent DNA was the fragments. On chasing, the fragments were converted into the higher-molecular DNAs (Fig.1b), indicating that the fragment seems to be a direct structural constituent for the higher-molecular DNA of the closed-circular replicating molecule. In addition, most of the nascent DNA taken from the open-circle fraction sedimented in the higher-molecular DNA region.

As the sedimentation characteristic of the mtDNA duplex in alkaline CsCl density gradient is to separate into heavy(H) and light(L) strands due to the difference in guanine and thymidine content (18), and the H/L ratio of radioactive thymidine incorporated into two separated strands was reported to be 1.24-1.4 (13,19). The higher-molecular DNA fraction was collected and subjected to the equilibrium centrifugation in alkaline CsCl. As can be seen in Fig.2a, the distribution of these DNAs after 10 min of labeling appeared to be bimodal and slightly asymmetric (H/L=1.5). When labeled for 45 min, this asymmetry in the

580

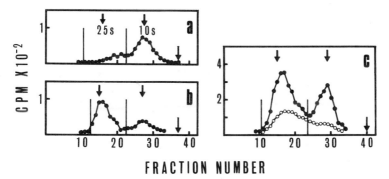

Fig. 1. Band velocity sedimentation of [3H]mtDNA of
the closed-circle fraction in alkaline CsCl.

After band velocity sedimentation in CsCl-dye, [3H] mt-
DNA collected from the closed-circle fraction was subjected
to the band velocity sedimentation in alkaline CsCl for 5h
at 38,000 rev./min. The radioactive profiles of [3H]mtDNA
are presented in (a) 10 min labeling, (b) 10 min labeling
followed by 35 min chasing in the presence of 1mM non-radio
active thymidine, (c) 45 min labeling. Arrows indicate the
positions of 25s and 10s, and the top of the gradient.

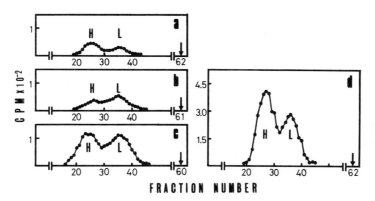

Fig. 2. Equilibrium centrifugation of [3H] higher-molecular DNAs in alkaline CsCl.

The higher-molecular DNA fraction was separated as
indicated by two vertical lines in Fig. 1 and was, then,
subjected to the equilibrium centrifugation in alkaline
CsCl for 70 h at 34,000 rev./min. The radioactive profiles
of the higher-molecular DNAs are presented in (a) from the
closed-circle fraction of 10 min labeling, (b) from the
open-circle fraction of 10 min labeling, (c) from the
closed-circle fraction of 10 min labeling followed by 35
min chasing and (d) from the closed-circle fraction of 45
min labeling. Arrows indicate the top of the gradient.

H/L ratio was also observed to be 1.5(Fig.2d). On the
other hand, Fig.2b shows that the higher-molecular DNAs
from the open-circle fraction were separated into two respective bands, however, the light one was more heavily la-
beled(H/L=0.6) probably resulted from an additional incorporation of [3H]thymidine into the light strand by gap-
filling synthesis (5). Furthermore, a substantial increase
of the higher-molecular DNAs was found in both L and H
bands after chase incubation(Fig.2c), however, the radioactivity rather symmetrically appeared in both L and H
bands(H/L=1.0). These observations indicate that the fragments are primarily synthesized on both template strands
and then converted into both light and heavy chains of the
higher-molecular DNA under appropriate mode, in which an
extension of the heavy chain seems to be preceding the
light one.

DNA-DNA Hybridization of the Fragments: Since the previous
results are indicative that the fragments consist of complementary copies to the parental strands, the fragments
were separated from the closed-circle fraction by the
alkaline CsCl band sedimentation. The hybridization reaction and the batch method of hydroxyapatite chromatography
were performed for the detection of DNA-DNA hybrid (15).
Experimental results are summarized as follows. When heated to 100°C for 10 min followed by rapid cooling, more than
95% of the labeled fragments were eluted at 0.12M phosphate
concentration as single-stranded DNA. After annealing of
20h, 72h and 90h at 67°C, 23%, 43% and 50% of the labeled
fragments were eluted at 0.4M phosphate concentration as
double-stranded DNA, respectively. Consistent with the
previous data obtained by Flavell et al(20), a "hairpin"
structure was hardly obtainable under the conditions used.
Present data likely indicate that a half of the fragments
at least are copies from complementary parts of the parental strands.

Electron Micrographical and Agarose Gel Electrophoretic
Analyses of EcoRI Restriction Segments of mtDNA: The in
vivo closed circular mtDNA, fractionated by the CsCl-dye
equilibrium centrifugation(12), was extensively cleaved
with EcoRI restriction endonuclease (16). The restriction
products were examined by electron microscopy. They were
photographed at random and the size was measured. After
EcoRI digestion, 100% of the circular molecules were con-

Table 1.   Length Analysis of EcoRI Segments

Segment	Number of Molecules Classified	Length in Microns
S1	85	2.01 ± 0.07
S2	84	1.28 ± 0.05
S3	120	0.95 ± 0.06
S4	84	0.60 ± 0.04
S5	99	0.29 ± 0.07
S6		
S1 + S2 + S3 + S4 + S5	472	5.13 ± 0.06
Circular Genome	38	5.16 ± 0.09

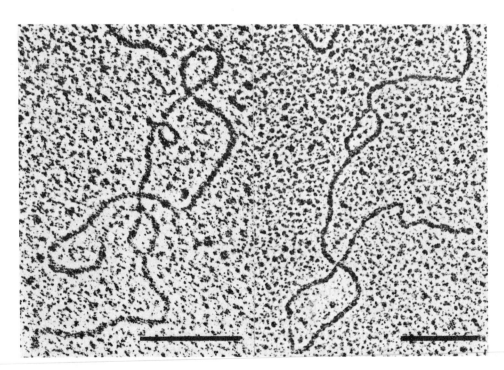

Fig. 3.   Electron micrographs of mtDNA S1 segment with a small eye after treatment with endonuclease EcoRI.
The linear replicating S1 segments generated by endonuclease EcoRI treatment of _in vivo_ closed circular mtDNA are shown.   The replicated region can be seen as a small eye.   The scale represents 0.2 μ.

verted to linear duplex forms. As shown in Table 1, length distribution for the populations of linear segments display at least five major components which have average length comprising $S1=2.01\pm0.07$ μ, $S2=1.28\pm0.05$ μ, $S3=0.95\pm0.06$ μ, $S4=0.60\pm0.04$ μ and $S5=0.29\pm0.07$ μ. The sum of these five segments is very slightly shorter than the contour length, 5.16 ± 0.09 μ, determined for rat mtDNA before EcoRI cleavage. In addition, short pieces, S6, were present with lengths of <400 nucleotides. These results indicate that EcoRI cleavage sites occur in at least five unique regions of the rat mtDNA genome. The restriction products were also examined by the dye-1.4% agarose gel electrophoresis (17). Electrophoresis pattern revealed that at least five discrete bands of mtDNA segments were detectable. This result is quite consistent with the electron micrographic data.

Position of the Replicative Eye: EcoRI cleavage of rat mtDNA provided for a basis for examination of the initiation site of DNA replication. When the in vivo closed circular mtDNA was digested with EcoRI, Fig.3 shows that the small replicative eye was found exclusively on the largest linear structure(2.01 ± 0.07 μ). Before enzyme digestion, the closed circular mtDNA contained $21.5\pm3.2\%$ of D-loop molecules among the population. After cleavage with EcoRI the frequency of resulting linear duplex molecule containing a replicative eye became substantially lower than the initial frequency. This suggests that the replicative eye in the linear form of replicating mtDNA is probably lost through the process of rewinding or branch migration during the EcoRI digestion. Alignment of these forms indicates that initiation of replication occurs on the largest linear segment of mtDNA at a position on the genome 0.45 ± 0.02 μ (1340 ± 70 nucleotide pairs) from the proximal restriction site. Discontinuous replication of rat mtDNA may proceed unidirectionally away from this initiation site on the largest EcoRI segment.

REFERENCES

1. Nass, M.M.K. Science 165, 25 (1969).
2. Borst, P. and Kroon, A.K. in Int. Rev. Cytol. Vol.26, 107 (1969) (Academic Press, New York).
3. Kirshner, R.H., Wolstenholme, D.R. and Gross, N.J. Proc. Natl. Acad. Sci. USA. 60, 1466 (1968)
4. Kasamatsu, H., Robberson, D. L. and Vinograd, J. Proc.

Natl. Acad. Sci. USA 68, 2252 (1971).

5. Robberson, D.L., Kasamatsu, H. and Vinograd, J. Proc. Natl. Acad. Sci. USA 69, 1737 (1972).
6. Kasamatsu, H. and Vinograd, J. Nature New Biol. 241, 103 (1973).
7. Koike, K. and Wolstenholme, D.R. J. Cell Biol. 61, 14 (1974).
8. Wolstenholme, D.R., Koike, K. and Cochran-Fouts, P. Cold Spring Harb. Symp. Quant. Biol. 38, 267 (1973).
9. Koike, K. and Kobayashi, M. Biochim. Biophys. Acta 324 452 (1973).
10. Koike, K., Kobayashi, M. and Fujisawa, T. Biochim. Biophys. Acta 361, 144 (1974).
11. Smith, C.A., Jordan, J.M. and Vinograd, J. J. Mol. Biol. 59, 255 (1971).
12. Radloff, R., Bauer, W. and Vinograd, J. Proc. Natl. Acad. Sci.USA 57, 1514 (1967).
13. Aloni, Y. and Attardi, G. J. Mol. Biol. 55, 251 (1971)
14. Davis, R.W., Simon, M. and Davidson, N. in Methods in Enzymology Vol.XXI, pp.413 (1971)
15. Fujinaga, K., Sekikawa, K. and Yamazaki, H. J. Virol. in press (1975).
16. Yoshimori, R N. PhD. Thesis, Univ. of Calif. at San Francisco Medical Center (1971).
17. Sharp, P.A., Sugden, B. and Sambrook, J. Biochemistry 12, 3055 (1973).
18. Vinograd, J., Morris, J., Davidson, N. and Dove, W.F. Jr. Proc. Natl. Acad. Sci. USA 49, 13 (1963).
19. Tibbetts, C.J.B. and Vinograd, J. J.Biol. Chem. 248, 3380 (1973).
20. Flavell, R.A., Borst,P. and Ter Schegget,J. Biochim. Biophys. Acta 272, 341 (1972).

THE REPLICATION OF mtDNA IN Tetrahymena

Piet Borst, Rob W.Goldbach, Annika C.Arnberg*
and Ernst F.J. Van Bruggen*

Section for Medical Enzymology and Molecular
Biology, Laboratory of Biochemistry, University
of Amsterdam, Eerste Constantijn Huygensstraat 20,
Amsterdam (The Netherlands) and *Biochemical
Laboratory, The University, Zernikelaan,
Groningen (The Netherlands)

ABSTRACT. The mtDNA of Tetrahymena pyriformis,
strain ST, is a linear duplex with a molecular
weight of 3×10^7 and a unique gene sequence.
Electron microscopy of replicating forms of this
DNA, identified by pulse-labelling, shows these
to be enriched in branched DNA molecules, all
linear and of the same end to end length as unbranched mtDNA. The major type of branched DNA
was a duplex with an internal "eye" of variable
size; "eyes" were largely duplex but often contained short single-stranded sections in one or
both forks.
    The data fit a replication model in which DNA
synthesis starts near the middle and proceeds bidirectionally to the ends via intermediates with
a largely duplex character. This replication mode
should yield linear DNA with one 3'-single-
stranded end, as pointed out by Watson. By electron microscopy of denatured mtDNA we have found
two structures that could be involved in converting this single-stranded end into a duplex:
Strain ST mtDNA contains a long (1.3 μm), imperfect terminal duplication-inversion; strain
GL mtDNA (which is very different from strain ST
mtDNA) contains a sub-terminal duplication-
inversion and a long (5 μm) palindrome on one
end. Possible models for synthesizing the ends
of linear DNA molecules are discussed.

586

Introduction

Our interest in the replication of mtDNA from Tetrahymena pyriformis has three sources:

1. This DNA is the only linear mtDNA known in nature (1) and all attempts to isolate it in a circular form have failed (see ref. 2). An analysis of replicative intermediates might provide us with a clue whether this linearity is an artefact or not.

2. In studying the structure of this DNA by electron microscopy we found displacement loops (D-loops) that differed from those in animal mitochondria in two respects: some monomeric molecules had more than one (up to 5) D-loops per molecule and these D-loops had no fixed position in relation to the molecular ends (3). This seemed difficult to fit into any conventional replication model.

3. In electron micrographs we found two molecules with a small, largely duplex "eye" in the middle (3). These molecules did not fit the highly asymmetric replication model, then in vogue for the replication of animal mtDNA (4).

On this basis we expected that an analysis of the replication mode of Tetrahymena mtDNA might not just provide us with a replica of results already obtained in the animal system. Fortunately, this was so. By a combination of biochemical experiments and electron microscopy we found (2,5,6) that:

a) All replicative intermediates were linear and of unit length.

b) The predominant replicative intermediate contained a replication bubble or "eye", which was mainly duplex in character.

c) All molecules with an eye could be arranged in a sequence in which replication starts within 0.05 fractional map unit from the middle and proceeds bi-directionally to the ends, both forks travelling approximately at equal speed (Fig. 1).

This replication mechanism differs from that currently accepted for animal mtDNA (4,7): largely single-stranded intermediates are absent

and replication is bi-directional rather than
uni-directional. These differences are not very
fundamental though and they mainly underline the
general philosophy emerging from the study of
DNA replication in viruses: the overall princi-
ples of the replication mechanism are similar,
but nature leaves room for originality when it
comes to details.

The model in Fig. 1 leaves us with two
questions:

1. Is <u>Tetrahymena</u> mtDNA really linear <u>in situ</u>
and, if so, how are the 5'-ends of the new
strands made? If DNA synthesis runs from 5' to
3' and if it requires an RNA primer, there is a
problem in filling in the last nucleotides at
the 5'-end of the new strand, as first pointed
out by Watson (8).

2. Where do the multiple D-loops fit in?

In this paper we will briefly summarize our
(meagre) progress in trying to answer these
questions.

The role of D-loops, if any

No progress has been made on this point,
because the D-loops have been elusive in recent
experiments. Attempts to pulse-label the short
displacing strand have been unsuccessful (2) and
we have not even seen D-loops in molecules recently spread for electron microscopy. Presumably they are lost, because of the metastable
character of D-loops in linear DNA.

In view of the fact that the D-loops previously observed were randomly spread over the DNA,
they cannot be linked in any obvious way to the
replication scheme in Fig. 1. Nevertheless, the
fact that these D-loops are uniform in size (3),
a size that happens to be identical with that of
D-loops in animal mtDNA (9), strongly suggests
that they are somehow related to replication. We
favour the idea that these D-loops represent prematurely synthesized Okazaki fragments (<u>cf</u>. ref.
10) in a discontinuous replication mode. Premature activation could be due to the rather un-
physiological conditions to which the mitochon-

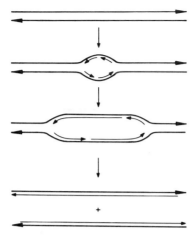

Fig. 1. Model for the replication of <u>Tetra-hymena</u> mtDNA (strain ST). The arrows indicate 3'-ends.

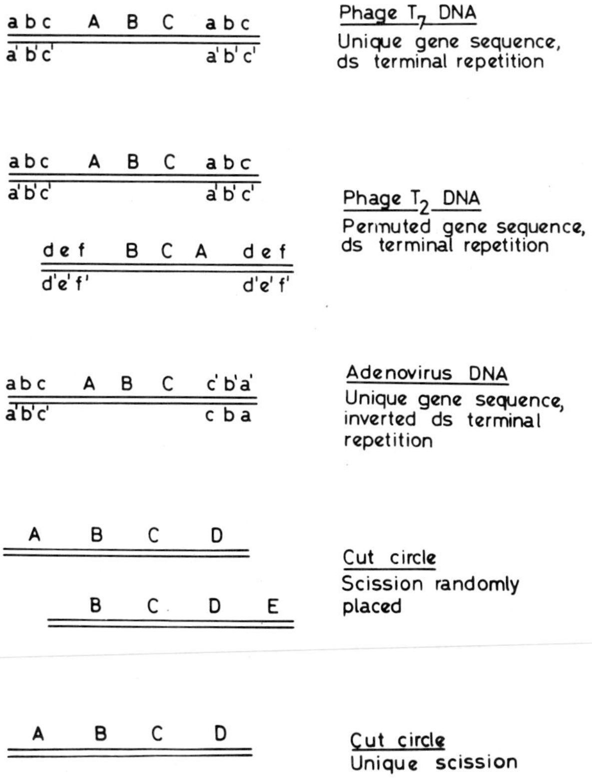

a b c A B C a b c		<u>Phage T$_7$ DNA</u>	
a' b'c' a' b' c'		Unique gene sequence, ds terminal repetition	

a b c A B C a b c			
a'b'c' a' b' c'		<u>Phage T$_2$ DNA</u>	
d e f B C A d e f		Permuted gene sequence, ds terminal repetition	
d'e' f' d'e' f'			

a b c A B C c' b'a'		<u>Adenovirus DNA</u>	
a'b'c' c b a		Unique gene sequence, inverted ds terminal repetition	

A B C D		<u>Cut circle</u>	
B C. D E		Scission randomly placed	

A B C D		<u>Cut circle</u>	
		<u>Unique scission</u>	

Fig. 2. Examples of linear genomes in nature. Modified after ref. 11.

dria are exposed during cell collection and mitochondrial isolation. Potentially this system remains interesting because it could throw some
light on the control of the initiation of
Okazaki fragment synthesis, once we learn how to
handle the system.

Copying of the ends of Tetrahymena mtDNA

As illustrated in Fig. 2, all authentic
linear DNAs found in nature have peculiar ends,
whereas broken circles should not have them. We
therefore, analysed whether Tetrahymena mtDNA
has a terminal repetition and/or a permuted gene
sequence. The former was found to be absent, but
standard denaturation-renaturation experiments
yielded a substantial fraction of full-length
duplex circles, at first suggesting that the
gene sequence of our isolated mtDNA was permuted
(12). Two facts, however, argued against this
conclusion:

1. None of the duplex circles formed after a
denaturation-renaturation cycle was perfect (i.e.
unambiguously traceable over all of its length)
and some were positively perverse, i.e. circles
with a duplex tail (12).

2. A partial denaturation map of the DNA --
although uninformative because of the even spread
of small denaturation bubbles over the length of
the DNA -- clearly showed that the ends of the
molecule were relatively under-denatured (13).
This excluded that this mtDNA as isolated is
randomly permuted.

This puzzling situation was resolved when we
tried to prove that the gene sequence was partially permuted by mapping the position of the
rRNA cistrons on this DNA (cf. ref. 14). For
this experiment we spread denatured DNA and immediately observed (12) that it contained a large
fraction of single-stranded circles held together by a duplex "pan handle" (Fig. 3). In
experiments with the mtDNA from strain ST only
up to 50% of all single-stranded DNA circularized (12). In more recent experiments with
strain T mtDNA, which is indistinguishable from

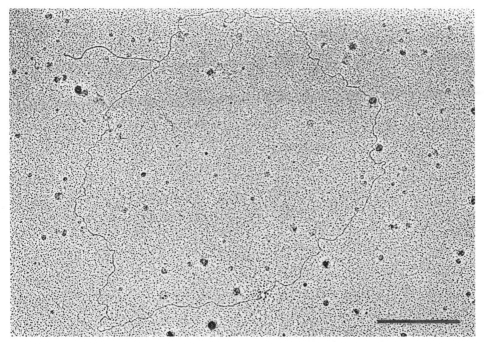

Fig. 3. Electron micrograph of denatured
mtDNA from <u>Tetrahymena</u> (strain T), spread from
formamide in a protein monolayer. The bar is 1
µm. See ref. 12 for details.

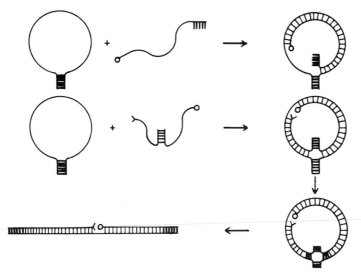

Fig. 4. Speculative scheme explaining the
formation of imperfect circles when denatured
<u>Tetrahymena</u> mtDNA (strain ST) is renatured.
From ref. 12.

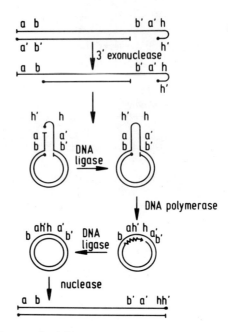

Fig. 5. Speculative scheme for the final stages of the replication of Tetrahymena mtDNA (strains ST and T). The long duplication-inversion is indicated by a b, a'b'; hh' is a short duplication-inversion, assumed to exist on one terminus, but not visible by electron microscopy. 3'-Ends are indicated by dots, 5'-ends by bars. The terminal sections are drawn too long in relation to the remainder of the molecule. Note that there is as yet no experimental evidence for the existence in mitochondria of the nucleases postulated in this scheme, nor for the additional duplication-inversion hh'.

ST mtDNA in its fragmentation by restriction endonucleases $Hind_{II+III}$ and $EcoR_I$, more than 90% circularization was observed. This proves that these <u>Tetrahymena</u> mtDNAs contain a terminal duplication-inversion, as first described in adenovirus DNA (15,16).

At first sight this finding appears to account for most of our problems: the terminal duplication-inversion readily explains why a denaturation-renaturation cycle leads to duplex pseudo-circles, even though the gene sequence of <u>Tetrahymena</u> mtDNA is unique rather than permuted (Fig. 4). The high GC content of the ends (deduced from the denaturation map) could slow down the branch migration leading to linear molecules. The duplication-inversion also provides a possible mechanism to complete the 5'-ends of newly synthesized strands, as illustrated in the highly speculative scheme in Fig. 5. Finally, the terminal duplication-inversion brings <u>Tetrahymena</u> mtDNA in line with other authentic linear DNAs in nature (Fig. 1) and, therefore, increases the probability that this linearity is not an isolation artefact.

Still one snag remained. Why is this duplication-inversion so long (1.3 µm = about 3000 base pairs = 14% of the total DNA, see ref. 12)? This is much longer than the approximately 100 base pairs found in the terminal duplication-inversion of adenovirus DNA (15,16) and adeno-associated virus DNA (17,18) and excessive in relation to its postulated circularization function for which the 12 base pairs of phage lambda DNA are already sufficient (19). Before going into the DNA sequence analysis of the ends to verify the predictions of Fig. 5, we therefore first looked at the mtDNAs from a few other <u>Tetrahymena</u> strains. The rationale behind this approach is that <u>Tetrahymena pyriformis</u> is a rather heterogeneous species, that includes subspecies or syngens that do not cross-breed, whereas the nuclear DNAs of some strains may cross-hybridize for less than 50% under standard conditions of DNA-DNA hybridization (20). Since

the evolution of mtDNA sequences is faster than
of comparable nuclear DNA sequences, at least in
yeast (21), one might hope to collect a number
of very different mtDNAs by screening sufficient
Tetrahymena strains. If all these mtDNAs share a
terminal duplication-inversion, this would support the idea that this duplication-inversion is
essential rather than incidental.

The complex ends of GL mtDNA

A Tetrahymena mtDNA, that we have found to be
very different from those of strains ST and T,
was isolated from a strain GL*. It sediments
ahead of ST mtDNA through neutral and alkaline
sucrose gradients, corresponding to a 20% difference in molecular weight. The fragmentation patterns of ST and GL mtDNAs obtained with restriction endonuclease $Hind_{II+III}$ or $EcoR_I$ share no
fragments of the same molecular weight. Preliminary DNA-DNA hybridization experiments give only
about 30% cross-hybridization at T_m - 25°C.

The analysis of the termini of GL mtDNA is
not yet complete, but some preliminary results
are summarized in Figs 6-10. Although denatured
GL mtDNA also contains single-stranded molecules
with a duplex pan handle, this pan handle appears
split at the end (Fig. 6). Moreover, two other
molecular species are present that are not found
in ST mtDNA: a linear single-stranded molecule
with a duplex end (hairpins, see Fig. 7), which
is the predominant species on our grids, and a
circle with a duplex portion, punctuated with a
single-stranded tail (Fig. 8). As a further complication some of the molecules are able to cir-

---

* The strain designation of Tetrahymena is a
  mess, probably because in the course of de-
  cades of laboratory handling the classical,
  amicronucleate strains have been repeatedly
  mis-labelled and mixed up (see ref. 22). The
  result is that GL strains in different laboratories may be very different, whereas some GL
  strains may well be identical to T.

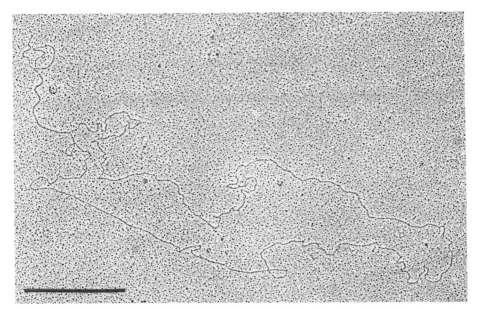

Fig. 6. A "pan handle" in denatured GL
mtDNA. The bar is 1 μm. See legend to Fig. 3
for details.

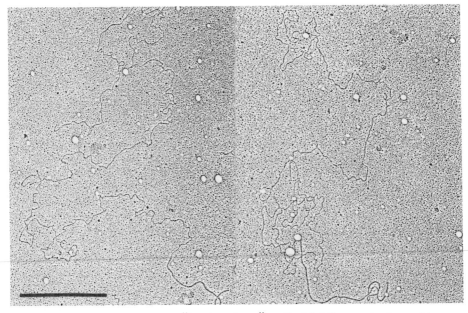

Fig. 7. Two "hairpins" of different lengths
in denatured GL mtDNA. The bar is 1 μm. See
legend to Fig. 3 for details.

cularize after a partial digestion of the native
duplex molecules with exonuclease III from Escherichia coli (Fig. 9).

An embarrassing feature of all these molecules
is the variable length of the duplex part. For
instance, the duplex part of the hairpin molecules varies from 0.5 to 5.0 μm. This variability
could be explained by a limited terminal degradation, as shown in Fig. 10, which presents our
present picture of this DNA, as isolated.

The sub-terminal position of the duplication-
inversion in GL mtDNA does not support the model
in Fig. 5 and alternative proposals for the synthesis of 5'-ends of DNA molecules should, therefore, be considered. One possibility is to leave
the putative terminal 5'-RNA primer in place.
This is not unreasonable: if the RNA primer is
removed by a DNA polymerase, acting as exonuclease (cf. ref. 10), the 5'-terminal RNA primer
will not be touched. The RNA end should be able
to act as template in the next replication round,
because DNA polymerases can use RNA templates
(cf. ref. 10). A disadvantage of this mechanism
is that erroneous sub-terminal starts of the RNA
primer are uncorrectable. Two alternative proposals, relying on terminal palindromes, are depicted in Fig. 11. The Bateman model has the disadvantage that the separation of daughter molecules would require a special site-specific unwinding protein; but it readily accounts for the
recent observation that vaccinia DNA behaves as
if terminally cross-linked (25). An interesting
feature of both models is that the terminal
palindrome could easily get larger if a new site
for the postulated nicking enzyme arises further
away from the end. This would be one way to account for the over-size terminal palindrome of
GL mtDNA.

From these results it is clear that we still
cannot decide whether the duplication found in
Tetrahymena mtDNA is essential for replication.
If it is, however, the variation in this structure in different Tetrahymena strains may well
provide a clue how a duplication-inversion helps

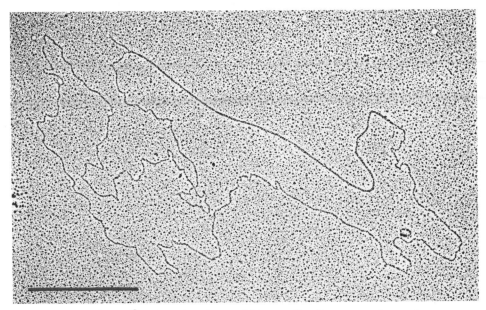

Fig. 8. A circle with a duplex part and a
single-stranded tail in GL mtDNA. The bar is 1
μm. See Fig. 3 for details.

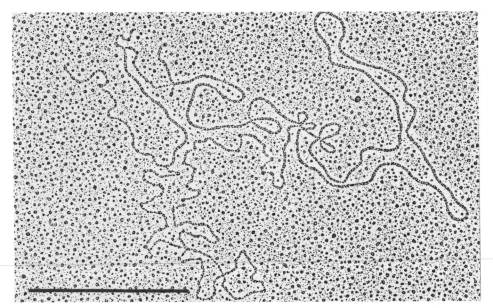

Fig. 9. Circularized molecule in native GL
mtDNA after digestion with exonuclease III, as
described in ref. 12. The bar is 1 μm. See Fig.
3 for details.

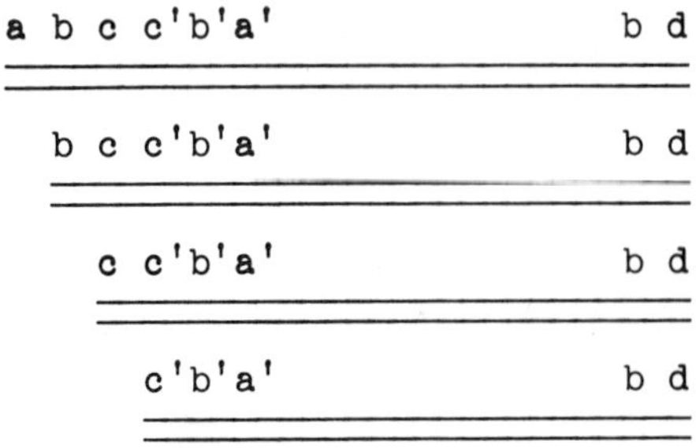

a b c c'b'a'   b d

b c c'b'a'   b d

c c'b'a'   b d

c'b'a'   b d

Fig. 10. Tentative model for the structure of GL mtDNA, as isolated.

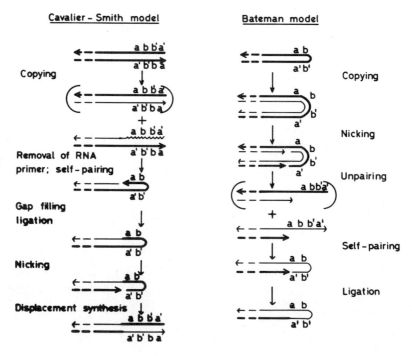

Cavalier – Smith model   **Bateman model**

Fig. 11. The palindromic models of Cavalier-Smith (23) and Bateman (24) for the synthesis of 5'-termini of linear DNA. Thick lines, old DNA; thin lines, new DNA; wavy line, RNA primer. The arrows mark 3'-ends.

in replicating DNA.

ACKNOWLEDGEMENTS

We thank Mrs J.E.Bollen-de Boer for expert technical assistance, Mr K.Gilissen for printing, the photographs and Mrs M.A.Tilanus for her help in constructing this manuscript. This work was supported in part by a fellowship (A.C.A.) and a grant (P.B.) from The Netherlands Foundation for Chemical Research (S.O.N.) with financial aid from The Netherlands Organization for the Advancement of Pure Research (Z.W.O.).

REFERENCES

1. Suyama, Y. and Miura, K. Proc.Nat.Acad.Sci. (Wash.) 60, 235-242 (1968).
2. Clegg, R.A., Borst, P. and Weijers, P.J. Biochim.Biophys.Acta 361, 277-287 (1974).
3. Arnberg, A.C., Van Bruggen, E.F.J., Schutgens, R.B.H., Flavell, R.A. and Borst, P. Biochim.Biophys.Acta 272, 487-493 (1972).
4. Robberson, D.L., Kasamatsu, H. and Vinograd, J. Proc.Nat.Acad.Sci.(Wash.) 69, 737-741 (1972).
5. Arnberg, A.C., Van Bruggen, E.F.J., Clegg, R.A., Upholt, W.B. and Borst, P. Biochim. Biophys.Acta 361, 266-276 (1974).
6. Upholt, W.B. and Borst, P. J.Cell Biol. 61, 383-397 (1974).
7. Wolstenholme, D.R., Koike, K. and Cochran-Fouts, P. J.Cell Biol. 56, 230-245 (1973).
8. Watson, J.D. Nature New Biol. 239, 197-201 (1972).
9. Arnberg, A., Van Bruggen, E.F.J., Ter Schegget, J. and Borst, P. Biochim.Biophys.Acta 246, 353-357 (1971).
10. Kornberg, A. (Ed.), DNA Synthesis, Freeman, San Francisco, 1974.
11. Thomas Jr., C.A., Kelly Jr., T. and Rhoades, M. Cold Spring Harbor Symp.Quant.Biol. 23, 417-424 (1968).
12. Arnberg, A.C., Van Bruggen, E.F.J., Borst, P.

Clegg, R.A., Schutgens, R.B.H., Weijers, P.J. and Goldbach, R.W. Biochim.Biophys. Acta <u>383</u>, 359-369 (1975).

13. Schutgens, R.B.H. Ph.D.Thesis, University of Amsterdam, Quadriga Press, Amsterdam (1973).

14. Schutgens, R.B.H., Reijnders, L., Hoekstra, S.P. and Borst, P. Biochim.Biophys.Acta <u>308</u>, 372-380 (1973).

15. Garon, C.F., Berry, K.W. and Rose, J.A. Proc.Nat.Acad.Sci.(Wash.) <u>69</u>, 2391-2395 (1972).

16. Wolfson, J. and Dressler, D. Proc.Nat.Acad. Sci.(Wash.) <u>69</u>, 3054-3057 (1972).

17. Koczot, F.J., Carter, B.J., Garon, C.F. and Rose, J.A. Proc.Nat.Acad.Sci.(Wash.) <u>70</u>, 215-219 (1973).

18. Berns, K.I. and Kelly Jr., T.J. J.Mol.Biol. <u>82</u>, 267-271 (1974).

19. Wu, R. and Taylor, E. J.Mol.Biol. <u>57</u>, 491-511 (1971).

20. Allen, S. and Gibson, I. Biochem.Gen. <u>6</u>, 293-313 (1972).

21. Groot, G.S.P., Flavell, R.A. and Sanders, J.P.M. Biochim.Biophys.Acta <u>378</u>, 186-194 (1975).

22. Borden, D., Whitt, G.S. and Nanney, D.L. J.Protozool. <u>20</u>, 693-700 (1973).

23. Cavalier-Smith, T. Nature <u>250</u>, 467-470 (1974).

24. Bateman, A.J. Nature <u>253</u>, 379-380 (1975).

25. Geshelin, P. and Berns, K.I. J.Mol.Biol. <u>88</u>, 785-796 (1974).

Part VIII
Initiation and Control
of the Replication Cycle
in *Escherichia coli*

REGULATION OF THE INITIATION OF DNA REPLICATION IN <u>E. COLI</u>.
ISOLATION OF I-RNA AND THE CONTROL OF I-RNA SYNTHESIS.

Walter Messer, Ludolf Dankwarth, Ruth Tippe-Schindler,
John E. Womack and Gabriele Zahn
Max-Planck-Institut für molekulare Genetik
Abt. Trautner
1 Berlin 33, Germany

ABSTRACT. An RNA species required for the initiation of
DNA replication (I-RNA)[*] was isolated as a copolymer of RNA
covalently bound to high molecular weight DNA. The dnaC
product is not required for the synthesis of I-RNA, but for
its function as a primer. I-RNA synthesis is negatively
controlled with the dnaA product being involved in this
regulation. This suggests the existence of an I-RNA
operon.

INTRODUCTION

The initiation of DNA replication is one of the most
important events in the bacterial life cycle. The
frequency of initiation determines the overall rate of DNA
synthesis, the amount of DNA per cell and the frequency of
cell division (1). It is because of this importance but
also because of the intellectual challenge that the
regulation of a cyclic event such as initiation might
reveal hitherto unknown regulation phenomena that
initiation of replication found widespread interest.

A number of models for the regulation of initiation
have been proposed. Models for positive control suggested
that the time of initiation could be determined by the
position of a gene coding for an initiator protein on the
genome (2) or that initiation occurred when initiation
protein(s) were accumulated beyond a threshold level (3).
Systems of negative control[**] suggested the synthesis of an
inhibitor at the time of initiation which is successively
diluted by increase in volume (4, 5). In the auto-
repressor model (6) similar assumptions are made. Other
models relate initiation to cell division or membrane
synthesis (7 - 10).

* O-RNA, Origin-RNA
** Negative control of initiation

The inability of DNA polymerases to start new chains led to the search for a natural primer for DNA synthesis. RNA as a primer for DNA synthesis has been demonstrated in phages (11 - 19). DNA-RNA copolymer was also found in plasmids (20 - 23) and several eucaryotic systems (24 - 27).

RNA is also directly involved in the initiation of replication in bacteria. Rifampicin, an antibiotic which specifically inhibits the binding of RNA polymerase to DNA, blocks the initiation of replication under conditions where initiation cannot be inhibited with chloramphenicol (28 - 30).

In this paper we describe the isolation of the RNA required for initiation (I-RNA) in E. coli and the regulatory aspects which lead to the postulate of a negatively controlled I-RNA operon.

METHODS

E. coli WM 301 is B/r, F‾, leu, pro, lac, gal, try, his, arg, thy A, str, met, drm, hss-K12 (29). The dnaA5 allele was transferred into this strain by P1 transduction from E. coli K12 PC5 (31), and dnaA46 from E. coli K12 CRT 4634 (32). DnaA203, dnaA204, dnaA211, dnaC201, dnaC215, and dnaI208 are mutants induced in E. coli WM 301 (33).

Minimal glucose medium, measurement of the rate of DNA synthesis and density shift experiments were done as described previously (34).

For the preparation of I-RNA cells were lysed by the addition of 2 volumes of boiling lysis mix (0.02 M Tris pH 7.6; 0.01 M EDTA; 0.1 M NaCl; 0.5 % SDS) and kept at 100° C for 2 min. Nucleic acids were extracted by phenolization, heated for 5 min to 100° C, and quickly chilled. The denatured nucleic acid mixture was applied to a Sepharose 2B column (40 cm x 2.5 cm diameter) and eluted with TENS buffer (5 mM Tris pH 8.0; 10 mM EDTA; 50 mM NaCl; 0.1 % SDS; 1 mM NaN_3) at a rate of 10 ml/hr. DNAse treatment was in 20 mM Tris pH 7.4, 10 mM $MgCl_2$, 2 mM $CaCl_2$, 20 µg/ml DNase. (Worthington, RNase-free).

RESULTS

Isolation of Initiation RNA (I-RNA):   RNA which is used as
a primer in the initiation of DNA replication can be
expected to be covalently linked to DNA by a phosphodiester
bond. When the nucleic acids of cells are separated
according to their molecular weight I-RNA must therefore
be associated with high molecular weight DNA, clearly
separated from all other, smaller RNA-species.

To isolate I-RNA a culture of E. coli WM 301 is pulse
labelled with ^{3}H-uridine (typically 50 μCi/ml, 40 Ci/mM,
4 min). Cells are lysed as described in Methods. Nucleic
acids are extracted by phenolization, denatured by heating
to 100° C for 5 min followed by quick chilling, and
subjected to gel filtration on a Sepharose 2B column. To
exaggerate the size difference between an RNA-DNA copolymer
and RNA, denaturation was achieved in some experiments by
treatment with 0.1 M NaOH at 0° for 5 min. This procedure
hydrolyzes approximately one phosphodiester bond for every
1500 nucleotides in RNA.

An elution profile from the Sepharose 2B column is
shown in Fig. 1. A small peak of ^{3}H label is eluted
together with DNA in the void volume around fraction 50.

About 10 % of the ^{3}H label in the void volume peak
is RNA as judged by it being sensitive to RNase and OH⁻
and resistant to DNase. The rest is label in DNA, due to
the conversion of ^{3}H-uridine into ^{3}H-dCTP (35). For a
determination of the amount of I-RNA present fractions or
aliquots from the void volume peak were first treated with
DNase, then RNA was measured as alkali-labile material.

I-RNA is Covalently Bound to DNA. In order to show that
^{3}H-labelled RNA recovered from the void volume peak is
really part of a DNA-RNA copolymer and not the consequence
of unspecific trapping of RNA two sets of experiments were
done.

DNA-RNA copolymer was isolated as described above.
One half of the material served as a control and was
applied to a new Sepharose 2B column without further
treatment. The other half was denatured again by treatment

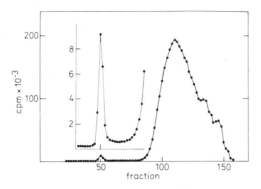

Fig. 1. Elution profile of ^{3}H-uridine labelled
nucleic acids from a Sepharose 2B column.

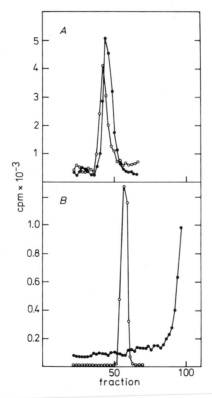

Fig. 2. A. Rechromatography of DNA-RNA copolymer
denatured again in 3 % formaldehyde. cpm ^{3}H-uridine in
RNA: HCHO-denatured (o——o), control (●——●).

B. Elution profile of ^{14}C-DNA and [3]H-
uridine labelled cell lysate treated with DNase.
cpm ^{14}C-thymine (o——o); cpm ^{3}H-uridine (●——●).

at 80° C for 10 min in the presence of 3 % formaldehyde
(36) and then applied to a Sepharose 2B column. As shown
in Fig. 2A about the same amount of labelled RNA is eluted
from both columns, still associated with DNA. Since
formaldehyde effectively inhibits the hybridization of RNA
and DNA this experiment shows that the RNA isolated from
the void volume is not bound to DNA via hydrogen bonds.

In a second set of experiments the nucleic acid
mixture from a culture which was pulse labelled with
^{3}H-uridine was treated with DNase. Further DNase activity
was inhibited with EDTA and a lysate from a ^{14}C-thymine
labelled culture was added. This mixture was then
subjected to denaturation (100° C, 5 min) and applied to
a Sepharose 2B column. The elution pattern is shown in
Fig. 2B. The void volume peak shows exclusively ^{14}C-
thymine label and no ^{3}H-labelled RNA (see also 35). This
experiment shows that the original integrity of the RNA-
DNA copolymer is required for the elution of RNA in the
void volume.

Size of I-RNA: DNA-RNA copolymer from a culture of
E. coli WM 301 pulse labelled with ^{3}H-uridine was isolated
and the DNA part was digested with DNase. The resulting
I-RNA was subjected to electrophoresis in acrylamide gels
(37). As shown in Fig. 3 the I-RNA migrated to a position
characteristic for RNA with a sedimentation coefficient of
about 8S. This corresponds to a length of approximately
500 nucleotides. Since a few deoxyribonucleotides might
still have been present in the preparation this is an
upper estimate for the size of I-RNA.

Synthesis of I-RNA in Initiation Mutants. DnaA and dnaC
mutants are temperature sensitive for gene products which
are required for initiation. Some of the mutants (dnaA5,
dnaA46, dnaC2, dnaC28, dnaC201, dnaC215, dnaC325; 38 - 41)
synthesize a thermolabile protein at non-permissive
temperature, 42° C, which upon a shift to the permissive
temperature, 30° C - 33° C, can acquire the active
conformation. Reinitiation therefore occurs under these
conditions in the absence of protein synthesis (i. e. in
the presence of chloramphenicol). Moreover when these
mutant strains are incubated at restrictive temperature
for some time replication resumes after a shift to

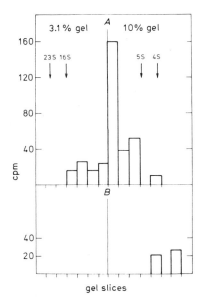

Fig. 3. Acrylamide gel electrophoresis of I-RNA.
DNA-RNA copolymer was isolated as described in Methods.
The copolymer was subjected to extensive DNase treatment
and electrophoresis was done in a 3.1 %/10 % step gel
together with marker RNA indicated by arrows.
A: I-RNA; B: I-RNA after degradation with OH⁻.

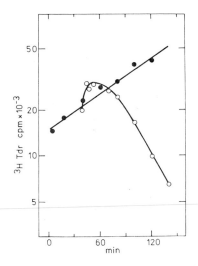

Fig. 4: Rate of DNA synthesis at 33° in E. coli
WM 301 dnaA5 after addition of chloramphenicol (30 µg/ml)
+ chloramphenicol (o——o); - chloramphenicol (●——●).

permissive temperature even in the presence of rifampicin
(40, 41). These experiments indicate that all proteins
required for initiation can be synthesized at 42° C and
stored for future use. Also I-RNA might be synthesized
at 42° C and stored. It is also possible, however, that
I-RNA is not actually synthesized, but that a rifampicin
resistant complex between RNA-polymerase and a promoter
for an I-RNA gene is formed which allows RNA synthesis
after a shift to 30° C in the presence of rifampicin.

The isolation of DNA-RNA copolymer in these mutants
should give information on the actual site of action of
the mutant products and at the same time demonstrate that
the isolated copolymer really contains I-RNA, i. e. an
RNA sequence which is required for the initiation of
replication.

If ^{3}H-uridine is incorporated at 42° C into dnaA or
dnaC mutants we would not expect to find I-RNA when the
cells are lysed immediately after the uridine pulse since
the isolation procedure relies on I-RNA being covalently
linked to DNA. If, however, the ^{3}H-uridine pulse is
followed by a chase at permissive temperature I-RNA should
be found in our standard isolation procedure described
above if I-RNA had been synthesized at 42° C.

In the experiments shown in Table 1 wild type E. coli
WM 301 and dnaA and dnaC mutants in this strain were
incubated at 42° C for 40 min. A 4 min pulse of ^{3}H-uridine
was given at 42° C. Radioactive uridine was removed and
one half of the culture was lysed immediately (- chase).
The other half was incubated at 30°C for 20 min in the
presence of 150 µg/ml rifampicin (+ chase). All mutants
showed only very low recovery of ^{3}H-uridine in I-RNA
(40 - 80 cpm/10^8 cells) when I-RNA was prepared immediately
after the pulse. After a chase at permissive temperature
I-RNA was recovered from the mutant dnaC201 with about
the same efficiency as from wild type. In both dnaA
mutants (dnaA5 and dnaA46) the recovery of I-RNA was low
even after a chase at 30° C.

Several conclusions can be drawn from these
experiments:

TABLE 1

SYNTHESIS OF I-RNA IN INITIATION MUTANTS

	I-RNA (cpm x 10^{-2}/10^8 cells)	
	Exp. 1	Exp. 2
Control (wild type)		
− chase	5.9	3.8
+ chase (30°)	18.7	9.3
dnaC201		
− chase	0.7	0.4
+ chase (30°)	12.3	12.9
dnaA5		
− chase	0.6	0.5
+ chase (30°)	3.0	1.0
dnaA46		
− chase	0.5	0.8
+ chase (30°)	2.8	1.0

TABLE 2

INCREASE OF THE AMOUNT OF DNA AFTER CHLORAMPHENICOL,
RIFAMPICIN OR 42° TREATMENT

	42°	CM	RIF
wild type (E. coli WM301)	−	1.4-1.5[+]	1.4-1.45
dnaA5	1.4	1.9-2.4	1.7-1.9
dnaA46	1.4	2.3	1.4

[+]indicates the range of increase factors in different
experiments. Temperature was 37° for wild type, 33°
for dnaA5, and 30° for dnaA46.

609

1. The RNA which is isolated in the copolymer actually is an RNA species required for initiation, it can justly be called I-RNA.
2. I-RNA is stable for at least 20 min.
3. I-RNA is synthesized in dnaC mutants at 42° C. The product of the dnaC gene is therefore required for the transition from ribonucleotide to deoxyribonucleotide synthesizing activity.
4. Only small amounts – if any – of I-RNA are synthesized in dnaA mutants at 42° C. Since in the dnaA5 mutant DNA synthesis occurs in the presence of rifampicin at 30° C after preincubation at 42° C the dnaA gene product must exert its function before the synthesis of I-RNA but after the RNA polymerase binds to the I-RNA promoter.

Negative Control of Initiation: Chloramphenicol was added to an exponentially growing culture of E. coli WM 301 dnaA5 at permissive temperature (33° C) at a concentration sufficient to block all protein synthesis (30 µg/ml). The rate of DNA synthesis, as measured with pulses of ^{3}H-thymidine, increased immediately after addition of the drug and reached a maximum after 10 – 20 min. After this time the rate of synthesis decreased due to termination of chromosome replication (Fig. 4). The initial stimulation of the rate of DNA synthesis could either be due to an increase in the replication velocity of existing replication points or to an increase in the number of replication points, i. e. to a stimulation of initiation. One can discriminate between these possibilities by measuring the amount of DNA which can be synthesized after chloramphenicol addition. As shown in Table 2 considerably more DNA is synthesized in the presence of chloramphenicol in the mutants dnaA5 and dnaA46 than in the wild type or in the same mutants at 42° C. These data show that indeed more replication points are induced.

A stimulation in the rate of DNA synthesis can also be induced with amino acid starvation in E. coli WM 301 dnaA5, with chloramphenicol in E. coli WM 301 dnaA46, and with rifampicin in E. coli WM301 dnaA5 but not in E. coli WM 301 dnaA46. This is also reflected in the residual DNA synthesis pattern shown in Table 2.

DnaA5 and dnaA46 were the only mutants in which such a stimulation of the initiation could be observed. No stimulation occurred in the wild type strain, in two dnaC mutants (dnaC201, dnaC215), in dnaI208, and in four other dnaA mutants (dnaA203, dnaA204, dnaA205, and dnaA211).

The observation that the inhibition of protein synthesis can result in the stimulation of initiation can only mean that a negative control element exists for initiation. A similar conclusion was reached from experiments in which initiation could be stimulated with subinhibitory concentrations of chloramphenicol (42) and for plasmid replication (43). Such a negative control element would have to have a high turnover rate such that a sufficient concentration of inhibitor or repressor is maintained by continuous synthesis.

Negative control occurs at the site of action of the dnaA product. Growth of E. coli WM 301 dnaA5 at temperatures intermediate between permissive and restrictive results in a progressive increase with temperature in the ability to be stimulated by chloramphenicol. The rate of DNA replication after the addition of chloramphenicol was determined in an analogous experiment to the one shown in Fig. 4. The maximal rate of synthesis, relative to the control rate, is given in Table 3. A somewhat smaller increase in the rate of DNA synthesis is also observed when a culture grown at 35° C or 37° C is shifted to 33° C in the absence of chloramphenicol. This suggests that a partially active dnaA product might be responsible for the ability to stimulate the initiation with chloramphenicol. The stimulation is higher the more inactive dnaA product is present. However, at 42° C where all dnaA product is denatured no stimulation can be observed any more.

Two properties distinguish the two dnaA mutants dnaA5 and dnaA46 from dnaA mutants which cannot be stimulated. Both mutants produce a product which can renature at 30° C after thermo-denaturation at 42° C. The dnaA product of the other four dnaA mutants tested (dnaA203, dnaA204, dnaA205, dnaA211) is permanently denatured at 42° C. The other property can best be described with the term "delayed initiation".

611

TABLE 3

EFFECT OF INTERMEDIATE TEMPERATURES IN <u>dnaA5</u>

Temp. °C	relative stimulation of the rate of DNA synthesis by chloramphenicol	average delay of initiation in min relative to wild type
control (wild type)	1.0	0
33	1.4	7
35	2.3	15
37	3.0	33
39	3.5	−

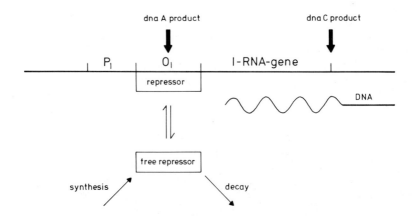

Fig. 5. The I-RNA Operon

We measured the average time it takes to initiate a new round and to replicate one length of chromosome. Cultures of the dnaA5 mutant grown at 33° C, 35° C, or 37° C were pulse labelled with ^{3}H-thymidine. They were then transferred to heavy medium (^{13}C-glucose, ^{15}N H $_4$Cl), samples were taken after different times and analyzed in CsCl density gradients. A progressively increasing time required for the conversion of 50 % of the pulse labelled DNA to hybrid density was observed with increasing temperature. This is expressed as minutes delay compared to a wild type control in Table 3, last column. Since the initial rate with which continuously ^{14}C-thymine labelled DNA became hybrid was much less effected by temperature we conclude that it is the event of initiation which is delayed, i. e. the delay is not due to decreased repli- cation velocity. This delay obviously is also due to partially inactive dnaA product at intermediate temperatures.

A partially inactive dnaA product must create an unbalanced situation. All other initiation proteins will continue to be synthesized, accumulate and "wait" for enough active dnaA product to become available. The "waiting position" for the initiation complex must be at the level at which the dnaA product acts, namely prior to the synthesis of I-RNA.

This is also the level at which the negative control of initiation must act. It seems therefore to be a control system for I-RNA synthesis. Inhibition of protein synthesis allows the synthesis of I-RNA, providing the substrate for the "waiting" initiation complex.

A stimulation of I-RNA synthesis can be directly demonstrated. I-RNA was isolated after a 4 min pulse with ^{3}H-uridine as described above from a culture of E. coli WM 301 dnaA5 growing at 37° C. Chloramphenicol was added to one part 2 min prior to the beginning of the pulse, the other part was without chloramphenicol. The amount of I-RNA found in the presence of chloramphenicol was increased 1.8 fold. The stimulation of the rate of DNA synthesis in the same time interval was 2 fold.

DISCUSSION

The dnaA product and a negative control system are effective after RNA polymerase attaches to an initiation promoter but before I-RNA is synthesized. I-RNA probably serves as a primer for DNA polymerase. The dnaC product is required for the transition from ribonucleotide to deoxyribonucleotide polymerization.

These data are very suggestive for an operon structure for the synthesis of I-RNA. In the model shown in Fig. 5 initiation repressor bound to DNA is in equilibrium with free repressor which is regulated by matching synthesis and decay rates. The dnaA product might either be required for the release of the repressor or might be the repressor protein itself. A detailed discussion (40) will be published elsewhere.

This could provide a very simple system for the temporal control of initiation. When the number of copies of free repressor is maintained by synthesis and decay the concentration of repressor would decrease when a cell increases in volume. I-RNA synthesis could then occur when the repressor concentration falls below a threshhold level. Such a model for the temporal control of initiation would be very similar to the one proposed by Pritchard et al. (4).

The negative control of the I-RNA operon is obviously not the only control system involved in initiation. The accumulation of initiation complex was required in order to be able to detect an effect on DNA synthesis due to derepression (or to the release of inhibition). This accumulation seems to be also tightly controlled since under normal conditions the accumulation of initiation complex and the synthesis of I-RNA are coordinated. No surplus in I-RNA exists, since rifampicin inhibits initiation in the wild type without delay (29). The surplus in initiation complex in dnaA5 decreases with decreasing temperature, suggesting that it approaches zero in the wild type.

The regulation of the synthesis – and possibly assembly – of the initiation complex may well be positive.

Thus both, positive and negative control mechanisms can be active in the initiation of DNA replication.

ACKNOWLEDGEMENT

We thank T. A. Trautner and the Wednesday Night Seminar Club for many stimulating discussions. The excellent technical assistance of G. Krug, C. Kurz and I. Wolf is gratefully acknowledged.

REFERENCES

1. Helmstetter, C. E., Cooper, S., Pierucci, D. and Revelas, E. Cold Spring Harbor Symp. Quant. Biol. 33, 809 (1968).
2. Jacob, F., Brenner, S. and Cuzin, F. Cold Spring Harbor Symp. Quant. Biol. 28, 239 (1963).
3. Maaløe, O. and Kjeldgaard, N. O. Control of Macro-molecular Synthesis. Benjamin, Inc., New York, 1966.
4. Pritchard, R. H., Barth, P. T. and Collins, J. Symp. Soc. Gen. Microbiol. 19, 263 (1969).
5. Rosenberg, B. H., Cavalieri, L. F. and Ungers, G. Proc. Nat. Acad. Sci (USA) 63, 1410 (1969).
6. Sompayrac, L. and Maaløe, O. Nature N.B. 241, 133 (1973).
7. Marvin, D. A. Nature 219, 485 (1968).
8. Donachie, W. D., Jones, N. C. and Teather, R. Symp. Soc. Gen. Microbiol. 23, 9 (1973).
9. Helmstetter, C. E. J. Mol. Biol. 84, 21 (1974).
10. Gudas, L. and Pardee, A. J. Bacteriol. 117, 1216 (1974).
11. Wickner, W., Brutlag, D., Schekman, R. and Kornberg, A. Proc. Nat. Acad. Sci. (USA) 69 (1972).
12. Keller, W. Proc. Nat. Acad. Sci. (USA) 69, 1560 (1972).
13. Staudenbauer, W. L. and Hofschneider, P. H. Proc. Nat. Acad. Sci. (USA) 69, 1634 (1972).
14. Buckley, P. J., Kosturko, L. D. and Kozinski, A. W. Proc. Nat. Acad. Sci. (USA) 69, 3165 (1972).
15. Speyer, J. F., Chao, J. and Chao, L. J. Virology 10, 902 (1972).
16. McNicol, L. A. J. Virology 12, 367 (1973).
17. Dove, W. F., Inokuchi, H. and Stevens, W. F. In (A. D. Hershey, ed.) The Bacteriophage Lambda, p. 747 Cold Spring Harb. Lab., Cold Spring Harbor, 1971.

18. Hayes, S. and Szybalski, W. Fed. Proc. 32, 529 Abs. (1973).
19. Hayes, S. and Szybalski, W. In (B. A. Hamkalo and J. Papaconstantinou, eds.) Molecular Cytogenetics p. 277, Plenum Press, New York, 1973.
20. Blair, D. G., Sherratt, D. J. Clewell, D. B. and Helinski, D. R. Proc. Nat. Acad. Sci (USA) 69, 2518 (1972).
21. Williams, P. A., Boyer, H. W. and Helinski, D. R. Proc. Nat. Acad. (USA) 70, 3744 (1973.
22. Sakakibara, Y. and Tomizawa, J. Proc. Nat. Acad. Sci. (USA) 71, 1403 (1974).
23. Sugino, Y., Tomizawa, J. and Kakefuda, T. Nature 253, 652 (1975).
24. Barker, S. T., Kurtz, H., Taylor, B. A. and Ackermann, W. W. Biochem. Biophys. Res. Commun. 50, 1068 (1973).
25. Faras, A. J., Taylor, J. M., Levinson, W. E., Goodman, H. M. and Bishop, J. M. J. Mol. Biol. 79, 163 (1973).
26. Magnusson, G., Pigiet, V., Winnacker, E. L., Abrams, R. and Reichard, P. Proc. Nat. Acad. Sci (USA) 70, 412 (1973).
27. Waqar, M. A. and Huberman, J. A. Biochem. Biophys. Res. Commun. 51, 174 (1973).
28. Lark, K. G. J. Mol. Biol. 64, 47 (1972).
29. Messer, W. J. Bacteriol. 112, 7 (1972).
30. Laurent, S. J. Bacteriol. 116, 141 (1973).
31. Carl, P. L. Mol. Gen. Genet. 109, 107 (1970).
32. Hirota, Y., Ryter, A. and Jacob, F. Cold Spring Harbor Symp. Quant. Biol. 33, 677 (1968).
33. Beyersmann, D., Messer, W. and Schlicht, M. J. Bacteriol. 118, 783 (1974).
34. Marunouchi, T. and Messer, W. J. Mol. Biol. 78, 211 (1973).
35. Womack, J. E. and Messer, W., this volume.
36. Boedtker, H. J. Mol. Biol. 35, 61 (1968).
37. Wrede, P. and Erdmann, V. A. FEBS Letters 33, 315 (1973).
38. Schubach, W. H., Whitmer, J. D. and Davern, C. I. M. Mol. Biol. 74, 205 (1973).
39. Wolf, B. Genetics 72, 569 (1973).
40. Tippe-Schindler, R. Ph.D. thesis, Freie Univ. Berlin, W-Germany, 1974.
41. Hanna, M. H. and Carl, P. L. J. Bacteriol. 121, 219 (1975).

42. Blau, S. and Mordoh, J. Proc. Nat. Acad. Sci (USA) 69, 2895 (1972).
43. Goebel, W. Biochem. Biophys. Res. Commun. 51, 1000 (1973).

Note added in proof:

There was an agreement reached between the participants of this meeting to call the RNA which is required for initiation at the origin of a replicon origin-RNA (o-RNA).

SPECIFIC LABELLING OF INITIATION RNA IN E. coli.

John E. Womack and Walter Messer
Max-Planck-Institut für molekulare Genetik
Abt. Trautner, 1 Berlin 33, Germany

ABSTRACT. We have constructed a strain which allows us
to independently label RNA and DNA, although RNA is
labelled at a reduced rate. The reduction in the rate
of labelling or RNA using ^{3}HUR precludes using this
strain for rapid pulse-chase experiments; however, the
labelling rates under special conditions allow us to
label the I-RNA piece and isolate it while it is bound
to relatively unlabelled DNA. This greatly facilitates
the physical studies on this important piece of RNA.

INTRODUCTION

In order to isolate the piece of RNA that initiates
the replication of the Escherichia coli chromosome (I-RNA)[*],
we devised a technique based on the property that I-RNA is
covalently bound to high molecular weight DNA (1). Thus
by column chromatography, we are readily able to isolate
this piece of RNA, for the DNA and thus the I-RNA moves
in the void volume; whereas, the bulk RNA is retarded on
these columns and does not come off until the later
fractions. When isolating small amounts of RNA
associated with DNA, it is important to realize that even
in thyA strains which do not convert dUMP to dTMP a size-
able fraction of uridine radioactivity enters into DNA
(Table 1, see also 1).

Since the I-RNA piece is isolated in the fractions
containing DNA, we decided to construct a strain which
would allow us to specifically label RNA with little or
no spillover into DNA (Fig. 1). In the thyA strain,
radioactivity - given as uracil or uridine - enters RNA
from both the cytidine triphosphate (CTP) and the uridine-
triphosphate (UTP) pools. However, the label also enters
DNA via the dCTP pools (for a recent review see 2).

By selecting the proper mutants, one can separate
the three nucleotide triphosphate pools (3). Previously
these strains have only been used to study the regulation

* The naming of this peice of RNA was changed at the Squaw
Valley meetings. From now on we intend to call the peice
of RNA that initiates DNA synthesis "Origin-RNA" or "O-RNA."

of pyrimidine biosynthesis (4), however, a silent feature
of these mutants is that they allow the specific labelling
of RNA. The separation of uracil, cytidine, and thymine
derived pools requires two mutations in addition to the
thyA, drm, or dra mutations normally used in DNA labelling.
These are in the genes cdd |encoding for cytidine
(deoxycytidine) deaminase (E.C.3.5.4.1) (5)| and pyrG
|encoding for cytidine-triphosphate synthetase (E.C.6.3.4.
2.) (3)|. The cytidine deaminase rapidly deaminates CR to
UR, and a mutant in pyrG requires CR for growth and no
other nucleoside will support growth of these strains (3).
The pyrG mutation blocks the conversion of UTP to CTP which
allows one to add radioactive uridine to the growth medium
and to label only RNA.

The enzyme cytidine deaminase not only deaminates its
natural substrates, but also their 5-fluoronated analogues.
These are highly toxic to the cell after deamination and
phosphorylation to FUMP and FUTP, but not in the form of
5-fluorocytidine (FCR) or 5-fluorodeoxycytidine (FCdR) (6).
Thus one can select a cdd strain based on resistance to
FCdR. The pyrG mutation confers an absolute requirement
for CR and is therefore selected by penicillin counterselection (3). These genotype-phenotype relationships
are shown in Fig. 1.

The results of this work show that with a cdd, pyrG,
thyA, drm strain one can not incorporate as much ^{3}HUR into
RNA as in the parent strain; however, the label present is
predominantly in RNA, and one can vastly improve the
precision of labelling RNA associated with DNA.

METHODS

The bacterial strains employed were E. coli WM 301,
which is thyA,drm (other markers are listed in 1). The
other strain E. coli JM463 was isolated in two steps from
WM301. First a cdd (Fig. 1) strain JM460 was isolated,
using EMS mutagenesis, by its resistance to FCR. The
enzyme cytidine deaminase was assayed for and not found.
The pyrG strain was then isolated from JM460 by EMS
mutagenesis and penicillin counterselection. These
strains were tested for their dependence on CR for growth.
One isolate JM463 showed an absolute dependence and was

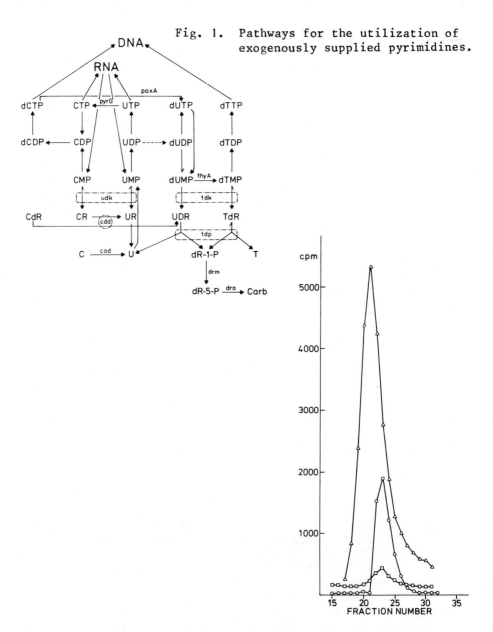

Fig. 1. Pathways for the utilization of exogenously supplied pyrimidines.

Fig. 2. DNA containing fractions from a sepharose 2B column (see text for details). WM301 ^{3}H cpm (Δ); JM463 ^{3}H cpm (◻); JM463 ^{14}C cpm (o).

chosen for further work.

Bacteria were grown in the minimal glucose medium (1) at 37° C. For growth of JM463 40 µg/ml CR was also added. ^{3}H-Uridine labelling was 4 min in WM301 and 20 min in JM463, both at 1x10^8 cells/ml. ^{14}C thymine labelling was done continuously.

The cell lysis, preparation of nucleic acids, column chromatography and I-RNA determination for these studies are detailed in Messer et al. (1). The amount of RNA present was determined by boiling in 0.3 M NaOH for 1 hr and measuring the loss of radioactivity.

RESULTS

WM301 and JM463 were labelled with ^{3}HUR. The pyrG strain JM463 would be expected to convert very little if any ^{3}HUR to dCTP. Thus this strain should synthesize very little radioactive DNA when given labelled UR.

The data shown in Table 1 are completely in agreement with this expectation. After labelling the cells were then washed with fresh media, lysed, and phenolyzed (1). Two equal samples were taken - one for direct TCA precipitation and one for alkaline hydrolysis. The parent strain incorporated 40 times more ^{3}HUR into RNA and DNA than does JM463. This is very similar to the incorporation rate of the parent strain when labelled in the presence of 40 µg/ml CR (data now shown). CR has been shown to enter the cell via the same permease as UR (7). It has also been shown in S. typhimurium that uridine kinase phosphory-lates both UR to UMP and CR to CMP (6). This phenomenon is probably true for E. coli as well. Thus, the labelled uridine faces competition by the CR at two reactions. It is, therefore, not surprising that our strain JM463 incorporates label at a very low rate.

Also from Table 1, one learns that strain JM463 incorporates label from UR into DNA (via dCTP) at a rate of less than 10 % that of the pyrG$^+$ parent. This can not be accounted for by the low level of total label in JM463 (see Table 1). We can therefore safely assume that JM463 has a tight mutation in pyrG.

TABLE 1

FATE OF LABELLED URIDINE IN A _pyrG_ STRAIN AND ITS PARENT

Strain	^{3}H in RNA	^{3}H in DNA	% ^{3}H in DNA	I-RNA	% ^{3}H in I-RNA
WM301	93,179	10,702	11.5	599	0.64
JM463	21,187	71	0.30	142	0.67

a. All labelling data normalized to 1 x 10^8 cells.

b. The ^{3}H-label in DNA varies with labelling conditions.

Figure 2 shows the results of the Sepharose 2B column chromatography of the nucleic acids from the two strains (only fractions containing DNA shown). Strain WM301 is labelled with ^{3}HUR ($\triangle$--$\triangle$) and strain JM463 is labelled with both ^{14}Cthy (O--O) and ^{3}HUR (□--□). The large difference in tritium radioactivity reflects the absence of tritium in DNA from strain JM463 (labelled with ^{3}HUR). Thus the majority of the ^{3}H-label, at least 66 % (Table 1), is the result of ^{3}H-uridine in RNA directly associated with DNA.

These data demonstrate that the cdd and pyrG mutations in JM463 greatly facilitate the detection and isolation of I-RNA or any other small RNA associated with DNA.

ACKNOWLEDGEMENT

We thank Ingrid Wolf and Kayron Dube for their excellent technical assistance, Nancy Lou Womack for proof reading of the manuscript and T. A. Trautner and the Wednesday night discussion group for their valuable suggestions. We also thank Drs. Hennes and Kapp of Hoffmann-LaRoche for kindly providing us with the 5-fluoro-nated analogues of cytidine. One of us, J.E.W. acknow-ledges the Deutsche Forschungsgemeinschaft for a grant to attend these meetings.

REFERENCES

1. Messer, W., Dankwarth, L., Tippe-Schindler, R., Womack, J. E. and Zahn, G. this volume.
2. O'Donovan, G. A. and Neuhard, J. Bacteriol. Rev. 34, 278 (1970).
3. Neuhard, J. and Ingraham, J. L. J. Bacteriol. 95, 2431 (1968).
4. Williams, J. C. and O'Donovan, G. A. J. Bacteriol. 115, 1071 (1973).
5. Wang, T. P. In: Method in Enzymology (C. P. Colowick and O. Kaplan, eds.) Academic Press Inc., New York, p. 478, 1955.
6. Beck, C.F., Ingraham, J. L., Neuhard, J. and Thomassen, E. J. Bacteriol. 110, 219 (1972).
7. Peterson, R. N., Boniface, J. and Koch, A. L. Biochem. Biophys. Acta 135, 771 (1967).

Note added in proof:

There was an agreement reached between the participants of this meeting to call the RNA which is required for initiation at the origin of a replicon origin-RNA (o-RNA).

CRYOLETHAL SUPPRESSORS OF
THERMOSENSITIVE dnaA MUTATIONS

James A. Wechsler and Malgorzata Zdzienicka
Department of Biological Sciences
Columbia University
New York, New York 10027

ABSTRACT. A series of suppressors of thermosensitive dnaA
mutations has been isolated. The suppressors chosen for
study simultaneously reverse the thermosensitivity of the
original mutant and render the cell inviable at low temperature, 25 C. The suppressors can be separated into at
least five genetic loci, only one of which could be identical to dnaA. Isotope incorporation experiments show
that all of these strains synthesize protein normally at
25 C. Incorporation of thymine into DNA at 25 C is either
similar to that observed with wild-type strains or faster.
All of the suppressed strains appear to be defective in
nuclear organization, chromosome segregation, or septation.

INTRODUCTION

As noted by Watson and Crick, the double-helical
structure of DNA provides the molecular logic for understanding DNA replication (1). The actual mechanism required to replicate a chromosome appears, however, to be
quite complex. Presumably, the constraints imposed by the
structure of the chromosome and the requirements that
chromosome replication be both fastidious and temporally
adjusted to the cell cycle, necessitate this mechanical
complexity. In the past few years our understanding of
DNA replication has progressed because of the development
of several in vitro assay systems and the ability to correlate in vitro results with gene products (2-8).
This combination of genetics and biochemistry has resulted in the isolation of many replicative proteins and
in some degree of understanding of their roles in DNA
chain elongation (2-6).

Initiation of DNA replication also appears to be complex. The determination of the components involved in DNA initiation has been refractory to study because of the inability to obtain chromosomal DNA initiation in vitro. In vitro systems which assay the replication of bacteriophage templates are not dependent on all of the known dna initiation gene products (4) and, thus, do not represent adequate models for the biochemical study of chromosomal initiation.

The possible interaction between various gene products involved in DNA replication currently seems more amenable to in vitro than to in vivo study (9). A method for determining the in vivo interactions of the various DNA replication proteins was one of the motivations for this study.

The control of DNA replication presumably depends on the ability to initiate rounds of DNA synthesis. Additional replication forks are observed both in fast growing cells, compared to those growing more slowly, and under conditions which significantly reduce the normal rate of DNA chain elongation (10,11). Physiological experiments have shown that there are two chloramphenicol sensitive steps and an RNA requirement involved in normal initiation (12,13). In addition, proteins which effect DNA chain elongation may be intimately involved in DNA initiation and control, as may a variety of other cellular processes, crudely represented by cell mass (14).

In an attempt to define the complexity of DNA initiation and its control and to determine the interaction of gene products involved in chain elongation, we have been analyzing suppressors of the known classes of dna mutations. This report summarizes our current results with suppressors of initiation-defective dnaA mutations. These suppressors are given the genotype designation das for "dnaA suppressor".

The initial selection for suppressors of the thermosensitive dnaA mutation was for thermoresistant revertants which were expected to be of two classes: 1) mutations within the dnaA gene (either true revertants or second-site revertants) or 2) extragenic suppressors. It is the latter class which is of interest to us.

If the suppressor confers only the single phenotype of thermoresistance, both genetic and eventual biochemical analyses would be difficult. To facilitate analysis we

625

have imposed a secondary selection, the inability to grow
at low temperature, on the initially selected thermoresistant revertants. We have termed this general class of
suppressors "cryolethal".

An example of the logic involved in the selection of
cryolethal suppressors is as follows: the thermosensitive
dnaA protein is normal at 30 C and is denoted as A. It is
assumed that A and an unknown component, X, interact at 30
C to allow DNA synthesis. At 42 C A becomes A*. A* and X
do not interact and the strain is inviable at 42 C. After
mutagenesis X is altered to a new form X*. A* and X* interact at 42 C and thus X* suppresses the thermosensitive
dnaA mutation. With a shift from 42 C to 25 C, A* changes
to A but X* remains X*. A and X* do not interact and the
cell is inviable at 25 C. Thus, the mutation which
changes X to X* results in cryolethal suppression.

METHODS

Bacteria and bacteriophage: The parental strains for
all but one of the suppressor mutations described was
strain E508: thr leu thy thi dra lac tonA str supE
dnaA508 (15). The single suppressor, das-1002, was isolated from N167: HfrH thy met dnaA167 (16). Marker
notation is according to Taylor and Trotter (17). The
suppressor carrying strains are denoted in the text by
their suppressor genotype only, e.g. das-61 means dnaA508
das-61. P1virS was used for all transductions.

Growth media: Liquid media were M9 salts supplemented
as necessary or Oxoid Nutrient Broth No. 2 as described
(15). Solid media were M9 minimal or Difco Antibiotic
Medium No. 3 solidified with 1.5% agar.

Labelling media: Labelling medium for E508 and its
derivatives was as previously described (15). The das-1002
strain was labelled in M9-minimal media containing 0.1%
vitamin-free casamino acids (Difco), thiamine 0.35 µg/ml
and thymine, 10 µg/ml. ^{3}H-thymine was added to give a
final specific activity of 30 µCi/µg.

Genetic techniques: Transductions and conjugations
were done as described previously (15).

Mutagenesis: The suppressors obtained in strain E508
were the result of directed mutagenesis. The dnaA mutant
was grown into log phase in nutrient medium at 30 C
(generation time approximately 1 hour) and shifted to

42.5 C for 1.2 generation time equivalents, by which time all DNA synthesis had terminated. The culture was diluted into a volume of nutrient broth containing N-methyl-N'-nitro-N-nitrosoguanidine (NTG) such that the temperature of the culture was instantaneously lowered to 30 C at an NTG concentration of 100 μg/ml. After either 30 s or 60 s the culture was filtered through green membrane filters, 0.45 μm pore size (Gelman Instrument Co.), washed with several volumes of buffer at 42 C, and the filters were placed on agar plates at 42 C. Revertant colonies were picked from the surface of the filter after 2 days. One suppressor, das-1002, was derived from strain N167, by a different method. N167 was grown in log phase at 30 C and mutagenized with diethylsulfate for 45 minutes according to the method of Martin (18). The mutagen treated culture was filtered through a polycarbonate membrane, 0.45 μm pore size (Nuclepore Corp.), resuspended in nutrient broth and grown for two generations at 30 C. This culture was filtered through a green membrane filter and placed at 42 C as above.

Reversion: Independent revertants of the cryolethal suppressors were isolated by inoculating single colonies into nutrient broth and growing these cultures to stationary phase at 42 C. These cultures were then plated at 25 C. Revertant colonies were picked and purified at 25 C. Each revertant used was from a separate, original single-colony inoculum.

Nuclear stain: Bacteria were stained according to Piechaud (19) and observed in a Zeiss Photomicroscope II.

Reagents: Chemicals and their suppliers were: chloramphenicol, Sigma Chemical Co.; rifampicin, Calbiochem; NTG, Aldrich Chemical Co. Inc.; diethylsulfate, Fisher Scientific Co.; [methyl-^{3}H]thymine was from Schwarz/Mann. L-[1-^{14}C]methionine were from Amersham/Searle. All other chemicals were reagent grade.

RESULTS

The first criterion for extragenic suppression is that the original dna allele still be present in the strain. Table 1 shows that for seven of the nine cold-sensitive, thermoresistant revertants, the dnaA508 allele could be cotransduced with ilv$^+$ to an Ilv$^-$ recipient strain. The suppressors obtained in dnaA167 have not been tested in

TABLE I

Transduction of dnaA508 and das alleles with ilv

Donor Genotype	Transduction at 30 C		Transduction at 42 C	
	ilv⁺	% dnaA	ilv⁺	% das
dnaA das-1 ilv⁺	200	11	100	0
dnaA das-46 ilv⁺	136	1.5	100	0
dnaA das-61 ilv⁺	150	0	180	5
dnaA das-107 ilv⁺	106	0	104	6
dnaA das-119 ilv⁺	70	13	188	3
dnaA das-122 ilv⁺	100	5	100	4
dnaA das-411 ilv⁺	100	4	100	0
dnaA das-416 ilv⁺	100	7	100	0
dnaA das-419 ilv⁺	64	13	46	0

this way as yet because this dnaA allele either cannot be transduced into, or is suppressed in, some recipient strains.

Since initiation of rounds of DNA replication has been shown to begin near the ilv locus and the mutagenesis was specifically directed at the region near the replication origin, it was expected that many of the suppressor mutations might also be cotransducible with ilv. These results are also shown in Table 1.

From the data in Table 1 three classes of dnaA suppressors are apparent. One class shows cotransducibility of both thermosensitivity and cold sensitivity with ilv (das-119 and das-122). In the second class only thermosensitivity is cotransducible with ilv (das-1, das-46, das-411, das-416, and das-419). The das-61 and das-107 strains show only cold sensitivity cotransducible with ilv.

For the strains which do show cotransducibility both of thermosensitivity and of cold sensitivity with ilv, the most likely map position for the suppressor is on the opposite side of ilv from dnaA. Furthermore, their map position with respect to ilv is probably correctly reflected by the 5-13% cotransduction frequency. If the suppressors were either 1) very close to dnaA or 2) very close to ilv and only conferred cold sensitivity in the presence of dnaA, the cotransduction frequency of dnaA thermosensitivity and ilv should be radically altered, which it is not.

The two suppressor mutations (das-61 and das-107) where only cold-sensitivity is cotransducible with ilv could be 1) in the dnaA gene, 2) very close to dnaA on either side so that the cotransduction frequency reflects the relative map position of the suppressor, or 3) if the das alleles confer cold-sensitivity only in the presence of dnaA508, they could map extremely close to ilv on the opposite side from dnaA or anywhere between ilv and dnaA, and the cotransduction frequency would reflect the position of dnaA508 rather than that of the suppressor.

The map position of the suppressors in strains where only thermosensitivity is cotransducible with ilv is generally unknown except that they are not near dnaA. Conjugation experiments place the das-1 suppressor between 61 and 72 minutes on the chromosome map and, thus, specify a fourth locus. The das-1002 mutation, which was

629

isolated in the dnaA167 strain, maps between 77 and 81
minutes but has not been carefully analyzed because the
original strain is an Hfr and the dnaA mutation is not
easily transducible into all ilv recipients. The map position of das-1002 does, however, define a fifth tentative
class of dnaA suppressor.

A causal relationship between the dnaA suppressor and
cryolethality is the primary requirement for establishing
the validity of these suppressor mutations. Reversion of
the suppressed strains to cold-resistance should result in
substantial restitution of thermosensitivity if both
suppression and cold-sensitivity are due to a single
lesion. Results of reversion studies for several of the
suppressed strains are shown in Table 2. A substantial
proportion of the cold-resistant revertants of all the das
mutations, except das-416, regain thermosensitivity. The
appearance of thermosensitivity concomitant with the
reversion to cold-resistance demonstrates that dnaA suppression and cold-sensitivity are the result of a single
mutation.

The 100% restitution of thermosensitivity in revertants of das-411 implies that the possible ways of obtaining reversion, either intragenic or extragenic, are
restricted. Conversely das-416 reverts at a much greater
rate than any of the others so that there may be many ways
to phenotypically revert this suppressor and the class
that leads to restitution of thermosensitivity may be a
minor one. The fact that one of the 43 cold-resistant
revertants of das-416 is thermosensitive is probable
evidence for this also being a true cryolethal suppressor.

It is not possible to determine if the suppressors
das-61 and das-107 map within the dnaA gene from these
reversion rates as the analysis just applied to the reversion results with das-411 and das-416 applies equally
to intragenic or extragenic suppressors. The kinetic results in the next section, however, imply that neither
das-61 nor das-107 are dnaA mutations.

Analysis of the kinetics of DNA and protein synthesis
following a shift from 42 C to 25 C divide the suppressed
dnaA strains into two general classes.

The first class is exemplified by das-1 and is shown
in Figure 1. Protein and DNA are synthesized at a
reduced rate following the shift to 25 C. This is similar to the kinetics observed with wild-type strains.

TABLE 2

Reversion of Cryolethal Suppressors

Suppressor Allele	No. Revertants Tested	No. ts	% ts
das-61	64	49	80
das-107	35	5	15
das-122	38	19	50
das-411	40	40	100
das-416	43	1	2.3
das-419	32	16	50

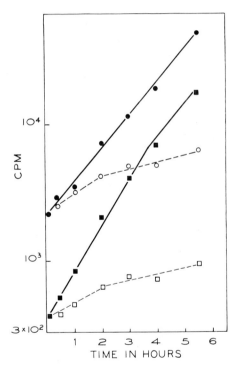

Fig. 1. The dnaA508 das-1 strain was pre-labelled with ^{14}C-leucine and ^{3}H-thymine for approximately four generations at 42 C. A log phase culture in labelled medium was divided into flasks at 42 C and 25 C. Samples, 50 μl, were taken at the times indicated.
^{3}H-thymine: ●, at 42 C; O, at 25 C. ^{14}C-leucine: ■ , at 42 C; □ , at 25 C.

TABLE 3

Summary of Suppressor Strain Characteristics

Strain Relevant Genotype	Group	Map Position[a]	DNA[b] 25 C	Protein[b] 25 C	Morphology 25 C
dnaA508 das-61	I	73.5 - 74.5' (dnaA-ilv)	++	+	filaments, many nuclei
dnaA508 das-107			+	+	cells distended
dnaA508 das-119	II	74.5 - 76'	++	+	very long filaments, many nuclei
dnaA508 das-122			+	+	long filaments, few nuclei
dnaA508 das-1	III	61' - 72'	+−	+−	short filaments, large nuclei
dnaA167 das-1002	IV	77' - 81'	++	+	not tested
dnaA508 das-46	V	unknown	+	+	nuclei at poles some DNAless
dnaA508 das-411			++	+	filaments, few nuclei
dnaA508 das-416			+	+	not tested
dnaA508 das-419			+	+	filaments, few nuclei

a. Map position is in minutes on standard E. coli chromosome (17).
b. + = synthesis as wild-type; ++ = synthesis greater than wild-type.

Though the das-1 strain is somewhat slow-growing and shows a reduced rate of synthesis of both protein and DNA at 42 C and at 25 C when compared with wild-type, no selective alteration in the kinetics of macromolecular synthesis is observed. Many mutants fall into this normal class though most grow faster than the das-1 strain.

Figure 2 shows the kinetics of DNA synthesis in the das-61 strain following the shift to 25 C. In this strain, protein synthesis (not shown) responds to the temperature shift as it does in the wild-type strain, but DNA synthesis continues at the 42 C rate for the first 2 to 3 hours at 25 C. After 2 to 3 hours at 25 C DNA synthesis continues at a greatly reduced rate for many hours (at least overnight), even in the presence of chloramphenicol. As shown in Table 3, many das carrying strains exhibit an increased rate of thymine incorporation following a shift to 25 C.

To test if this rapid rate of DNA synthesis at 42 C is due to the presence of an extra replication fork, chloramphenicol was added to half of the culture at the time of the temperature shift and to half of the control at 42 C. Figure 2 shows that the effect of chloramphenicol on the 42 C culture is as expected for an agent that blocks initiation. At 25 C, if one assumes that chloramphenicol is acting similarly, the amount of residual synthesis is somewhat less than that expected if all cells in the culture had acquired an extra growing fork. Rifampicin also blocks initiation, and if rifampicin rather than chloramphenicol is added to the culture, the incorporation at 25 C and 42 C are identical and inconsistent with an additional replication fork at 25 C (results not shown). The simplest explanation to resolve the different action of the two drugs is to argue that time and RNA synthesis, but not protein synthesis, is required for the induction of a new growing fork at 25 C and that rifampicin acts before the new growing fork is induced.

The primary action of the das-61 suppressor mutation is not suitably explained even if this argument is correct as the increased rate of thymine incorporation is short-lived and the ability to induce replication forks appears to be limited. Repeated transfers at 15 minute intervals followed by incubation in the presence of chloramphenicol do not lead to greater levels of

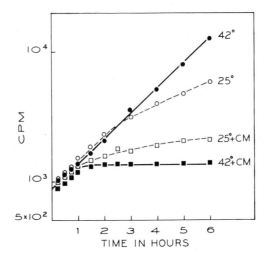

Fig. 2.   The dnaA508 das-61 strain was prelabelled with
³H-thymine for approximately 12 generations at 42 C.   At
the beginning of the experiment, the culture was divided
into flasks at 25 C and 42 C with or without chloramphenicol (CM).   Samples 50 µl, were taken at the times indicated.

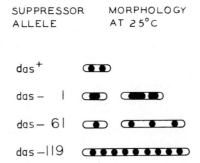

Fig. 3.   A schematic diagram of several das strains and
the wild-type dnaA⁺ das⁺ strain, CR34, stained according
to Piechaud (19) after extended periods of growth at 25 C.
Solid areas represent stained nuclear regions.

synthesis (data not shown). Also, it is clear from recent results with rep mutants that cells can tolerate extra growing forks (20). The extensive DNA synthesis which continues for many hours at a low rate in the presence of chloramphenicol does, however, imply a defect in DNA synthesis control.

All of the das strains appear to be defective in nuclear organization, segregation, or septation at low temperature. The morphology of several strains are schematically diagrammed in Figure 3. It is obvious that membrane alterations could explain both the suppression of the initiation-defective phenotype and the inability to form colonies at low temperature. At least two of the das strains are unusually sensitive to low deoxycholate concentrations, but since wild-type strains differ in their sensitivities to deoxycholate, we do not yet know if deoxycholate-sensitivity is due to the presence of these das mutations.

DISCUSSION

A series of dnaA suppressors which render the cell inviable at low temperature has been described. Though mutagenesis was directed toward the origin of DNA replication and dnaA maps very close to the origin, none of the cryolethal suppressor carrying strains display a classical initiation defective phenotype. As cold-sensitive initiation-defective mutants which map at the dnaA locus have been obtained by conventional techniques (21), the lack of such a phenotype among the suppressors is surprising.

Of the five tentative genetic classes of suppressors, only one could be coincident with dnaA. As the two suppressors, das-61 and das-107, do not result in a standard initiation-defective phenotype at low temperature, they are either not dnaA lesions or alteration of the dnaA gene product can result in inviability without specifically blocking DNA initiation. Tentative results with das-61 imply that it is recessive to das⁺.

The reversion results in Table 2 show that both cold-sensitivity and dnaA suppression result from a single lesion in each of the das strains.

We do not yet know if the das mutations suppress all dnaA lesions or are allele specific; nor do we know if they all confer cold-sensitivity in the absence of a dnaA mutation. We expect to answer these questions and to test the ability of the das mutations to suppress dna alleles at other known loci.

Many thermosensitive dna mutations are suppressed by high salt or high sugar concentrations (22). As such phenotypic reversal is not uncommonly observed with temperature sensitive mutations of all types, it is possible that the effect of high salt or sugar concentrations could be mimicked by mutation. It is, thus, necessary for us to exclude this non-specific suppression mechanism as an explanation for the dnaA suppressors. The das mutants, themselves, are not suppressed by the presence of high salt which is known to inhibit the growth of dnaA mutants at the permissive temperature (22).

It should be noted that all but one of the suppressors reported here resulted from mutagenesis directed toward the origin of chromosomal DNA replication. From our preliminary results on dnaA suppressors obtained by random mutagenesis, we believe that the majority of das loci are fortuitously in this region of the chromosome map. As the suppressors described were isolated by directed mutagenesis, however, no conclusion about the possible association of the dnaA product with other DNA replication gene products is possible.

Isotope incorporation divides the mutants into two groups which do not correlate with the genetic classification (Table 3). Members of one group show no gross alteration in macromolecular synthesis, while those in the other show enhanced DNA synthesis for several or many hours. Several dnaA suppressors obtained by random, rather than directed, mutagenesis and with allowance for a period of expression following mutagen treatment also separate into these two kinetic groups.

The low rate of DNA synthesis observed in the das-61 strain in the presence of chloramphenicol is not unlike that seen in one dnaE mutant (15, Kurtz and Wechsler, unpublished) or when protein synthesis is inhibited following a period of thymine starvation or nalidixic acid treatment (23). The fact that the only aberrant macromolecular synthesis observed in das mutants is enhanced, rather than deficient, DNA synthesis lends some support to negative control models of DNA initiation (24-27). However, simple models with only one level of negative control do not easily explain our data.

636

A more probale explanation of the cryolethality of the das mutations comes from the preliminary observations of cellular morphology of these mutants at 25 C (Figure 3). All of the das mutants appear to be defective in septation, chromosome segregation, or both. Abnormal cell morphologies could be secondary consequences of the das allele or could be the primary result of an altered membrane component. As several of the das strains are sensitive to deoxycholate and the das-107 strain has an unusual shape even at 42 C, we favor the second possibility.

The postulation of membrane chromosome interaction for normal DNA metabolism (28) has received considerable support in recent years. Perturbation of membrane fatty acid composition affects the phenotypic properties of dnaA and dnaC mutants (29), and a study of conditional mutants specifically defective in phospholipid synthesis suggests that this synthesis is coupled to protein, RNA, and DNA synthesis (30). Several reports of membrane detachment of the folded E. coli chromosome after inhibition of rounds of DNA replication imply a specific role for membrane attachment in DNA synthesis (31, 32). Finally, in slow-growing E. coli the observation that visible cross-wall formation correlates with DNA initiation implies that these two processes may be intimately related (32).

If the suppressor action of the das mutations results from specific membrane alterations, it may be possible to identify the membrane components involved in DNA initiation.

ACKNOWLEDGEMENTS

The aid of David Turkewitz and Harvey Stern with characterization of the das-1 suppressor and the excellent technical assistance of Mrs. Leei-Hu Yang is gratefully acknowledged. This work was supported by grants GB32417 and BMS75-02300 from the National Science Foundation and grant CA-12590 from the U.S. Public Health Service. M.Z. is a Fellow of the American Association of University Women.

REFERENCES

1. Watson, J.D. and Crick, F.H.C. Nature 171, 737 (1953).
2. Gefter, M.L., Hirota, Y.,Kornberg, T., Wechsler, J. and Barnoux, C. Proc. Nat. Acad. Sci. 68 3150 (1971).
3. Lark, K.G. Nature N. Biol. 240, 237 (1972).
4. Wickner, S. and Hurwitz, J. Proc. Nat. Acad. Sci. 71, 4120 (1974).
5. Geider, K. and Kornberg, A. J. Biol. Chem. 249, 3999 (1974).
6. Lark, K.G. and Wechsler, J.A. J. Mol. Biol. (1975) in press.
7. Schaller, H., Otto, B., Nusslein, V., Huf, J., Herrmann, R. and Bonhoeffer, F. J. Mol. Biol. 63, 183 (1972).
8. Wickner, W.T., Brutlag, D., Schekman, R. and Kornberg, A. Proc. Nat. Acad. Sci. 69, 965 (1972).
9. Wickner, S. and Hurwitz, J. Proc. Nat. Acad. Sci. (1975) in press.
10. Oishi, M., Yoshikawa, H. and Sueoka, N. Nature 204, 1069 (1964).
11. Pritchard, R.H. and Zaritsky, A. Nature 226, 126 (1970).
12. Lark, K.G. and Renger, H. J. Mol. Biol. 42, 221 (1969).
13. Lark, K.G. J. Mol. Biol. 64, 47 (1972).
14. Donachie, W.D. Nature 219, 1077 (1968).
15. Wechsler, J.A. and Gross, J.D. Molec. Gen. Genetics 113, 273 (1971).
16. Abe, M. and Tomizawa, J.-I. Genetics 69, 1 (1971).
17. Taylor, A.L. and Trotter, C.D. Bacteriol. Rev. 36, 504 (1972).
18. Martin, R.C. J. Mol. Biol. 31,127 (1968).
19. Piechaud, M. Ann. Inst. Pasteur 86, 787 (1954).
20. Lane, H.E.D. and Denhardt, D.T. J. Bacteriol. 120, 805 (1974).
21. Wehr, C.T., Waskell, L. and Glaser, D.A. J. Bacteriol. 121,99 (1975).
22. Gross, J. Curr. Top. Microbiol. Immunol. 57, 39 (1972).
23. Kogoma, T. and Lark, K.G. J. Mol. Biol. 52, 143 (1970).
24. Pritchard, R.H., Barth, P.T. and Collins, J. Symp. Soc. Gen. Microbiol. 19, 263 (1969).
25. Rosenberg, B.H., Cavalieri, L.F. and Ungers, G. Proc. Nat. Acad. Sci. 63, 1410 (1969).

26. Blau, S. and Mordoh, J. Proc. Nat. Acad. Sci. <u>69</u>, 2895 (1972).
27. Goebel, W. Biochem. Biophys. Res. Commun. <u>51</u>, 1000 (1973).
28. Jacob, F., Brenner, S. and Cuzin, F. Cold Spring Harb. Symp. Quant. Biol. <u>28</u>, 329 (1963).
29. Fralick, J.A. and Lark, K.G. J. Mol. Biol. <u>80</u>, 459 (1973).
30. Glaser, M., Bayer, W.H., Bell, R.M. and Vagelos, R. Proc. Nat. Acad. Sci. <u>70</u>, 385 (1973).
31. Worcel, A. and Burgi, E. J. Mol. Biol. <u>82</u>, 91 (1974).
32. Jones, N.C. and Donachie, W.D. Nature <u>251</u>, 252 (1974).
33. Gudas, L.G. and Pardee, A.B. J. Bacteriol. <u>117</u>, 1216 (1974).

TRITON X-100 DEPENDENT IN VITRO DNA SYNTHESIS
IN AN E. COLI dnaC MUTANT

Michael H. Hanna, L. Stephanie Soucek and Philip L. Carl

Department of Microbiology
University of Illinois
Urbana, Illinois 61801

ABSTRACT. When an E. coli dnaC mutant is returned from
nonpermissive to permissive conditions, DNA synthesis
reinitiates in vivo but not in a toluene-treated in
vitro system. However, Triton X-100 added to the toluene-
treated cells stimulates DNA synthesis more than 15 fold.
This DNA synthesis is ATP-dependent, semiconservative, and
resistant to chloramphenicol or rifampin added in vitro,
but sensitive to chloramphenicol added in vivo. In vitro
reinitiation does not appear to require the dnaC protein.

INTRODUCTION

Strains carrying the dnaC mutation appear to be specifically blocked in initiation of rounds of chromosome replication (1,2,3). When E. coli cells whose chromosome
replication has been terminated by a dnaC initiator mutation are returned from nonpermissive to permissive
conditions, DNA synthesis can resume in vivo even in the
presence of high concentrations of chloramphenicol or
rifampin (4). Apparently, all the components of the
initiation apparatus can accumulate in the absence of
functional dnaC protein, and, upon renaturation of the
inactive protein, DNA synthesis can reinitiate in vivo.
Cells which possess the potential to reinitiate DNA
synthesis in vivo seemed to provide an attractive system for studying the initiation of bacterial DNA synthesis in vitro. The toluene-treated cell system faithfully reproduces the elongation reactions of the bacterial chromosome (5). However, we and others (6) have
been unable to observe reinitiation of DNA synthesis

in vitro in this system. We were therefore intrigued by
the observation of Moses (7) that stationary phase cells,
which may contain terminated chromosomes (8), do not synthesize DNA in the toluene-treated cell system unless the
non-ionic detergent Triton X-100 is added to the assay
mixture.

We report here that Triton X-100 can also reactivate
DNA synthesis in a toluene-treated cell system derived
from cells carrying a dnaC mutation, whose chromosome
replication has been terminated in vivo by prior incubation under nonpermissive conditions. The characteristics
of this Triton-dependent synthesis are described.

METHODS

Bacterial strains: D110 (pol A1, end⁻, thy⁻, dra⁻) is
the wild type strain for this study. PC233 (pol A1, end⁻,
thy⁻, dra⁻, dnaC2) is an isogenic derivative of this
strain constructed by P1 transduction.

Growth and toluene treatment: Cells were grown at
30°C in a minimal glucose-salts medium supplemented with
casamino acids and thymine. dnaC aligned cells were produced by shifting such cultures to 40°C for 75 minutes.
Under these conditions chromosome replication in vivo
terminates after an increase in DNA content of approximately 33% relative to the amount of DNA present at the
time of the shift. KCN (0.03M) was then added to prevent
any reinitiation in vivo during harvesting. Cultures
were iced, washed once in 50 mM potassium phosphate pH
7.4 and resuspended in the same buffer plus 1% toluene
at about 1×10^{10} cells/ml. The cells were vortexed at
4°C for 2 min, spun down, the pellets rinsed, and finally
resuspended in buffer at about 10^{10} cells/ml. They were
then frozen in liquid N_2 and stored at -80°C.

Assay conditions: DNA synthesis was measured in 0.3
ml reaction mixtures containing 1.3 mM ATP, 33 µM each
dATP, dCTP, dGTP and ^{3}H-TTP (30-60 cpm/pmole), 13 mM
MgCl, 100 mM KCl, 0.1 mM EDTA and 20 mM MOPS buffer pH 8.
Toluene-treated cells were added to 3×10^8 cells/assay
and Triton X-100 (Beckman Instuments) to 1% final concentration. RNA synthesis was assayed as described by

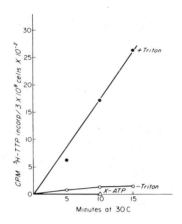

Fig. 1.   PC233 was grown at 30°C and transferred to
40°C for 75 min.   The cells were harvested and toluene-
treated as in Methods.   The assays were done at 30°C
with or without Triton X-100.   Samples were removed at
5 min. intervals, precipitated and counted as in Methods.
One pmole of ³H-TTP corresponds to 50 cpm.

TABLE 1

DNA Synthesis in a toluene-treated aligned dnaC mutant.

Complete	100%
-cells	5
-Triton	1
-KCl	78
-Mg++	3
-ATP	8
-dATP, dCTP, dGTP	8
+CTP, GTP, UTP (1mM each)	119
+CAP (150 µg/ml) in vivo	3
+CAP (150 µg/ml) in vitro	95
+RIF (200 µg/ml) in vitro	167
+NEM (13mM)	0

dnaC aligned cells were harvested, toluene-treated and
assayed as in Methods.   The assay was at 30°C for 10 min.
The activity is expressed as percentage of the activity
of a complete reaction mixture.

642

Peterson et al (9) using ^{3}H-UTP (4 cpm/pmole). Acid precipitable counts were determined by filtration on glass
fiber filters.

Density gradient centrifugation: The procedure was
essentially that of Burger (10) with minor modifications.

RESULTS

Figure 1 shows the kinetics of in vitro DNA synthesis
in the Triton-toluene system prepared from dnaC aligned
cells. DNA synthesis is stimulated more than 15 fold
by the addition of the non-ionic detergent Triton X-100
to the toluene-treated cells. The Triton-dependent
synthesis is almost completely ATP dependent. The requirements for the reaction (Table 1) are similar to
those previously reported (7). Note that Triton-dependent DNA synthesis is resistant to high concentrations of
rifampin and chloramphenicol added in vitro, just as is
the reinitiation seen in vivo (4). However, if chloramphenicol is added to the cells in vivo at the moment of
transfer to 40°C, subsequent in vitro DNA synthesis is
almost completely abolished. Thus, similar to the reinitiation seen in vivo, one or more components needed for
DNA synthesis in vitro must accumulate during the period
the cells are incubated at the nonpermissive temperature.

As shown in Table 2, in vitro RNA synthesis in toluene-
treated cells prepared from dnaC aligned cultures is
essentially unaffected by Triton X-100. This RNA synthesis depends on externally supplied ribonucleoside triphosphates. Thus dnaC aligned cells do not require Triton to
become permeable to nucleoside triphosphates.

The Triton-dependent DNA synthesis in dnaC aligned
cells is normal semiconservative DNA replication as indicated by the density gradient analysis presented in Fig.
2. The DNA synthesized in vitro bands at the density
of authentic hybrid DNA. Approximately 1-3% of the
chromosome has been replicated in vitro.

We next asked whether the Triton-dependent synthesis
required the dnaC gene product. dnaC aligned cells were
assayed in the Triton-toluene system at 30°C and at 40°C,

TABLE 2

RNA synthesis in a toluene-treated aligned dnaC mutant.

Complete	100%
-Triton	120
-GTP, CTP, ATP	12

The growth conditions and toluene-treatment were the same as in Table 1. The cells were assayed for RNA synthesis as in Methods, for 15 min. at 30°C. The activity is expressed as percentage of the activity of a complete reaction mixture.

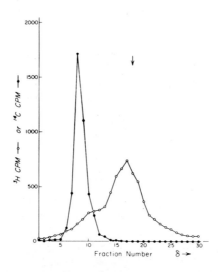

Fig. 2. PC233 was grown and toluene-treated as in Fig. 1. The cells were assayed as in Methods except BrdUTP and ^{3}H-dATP were used in place of TTP and dATP. The assay was for 10 min. at 30°C. The cells were prepared for density gradient analysis as in Methods. ^{14}C-light DNA was added as marker. The arrow shows the position expected for hybrid DNA. ^{14}C- ●-●, ^{3}H- o-o.

where the dnaC protein should have been inactive. The
results shown in Table 3 show a slight temperature sensitivity of the ATP-dependent synthesis at 40°C. As shown
in Figure 3, the residual synthesis at 40°C is largely
semiconservative. We have been unable to reduce the a-
mount of residual 40°C synthesis by preincubating the
cells for up to 15 minutes at 40°C prior to the start of
the assay (data not shown). Although exponential phase
wild type cells are usually slightly more active at 40°C
than at 30°C, stationary phase wild type cells, like dnaC
aligned cells, are usually slightly less active (Table 3).
Our inability to obtain conditions in which the activity
of aligned cells is totally absent and the activity of
wild type cells still measureable, leads us to conclude
that the aligned cells probably do not require the dnaC
protein for DNA synthesis in vitro.

DISCUSSION

Our data is consistent with the notion that the Triton-dependent DNA synthesis observed in vitro is similar
to the reinitiation seen in vivo: (a) It is normal semiconservative replication. (b) It is resistant to chloramphenicol and rifampicin added at the time of return to
permissive conditions. (c) Both in vivo and in vitro reinitiation are eliminated by adding chloramphenicol in
vivo during the incubation at the nonpermissive temperature, implying that a period of protein synthesis in the
absence of initiation of DNA synthesis is required to observe both in vivo and in vitro reinitiation. On the
other hand, Triton-dependent in vitro reinitiation seems
to show only slight evidence of temperature sensitivity
in a dnaC strain. Stationary phase wild type cells also
show a slight temperature sensitivity compared to wild
type exponential cells. Since the amount of in vitro DNA
synthesis is relatively low in dnaC aligned cells even
under optimum assay conditions, it is hard to decide
whether Triton-dependent synthesis is generally temperature-sensitive or whether the temperature-sensitivity is
due to the dnaC gene product. However, the fact that we
cannot reduce the amount of in vitro DNA synthesis at 40°
C by preincubation of the cells at 40°C, makes it seem
likely that in vitro reinitiation in our system does not
require the dnaC protein.

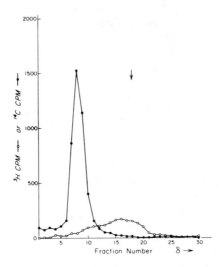

Fig. 3. PC233 was grown and treated as in Fig. 2 for density gradient analysis, except the assay was for 5 min. at 40°C. ^{14}C- ●-● and ^{3}H- o-o.

TABLE 3

Comparison of in vitro temperature-sensitivity in exponential, dnaC aligned, and stationary phase cells.

	30° assay	40° assay	Ratio 40/30
dnaC$^+$ exponential cells	100%	141%	1.4
dnaC aligned cells	22	8	0.4
dna C$^+$ stationary cells	12	11	0.9

Exponential cells are from late exponential phase cultures, stationary phase cells from cultures grown 10 hours
from the end of exponential growth, and dnaC aligned cells
as in Methods. All cultures were at 40°C before being
toluene-treated and assayed as in Methods. The data are
normalized to the amount of DNA synthesized in 5 min. in
dnaC$^+$ exponential phase cells assayed at 30°, taken as
100%.

ACKNOWLEDGEMENTS

This study was conducted under Grant #HEW PHS GM 21536 from the National Institutes of Health, U.S. Public Health Service and Grant #GB 38334 from the National Science Foundation.

REFERENCES

1. Carl, P., Molec. Gen. Genetics <u>109</u>, 107 (1970).
2. Schubach, W., Whitmer, J. and Davern, C., J. Mol. Biol. <u>74</u>, 205 (1973).
3. Wolf, B., Genetics <u>72</u>, 569 (1973).
4. Hanna, M. and Carl, P., J. Bacteriol. <u>121</u>, 219 (1975).
5. Moses, R. and Richardson, C., Proc. Natl. Acad. Sci. <u>67</u>, 674 (1970).
6. Moses, R., Campbell, J., Fleischman, R., Frenkel, G., Mulcahy, H., Shizuya, H. and Richardson, C., Fed. Proc. <u>31</u>, 1415 (1972).
7. Moses, R., J. Biol. Chem. <u>247</u>, 6031 (1972).
8. Sueoka, N. and Yoshikawa, H., Cold Spring Harbor Symp. Quant. Biol. XXVIII, 47 (1963).
9. Peterson, R., Radcliffe, C. and Pace, N., J. Bacteriol. <u>107</u>, 585 (1971).
10. Burger, R., Proc. Natl. Acad. Sci. <u>68</u>, 2124 (1971).

Part IX
Topology and Replication
Control in Bacteria

BIDIRECTIONAL REPLICATION IN *BACILLUS SUBTILIS*

R.G. Wake
Department of Biochemistry
University of Sydney
Australia

ABSTRACT. The early conclusion, which has now been
rejected, that replication of the *B. subtilis* chromosome
is a unidirectional process was based largely on certain
features of the genetic map constructed from a combination
of replication order and linkage data on a restricted
number of markers. While several discrepancies between
the map as constructed and some experimental data were
apparent at the time they were probably considered minor.

Autoradiographic experiments on replication fork
movements established directly that replication is bidirectional over the whole chromosome. This has led to a
reconsideration of the earlier discrepancies and more
extensive experimentation into replication order and
linkage of markers. As a result, a new map (Harford-
Lepesant), which is consistent with bidirectional replication over the whole chromosome, has been constructed. It
will be seen how the misplacement of a relatively small
number of genetic markers, with respect to replication
order, was the major factor leading to the earlier incorrect arrangement of established linkage groups. However,
there is genetic evidence from Sueoka's laboratory to
support the contention that replication is only partially
bidirectional. It is argued here that the data on which
this proposition is based may reflect some unusual feature
of the particular experimental system used to achieve
synchrony of chromosome replication.

Some of the consequences of bidirectionality with
respect to various aspects of the overall replication
process are also considered.

650

INTRODUCTION

It is doubtful if anyone would consider that bidirectional, in contrast to unidirectional, replication of the bacterial chromosome has not been definitively established. The first indications of such a mode of replication appeared about 4 years ago from quite different approaches in two bacterial species, *E. coli* and *B. subtilis*. Since then much evidence to substantiate these initial indications has been forthcoming. We will concern ourselves here almost exclusively with the situation in *B. subtilis*. Most significantly, the finding of bidirectional replication has led to a major rearrangement of the genetic map of this organism.

A common question has been, what led to the almost universal acceptance of the unidirectional mode of replication? The reasons for this will be set out, and the crucial findings leading to its rejection and replacement by bidirectional replication will be described. It will then be possible to assess in a more meaningful way the conflicting data which currently argue for complete bidirectional replication on the one hand and partial bidirectional replication on the other. The consequences of bidirectionality with respect to the overall replication process are several and these will be considered towards the end of this presentation.

ORDERED REPLICATION

It was Cairns, in 1963, who showed that the *E. coli* chromosome consists of a single piece of double stranded DNA with no free end, i.e. it is circular (1,2). During replication this molecule, $2-3 \times 10^9$ daltons, becomes split into two arms over part of its length (see Fig. 1). The length over which the chromosome could be split into two was found to vary considerably and Cairns interpreted this region, or loop, to represent the replicated portion of the chromosome. The loop expands, through the Meselson-Stahl semiconservative mechanism, as a round of replication progresses. Eventually, two complete daughter chromosomes are generated and these separate from one another.

From a consideration of the lengths of DNA replicated upon pulse-labelling with ³H-thymidine, the

length of the whole chromosome and the generation time of
the bacterial culture, Cairns concluded that "one or at
the most two regions of the chromosome are being
duplicated at any moment." For reasons that are not
entirely clear he preferred the apparently simpler
situation of only one replication region per chromosome.
Thus, he interpreted the intermediate structure shown in
Fig. 1 as having arisen by the unidirectional movement of
a replication fork, Y, away from the origin, X, generating
the two daughter arms (comprising the replication loop) in
the process.

At the same time as Cairns' illuminating studies on
E. coli, Sueoka and Yoshikawa were investigating the
process of bacterial chromosome replication in *B. subtilis*
through a different but equally elegant approach (3).
They wanted to know if there was a fixed order of
replication of genetic markers, assayable by transformation, during the replication cycle. From a comparison
of the relative frequencies of various markers in the
exponential and stationary growth phases of strain W23
they concluded that indeed there was a fixed order of
replication. Of the markers analysed at that time it was
considered that *purB* was replicated very early in the
cycle and *metB* very late. Others, e.g. *thr* and *leu*
replicated at intermediate stages. The simplest model to
explain such an ordered replication of markers, and the
one proposed by Sueoka and Yoshikawa, was that replication always started at one end (the origin) of the
chromosome and proceeded in a continuous fashion towards
the other (the terminus)*. However, they made it quite
clear that their data did not rule out a "subunit-model"
of replication, into which bidirectional replication of a
circular chromosome (to be considered below) would fit.
But because of Cairns' preference for only one replication fork per *E. coli* chromosome the "subunit-model" was
considered unlikely.

The replication order map constructed from measurements of the relative frequencies of various markers was
confirmed most convincingly by the direct determination of
the time of replication of markers (transfer from

---

*At that time circularity of the *B. subtilis*
chromosome had not been established and a linear
structure was the simplest to assume.

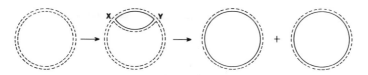

Fig. 1. Stages in the replication of a circular bacterial chromosome. The intermediate, partially replicated structure, with forks at X and Y, is shown in the middle. The broken line represents the parental DNA strand, and the solid line the newly synthesized one.

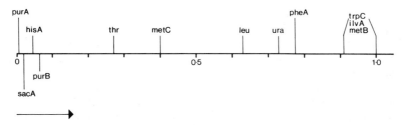

Fig. 2. Replication order map of *B. subtilis* according to O'Sullivan and Sueoka (5). Not all markers are shown. See refs. 30 and 33 for gene designations.

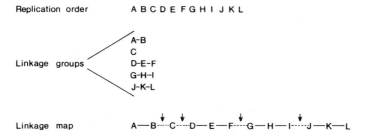

Fig. 3. Constructing a linkage map of *B. subtilis*. The replication order of only two markers within a single linkage group is needed to orientate it. The arrows point to the gaps.

653

unreplicated to replicated DNA in a density shift
approach) during the synchronous germination of spores
(4,5). The map constructed by such an approach is shown
in Fig. 2. *purA* is shown as the earliest replicating
marker and is estimated to lie within 2-3% of the origin.

EARLY LINKAGE MAPS - SUPPORT FOR
UNIDIRECTIONAL REPLICATION

Dubnau and coworkers were the first to attempt the
construction of a genetic linkage map of *B. subtilis* (6).
In the absence of a conjugation system, which had proven
so invaluable in the case of *E. coli* (see ref. 7), these
workers set about the task as follows. The early replication order data of Sueoka and his colleagues (3,4,8,9),
was extended by density transfer experiments so that no
major portions of the map were lacking markers. Linkage
of selected pairs of markers was then tested by cotransduction using the generalized transducing phage PBS1.
Every marker studied, except *purB*, could be placed into
one of four linkage groups. Assuming overall unidirectional replication the composite map was obtained by
ordering the linkage groups according to the replication
order of key markers. A diagrammatic representation of
the approach is shown in Fig. 3. Assuming a single
chromosome in *B. subtilis*, the map constructed contained
gaps between the various linkage groups. But, hopefully,
with the accumulation of more data these would eventually
be filled. However, the map as put together, even at this
relatively early stage, tended to confirm the assumption
of unidirectional replication. Thus, when markers were
found to replicate in the order DEF say (see Fig. 3), and
D was linked to F, the other linkages D-E and E-F were
also always found. As more data gradually became
available from various laboratories it appeared that some
of the gaps were being filled. In 1970 Dubnau put most of
the available data together to give a much more detailed
map, made up now of only three linkage groups (10). An
abbreviated form of it is shown in Fig. 4. The
construction of this map was invaluable to geneticists
and acted as a catalyst, generating more intensive efforts
to identify and fix the position of new markers and to
fill in the various gaps. The most significant aspect of
this map with respect to the replication process is that

the largest linkage group accounts now for at least 70% of the chromosome. This would mean that replication is unidirectional at least over this portion. And such a conclusion would argue for a completely unidirectional process, the direction of which is shown by the arrow. The arrangement of the linkage groups is seen to be circular. This was a reasonable assumption although convincing evidence for it had not yet been forthcoming. Certainly most workers in the *B. subtilis* field would have agreed with this map although there are five particular difficulties, probably considered by most to be minor at the time, which deserve comment.

(a) *sacA* is placed closer to the origin than *purA*. This positioning is based on transductional mapping analysis of sucrase defective mutants (11), and is at variance with the data of O'Sullivan and Sueoka (see Fig. 2).

(b) *hisA* is placed after *thr*. This is based largely on the replication order data of Dubnau *et al* (6). But O'Sullivan and Sueoka (see Fig. 2) and Huang *et al* (12) found the opposite.

(c) *leu* is placed after *ura*. This is based on a combination of replication order and linkage data (6). O'Sullivan and Sueoka found *leu* to replicate before *ura* (see Fig. 2).

(d) Some of the linkage data considered to close the gaps remaining in the original map of Dubnau and co-workers (6) is not entirely convincing. Thus, some workers maintained that the largest group, numbered III in Fig. 4, was made up of three smaller ones, IIIa, IIIb and IIIc, yet to be shown linked (13).

(e) Despite intensive efforts no linkage between the earliest and latest replicating markers, i.e. across the proposed origin-terminus region of the circular chromosome could be demonstrated (see ref. 13).

BIDIRECTIONAL REPLICATION NEAR THE ORIGIN

That the *B. subtilis* chromosome can exist in a physically circular form has now been established (14). Prior to this demonstration there were suggestions from two different approaches that replication may be bi-directional, rather than unidirectional, at least in the vicinity of the chromosome origin.

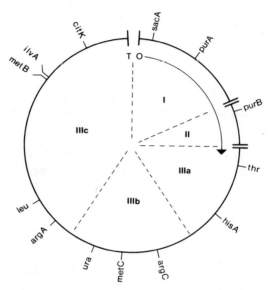

Fig. 4. An abbreviated form of the linkage map of
B. subtilis compiled by Dubnau (10). The arrow shows the
direction of replication, from the origin (O) to the
terminus (T). The orientation of linkage group II was
not defined. Linkage group III is shown as being made up
of 3 smaller ones, IIIa, IIIb and IIIc (see ref. 13).

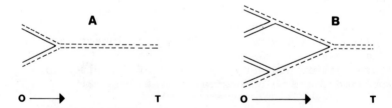

Fig. 5. Diagrammatic representation of the configuration of a reinitiated chromosome. In A (non-
reinitiated) the origin (O) has replicated once; in B
(reinitiated) it has replicated twice. The arrows show
the direction of replication, origin to terminus (T).

One suggestion came from the elucidation of the structure of the replicated portion of reinitiated (dichotomous) chromosomes. In another major contribution to our knowledge of the process of chromosome replication in *B. subtilis* Sueoka and his colleagues had shown that, in a rich medium, reinitiation of replication from the normal origin occurs before the round in progress is complete (15). A diagrammatic representation of the configuration of a symmetrically reinitiated chromosome is shown in Fig. 5. When spores of *B. subtilis*, containing completed chromosomes, were induced to undergo dichotomous replication in the presence of ^{3}H-thymine, and the replicated portion of such chromosomes examined by autoradiography, a rather unexpected result was obtained (16). Many of the reinitiated structures were displayed as fully closed, multiforked forms, actually large loops split into two arms at two positions (Fig. 6). The smaller loops within any single structure were always of approximately the same size. Furthermore, one arm of each smaller loop always had a higher grain density than the rest of the structure, and this allowed their identification as that part of the chromosome that had undergone a second cycle of replication (see Fig. 5). The interpretation put forward to account for the general form of these structures is shown in Fig. 7. It involves the initial movement of replication forks in both directions away from the origin. A second round of replication starting soon afterwards, in the same bidirectional manner from the two daughter origins, would generate the types of reinitiated structures visualized by autoradiography. From the lengths of these structures it was estimated that replication is bidirectional to an extent of at least 10% on both sides of the origin.

The other suggestion that replication may be bidirectional in the vicinity of the chromosome origin came from a consideration of the apparent contradiction between density transfer and transduction data in positioning *sacA* relative to *purA* (see previous Section). While this contradiction had been apparent for some time it was not until the more extensive mapping of mutations affecting sucrose metabolism in *B. subtilis* that Lepesant and coworkers pointed out that the difficulty was resolved if replication commenced in a bidirectional manner from an origin close to *purA* (17). This would explain the earlier

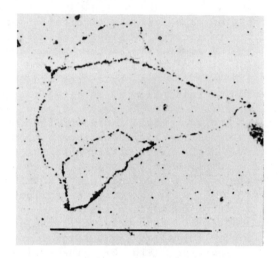

Fig. 6. Autoradiographic visualization of the replicated portion of a symmetrically reinitiated (dichotomous)
chromosome. The scale shows 100μm (from ref. 16).

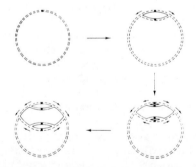

Fig. 7. Scheme for the formation of a reinitiated
(dichotomous) chromosome from a circular structure
containing a single initiation site or origin (●). The
arrows indicate the directions of movement of replication
forks. The broken line represents the parental DNA strand,
and the solid line the newly synthesized one (taken from
ref. 16).

replication of *purA* (see Fig. 2) and allow the positioning of *sacA* to the left of it and other later replicating markers to the right. Sueoka *et al* (18) quite independently pointed out the same resolution of the apparent discrepancy.

AUTORADIOGRAPHIC STUDIES ON THE EXTENT OF BIDIRECTIONAL REPLICATION

There was certainly no major reason, obvious at the time, to cast doubt on the general features of the map compiled by Dubnau. When it became apparent that replication might be bidirectional in the vicinity of the chromosome origin there was a strong feeling amongst workers that the extent of bidirectionality was probably limited, and restricted to the origin region. Actually only partial bidirectional replication is needed to explain nicely, at the gross level, how reinitiation of chromosome replication is achieved. Young and Wilson in their very comprehensive and valuable extension of Dubnau's linkage map would have expressed the conviction of most in stating that "in view of the relatively long regions which contain only a few markers, it would be possible to overlook short stretches of bidirectional replication" (19). When one considers that at the time of this statement there was good evidence for extensive bidirectional replication in mammalian cells (20), phage lambda (21) and *E. coli* (22) the strength of the general feeling that chromosome replication in *B. subtilis* was a largely unidirectional process is evident.

Certainly the most direct and convincing evidence for extensive bidirectional replication in *B. subtilis* comes from an autoradiographic study of the sites of replication loop expansion during the first round of replication following germination of spores (14,23). Thymine-requiring spores of strain 168 were germinated in the presence of ^{3}H-thymidine for sufficient time to allow the first round to start, the specific radioactivity was then increased three-fold and incubation continued for another 20 minutes before stopping replication and preparing samples for autoradiography. Chromosomal loop structures containing two symmetrically situated regions of higher grain density than the rest of the visible structure would be indicative of bidirectional replication, the two

higher grain density regions representing the sites of DNA
chain growth during the last 20 minutes of incubation.
The majority of loop structures with two levels of grain
density were indeed indicative of bidirectional replication (Fig. 8). The longest such loops accounted for replication of at least 50% of the whole circular chromosome
of length 900-1100μm (14). Within experimental error the
extent of growth during the last 20 minutes at each of the
two regions within a single loop was the same. The
overall conclusion from these studies was that, upon
initiation, the two forks generated at the chromosome
origin move apart at approximately the same rate until at
least 50% of the chromosome is replicated. In the
experiments carried out the overall sizes of the loops
varied considerably, and a very significant result was
obtained when the combined lengths of the heavy regions
(amount of chain growth during the last 20 minutes) were
plotted against loop size (extent of first round replication)(Fig. 9). The two features to note, with the
conclusions that can be drawn from them, are as follows:
    (a) All points fall onto a common straight line.
Thus, the bidirectional loops are related in the sense
that they form part of a single series. This would argue
for only one type of chromosome in B. subtilis, but of
course it does not prove it. In addition, the fact that
the longest loops, on which is based the estimate of at
least 50% bidirectional replication, form part of this
series argues for their validity.
    (b) The line has a very significant positive slope.
From its value it is obvious that the rate of replication
fork movement within any single chromosome increases by a
factor of 2-3 during the first half of the round of
replication. This linear increase in rate is possibly
unique to the germinating spore system, reflecting the
gradual build-up in the capacity for DNA synthesis during
the period of conversion from the dormant spore to fully
vegetative state. In this connection it is interesting to
note that there is a linear increase, by a factor of about
7-8, in the total deoxynucleoside triphosphate level
during the first round of significant (stage II) replication following germination of B. megaterium spores (24).
    Attempts to extend these experiments with germinating
spores to more than the first half of a round of replication were unsuccessful because of the large extent of

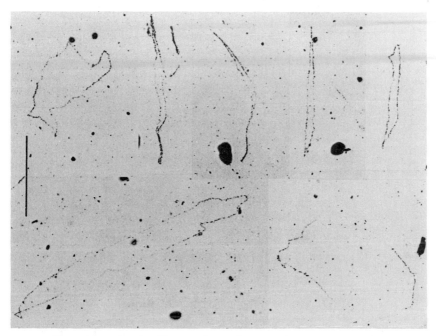

Fig. 8. Autoradiographs of chromosomal replication
loops indicative of bidirectional replication. In all
cases, except the smallest one, two regions of higher
grain density (DNA chain growth) are distinguishable. The
scale shows 100µm (from ref. 23).

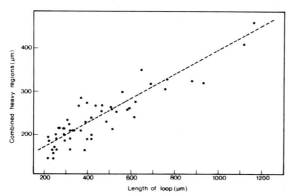

Fig. 9. The relationship between the rate of DNA
chain growth (combined heavy regions) and extent of first
round replication (length of loop) following germination
of *B. subtilis* spores (from refs. 14 and 23).

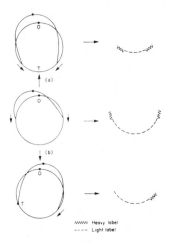

Fig. 10. Two extreme possibilities for the manner of approach of replication forks during the second half of a round of replication. See text for full explanation (O, origin; T, terminus)(from ref. 26).

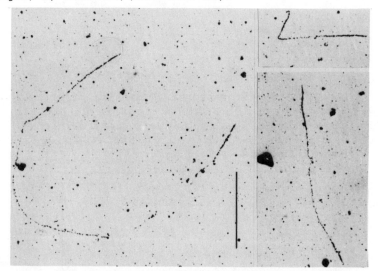

Fig. 11. Autoradiographic visualization of the process of replication fork movement during termination of a round of replication. The heavy regions (at ends) represent the DNA chain growth during the period immediately before transfer of *dna-20*(Ts) at 45°C to lower specific activity medium in which termination occurred. The scale shows 100µm (from ref. 26).

chromosome breakage. To investigate the process of
replication fork movement during the second half of a
round, therefore, use was made of a mutant of *B. subtilis*
strain W23, temperature-sensitive for initiation of
replication. It had been shown already that new
initiations are completely blocked when this mutant,
dna-20(Ts), is transferred to $45^{o}C$, and the rounds in
progress terminate completely (25). With respect to replication fork movement during the second half of a round
it was relevant to distinguish between the following two
extreme possibilities (see Fig. 10 and ref. 26):

(a)   the continued movement of both replication forks
at approximately equal rates until they meet at a
termination site or region that is approximately diametrically opposite the origin; and

(b)   the cessation in movement of one of the replication forks at a site far removed from the antipodal
position and the continued progression of the other until
it eventually reaches this site.

To distinguish between these two possibilities the
following procedure was used: *dna-20*(Ts) growing at $34^{o}C$
was transferred to $45^{o}C$. After a short time, ^{3}H-thymidine
was added and after a further 5 minutes the specific
radioactivity was lowered by a factor of 3. Rounds of
replication were then allowed to complete under these nonpermissive conditions. In the case of complete bidirectional replication,(a),one would expect to observe by
autoradiography linear chromosomal structures with both
ends more heavily labelled (higher grain density) than the
internal region (see Fig. 10). This should be true of
even the shortest structures. In the case of partial bidirectional replication,(b),the linear structures containing heavy label, introduced after one fork had stopped,
should be heavily labelled at only one end. More
specifically, if the termination site were positioned 15%
of the chromosomal length from the antipodal position all
structures accounting for 30% of the chromosome (300μm) or
less should be heavily labelled at only one end.

The result of the experiment was that many linear
structures with both ends heavily labelled were observed
(Fig. 11). These accounted for 36% of all structures with
two levels of labelling, and they ranged in size from 100
to 500μm, i.e. 10-50% of the whole chromosome. Well over
half of them were <250μm, with a significant proportion

< 150μm. From an analysis of these double-heavy
structures it was concluded that both replication forks
can continue to move at approximately the same rate (within a factor of 2) until they are separated by < 10% of the
length of the whole chromosome. Most of the other
structures with two levels of labelling were heavy at only
one end. However, they were not restricted to the smaller
size class and it was considered most likely that they
arose by breakage of those with both ends heavily
labelled. A clear example of such breakage is seen in the
largest structure in Fig. 11.

The overall picture of chromosome replication, at the
level of what can be observed by autoradiography, emerges
therefore from the use of two systems. Germinating spores
of strain 168 have given information on the first half of
a round of replication, and vegetative cells of strain W23
on the second half. Marker frequency and density transfer
studies have shown no significant differences in the
pattern of replication in germinating spores *versus* vegetative cells nor in strain 168 *versus* W23. The overall
picture then is that following initiation of chromosome
replication two forks are generated. They proceed to move
in opposite directions, and at approximately equal rates,
away from the origin. Both forks can continue in this
manner until they approach one another to within <10% of
the length of the chromosome in a region that is approximately diametrically opposite the origin.

RECENT GENETIC STUDIES ON THE EXTENT
OF BIDIRECTIONAL REPLICATION

The autoradiographic data described in the previous
Section strongly supports a process of replication fork
movement in *B. subtilis* that is completely bidirectional.
But what of the linkage map, constructed by Dubnau, which
suggests a largely unidirectional process? Recently there
have been two very significant developments that relate
directly to this question. One has been the extensive
remapping accomplished by Harford in Brussels in
collaboration with the Lepesants and their colleagues in
Paris. The new map is based on, and indeed very strongly
supports a fully bidirectional mode of replication. On
the other hand Sueoka and his colleagues, through a somewhat similar approach, have concluded that replication is

only partially bidirectional, there being a termination
site located at a distance of about 25% of the length of
the chromosome from the origin.

The Harford-Lepesant Map: In a very detailed study of the
order of replication of many markers, by density transfer
using germinating spores of strain 168, Harford provided
evidence that was not compatible with previous suggestions
of unidirectional or even partial bidirectional replication.   Included in his study were many of the markers
mapped in the same way by O'Sullivan and Sueoka (5) and
the results of the two studies, with respect to these
markers, were in close agreement.   Assuming bidirectional
replication of a circular chromosome and with the aid of
some new linkage data of his own and especially Lepesant-
Kejzlarova *et al* (28,29) Harford suggested a new arrangement of the well established linkage groups.   In a
reciprocal approach, Lepesant-Kejzlarova *et al* using their
new linkage data as the basis of their argument, and
helped by Harford's replication order findings, were led
to the same rearrangement (29).   Fig. 12 shows how it was
achieved.   Two of the major factors with respect to replication order that led to it were the confirmation that
hisA replicates ahead of *thr* and *leu* ahead of *ura*, not the
reverse as shown in the map compiled by Dubnau (see Fig.
4).   The new arrangement shown in Fig. 12 eliminates or
enables the resolution of all 5 difficulties associated
with the unidirectional map as discussed previously.
    The closure of the gaps created by the new arrangement
was achieved largely by Lepesant-Kejzlarova *et al* (29).
The most recent map with all gaps closed was described
late last year and is reproduced in Fig. 13 (30).
    While there are some weak linkages in this map - some
of them have already been strengthened by new data
(Lepesant, private communication) - a very strong argument
for its validity is the fact that no significant discrepancies have emerged between the very extensive replication order and linkage data.   Also, from totalling transduction map distances the two arms of the chromosome,
origin to terminus in each direction, are of approximately
equal length.   In addition, the very recent replication
order and map position discrepancies reported by Copeland
(31) can be explained nicely by the new arrangement.   The
confidence of geneticists in the Harford-Lepesant map is

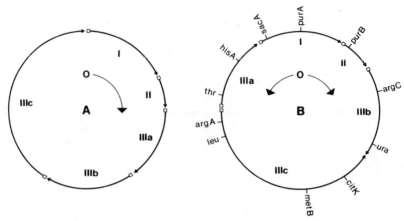

Fig. 12. Rearrangement of the genetic linkage groups of *B. subtilis*. A shows the old arrangement (see Fig. 4), and B the new one based on bidirectional replication (see refs. 27 and 29).

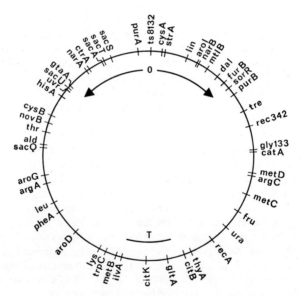

Fig. 13. The Harford-Lepesant map of the *B. subtilis* chromosome. Replication is completely bidirectional, from the origin (O) to the terminus (T) (from ref. 30, courtesy of Dr. J.A. Lepesant).

reflected in its adoption as the basis for updating the previous most detailed map compiled (32,33).

Genetic Evidence for Partial Bidirectional Replication and a Unique Terminus: O'Sullivan *et al*, through more extensive studies by the synchro-density transfer technique of markers near the origin, have confirmed that replication is bidirectional in this region (34). They also reaffirmed that *hisA* replicates ahead of *thr*. Their picture of the situation agrees with the Harford-Lepesant map over the region, *narA* to *lin* (see Fig. 13). However, in one experiment, they found one of Harford's late markers, *gltA*, to replicate very early, actually well ahead of *purB*. It should be emphasized that this particular experiment, which gave the clearest indication of early replication of *gltA*, utilized not germinating spores but the restart in replication upon transfer of the 168 Ts initiation mutant, *dna-1* (35) from 45°C back to 35°C. In other experiments, using the same temperature shift approach on a derivative of *dna-1* as well as synchronously germinating spores of strain W23, *gltA* was found to replicate only somewhat early, well after *purB* but before *leu*. In all experiments *citK* was found to replicate very late, in agreement with Harford. In separate transduction experiments, they confirmed the earlier report (13) of linkage between *gltA* and *citK*. To account for these findings *viz* early replication of *gltA* versus late replication of *citK* and linkage between them, it was proposed that the extent of bidirectional replication is limited to about the first 50% of the chromosome. The limitation is achieved through a block in the movement of one of the replication forks at a locus (termination site) between the two linked markers, *citK* and *gltA*. The situation is depicted in Fig. 14.

Thus with respect to replication fork movement there is substantial agreement that it is a bidirectional process for the first 50% of the chromosome. There is both autoradiographic and genetic evidence for this. With respect to the second half of the round of replication there is disagreement. Autoradiographic data (26) and the Harford-Lepesant map (27,29,30) strongly support a completely bidirectional mode of fork movement. O'Sullivan *et al*, however, have concluded that movement of one of the forks is blocked around the time of 50% overall replication (34).

It is worthwhile to look more closely at the precise nature of the discrepancy within the genetic studies relating to the extent of bidirectional replication. It arises because of the different findings with respect to the time of replication of the *gltA* marker. Harford finds it to replicate very late while O'Sullivan *et al* have shown that it can replicate early. There are two reasons why one cannot explain this discrepancy by invoking a difference in the *gltA* markers used by the two groups. Firstly, the recipient strains used for transformational analysis and carrying *gltA1* (Harford) and *gltA14* (O'Sullivan *et al*) both originated from Hoch, who has shown them to be alleles of the same gene and linked by recombination index analysis in transformation (Hoch, personal communication). Secondly, in the experiment of O'Sullivan *et al* where *gltA14* was found to replicate earliest, *citB17* was found to replicate ahead of it. Using exactly the same recipient for transformation, Harford, on the other hand, found *citB17* to replicate very late. Thus one can only conclude that the very early replication of *gltA* observed by O'Sullivan *et al* is valid in the sense that it reflects replication of the same *citB-gltA* region that Harford had found to replicate late. However, the experimental systems used by the two groups to achieve synchrony in chromosome replication were different and it is possible that the source of the discrepancy lies here. One reason for suspecting this is the fact that with the different approaches used by O'Sullivan *et al* considerable variability in the time of replication of *gltA* with respect to other markers was found. These workers also imposed a period of thymine starvation, before adding density label, to improve the initiation synchrony in all their experiments. Harford, on the other hand, followed the well-tried procedure of germinating spores in density-labelling medium (no thymine starvation) to fix the order of replication of more than 25 markers uniformly dispersed over the chromosome. Where the replication order of markers had been studied carefully using the same approach before (5,12) there is virtually complete agreement with Harford's results. Taking all these facts into consideration one cannot help but suspect that, normally, *gltA* replicates late, as Harford finds, and that there is some unusual feature of the system or

668

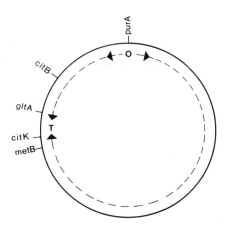

Fig. 14. Partial bidirectional replication of the
B. *subtilis* chromosome as proposed by O'Sullivan *et al*
(34). Replication commences in a bidirectional manner
from the origin (O). The leftward fork is blocked upon
reaching a unique termination site (T) located between
gltA and *citK* and about 25% of the chromosome from the
origin.

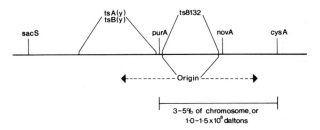

Fig. 15. The region of the B. *subtilis* chromosome
containing the origin of replication. Both the origin and
ts8132 are located between *purA* and *novA*, but it is not
known how they are positioned with respect to one another.
The map is to scale and has been compiled from the data in
refs. 17,27,30,36 and 44. The arrows show the directions
of replication fork movement.

systems used in the synchro-density transfer experiments
of O'Sullivan *et al* which leads to the early replication
of *gltA*. But it would be unwise, on the basis of past
experience, not to persevere until these apparent discrepancies in the genetic data are resolved by direct
experimentation.

It should be mentioned that Hara and Yoshikawa also
showed that replication is bidirectional in the origin
region (36). In addition they concluded that the movement
of one fork is blocked after traversing no more than 20%
of the chromosome, but there was no direct experimental
evidence to support this conclusion.

THE CONSEQUENCES OF BIDIRECTIONAL REPLICATION

Reinitiation of Replication: It was Alberts who first
appreciated the advantages of bidirectional replication in
accounting for the structure of reinitiated (dichotomous)
chromosomes (37). It was put forward as a concrete
suggestion to overcome the difficulty of linking 4 origins
to one terminus in a reinitiated circular chromosome
undergoing unidirectional replication. And it was the
direct investigation into the structure of reinitiated
chromosomes by autoradiography that provided the first
evidence for bidirectional replication in *B. subtilis*
(16). Alberts presented a molecular model, based on
template priming at breaks, to account for the generation
of replication forks leading to bidirectional replication
and subsequent reinitiation. The template priming
aspect of the model was suggested by the data of Yoshikawa
supporting the existence of physical linkage at the
initiation site between daughter and parental strands in
B. subtilis (38,39). However, Yoshikawa's proposition of
template priming was made well before the discovery of RNA
as a primer for DNA synthesis (40) and was almost certainly encouraged by the consensus at the time that replication in *B. subtilis* was essentially unidirectional. The
discovery of bidirectional replication, RNA priming of DNA
in some systems and discontinuous DNA synthesis at replication forks raises the possibility of alternative and much
less complex mechanisms for the initiation and reinitiation of rounds of chromosome replication than those put
forward by Alberts. The first initiation could involve
the simple "opening-out" or "denaturation" of the origin

region without the immediate introduction of strand breaks. The generation of two readily accessible daughter origins through the bidirectional movement of replication forks would then allow a straightforward repetition of the process to accomplish reinitiation.

Segregation of Chromosomes: There is now considerable evidence to substantiate the original finding (41) that the origin region of the B. *subtilis* chromosome is attached in some special way to the cell membrane. Indeed, such attachment offers an extremely attractive mechanism by which chromosome segregation could be achieved through membrane expansion between such sites of attachment (see ref. 42). The finding of bidirectional replication makes such a proposition even more attractive as, clearly, the process of segregation of daughter chromosomes could commence immediately or very soon after replication of the origin regions. And such a process would be most efficient if the replication process were completely rather than partially bidirectional. The advantage of complete bi-directional replication in this respect is particularly evident when one considers the ease with which one can order the segregation of multiply reinitiated circular chromosomes through the separation of membrane bound origins (43).

Fine Structure of the Origin Region: As already pointed out, one of the early observations suggesting that replic-ation in B. *subtilis* may be bidirectional came from the apparent discrepancy between replication order and linkage data concerning markers near the origin. The confirmation of bidirectional replication has led to a more confident mapping of markers in this region. And it is certain that substantial information on the mechanism of initiation of bacterial chromosome replication will ensue from more refined mapping still. For example, it will be important to establish the precise location of the site from which polynucleotide chain growth starts. While this is a formidable task B. *subtilis*, because it can be readily transformed and induced to give synchronous initiation of chromosome replication, offers advantages over other bacterial systems for such studies.

Hara and Yoshikawa have already exploited these features of B. *subtilis* to gain more detailed information

671

on the structure of the origin region (36). They have
developed an ingenious technique for the selective
mutation of genes near the origin. This has led to the
identification of a marker, *ts8132*, which replicates ahead
of *purA*. The mutant containing this marker is unusual in
that it shows some residual DNA synthesis at the nonpermissive temperature without any substantial change in
the origin:terminus marker ratio. A three point cross by
transformation placed *ts8132* between *purA* and *novA*, the
latter having been mapped previously at approximately
mid-way between *purA* and *cysA* (44). The location of
ts8132 between *purA* and *cysA* has been confirmed (30).
Because *purA* has been found by various workers to replicate
before *cysA*, and *ts8132* lies between them, it follows that
the origin of replication also occurs in this region.
Furthermore, if fork movement is symmetrical right from the
start it means that the origin lies in the *purA-novA*
portion, as does *ts8132* (see Fig. 15). From published data
it can be calculated that the distance separating *purA*
and *cysA* accounts for 3-5% of the chromosome or
$1.0-1.5 \times 10^8$ daltons of DNA. Thus the *purA-novA* region,
containing both *ts8132* and the origin is still fairly
large, 1.5-2.5% of the chromosome, $5-8 \times 10^7$ daltons or 1-3
times the size of the lambda chromosome. Obviously much
more work is needed to fix precisely the positions of the
origin and *ts8132* and it will be interesting to know just
how close to one another they are. The additional genes
for DNA synthesis identified by Hara and Yoshikawa, *tsA(y)*
and *tsB(y)*, lie to the left of *purA* as shown in Fig. 15.

Prior to the identification of *ts8132* Borenstein and
Ephrati-Elizur (45) had shown the *dnaC30* marker (immediate
stop in DNA synthesis) of Karamata and Gross (46) to replicate before *purA*. This has been confirmed by O'Sullivan
and Sueoka (personal communication) who have suggested
that *dnaC30* is located at a position similar to that shown
for *ts8132* above. Also, by PBS1 transduction, two other
immediate-stop DNA mutants, *dnaG* and *H*, have been mapped
between *purA* and *cysA* (46). The possible relationship
between *ts8132* and *dnaC,G* and *H* has not yet been investigated. However, the clustering of at least 3 distinct
genes for DNA synthesis between *purA* and *cysA* and within
3-5% of the origin of replication could be significant.
Although none of them has been classified as an initiation
gene it is interesting to speculate that their action in

DNA synthesis could be mediated through events associated
with or subsequent to the act of initiation itself.

Fork Movement at the Terminus: Bidirectional replication
commencing from a genetically defined origin on a circular
chromosome, involves the gradual separation of two replication forks. Eventually, the two forks must come back
together to effect termination and allow the segregation
of daughters. [As pointed out by Dingman (43) the two
replication forks do not necessarily separate in space.
It could be that they are held close to one another in a
single complex capable of effecting chain growth at both
forks]. With respect to the movement of the two replication forks at termination the question that arises
immediately is, do they meet at a specifically defined
site on the chromosome? For *B. subtilis* it is possible to
give only a partial answer at this stage. There is agreement from the replication order studies of Harford (27)
and O'Sullivan *et al* (34) that the forks meet in the
vicinity of the *citK* and *gltA* markers, which could be
separated by up to 5% of the chromosome. But is this
because these markers happen to be located approximately
diametrically opposite the origin, as suggested by Harford,
and the replication forks moving at approximately equal
rates happen to meet in the vicinity of this region on
most occasions? Or is there a very specific site between
citK and *gltA*, as suggested by O'Sullivan *et al*, which is
capable of blocking replication fork movement? Assuming
that the *citK-gltA* region is located approximately diametrically opposite the origin, the autoradiographic
studies (26) and the genetic studies of Lepesant-
Kejzlarova *et al* (30) cannot distinguish between the two
possibilities. On the other hand, the evidence presented
by O'Sullivan *et al* (34) for a very specific termination
site between *citK* and *gltA* rests very heavily on their
data suggesting that this region is located far from the
position opposite the origin, actually at only about 25%
of the chromosome from the origin. The problem of recon-
ciling this conclusion with the findings of Wake (26) and
Lepesant-Kejzlarova *et al* (30) has already been considered.
While this problem certainly reduces the strength of the
overall argument for the specific termination site it does
not eliminate it. Thus, it has been argued that the very
early replication of the *citB-gltA* region, observed by

O'Sullivan *et al* when the temperature sensitive initiation mutant *dna*-1 restarts replication at 35°C, may reflect some unusual feature of this experimental system. But the fact that one can get conditions that allow the two regions, *citK* and *citB-gltA*, to replicate independently even though they are situated next to one another on the chromosome still raises the possibility of a site, with some special property, located between them. And this could be the specific termination site with the blocking properties suggested by O'Sullivan *et al*.

ACKNOWLEDGEMENTS

In the preparation of this manuscript I have been helped considerably, through general and specific comment and/or the communication of unpublished material, by Drs. J.C. Copeland, N. Harford, J.A. Hoch, J.A. Lepesant, N. Sueoka and F.E. Young. The financial support of the Australian Research Grants Committee and the University of Sydney Cancer Research Fund is gratefully acknowledged.

REFERENCES

1. Cairns, J., J. Mol. Biol. *6*, 208 (1963).
2. Cairns, J., Cold Spr. Harb. Symp. Quant. Biol. *28*, 43 (1963).
3. Sueoka, N. and Yoshikawa, H., Cold Spr. Harb. Symp. Quant. Biol. *28*, 47 (1963).
4. Oishi, M., Yoshikawa, H. and Sueoka, N., Nature *204*, 1069 (1964).
5. O'Sullivan, A. and Sueoka, N., J. Mol. Biol. *27*, 349 (1967).
6. Dubnau, D., Goldthwaite, C., Smith, I. and Marmur, J., J. Mol. Biol. *27*, 163 (1967).
7. Hayes, W., *The Genetics of Bacteria and their Viruses*, 2nd ed., Blackwell Scientific Publications, Oxford and Edinburgh (1968).
8. Oishi, M. and Sueoka, N., Proc. Nat. Acad. Sci. U.S.A. *54*, 483 (1965).
9. Oishi, M., Oishi, A. and Sueoka, N., Proc. Nat. Acad. Sci. U.S.A. *55*, 1095 (1966).
10. Dubnau, D., In *Handbook of Biochemistry* (Sober, H.O. and Harte, R.A. eds.), 2nd ed., pp. 39-45, Chemical Rubber Co., Cleveland, Ohio (1970).

11. Lepesant, J.A., Kunst, F., Carayon, A. and Dedoner, R., Compt. Rend. Acad. Sci. Paris, Ser. D. *269*, 1972 (1969).
12. Huang, P.C., Eberle, H., Boice, L.B. and Romig, W.R., Genetics *60*, 661 (1968).
13. Hoch, J.A. and Mathews, J., In *Spores V* (Halvorson, H.O., Hanson, R. and Campbell, L.L. eds.) pp. 113-116, Amer. Soc. Microbiol., Wash. D.C. (1972).
14. Wake, R.G., J. Mol. Biol. *77*, 569 (1973).
15. Yoshikawa, H., O'Sullivan, A. and Sueoka, N., Proc. Nat. Acad. Sci. U.S.A. *52*, 972 (1964).
16. Wake, R.G., J. Mol. Biol. *68*, 501 (1972).
17. Lepesant, J.A., Kunst, F., Lepesant-Kejzlarova, J. and Dedoner, R., Mol. Gen. Genet. *118*, 135 (1972).
18. Sueoka, N., Matsushita, T., Ohi, S., O'Sullivan, A. and White, K., In *DNA Synthesis in Vitro* (Wells, R.D. and Inman, R.B. eds.), pp. 385-404, University Park Press, Baltimore, Maryland (1972).
19. Young, F.E. and Wilson, G.A., In *Spores V* (Halvorson, H.O., Hanson, R.R. and Campbell, L.L., eds.), pp. 77-106, Amer. Soc. Microbiol., Wash. D.C. (1972).
20. Huberman, J.A. and Riggs, A.D., J. Mol. Biol. *32*, 327 (1968).
21. Schnos, M. and Inman, R.B., J. Mol. Biol. *51*, 61 (1970).
22. Masters, M. and Broda, P., Nature New Biol. *232*, 137 (1971).
23. Gyurasits, E.B. and Wake, R.G., J. Mol. Biol. *73*, 55 (1973).
24. Setlow, P., J. Bacteriol. *114*, 1099 (1973).
25. Upcroft, P., Dyson, H.J. and Wake, R.G., J. Bacteriol. *121*, 121 (1975).
26. Wake, R.G., J. Mol. Biol. *86*, 223 (1974).
27. Harford, N., J. Bacteriol.(in press).
28. Lepesant-Kejzlarova, J., Walle, J., Billault, A., Kunst, F., Lepesant, J.A. and Dedoner, R., Compt. Rend. Acad. Sci. Paris, Ser. D. *278*, 1911 (1974).
29. Lepesant-Kejzlarova, J., Lepesant, J.A., Walle, J., Billault, A. and Dedoner, R., J. Bacteriol. (in press).
30. Lepesant-Kejzlarova, J., Harford, N., Lepesant, J.A. and Dedoner, R. In *Spores VI*, Amer. Soc. Microbiol., Wash. D.C. (in press).
31. Copeland, J.C., Genetics (in press).

32. Young, F.E. and Wilson, G.A., In *Handbook of Genetics*, Vol. 1, pp. 69-114, Plenum Pub. Corp., New York (1974).
33. Young, F.E. and Wilson, G.A., In *Spores VI*, Amer. Soc. Microbiol., Wash. D.C. (in press).
34. O'Sullivan, A., Howard, K. and Sueoka, N., J. Mol. Biol. *91*, 15 (1975).
35. White, K. and Sueoka, N., Genetics *73*, 185 (1973).
36. Hara, H. and Yoshikawa, H., Nature New Biol. *244*, 200 (1973).
37. Alberts, B.M., Fed. Proc. *29*, 1154 (1970).
38. Yoshikawa, H., Proc. Nat. Acad. Sci. U.S.A. *58*, 312 (1967).
39. Yoshikawa, H., J. Mol. Biol. *47*, 403 (1970).
40. Brutlag, D., Schekman, R. and Kornberg, A., Proc. Nat. Acad. Sci. U.S.A. *68*, 2826 (1971).
41. Sueoka, N. and Quinn, W.G., Cold Spr. Harb. Symp. Quant. Biol. *33*, 695 (1968).
42. Jacob, F.S., Brenner, S. and Cuzin, F., Cold Spr. Harb. Symp. Quant. Biol. *28*, 329 (1963).
43. Dingman, C.W., J. Theor. Biol. *43*, 187 (1974).
44. Harford, N. and Sueoka, N., J. Mol. Biol. *51*, 267 (1970).
45. Borenstein, S. and Ephrati-Elizur, E., J. Mol. Biol. *45*, 137 (1969).
46. Karamata, D. and Gross, J.D., Mol. Gen. Genet. *108*, 277 (1970).

THE ROLE OF SEMICONSERVATIVE DNA
REPLICATION IN BACTERIAL CELL DEVELOPMENT

Terrance Leighton,* George Khachatourians,[†]
and Neal Brown[††]

ABSTRACT. Bacillus subtilis sporulation is initiated
when vegetative bacteria cease exponential growth.
Chemical and radioisotopic measurements of DNA synthesis
suggest that during the first two hours of sporulation,
there is a postexponential doubling of cellular DNA content.
During this same time period there is a substantial
increase in cell mass. Microscopic and electronic particle
size observations indicate that the termination of DNA
synthesis is followed by a conventional cell division and
a concomitant internal asymmetric septation of mother cell
and forespore chromosomes. The observed DNA synthesis is
evidently semi-conservative as it is totally inhibited by
the DNA polymerase III directed drug, 6-(p-hydroxyphenylazo)-uracil (HP-Ura). Addition of 50 μM HP-Ura to cultures
of wild type B. subtilis causes an inhibition of sporulation during the first two hours (t0-t2) of the developmental process. HP-Ura addition has no effect on sporulation when added after t2.5. Similarly, 50 μM HP-Ura
addition, at any time, to an isogenic strain of B. subtilis
containing a single site HP-Ura resistant DNA polymerase
III mutation has no effect on the sporulation sequence.
From experiments where HP-Ura is added at various times
during the t0-t2 increment it is apparent that the postexponential cell division is coupled to semi-conservative
DNA replication. DNA synthesis and cell size analysis of
several early blocked temperature sensitive sporulation
mutants (including possible DNA polymerase III mutations)

*Department of Bacteriology and Immunology, University
of California, Berkeley, California.
  †Department of Microbiology, University of
Saskatchewan, Saskatoon, Saskatchewan.
  ††Department of Pharmacology, University of Massachu-
setts Medical School, Worcester, Massachusetts.

also supports the association of correctly scheduled DNA synthesis and cell division with the normal developmental sequence. Although HP-Ura is capable of blocking sporulation selectively at early stages, the drug seems to have no effect on early stage sporulation specific transcription or translation. The ability of a small host dependent virus (ϕ29) to replicate in sporulating cells is positively correlated with the time period of host semi-conservative DNA synthesis.

INTRODUCTION

Bacterial sporulation is characterized by the sequential appearance of sporulation-specific and sporulation-associated proteins (1, 2). These temporal molecular changes are accompanied by a well defined series of cellular ultrastructural differentiations (3). Hence this process represent a suitable model system for investigating the control of normal and abnormal cell development.

It has been demonstrated previously that sporulation is specifically regulated by alterations in the transcriptional (4,5,6,7,8), translational (9), and posttranslational (10,11) capabilities of Bacillus subtilis cells. Aubert, Ryter and Schaeffer (12) have reported that DNA synthesis is limited to the early periods of bacterial cellular differentiation. This observation has not been subsequently extended to determine whether the observed DNA synthesis was semi-conservative or repair type polymerization, what factors might be responsible for the timing and shut-off of DNA synthesis, and whether the observed doubling of DNA mass was essential to the developmental program. The existence of a large collection of sporulation mutants (3,4,7,8,9,10), a drug 6-(p-hydroxyphenylazo)-uracil (HP-Ura), which specifically inhibits B. subtilis DNA polymerase III mediated semi-conservative DNA synthesis (13,14,15), and a variety of mutants affecting various components of the DNA synthesizing apparatus (16) have prompted us to investigate further the role of DNA synthesis in B. subtilis sporulation. Our preliminary results will be described in this paper.

METHODS

Experimental Conditions: The growth and sporulation of
Bacillus subtilis strains W168 (17), ts-14 (4), WB746 (10),
and ts-5 (10) in a glucose-supplemented nutrient broth
medium have been described. B. subtilis strains HA101F
(18), HA101FT12 (an isogenic DNA polymerase III HP-Ura
resistant mutant; 15, 19), JB1-49(59) (18), 1306 (18),
MN13 (18), N5-10 (18), SB1060 (20), and W168 spoA12 (21)
were grown as described above. Previously published
procedures were employed for the estimation of proteolytic
activity (10), forespore refractility (17), spore heat
resistance (17), cell viability (17), and the incorporation
of ^{3}H-adenine into DNA (22). DNA was quantitated by the
method of Burton (23) utilizing salmon sperm DNA as a
standard. B. subtilis DNA polymerase I, II and III
activities were quantitated using a procedure similar to
that described by Gass and Cozzarelli (18). Bacterial
cell numbers and size distributions were determined as
described previously (24) utilizing a Coulter Counter
equipped with a pulse-height analyzer and data storage-
retrieval system.

RESULTS

Figure 1 depicts the time course of growth and
sporulation under our experimental conditions. B. subtilis
W168 cells grew with a 28 min generation time for 2 hrs at
37°C. The developmental sequence was initiated at the
time when exponential growth ceased (t_0). By t_{12} 95% of
the cells present at t_0 had produced heat resistant
endospores. The acquisition of refractility by the forespore is a late sporulation event (1,3). Proteolytic
enzyme synthesis (an early sporulation event; 1,3,10,11)
was initiated at t_0. DNA polymerase I deficient mutant
cells (HA101F, JB1-49(59), 1306, MN13, N5-10, and SB1060)
sporulated with the same time course as W168 cells. An
isogenic derivative strain of HA101F containing an HP-Ura
resistant DNA polymerase III (HA101FT12) sporulated with
the same time course as HA101F.

Chemical measurements of postexponential DNA
accumulation are illustrated in Figure 2. Similar results
were obtained over the same time increment by quantitating

679

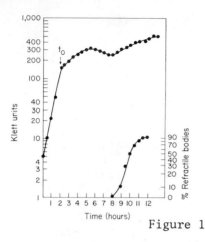

Figure 1

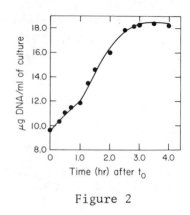

Figure 2

Fig. 1.   Growth and sporulation of B. subtilis. t_0 is the start of the sporulation sequence.

Fig. 2.   Postexponential accumulation of DNA by B. subtilis cells.

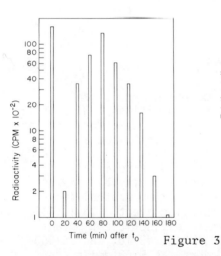

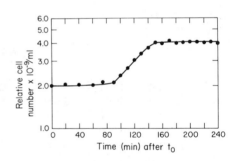

Figure 4

Figure 3

Fig. 3.   Pulse label incorporation of ^{3}H-adenine into DNA during postexponential time periods. Each time point represents a 5 min pulse of ^{3}H-adenine (25 μCi/ml ^{3}H-adenine, 5 μg/ml carrier adenine). Zero time backgrounds (cells + TCA + ^{3}H-adenine) have been subtracted.

Fig. 4.   Postexponential cell division in B. subtilis. Relative cell numbers were determined with a Coulter Counter. Coulter Counter values were two fold higher than the actual number of viable cells determined by plate counts.

the long-term accumulation of ^{3}H-adenine into DNA. The pattern of postexponential pulse label incorporation into DNA is shown in Figure 3. There are approximately 1 x 10^9 cells/ml present at t_0, and approximately 2 x 10^9 cells/ml present at t_3. Therefore, at t_0 or t_3 each cell contained 1.0 x 10^{-14} g of DNA, or an average of 2 genome equivalents (chromosomes) per cell. Just prior to the postexponential cell division (see Figures 4 and 7) each sporangium contained 4 chromosomes.

Figure 4 depicts the time course of B. subtilis W168 postexponential cell division. Conditions which resulted in asynchronous sporulation also produced asynchronous cell division patterns (see ref. 17). Postexponential cell division was observed at the nonpermissive temperature (46°C) in B. subtilis sporulation mutants ts-5 (a stage 0 block), ts-14 (a stage II block), and at 37°C in another stage 0 mutant W168 spoA12. Sporulation stages 0-VII are equivalent to t_0-t_{7-8} at 37°C. For a complete description of the morphological changes accompanying bacterial sporulation see Schaeffer (3).

B. subtilis W168 postexponential cell volume changes are illustrated in Figure 5. Under our experimental conditions vegetative and postexponential volume profiles of strains W168, HA101F, HA101FT12, WB746 and ts-14 indicated that at least 90% of the population exists as 1X, 1X-2X, or 2X cells. These estimates of cell sizes, and the previously described cell division data were corroborated by sequential phase contrast microscopic observations of postexponential B. subtilis cells. Strain W168 spoA12 grew exponentially (28 min generation time at 37°C) as filamentous cells some 3X-5X as large as 1X W168 cells. These filamentous cells divided postexponentially into 1X size cells over a very prolonged time period (15 hrs at 37°C). The postexponential mass increase in the spoA12 strain also was extended over a very prolonged time course. WB746 ts-5 cells grown at the nonpermissive temperature exhibited normal exponential phase cell volume profiles. However, postexponentially the time course of DNA synthesis (measured chemically or isotopically) was prolonged while the postexponential mass increase occurred over the usual time increment. Hence postexponential cell division resulted in cells 2X-3X larger than 1X vegetative cells. Similar results were obtained when 100 μM HP-Ura was used to slow the time course of DNA synthesis, cell division and sporulation in HA101FT12 cells.

Figure 6 depicts the effects of HP-Ura and chloramphenicol on B. subtilis W168 cell division. Pulse label experiments demonstrated that 50 μM HP-Ura inhibited 95% of the ^{3}H-adenine incorporation into DNA. 150 μg/ml of chloramphenicol totally inhibits B. subtilis protein synthesis (17). HP-Ura inhibition of sporulation quantitatively followed the same time course as its effects on cell division (i.e., addition of HP-Ura at t_0 stopped all sporulation, addition at $t_{1.5}$ resulted in 50% sporulation, addition at $t_{2.5-8.0}$ resulted in 95% sporulation). 50 μM HP-Ura had little affect on cell viability. The largest decrease observed after 18 hrs exposure was a 40% decrease in cell numbers when the drug was added at t_0. The addition of 50 μM HP-Ura to strain HA101FT12, at any time, had no affect on the developmental sequence or cell viability. The addition of 100 μM HP-Ura to t_0 HA101FT12 cells delayed the time of appearance refractile bodies 3 hrs (37°C). The forespores and sporangia were significantly larger (1.5X-2X) when compared to an untreated culture.

The mechanism by which HP-Ura blocks sporulation does not involve the inhibition of early stage sporulation-specific transcription or translation. W168 cells exposed to 50 μM HP-Ura continued to accumulate serine and metal protease activity (early sporulation-specific and sporulation-associated proteins; 10, 11) at the rate of an untreated culture. The affect of 50 μM HP-Ura on sporulation in W168 cells was reversible. If cells were exposed to 50 μM HP-Ura between t_0 and t_1, and the inhibitor was removed by filtration (17), the cells proceeded through the complete developmental sequence.

DISCUSSION

It has been suspected for many years that certain aspects of DNA synthesis, i.e., the timing or quality of chromosome replication, may be critical to normal cellular developmental programs. Our results would indicate that this notion is correct in the case of B. subtilis

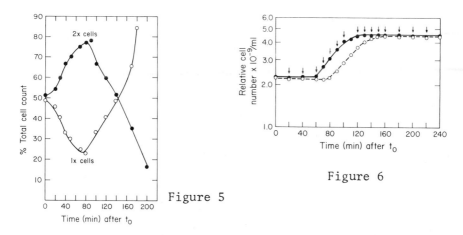

Figure 5

Figure 6

Fig. 5.   Postexponential cell volume changes in B. subtilis. Population size distributions were determined by a pulse height analyzer equipped with selective integration modes. 1X cells were equivalent in size to newly divided vegetative cells.

Fig. 6.   Inhibition of B. subtilis postexponential cell division by HP-Ura (50 μM) or chloramphenicol (150 μg/ml). Portions of a sporulating culture were exposed to either HP-Ura (●) or, chloramphenicol (O) at the times indicated by the arrows. The amount of terminal cell division was measured in each culture and the final relative cell numbers were plotted as data points.

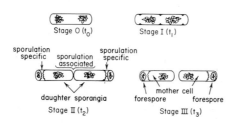

Fig. 7.   A model for B. subtilis postexponential cell division (nutrient broth medium). Septum sites are indicated. The sporulation-associated division site is a normal vegetative septum containing peptidoglycan. The sporulation-specific septum contains no peptidoglycan.

683

sporulation. We have been able to confirm the results of Aubert et al. (12) that there is a doubling of cellular DNA content during the early stages of sporulation. We find that the termination of DNA synthesis is followed by a single synchronous cell division. This cell division requires prior DNA and protein synthesis. A similar relationship of DNA synthesis, protein synthesis, and cell division has been observed in E. coli (25). The sensitivity of postexponential DNA synthesis to HP-Ura would suggest that it is semi-conservative and involves DNA polymerase III (13,14,15). Our data also would be consonant with the initiation of a single round of DNA replication shortly after t_0. Further studies are in progress to definitely establish whether these inferences are correct.

It would appear that the completion of the post-exponential round of DNA replication is an essential early developmental event. Inhibition of DNA replication precludes later morphogenesis without affecting the transcription or translation of stage O-I gene products. However, several stage O blocked mutants are capable of exhibiting postexponential DNA replication and cell division. These facts can be reconciled in the following fashion. At t_0 B. subtilis cells, growing rapidly in nutrient broth, contain two chromosomes per cell (26,27, 28). Between t_0 and $t_{2.0}$ the chromosome content of such cells doubles. These events are followed by a conventional cell division. However, concomitantly a sporulation-specific internal asymmetric cell division occurs. This asymmetric septation process, which divides mother cells from developing forespores, requires the action of sporulation-specific gene products (3,28,29), and is a priori a prerequisite for further cell differentiation. The same signal, i.e., the completion of a round of DNA replication, triggers both division events. However, the sporulation-specific asymmetric septation will not occur if mutations are present which prevent the expression of necessary gene products. Hence stage O mutants are capable of postexponential daughter cell separation (a vegetative process), without the production of a forespore septum (3,28,29). This model is diagramatically represented in Figure 7 and is consistent with electron microscopic studies of postexponential wild type and mutant cells

(28,29,30), and with the results of Callister and Wake (31) which demonstrate that a single completed chromosome is present in mature spores. Furthermore, it would appear that the absence of cell wall deposition in the forespore septum is actively developmentally regulated, since the RNA polymerase mutant ts-14 deposits cell wall material in this septum (30), which necessarily precludes the possibility of further differentiation. This interpretation of the available data would suggest that cell surface site-specific differentiation occurs during bacterial sporulation. As well as serving as a trigger for septation events DNA replication may also serve as a "clock" for early developmental processes since slowing the rate of DNA replication in HA101FT12 concomitantly delays the time of forespore appearance. In addition, we have found that some revertants from a temperature sensitive lethal dnaF mutation (KF-8, see ref. 16), which is allelic with the known DNA polymerase III drug resistance allele (19), demonstrate temperature sensitive sporulation, but are able to grow normally at the previously nonpermissive temperature. The fact that the stage 0 mutant spoA12.grows vegetatively as filamentous cells raises the possibility that this mutant phenotype could be due in part to a cell septation lesion which prevents the formation of a forespore septum. Clearly our model does not explain the series of events which would obtain following slow vegetative growth rates, i.e., the situation of one chromosome per cell at t_0. Experiments are in progress to determine whether a postexponential cell division occurs under such conditions.

We have begun to study the mechanisms, which could be responsible for the turn-off of DNA synthesis after $t_{2.5}$. Our results and those of Setlow (22) would indicate that after $t_{2.5}$ B. subtilis cells may become deficient in ribonucleotide reductase, and the ability to initiate new rounds of DNA replication. Consistent with the importance of a precursor defect we have found that t_3 sporulating cells contain vegetative levels of DNA polymerases I, II and III. In this context it is interesting that a small host dependent virus, $\phi29$, which is dependent on the host ribonucleotide reductase system, ceases to productively infect sporulating cells over the same time periods that postexponential DNA synthesis declines (32,33). However, large viruses such as SP01, which are independent of the host ribonucleotide reductase system, can productively infect sporulating cells over a much longer time course (32,33). These results would be consonant with the inability to detect significant repair type DNA synthesis after t_3 (33,34).

ACKNOWLEDGEMENT

This work was supported by grants from the National Science Foundation, National Institutes of Health, and American Cancer Society (California Division) to TL, and by a grant from the National Institutes of Health to NB. We are grateful to Donald Tipper for use of facilities during part of these experiments.

REFERENCES

1. Hanson, R. S., Peterson, J. A. and Yousten, A. A. Ann. Rev. Microbiol., 24, 53 (1970).
2. Dawes, I. W. and Hansen, J. N. Critical Rev. Microbiol. 1, 479 (1972).
3. Schaeffer, P. Bacteriol. Rev., 33, 48 (1969).
4. Leighton, T. J. Proc. Nat. Acad. Sci. U.S.A., 70, 1179 (1973).
5. Sumida-Yasumoto, C. and Doi, R. H. J. Bacteriol., 117, 775 (1974).
6. DiCioccio, R. A. and Strauss, H. J. Mol. Biol., 77, 325 (1973).
7. Sonenshein, A. L., Cami, B., Brevet, J. and Cote, R. J. Bacteriol., 120, 253 (1974).
8. Leighton, T. (submitted for publication).
9. Leighton, T. J. Mol. Biol., 86, 855 (1974).
10. Leighton, T. J., Doi, R. H., Warren, R. A. J., Kelln, R. A. J. Mol. Biol., 76, 103 (1973).
11. Leighton, T. and Simpson, D. (submitted for publication)
12. Aubert, J. P., Ryter, A. and Schaeffer, P., Spores IV 149 (1969).
13. Brown, N. C. J. Mol. Biol., 59, 1 (1971).
14. Brown, N. C., Wisseman, C. L. and Matsushita, T. Nature, 237, 72 (1972).
15. Clements, J. E., D'Ambrosio, J. and Brown, N. C. J. Biol. Chem., 250, 522 (1975).
16. Karamata, D. and Gross, J. D. Mol. Gen. Genet., 108, 277 (1970).

17. Leighton, T. J. Biol. Chem., 249, 7808 (1974).
18. Gass, K. B. and Cozzarelli, N. R. J. Biol. Chem., 248, 7688 (1973).
19. Love, E., D'Ambrosio, J., Brown, N. C. and Leighton, T. Manuscript in preparation.
20. Laipis, P. J. and Ganesan, A. T. J. Biol. Chem., 247, 5867 (1972).
21. Ito, J. and Spizizen, J. Spores V, 107 (1972).
22. Setlow, P. J. Bacteriol., 114, 1009 (1973).
23. Burton, K. Methods Enzymol., 12B, 163 (1968).
24. Khachatourians, G. and Tipper, D. J. Antimicrob. AG. Chemother., 6, 304 (1974).
25. Slater, M. and Schaechter, M. Bacteriol. Revs., 38, 199 (1974).
26. Sueoka, N. and Yoshikawa, H. Genetics, 52, 747 (1965).
27. Cooper, S. and Helmstetter, C. E. J. Mol. Biol., 31, 519 (1968).
28. Santo, L. Y. and Doi, R. H. The Cell, 6, 154 (1974).
29. Santo, L., Leighton, T. J. and Doi, R. H. J. Bacteriol., 111, 248 (1972).
30. Santo, L., Leighton, T. J. and Doi, R. H. J. Bacteriol., 115, 703 (1973).
31. Callister, H. and Wake, R. G. J. Bacteriol., 120, 579 (1974).
32. Kawamura, F. and Ito, J. Virology, 62, 414 (1974).
33. Leighton, T. Manuscript in preparation.
34. Sonenshein, A. L. Virology, 42, 488 (1970).

EFFECT OF AN ARYLAZOPYRIMIDINE ON DNA SYNTHESIS IN TOLUENE-TREATED Salmonella typhimurium

Charles P. Van Beveren* and Andrew Wright

Department of Molecular Biology and Microbiology
Tufts University School of Medicine
Boston, Massachusetts 02111

ABSTRACT. We have examined DNA synthesis in toluene-treated cells of Salmonella typhimurium. DNA synthesis in cells deficient in DNA polymerase I requires the four deoxynucleoside triphosphates and ATP, and is sensitive to sulfhydryl reagents and nalidixic acid. It is also sensitive to 6-(p-hydroxyphenylazo)-uracil. This is the first demonstration that an arylazopyrimidine can inhibit DNA synthesis in a Gram-negative aerobe.

INTRODUCTION

The arylazopyrimidines inhibit the growth of most Gram-positive bacteria (1) by blocking DNA replication. The reduced, hydrazino derivative of 6-(p-hydroxyphenyl-azo)-uracil (HPUra) inhibits DNA synthesis in toluene-treated cells of Bacillus subtilis (2); the drug was shown to specifically inhibit B. subtilis DNA polymerase III, the presumed replicase, but to have no effect on other B. subtilis DNA polymerases (3,4). The sum of present evidence indicates that the drug acts by forming a slowly dissociating ternary complex with DNA polymerase III and with cytosine residues in template DNA (5,6,7).

In contrast to its marked effect in Gram-positive bacteria, HPUra has no effect on the growth of Gram-negative aerobes. It is of interest to know whether the different sensitivities of these organisms reflects a difference in the DNA replication apparatus of Gram-positive and negative bacteria, or is rather due to a different permeability to the drug. DNA synthesis in toluene-treated cells of Escherichia coli is not inhibited by HPUra (7). The three purified DNA polymerases of E. coli are similarly resistant to HPUra (7). The resistance of at least one

---

*Present address: Department of Medicine, University of California at San Diego, La Jolla, California 92037.

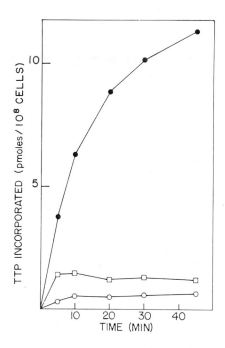

Fig. 1. Kinetics of ^{3}H-TTP incorporation into toluene-treated CV112 cells. Cells at a density of $3\text{-}4\text{x}10^9$/ml in 50 mM KPO$_4$ buffer, pH 7.4, were vigorously shaken with 1% toluene for 3 min at 37° (9). 0.05 ml cells was added to 0.25 ml of a mixture containing 21 µmole KPO$_4$, pH 7.4, 6 µmole MgCl$_2$, 0.6 µmole ATP (Calbiochem), and 10 nmole dGTP, dCTP, dATP (Schwarz-Mann) and ^{3}H-TTP (250 mCi/mmole; New England Nuclear). When NEM (Sigma) was used, cells were treated with 1 mM 2-mercaptoethanol (Sigma) in ice for 2 min before they were added to the reaction mixture. Reactions were carried out at 30°, and stopped with 1 ml cold 10% TCA-0.1 M sodium pyrophosphate. Precipitates were collected at 1500 g for 10 min, resuspended in 0.5 ml 0.5 N NaOH-1% sodium dodecyl sulfate, precipitated with 3 ml cold 10% TCA, and collected on 24 mm GF/A filter discs (Whatman). o, -ATP; •, +ATP; □, +ATP+NEM.

Gram-negative organism to an arylazopyrimidine does not, then, appear to be due merely to a permeability effect.

We have examined the sensitivity of DNA synthesis to HPUra in another Gram-negative organism, Salmonella typhimurium. We have isolated a DNA polymerase I-deficient mutant of S. typhimurium LT2, CV112 (8), and have used this to study DNA synthesis in toluene-treated cells.

RESULTS

The kinetics of DNA synthesis in this system are shown in Figure 1, and the characteristics are illustrated in Table 1. The reaction requires the four deoxynucleoside triphosphates and ATP, and is sensitive to N-ethylmaleimide (NEM) and to nalidixic acid. By analogy to similar systems in E. coli (9) and B. subtilis (10), the DNA synthesis seen here is probably replicative in nature. It may also be seen that, in contrast to its effect in toluene-treated cells of B. subtilis (2), NADPH strongly inhibits DNA synthesis in these cells.

The effect of increasing concentrations of HPUra on Salmonella DNA synthesis is shown in Table 2. 200 μM HPUra gave 34% inhibition in this experiment. In another experiment concentrations of HPUra higher than 200 μM did not increase the observed inhibition.

The extent of inhibition seen with a given concentration of HPUra varied widely. 100 μM HPUra caused from 10 to 75% inhibition in different experiments. The observed inhibition was, however, reproducible within a given experiment.

It has been shown that, in toluene-treated cells of B. subtilis, a reducing agent must be added to generate the active form of HPUra (2). As is shown in Table 3, though HPUra does inhibit Salmonella DNA synthesis in the presence of reducing agents, it also inhibits DNA synthesis in the absence of any exogenous reducing agent. Since an essential element in HPUra inhibition of B. subtilis DNA polymerase III involves hydrogen-bonding between cytosine residues and the reduced form of the drug, either the drug, supplied to the cells in the oxidized form, is reduced in toluene-treated cells of S. typhimurium and not B. subtilis, or HPUra inhibits DNA synthesis in Salmonella by a mechanism different from that demonstrated for B. subtilis.

One indication of whether the drug is acting by the

690

TABLE 1

CHARACTERISTICS OF DNA SYNTHESIS IN
TOLUENE-TREATED CV112 CELLS

System	% Activity
Complete	(100)
– ATP	6
– dCTP	5
+ nalidixic acid (100 or 200 μg/ml)	54
+ NADPH (2 mM)	40
+ sodium dithionite (0.3 mM)	110

CV112 cells were treated with toluene and incubated for 20
min as described in Figure 1. Nalidixic acid (Calbiochem),
NADPH (Sigma) and sodium dithionite (Eastman Chemical) were
included as indicated.

TABLE 2

EFFECT OF INCREASING CONCENTRATIONS OF HPUra
ON ^{3}H-TTP INCORPORATION
INTO TOLUENE-TREATED CV112 CELLS

HPUra concentration	% Inhibition
50 μM	13
100 μM	25
200 μM	34

CV112 cells were treated with toluene and incubated for 20
min as described in Figure 1. Sodium dithionite (0.3 mM)
was included in the incubation mixtures.

TABLE 3

EFFECT OF REDUCING AGENT ON HPUra
INHIBITION OF ^{3}H-TTP INCORPORATION
INTO TOLUENE-TREATED CV112 CELLS

Expt.	Reducing agent	HPUra conc.	% Inhibition
1	NADPH	75 µM	40
2	dithionite	80 µM	23
2	NADPH + dithionite	80 µM	22
3	dithionite	100 µM	51
3	none	100 µM	33

CV112 cells were treated with toluene and incubated for 20
min as described in Figure 1. NADPH (2 mM) and sodium dithionite (0.3 mM) were included as indicated.

TABLE 4

EFFECT OF dNTP'S ON HPUra INHIBITION
OF ^{3}H-TTP INCORPORATION
INTO TOLUENE-TREATED CV112 CELLS

Reducing agent	dNTP	dNTP conc.	% Inhibition
NADPH	none	——	64
	dCTP	0.9 mM	64
	dGTP	0.9 mM	36
dithionite	none	——	73
	dATP	1.0 mM	73
	dGTP	1.0 mM	59

CV112 cells were treated with toluene and incubated for 20
min as described in Figure 1. NADPH (2 mM) and sodium dithionite (0.3 mM) were included as indicated.

same mechanism in both organisms would be the ability of dGTP to specifically reverse HPUra inhibition of S. typhimurium DNA synthesis. In toluene-treated cells of CV112, 0.9-1.0 mM dGTP partially reversed the HPUra inhibition, whereas the same concentration of dCTP or dATP had no effect whatsoever (Table 4). Thus it would be reasonable to assume that HPUra inhibition of DNA synthesis in S. typhimurium does involve some interaction with cytosine residues.

DISCUSSION

Our results indicate that ATP-dependent DNA synthesis in toluene-treated cells of S. typhimurium is partially sensitive to HPUra. It is reasonable, therefore, to speculate that a DNA polymerase III-like activity of S. typhimurium is sensitive to the drug. The reversibility of HPUra inhibition specifically by dGTP suggests that the drug acts by the same mechanism in both S. typhimurium and B. subtilis. It must also be concluded that resistance of E. coli DNA replication to HPUra is not a characteristic shared by all Gram-negative aerobic bacteria.

ACKNOWLEDGMENTS

The authors wish to thank Robb Moses and Judith Campbell for assistance with the toluene system; and Neal Brown for generously supplying us with HPUra. This work was supported by Grant GM15837 from the National Institutes of Health.

REFERENCES

1. Brown, N.C. and Handschumacher, R.E., J. Biol. Chem. 241, 3083 (1966).
2. Brown, N.C., Bichem. Biophys. Acta 281, 202 (1972).
3. Bazill, G.W. and Gross, J.D., Nature New Biol. 240, 82 (1972).
4. Neville, M.M. and Brown, N.C., Nature New Biol. 240, 80 (1972).
5. Cozzarelli, N.R., Coulter, C.L., Low, R.L., Rashbaum, S.A. and Peebles, C.L., this Symposium (1975).
6. Gass, K.B., Low, R.L. and Cozzarelli, Proc. Natl. Acad. Sci. U.S.A. 70, 103 (1973).

7. Mackenzie, J.M., Neville, M.M., Wright, G.E. and Brown, N.C., Proc. Natl. Acad. Sci. U.S.A. 70, 512 (1973).
8. Van Beveren,C.P. and Wright, A., manuscript in preparation.
9. Moses, R.E. and Richardson, C.C., Proc. Natl. Acad. Sci. U.S.A. 67, 674 (1970).
10. Matsushita, T., White, K.P. and Sueoka, N., Nature New Biol. 232, 111 (1971).

Replication fork velocity in E. coli at various cell growth rates

M. Chandler, M. Funderburgh and L. Caro

Département de Biologie Moléculaire
Université de Genève, Switzerland

ABSTRACT. The relative frequencies of three chromosomal markers in a strain of E. coli K12 have been
measured, as a function of cell doubling time, by
DNA:DNA hybridisation. The results indicate that
the replication time of the chromosome is a
constant and independent of the doubling time of
the cells.

The pattern of chromosome replication in bacteria changes with the doubling time (τ) of the
cells. In fast growing cells DNA synthesis is continuous and the chromosome contains multiple
growing points (1). As the cell doubling time is
increased the mean number of growing points per
chromosome decreases and gaps in DNA synthesis
appear during the cell cycle (2,3,4). Most authors
agree that initiation of new rounds of replication
plays a major role in regulating DNA synthesis but
whether or not the velocity of the replication
fork remains constant at all cell growth rates has
been the object of some controversy. Kubitschek
and Freedman (5) have argued that it does, while
Helmstetter et al. (6) and Gudas and Pardee (7)
have concluded that, for cell doubling times
higher than 60 min., the replication time becomes
proportional to the cell doubling time.

In an attempt to resolve this question by a
new approach, we have evaluated the replication
fork velocity between selected markers in expo-

nentially growing cultures of E. coli K12, as a function of the doubling time of the cultures, by measuring the relative frequency of two chromosomal genes. This relative frequency is dependent on the mean number of growing points per chromosome (8) and on the relative position of the two genes on the genetic map. The relationship between the frequency of two chromosomal markers (a and b) which are located on the same replication unit of the chromosome, the cell doubling time (τ) and the mean replication time (C) is : $a/b = 2^{C\Delta/\tau}$ (9) where Δ is the fraction of the replication time C needed for a fork to go from a to b. If the average fork velocity is constant over the chromosome, Δ is equal to the distance between the two genes as a fraction of the replication unit.

We measured the relative frequency of two genes by the method of Bird et al. (10), in which the relative number of copies of two bacterio-phages (Mu-1 and λ) integrated into the host chromosome at known positions is determined by DNA:DNA hybridisation. The strains used were : Mx213 = E. coli K12 Wl485 thyA thi leu ilv (Mu-1) (λind⁻) in which Mu-1 is inserted in ilv (Fig. 1A) (11), Mx239 which is identical to Mx213 except that Mu-1 is in malA (Fig. 1B) (11), and Mx249 which is Mx213 made thy⁺ by transduction (9). DNA extracted from cultures growing at different rates in different media (L broth or M9 supplemented with glucose and casamino acids, glucose minimal, succinate, glycerol, aspartate, acetate, or proline) was immobilised on membrane filters and the Mu-1/λ ratios determined by hybridisation with a mixture of ³H-labeled Mu-1 DNA and ¹⁴C labeled λ DNA. The results are shown in Fig. 1 and Table 1. The points represent the ratio of the ³H (Mu-1) to ¹⁴C (λ) counts bound divided by the ratio obtained from parallel hybridisations with DNA from an amino acid starved culture. It has previously been shown that amino acid starvation of these strains results in a Mu-1/λ ratio of unity (10). This operation therefore acts as a normalisation in all hybrid-

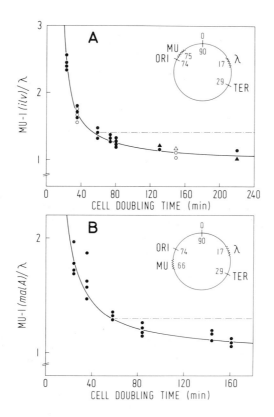

FIG. 1. Ratio of Mu-1/λ as a function of cell
         doubling time
         A. Strains Mx213 (●,▲) and Mx249 (○,△)
         B. Strain Mx239

Table 1. The replication time ($\bar{C}$) of the E. coli
chromosome as a function of cell doubling
time

Strain	τ (min)	No. of DNA samples	Mu-1/λ	$\bar{C}$ (min)
Mx213 (ilv⁻)	22	4	2.43 ± 0.05	39.0
	35	3	1.74 ± 0.04	38.7
	58	3	1.41 ± 0.04	39.8
	73	4	1.34 ± 0.02	42.7
	80	4	1.29 ± 0.01	40.7
	130	2	1.18 ± 0.02	43.0
	220	2	1.11 ± 0.07	45.9
Mx249 (ilv⁻)	35	3	1.70 ± 0.05	37.1
	150	3	1.11 ± 0.04	31.3
Mx239 (malA⁻)	25	4	1.79 ± 0.06	37.9
	36	4	1.64 ± 0.09	46.3
	59	4	1.33 ± 0.01	43.8
	85	4	1.20 ± 0.03	40.3
	145	3	1.16 ± 0.03	56.0
	162	3	1.08 ± 0.01	32.4

The mean replication time ($\bar{C}$) is calculated
from the equation : $\frac{Mu}{\lambda} = 2^{\Delta\bar{C}/\tau}$. The mean Mu/$\lambda$
ratios are given with the standard error of the
mean.

isation results. For details of the procedures
see Bird et al. (10) and Chandler et al. (9).

Two models are shown on the figures. The solid
curve represents, in each case, the theoretical
behavior of the Mu-l/λ ratio assuming that the
replication time is constant at 40 minutes and
that replication is bidirectional and symmetric,
with an origin close to the ilv region. The value
for Δ evaluated from the map of Taylor and Trotter
(12) is 0.72 for Mx213 and Mx249. For Mx239, where
Mu-l and λ are not on the same replicon, Δ is
evaluated by assuming that the origin of replication is at 74 min. on the map (10,11) and calculating distances from that point. This gives
Δ = 0.55. The dotted line shows the expected behavior of the Mu/λ ratio according to Helmstetter
et al. (6). The data are clearly in agreement with
the first model.

Some data obtained at a cell density of $A_{450} =$
0.1 (▲) gave results identical to those at cell
density of 0.2. The results obtained with Mx249
(0), similar to those of Mx213, (●) indicate that
the thyA⁻ mutation does not influence fork velocity under the conditions used.

In the experiments presented here we have considered the segment ilv-attλ, representing approximately 2/3 of the upper "replicon" of the bidirectionally replicating chromosome, and we have
measured the average number of growing forks present between the two markers in exponential cultures at various growth rates. The results show
that, in that region, the average fork velocity
remains constant over a wide range of cell growth
rates. The data from strain Mx239 give a strong
indication that the lower "replicon" is replicated
in a symmetrical manner. The data, thus, are
entirely consistent with the conclusion that the
replication time, in E. coli K12, is constant and
independent of cell doubling time. The assumptions
implicit in this conclusion, non-selective iso-

lation of DNA from various chromosomal regions, constancy of average fork velocity over the entire replicon, symmetry of the two forks produced in bidirectional replication, have been thoroughly tested only for some of the growth conditions used, and mostly for the fast growth rates (10).

M.C. was the recipient of a Royal Society European Fellowship. This work was supported by grant number 3.810.72 from the Fonds National Suisse. Some of the data presented (Fig. A) have been published elsewhere (9).

REFERENCES

1. Yoshikawa, H., O'Sullivan, A. and Sueoka, N. Proc. Nat. Acad. Sci. U.S. 52, 973 (1964).
2. Lark, C. Biochim. Biophys. Acta 119, 517 (1966).
3. Helmstetter, C.E. J. Mol. Biol. 24, 417 (1967).
4. Kubitschek, H.E., Bendigkeit, H.E. and Loken, M.R. Proc. Nat. Acad. Sci. U.S. 57, 1611 (1967).
5. Kubitschek, H.E. and Freedman, M.L. J. Bacteriol. 107, 95 (1971).
6. Helmstetter, C.E., Cooper, S., Pierucci, O. and Revelas, E. Cold Spring Harbor Symp. Quant. Biol. 33, 809 (1968).
7. Gudas, L.J. and Pardee, A.B. J. Bacteriol. 117, 1216 (1974).
8. Sueoka, N. and Yoshikawa, H. Genetics 52, 747 (1965).
9. Chandler, M., Bird, R.E. and Caro, L. J. Mol. Biol., in press.
10. Bird, R.E., Louarn, J., Martuscelli, J. and Caro, L. J. Mol. Biol. 70, 549 (1972).
11. Louarn, J., Funderburgh, M. and Bird, R.E. J. Bacteriol. 120, 1 (1974).
12. Taylor, A.L. and Trotter, C.D. Bacteriol. Rev. 36, 504 (1972).

Part X
Eucaryote DNA Replication Systems

TEMPERATURE SENSITIVE DNA⁻ MUTANTS OF A MAMMALIAN TISSUE CULTURE CELL LINE

Donald J. Roufa and Michael A. Haralson
Depts. of Biochemistry and Medicine
Baylor College of Medicine
Houston, Texas 77025

ABSTRACT. Several Chinese hamster lung (CHL) cell mutants that are not able to synthesize DNA at elevated temperature have been isolated. Mutants were generated from the parental cell line by treatment with ethyl methanesulfonate and were selected in culture for conditional resistance to the photosensitizing effects of BrdU. One class of mutants (TS-13, - 14 and - 41), which ceases protein and RNA synthesis as well as DNA biosynthesis at nonpermissive temperature, appears to have obtained its DNA⁻ phenotype via a genetic lesion directly affecting protein biosynthesis. Analysis of cell-free protein synthesizing extracts from TS-14 and of TS-14's polyribosomes after transfer to nonpermissive temperatures in culture indicate that the mutant's genetic defect results in the rapid cessation of protein synthesis. Together with the striking DNA⁻ phenocopy effected either by cycloheximide, puromycin or mixture of toxic amino acid analogs in the wild type cell lines, these observations suggest that continuous synthesis of one or more proteins is required by CHL cells to initiate or maintain programmed DNA replication.

INTRODUCTION

In order to recognize discrete biochemical functions that are required by Chinese hamster lung (CHL) cells for the replication of their DNA, and also to recognize regulatory mechanisms that coordinate DNA replication with the mammalian mitotic life cycle, we have been isolating temperature sensitive mutants that do not synthesize DNA at nonpermissive temperature. These somatic cell mutants are being obtained in tissue culture after treatment of the parental cell line with an alkylating reagent, ethyl methanesulfonate (1-3). At this time our collection of mutants consists of nine independently-isolated, temperature sensitive clones, seven of which abruptly reduce or cease DNA replication within six hours of being shifted to nonpermissive temperature. These mutants can be arranged into at least three phenotypic classes: clones that cease DNA replication only, clones that cease DNA, RNA and protein biosynthesis at high temperature and clones that reduce their rates of but do not cease DNA synthesis during the first 24 hrs. at high temperature (3). The CHL mutants are similar in character to the *E. coli* mutants mapped within the dna A-G loci (4). The *E. coli* mutants' biochemical properties have implicated several membrane functions (dna A and B) with bacterial DNA replication (5-8), and have assigned *E. coli* DNA polymerase III to a replicative function as assayed in cell-free extracts of dna E mutants (9). In much the same manner, we anticipate that biochemical analysis of the CHL mutants will be useful for recognizing particular mammalian enzymes that mediate chromosomal DNA replication.

METHODS

Commercial sources for tissue culture supplies, as well as the origin of our wild type CHL (V79) cell line have been described before (1-3). Radioactive nucleic acid bases, nucleosides and amino acids were purchased from New England Nuclear Corp. and Schwarz/Mann Biochemicals. Nonradioactive nucleic acids were obtained from Pabst Laboratories, Inc.; amino acids and firefly lantern extract (FLE-50) were from Sigma Chemical Co. Methods used for routine maintenance of CHL cells, mutagenesis with ethyl methanesulfonate (EMS) and for purifying mutant cell clones also have been described before (1, 2). Selection of temperature sensitive mutants from EMS-treated cultures was carried out by irradiating cultures with black light after the cells had been exposed to 10^{-5} M bromodeoxyuridine (BrdU) for 24 hrs. at nonpermissive temperature (3). Permissive temperature in these studies was 33°C and nonpermissive temperature, 39°C.

Efficiencies of colony formation were determined by plating 10^2, 10^3, 10^4, and 10^5 cells on 60 mm culture dishes. After 4 hrs. at 33°C, duplicate dishes were transferred to the appropriate temperature. Seven days later, when large colonies were apparent at permissive temperature, cells were fixed, stained and scored as described previously (1). Mitotic growth was monitored in asynchronous cultures in short experiments (< 12 hrs.) by the procedure of Puck and Steffen (10). The fraction of cells that enter mitotic metaphase at several times were scored in a series of culture dishes with media containing .05 μg/ml colcemid (GIBCO). At each time the fraction of cells accumulated in metaphase (mitotic index) was determined under a microscope, and the mitotic collection function, Log_{10} (1 + mitotic index), was plotted versus time. As shown by Puck and Steffen (10), the slope of that plot equals .301/T, where T is the cultures' generation time.

The abilities of cells to incorporate radioactive DNA, RNA and protein precursors were monitored by assaying acid-precipitable radioactivity that is extractable into 1% sodium dodecylsulfate (1, 3). For these studies we used 2 ml of culture medium supplemented either with 10^{-5} M [^{14}C] dT

(2.4 Ci/M), $10^{-5} \text{ M} [^{3}\text{H}]$ uridine (240 Ci/M), $10^{-4}\text{M} [^{3}\text{H}]$ proline (40 Ci/M), or a combination of $[^{14}\text{C}] \text{dT}$ and one ^{3}H-labeled substrate.

ATP was extracted from cultures of CHL cells (approx. 2×10^{6} cells) with ice-cold perchloric acid (6%) and was assayed with firefly lantern luciferase (11). Under the conditions of assay luminescence depends upon ATP concentration (0.25-80 μM), and was quantitated in a Beckman LS-100C scintillation counter. Protein concentrations contained in the perchloric acid precipitates were determined by the method of Lowry, *et al.* (12).

Extracts of wild type and mutant CHL cell lines that are capable of *in vitro* protein biosynthesis were prepared by a method similar to that of Aviv, Boime and Leder (13). A complete description of our cell-free preparations and their properties will be reported elsewhere (14). The extracts described in this paper were dialyzed 30,000 x g cytoplasmic supernatant fractions (S-30's). Reaction mixtures (.05 ml) contained 45 mM Tris HCI(pH7.5), 4 mM $\text{Mg(CH}_3\text{COO)}_2$, 128 mM KCl, 10 mM β-mercaptoethanol, 1 mM ATP, 0.1 mM GTP, and 0.6 mM CTP, $4 \times 10^{-5}\text{M}$ each of 19 non-radioactive amino acids and $2.5 \times 10^{-6} \text{ M} [^{3}\text{H}]$ leucine (SA = 44 Ci/M), and S-30 fraction (0.05 - 0.2 A^{260} units). Reaction mixtures were incubated at the specified temperatures and were terminated by addition of 0.1 ml of 0.1 N KOH. After 15 min. at 37°C, 1 ml of 10% Cl_3CCOOH was added and precipitates were collected on glass fiber filter discs (Whatman, GF/A). They were washed extensively with cold 5% Cl_3CCOOH and 0.01N HCl, dried and counted in a liquid scintillation spectrometer.

Polyribosomes contained by CHL cells growing in monolayer culture were analyzed on sucrose gradients. Cultures containing approximately 2×10^{6} cells were abeled with $[^{14}\text{C}]$ guanine (0.1 μC/ml, 50 Ci/Mole) for 14 hours at 33°C. Cultures then were rinsed with buffered physiological saline and pulsed for 30 min. with 60 μC $[^{3}\text{H}]$ alanine (53 Ci/mM) as they were shifted to 39°C or 90 min. after the temperature shift. After 120 min. at 39°C, 10^{-6} M cycloheximide was added to all cultures to stabilize the polysomes. Cells were scraped from dishes in buffer containing 10^{-6}M cycloheximide and were ruptured with a Dounce homogenizer (15). Homogenates were centrifuged (2000 XG for 5 min.) at 4°C. The supernatant fractions then were layered onto preformed 40 ml 15-30% sucrose gradients, and were centrifuged in a SW25.2 rotor at 25,000 RPM for 120 min. Gradients were fractionated from the bottom by collecting 40 drop (approximately 1 ml) aliquots. Bovine serum albumin (100 μg) and 1 ml of 10% Cl_3CCOOH were added to each fraction, and radioactive precipitates were collected on glass fiber discs and assayed in a liquid scintillation spectrometer.

TABLE 1.
TEMPERATURE SENSITIVE CHINESE HAMSTER MUTANTS SELECTED BY BRDU-BLACK LIGHT TREATMENT

Clone	Efficiency of Colony Formation	
	33°C(%)	39°C/33°C
HT-1 (Wild Type)	80.-100.	0.81
TS-12	72.2	<0.00009
TS-13	74.8	0.0007
TS-14	67.5	<0.0010
TS-38	15.9	<0.00036
TS-39	4.05	<0.0013
TS-40	18.2	<0.00056
TS-41	17.3	0.00061
TS-44	18.9	0.00011
TS-46	35.3	0.0034

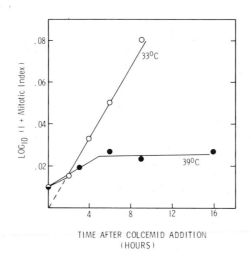

Figure 1. Growth of TS-13 at permissive and nonpermissive temperatures.

RESULTS AND DISCUSSION

Nine temperature sensitive Chinese hamster cell mutants have been isolated by conditional resistance to the photosensitizing effects of BrdU upon DNA replicated at 39°C. These mutants all are unable to grow at 39°C (Table 1). The few colonies which do grow from mutant cultures at high temperature appear to be revertant clones, since they plate equally well at both restrictive (39°C) and permissive (33°C) temperatures. While all of the mutants cease mitotic growth before forming visible colonies at 39°C, several of the mutants stop entering mitosis only a few hours after being transferred to high temperature. At 33°C mutant TS-13 grows with a generation time of 35 hours (Fig. 1), as compared to the wild type generation time of 20 hours, but ceases mitosis within six hours after a shift to nonpermissive temperature.

As expected on the basis of the selection procedure, mutant rates of dT incorporation at 39° are substantially less than their rates at 33°C. The kinetics of $[^{14}C]$ dT incorporation by TS-12, TS-13 and TS-14 are summarized in Figure 2. At 33°C the mutants incorporate thymidine into acid-insoluble polymer with rates ranging from 25 to 80% of the wild type. Rates of dT incorporation are either reduced (TS-12) or abolished (TS-13 and -14) shortly after transfer to nonpermissive temperature.

In order to identify mutants that contain altered DNA polymerases rapidly, we have surveyed crude cell-free extracts of the mutants for thermo-labile 4 and 6 S polymerases by procedures already described (15). CHL wild type 4 S DNA polymerase has a half-life of 1.0 - 1.5 min. at 47.5°C, and the mutants so far tested (TS-12, -13, -14, and -40) all have 4 S DNA polymerase with half-lives ranging from 0.5 - 1.5 min. Wild type CHL 6 S polymerase has a half-life of 3 min. at 50°C, and 6 S polymerase prepared from the above mutants have half-lives that range from 2-3.5 min. Since none of these DNA polymerases are temperature sensitive *in vitro*, we have sought other explanations for the mutants' temperature-dependent inhibition of dT incorporation.

One class of mutants, TS-13, -14 and -41, ceases not only DNA synthesis but also RNA and protein biosynthesis at 39°C (Fig. 3). Three types of models have been useful in considering the genetic basis of these mutants' complex phenotype:

1. Multiple, independent, temperature-sensitive mutations, each affecting one biosynthetic process, might be responsible for each clone's altered properties.
2. A single genetic lesion that affects intracellular levels of high energy nucleotides could abolish all biosynthetic processes that depend upon nucleoside triphosphates.
3. A single mutation that affects one biosynthetic process directly might affect the other two indirectly *via* a regulatory mechanism.

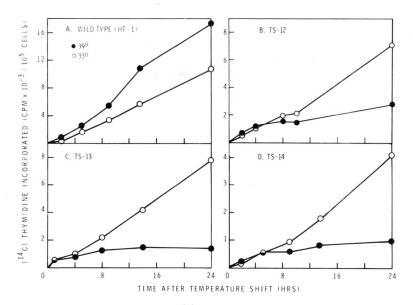

Figure 2. Incorporation of $[^{14}C]dT$ at permissive and nonpermissive culture temperatures.

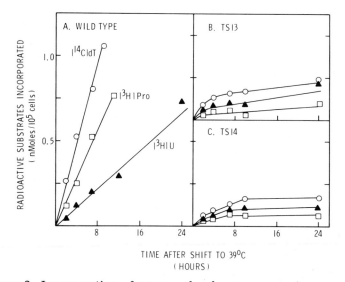

Figure 3. Incorporation of macromolecular precursors at nonpermissive temperature.

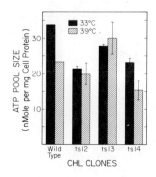

Figure 4. ATP levels in CHL TS clones.

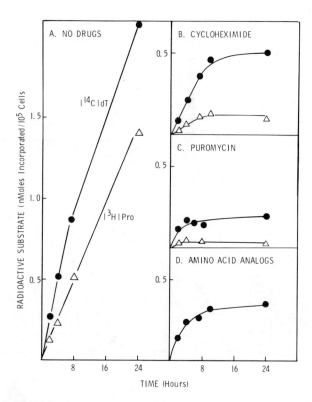

Figure 5. DNA⁻ phenocopy in wild type cells by protein biosynthetic inhibitors.

Revertants of TS-13 and -14 (see above) exhibit wild type kinetics of DNA, RNA and protein synthesis in culture, and thus suggest that inhibition of all three biosynthetic processes in the mutants probably relates to a single genetic alteration. Models based upon multiple mutations, such as Model 1 above, therefore, are not likely to account for the complex phenotype. The relationship between energy metabolism and the mutants' phenotype (model 2) has been examined by quantitating intracellular pools of ATP in mutants growing at 33°C and after exposure to 39°C for 20 hours (Figure 4). The data appear to eliminate models based upon the temperature-dependent reduction of cellular energy charge.

The third model described above, based upon a single mutation affecting protein biosynthesis directly and DNA and RNA synthesis indirectly, best accounts for the complex character of TS-13, -14, and -41. As shown in Figure 5, metabolic inhibitors that affect protein biosynthesis directly, i.e. 10^{-6}M cycloheximide and 10^{-4}M puromycin, produce a striking phenocopy of the mutant character in wild type cells. Mammalian protein and DNA biosynthesis are linked functionally either by a regulatory mechanism, such as bacterial stringency, or by a requirement for the continual synthesis of one or more protein factors necessary for programmed mammalian DNA replication. The latter appears to be the case, since a mixture of toxic amino acid analogs that are incorporated into mammalian proteins (10^{-3}M each of p-fluorophenylalanine, 6-fluoro-tryptophan and azetidine carboxylic acid) also promotes a mutant phenocopy (Fig. 5D).

In contrast to its DNA polymerase, mutant TS-14's protein biosynthetic apparatus is thermolabile in vitro. The kinetics of cell-free protein biosynthesis catalyzed by extracts of TS-14 and wild type cells were compared at 25 and 40°C (Fig. 6). At 25°C both extracts catalyzed protein biosynthesis for at least 60 min. in vitro (Fig. 6A). When mutant and wild type extracts possessing equal biosynthetic activity at 25°C were examined at 40°C, the mutant extract prematurely ceased protein synthesis after 15 min. (Fig. 6B). The wild type extract, in contrast, continued to polymerize [^{3}H]leucine for 60 min. at high temperature. As will be described elsewhere in detail, the defect in TS-14's protein biosynthetic machinery is associated specifically with a structural element of the mutant's ribosomes (14).

Analysis of TS-14's polysomes by sucrose gradient centrifugation also has indicated that protein biosynthesis in the mutant is affected rapidly by exposure to nonpermissive temperature. As shown in Figure 7, [^{14}C]polyribosomes contained by TS-14 (●-●) were not dissociated when cultures were exposed to 39°C for 120 min. When a pulse of [^{3}H]alanine (O-O) was given to a culture at the time of the temperature shift (Fig. 7A), ^{3}H-labeled peptides still were associated with the

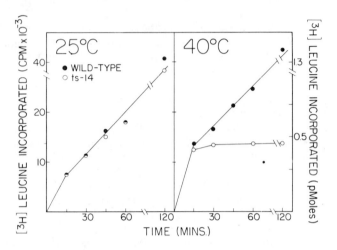

Figure 6. Protein biosynthesis in CHL cell-free extracts at permissive and nonpermissive temperatures.

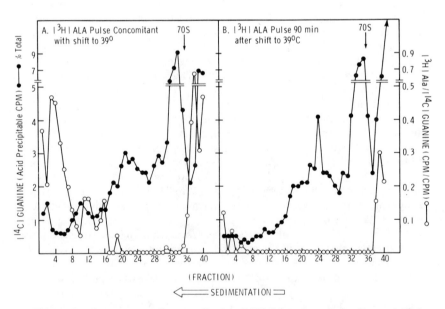

Figure 7. Sucrose gradient analysis of TS-14 polysomes after a shift to nonpermissive temperature.

polysomes after a 90 minute chase at restrictive temperature. When the pulse of [^{3}H]alanine was administered 90 min. after the temperature shift (Fig. 7B), protein synthesis already had ceased and no nascent peptides were labelled.

The phenotypic properties of mutants TS-13, -14 and -41, as well as biochemical analysis of TS-14, imply that one or more proteins must be synthesized continuously to permit animal cells to replicate their DNA. If synthesis of these proteins is inhibited, as is the case in mutant cultures at high temperature, or if defective proteins that contain nonfunctional amino acid analogs are synthesized, DNA replication ceases within 6-8 hours. Whether the required proteins serve a catalytic function in DNA replication, as recently suggested by other investigators (16), or whether they regulate DNA synthesis or coordinate it with the cell's mitotic cycle, such as proteins that affect cells' growth in culture (17-19) is not clear at present. It is interesting to note that a heat-labile factor that stimulates ATP-dependent DNA replication in S phase HeLa nuclei (20) and cytoplasmic 6 S CHL DNA polymerase (15) share several common properties. In addition, Otto and Reichard have found that cytoplasmic factor can be replaced by purified *E. coli* DNA pol I and DNA ligase in cell-free reactions that catalyze intranuclear polyoma DNA replication (16). Proteins which must be synthesized continuously to maintain mammalian cells in DNA synthesis also might function, at least in part, by regulating either the subcellular location of 6 S DNA polymerase or by facilitating the interconversion of 4 and 6 S mammalian DNA polymerase (21). Both of these models should be testable in the mutants by cell-free reactions that replicate intranuclear DNA *in vitro*.

ACKNOWLEDGEMENTS

These researches have been supported by grants from the American Cancer Society (NP135) and U.S. Public Health Service (GM 21070). The authors wish to thank Mrs. Susan Reed and Mrs. Marie Demetriades for their skillful technical aid.

REFERENCES

1. Gillin, F.D., Roufa, D.J., Beaudet, A.L. and Caskey, C.T., Genetics 72, 239 (1972).
2. Roufa, D.J., Sadow, B.N. and Caskey, C.T. Genetics 75, 515 (1973).
3. Roufa, D.J. and Reed, S.J. Genetics, in press (1975).
4. Hirota, Y., Mordoh, J., Scheffler, I. and Jacob, F. Fed. Proc. 31, 1422 (1972).
5. Inoue, M. and Gutherie, J.P. Proc. Nat. Acad. Sci. USA 64, 957 (1969).

6. Inoue, M. and Pardee, A.J. Biol. Chem. *245*, 5813 (1970).
7. Shapiro, B., Siccardi, A., Hirota, Y. and Jacob, F.J. Mol. Biol. *52*, 75 (1970).
8. Siccardi, A., Shapiro, B., Hirota, Y. and Jacob, F.J. Mol. Biol. *56*, 475 (1971).
9. Gefter, M.L., Hirota, Y., Kornberg, T., Wechsler, J.A. and Barnoux, C. Proc. Nat. Acad. Sci. USA *68*, 3150 (1971).
10. Puck, T.T. and Steffen, J. Biophys. J. *3*, 379 (1963).
11. Stanley, P.E. and Williams, S.G. Analyt. Biochem. *29*, 381 (1969).
12. Lowry, O.H., Rosebrough, N.J., Farr, A.L. and Randall, R.J. J. Biol. Chem. *193*, 265 (1951).
13. Aviv, H., Boime, I. and Leder, P. Proc. Nat. Acad. Sci. USA *68*, 2303 (1971).
14. Haralson, M.A. and Roufa, D.J., manuscript in preparation.
15. Roufa, D.J., Moses, R.E. and Reed, S.J. Arch. Biochem. Biophys. *167*, in press (1975).
16. Otto, B. and Reichard, P.J. Virol. *15*, 259 (1975).
17. Holley, R.W. and Kiernan, J.A. Proc. Nat. Acad. Sci USA *71*, 2908 (1974).
18. Gospodarowicz, D. Nature *249*, 123 (1974).
19. Dulak, N.C. and Temin, II.J. Cell. Physiol. *81*, 153 (1973).
20. Hershey, H.V., Stieber, J.F. and Mueller, G.C. Eur. J. Biochem. *34*, 383 (1973).
21. Hecht, N.B. and Davidson, D. Biochem. Biophys. Res. Commun. *51*, 299 (1973).

THE INHIBITION OF DNA STRAND ELONGATION BY CYCLOHEXIMIDE
IN *PHYSARUM POLYCEPHALUM*

Helen H. Evans, Thomas E. Evans, and Eugene N. Brewer
Division of Radiation Biology, Department of Radiology
Case Western Reserve University
Cleveland, Ohio 44106

ABSTRACT. The mechanism of the inhibition of nuclear DNA
replication by cycloheximide in *Physarum polycephalum* has
been investigated by means of alkaline sucrose density
gradient centrifugation techniques. The results suggest
that the rapid decrease in the incorporation of [^{3}H]thy-
midine into DNA after exposure to cycloheximide is caused
by an inhibition of the elongation of progeny strands. No
effect was detected on either the initiation of the synthesis of DNA replication units or on the ligation of DNA
fragments produced by ionizing radiation. It has been concluded that cycloheximide treatment inhibits the elongation of progeny strands within replication units, presumably by inhibiting the synthesis of short-lived proteins
necessary for this process.

INTRODUCTION

In many types of eukaryotic cells protein synthesis
appears to be necessary not only for the entry of cells
into the S period but also for the continuation of DNA
replication (1-7). It has been suggested that protein
synthesis may be necessary for the initiation of the synthesis of the many DNA replication units comprising the
eukaryotic genome (4, 8-10), whereas evidence for an inhibition of the elongation of growing DNA strands within
replication units has also been presented (11-14). We
have examined these alternatives using *P. polycephalum*, a
myxomycete in which each synchronous mitosis is followed
by the S period. Earlier studies have shown that cycloheximide inhibits both the initiation and continuation of DNA
replication in this organism (15), and that required pro-

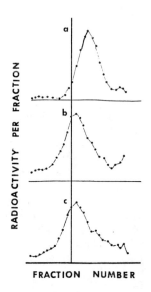

Fig. 1. Effect of cycloheximide on the elongation of progeny strands. Sectors of a single plasmodium were incubated with [3H]thymidine for the first 15 min. of the S period, then rinsed and incubated in non-radioactive medium containing unlabeled thymidine (1 x 10^{-4}M) for two hours. Cycloheximide was added to the medium at the times indicated by the open circles, and was present until harvest, which was at the end of the 120 min. chase period in all cases. Sectors were harvested, combined with sectors of a second plasmodium prelabeled with [14C]thymidine and harvested during the G$_2$ period, and nuclei were co-isolated, layered on alkaline sucrose gradients and centrifuged as described previously (18). Molecular weight values of the peak [3H] fractions were calculated, using the peak fraction of the [14C]-prelabeled non-replicating DNA as a marker (4 x 10^7 d).

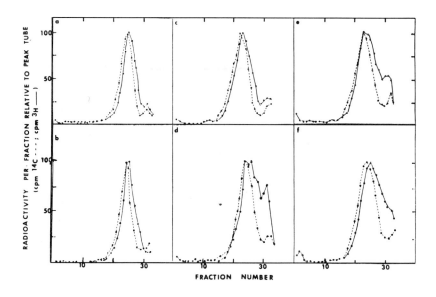

Fig. 2. The effect of cycloheximide on the initiation of
the synthesis of replication units. Sectors of plasmodia
were placed in medium containing [^{14}C]thymidine at various
times after mitosis and incubated for 15 min. The sectors
were then rinsed and immediately placed in medium containing [^{3}H]deoxyadenosine, non-radioactive thymidine ± cycloheximide, incubated another 15 min. and harvested. Nuclei
were isolated, lysed on top of alkaline sucrose gradients
and centrifuged. Sedimentation was from right to left.
(a). Incubated with [^{14}C]thymidine for 0-15 min. following
mitosis and with [^{3}H]deoxyadenosine in the following 15
min.; (b). Incubated as in (a) but with cycloheximide present during the [^{3}H]deoxyadenosine labeling period; (c).
incubated 30-45 min. following mitosis with [^{14}C]thymidine
and with [^{3}H]deoxyadenosine 45-60 min. following mitosis;
(d). Incubated as in (c) but with cycloheximide present
during the [^{3}H]deoxyadenosine labeling period; (e). Incubated with [^{14}C]thymidine 60-75 min. following mitosis and
with [^{3}H]deoxyadenosine 75-90 min. following mitosis; (f).
Incubated as in (e) but with cycloheximide present during
the [^{3}H]deoxyadenosine labeling period. Cpm/fraction for
the peak tubes were 200-1200 cpm ^{3}H and 130-400 cpm ^{14}C.

715

teins appear to be synthesized at defined times during the
S period (16).

METHODS

The organism was cultured as described previously
(16). For measurement of the extent of DNA synthesis,
macroplasmodia were prelabeled with [^{14}C] thymidine, and
at various times after mitosis sectors of single macroplasmodia were transferred to medium containing [^{3}H]thymi-
dine and incubated for 15 min. ± cycloheximide (10 µg/ml).
Samples were harvested and washed with cold 4% TCA in 50%
acetone, and then extracted at 85° two times with 0.6N
HClO$_4$. The ^{3}H/^{14}C ratio was used as a measure of the
amount of DNA synthesized per unit of DNA present in the
sample. For alkaline sucrose density gradient centrifugation, labeled portions of plasmodia were harvested and nuclei were isolated (17) and layered on top of alkaline
sucrose gradients and centrifuged according to methods
described previously (18).

RESULTS AND DISCUSSION

When sectors of plasmodia were transferred to media
containing [^{3}H]thymidine ± cycloheximide, incorporation
into DNA during the first 15 min. in the presence of the
inhibitor was 45-60% of the control. Very little incorporation occurred after 15 min. in the presence of the drug.
Treatment of the plasmodia with cycloheximide at any time
during the S period was found to inhibit elongation of
progeny strands pulse-labeled at the beginning of the S
period, as determined by alkaline sucrose density gradient centrifugation (Fig. 1). The initiation of the synthesis of replication units was indicated by the non-super-
imposibility of ^{3}H and ^{14}C density gradient profiles after
successive 15 min incubations in the presence of differentially labeled precursors (19). Cycloheximide was found
to have no detectable effect on initiations as determined
in this manner (Fig. 2). The results in Fig. 3 show that
single-strand breaks caused by γ-irradiation during S can
be repaired in the presence of cycloheximide, indicating
that the synthesis of protein is not required for ligation
of DNA fragments and that ligation *per se* is not directly
affected by cycloheximide. Thus, these results indicate

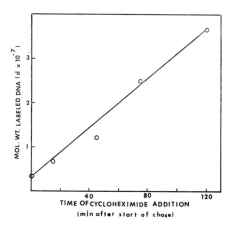

Fig. 3. Effect of cycloheximide on the repair of single-
strand breaks (produced by gamma irradiation) during the
S period. Plasmodia were prelabeled either with [3H]thy-
midine or [14C]thymidine. At the beginning of the S period, sectors of each were treated with 10,000 R [60Co]
gamma radiation. Sectors were then either harvested immediately or allowed to incubate for 3 hours in the presence
or absence of cycloheximide. Differentially labeled irradiated and non-irradiated nuclei were co-isolated, lysed on
top of alkaline sucrose gradients and centrifuged. The
vertical line indicates the sedimentation position of the
peak fraction of the DNA from unirradiated S-phase plasmodia. (a). Harvested immediately following irradiation;
(b). Irradiated and incubated for 3 hours in the absence
of cycloheximide; (c). Irradiated and incubated for 3
hours in the presence of cycloheximide. Cpm/fraction in the
peak tubes ranged from 700-1700. Sedimentation was from
right to left.

that in *P. polycephalum* cycloheximide prevents nucleotide addition, *i.e.* elongation within replicating units; presumably by preventing the synthesis of short-lived proteins necessary for this process.

ACKNOWLEDGEMENT

This investigation was supported by U.S. A.E.C. contract W-31-109-ENG-78 and NIH grant GM 19484. We are indebted to Miss Sandra Littman and Mrs. Pauline Ting for their excellent assistance.

REFERENCES

1. Mueller, G.C., Kajiwara, K., Stubblefield, E. and Rueckert, R. Cancer Res. 22, 1084 (1962).
2. Littlefield, J.W. and Jacobs, P.S. Biochim. Biophys. Acta 108, 652 (1965).
3. Taylor, E.W. Exp. Cell Res. 40, 316 (1965).
4. Young, C.W. Mol. Pharmacol. 2, 50 (1966).
5. Kim, J.H., Gelbard, A.S. and Perez, A.G. Exp. Cell Res. 53, 478 (1968).
6. Brown, R.F., Umeda, T., Takai, S. and Lieberman, I. Biochim. Biophys. Acta 209, 49 (1970).
7. Highfield, D.P. and Dewey, W.C. Exp. Cell Res. 75, 314 (1972).
8. Ensminger, W.D. and Tamm, I. Virology 40, 152 (1970).
9. Fujiwara, Y. Cancer Res. 32, 2089 (1972).
10. Hori, T. and Lark, K.G. J. Mol. Biol. 77, 391 (1973).
11. Weintraub, H. and Holtzer, H. J. Mol. Biol. 66, 13 (1972).
12. Gautschi, J.R. and Kern, R.M. Exp. Cell Res. 80, 15 (1973).
13. Hand, R. and Tamm, I. J. Cell Biol. 58, 410 (1973).
14. Gautschi, J.R. J. Mol. Biol. 84, 223 (1974).
15. Cummins, J.E. and Rusch, H.P. J. Cell Biol. 31, 577 (1966).
16. Muldoon, J.J., Evans, T.E., Nygaard, O.F. and Evans, H.H. Biochim. Biophys. Acta 247, 310 (1971).
17. Mohberg, J. and Rusch, H.P. Exp. Cell Res. 66, 305 (1971).
18. Brewer, E.N. J. Mol. Biol. 68, 401 (1972).
19. Brewer, E.N., Evans, T. and Evans, H.H. J. Mol. Biol. 90, 335 (1974).

STUDIES OF DNA SYNTHESIS
IN PERMEABILIZED MOUSE L CELLS

Nathan A. Berger and Elizabeth S. Johnson
Washington University Medical Center and
The Jewish Hospital of St. Louis
St. Louis, Missouri 63110

ABSTRACT. Mouse L cells, maintained in continuous suspension culture have been rendered permeable to deoxynucleoside phosphates by treating for 15 minutes at $4^{\circ}C$ with a buffer containing $.01\underline{M}$ Tris-HCl pH 7.8, $0.25\underline{M}$ Sucrose, $4\underline{mM}$ Mg Cl_2, $1\underline{mM}$ EDTA and $30\underline{mM}$ 2 mercaptoethanol. These cells use ^{14}C-UTP to synthesize RNA and ^{3}H-TTP to synthesize DNA. Subcellular distribution studies revealed that 98% of the DNA synthesized in these permeabilized cells was located in the nucleus. Cesium chloride gradient studies demonstrated that the newly synthesized DNA was the product of semiconservative replication. During the course of the DNA synthesis reaction, some of the ^{3}H-TTP used to measure DNA synthesis was degraded to TDP, TMP, Thymidine and Thymine. Double label experiments were performed to determine whether the nucleotides were incorporated directly into DNA or whether they underwent intermediate degradation and uptake as nucleosides. The results demonstrate that ^{3}H-TTP and $\alpha\,^{32}P$-TTP were incorporated into DNA at the same rate, indicating that in this system the nucleotides are incorporated directly into DNA. This was confirmed in a more rigorous fashion by demonstrating that L cells, deficient in Thymidine Kinase, can be permeabilized by the same technique and will then incorporate $\alpha\,^{32}P$-TTP and ^{3}H-TTP into DNA at the same rate. DNA synthesis requires the presence of NaCl, ATP, all four deoxynucleotides and Mg Cl_2. The reaction is inhibited by NEM, PHMB, and Actinomycin D. Hydroxyurea does not inhibit the reaction, nor does the nucleoside analog Cytosine Arabinoside. However, Cytosine Arabinoside Triphosphate does inhibit DNA synthesis and shows competitive inhibition with the deoxynucleotides.

719

INTRODUCTION

The development of techniques to incorporate deoxy-nucleoside triphosphates (dNTPs) into permeabilized bacteria provided new methods of analyzing the components involved in DNA synthesis (1-4). Techniques involving treatment with DEAE Dextran (5) or Hypotonic Buffer (6) have been developed to incorporate dNTPs into eukaryotic cells. We found that mouse L cells can be rendered permeable to dNTPs by a brief treatment with a near isotonic buffer (7). These cells retain their microscopic appearance as a single cell suspension. They use exogenously supplied dNTPs to carry out replicative DNA synthesis and have been used to study some of the factors affecting DNA synthesis in eukaryotic cells.

METHODS

Mouse L cells and Thymidine Kinase deficient L cells (TK$^-$) (8) are maintained in suspension culture in α Minimal Eagles Medium with 10% fetal calf serum at 37°C and 5% CO_2. Cells are collected by centrifugation at 1000xg for 10 minutes at 4°C. They are suspended at 2×10^6 cells/ml in permeabilizing buffer (P. Buff.) composed of .01$\underline{M}$ Tris-HCl pH 7.8, 0.25$\underline{M}$ Sucrose, 1m$\underline{M}$ EDTA, 30m$\underline{M}$ 2 mercaptoethanol (2ME), and 4m$\underline{M}$ MgCl$_2$ and incubated for 15 minutes at 4°C. This concentration of 2ME is optimal to insure reproducible results. The cells are collected by centrifugation and resuspended in P. Buff. at 2.5×10^7 cells/ml. The assay for DNA synthesis is performed in a final volume of 0.6 ml composed of 0.4 ml of cell suspension in P. Buff mixed with 0.2 ml of DNA synthesis reaction mix. The concentration of all components in the complete 0.6 ml DNA synthesis reaction is 1×10^7 cells, 6.6m$\underline{M}$ Tris-HCL pH 7.8, 0.16$\underline{M}$ Sucrose, 9.3m$\underline{M}$ MgCl$_2$, 0.66m$\underline{M}$ EDTA, 20m$\underline{M}$ 2-ME, 70m$\underline{M}$ NaCl, 33m$\underline{M}$ HEPES pH 7.8, 5m$\underline{M}$ ATP, 0.1m$\underline{M}$ dATP, 0.1m$\underline{M}$ dCTP, 0.1m$\underline{M}$ dGTP and 0.1m$\underline{M}$ dTTP with a radioactive label in one of the dNTPs. Components are combined in an ice water bath; the reactions are initiated by transfering to a 37°C shaking water bath and terminated by the addition of ice cold 10% TCA, 2% $Na_4P_2O_7$. Cell pellets are prepared for scintillation counting by sonicating, washing once with 5% TCA, 1% $Na_4P_2O_7$, and once with Ethanol. They are solubilized with 0.5 ml Protosol, then dissolved in 10 ml of Toluene Scintillation Fluid containing 6g PPO and 50mg POPOP per liter.

RESULTS AND DISCUSSION

Fig. 1 demonstrates that when untreated cells are incubated in the complete DNA synthesis mix there is essentially no incorporation of the exogenously supplied dNTPs into DNA during the first 15 to 20 minutes. A very low rate of incorporation into DNA occurs during the remainder of a 60 minute incubation period. When double label experiments are performed, with a reaction mix containing quantities of ^{3}H-TTP and α^{32}P-TTP combined to give equal cpm of each isotope, the untreated cells incorporate the ^{3}H label into DNA at 4 times the rate of the ^{32}P label. Dissociation of incorporation of the ^{3}H label from the ^{32}P label indicates that some of the dNTPs are hydrolyzed, transported as nucleosides and rephosphorylated before incorporation into DNA. In contrast, when permeabilized cells are incubated in the DNA synthesis mix, they use the exogenously supplied dNTPs to synthesize DNA at a linear rate for 30 minutes (Fig. 1). The permeabilized cells incorporate the ^{3}H and α^{32}P labels at identical rates, indicating that the exogenously supplied deoxynucleotides are incorporated directly into DNA without intermediate degradation to the nucleosides. The supernatants from the DNA synthesis reactions were analyzed by chromatography on PEI cellulose (9) and demonstrated the gradual degradation of ^{3}H-TTP. After 30 minutes of incubation the distribution in the supernatant of the ^{3}H added as ^{3}H-TTP was 57% in TTP, 27% in TDP, 6% in TMP, 7% in TdR, and 3% in Thymine.

Since some degradation of TTP to nucleosides occurred during the course of the reaction, studies were

721

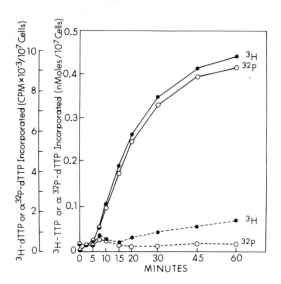

Figure 1. Incorporation of ^{3}H-TTP (●) and α^{32}P-TTP (o) into DNA of permeabilized (————) and nonpermeabilized cells (------). Permeabilized cells were suspended in P. Buff., nonpermeabilized cells were suspended in α MEM without serum. The complete DNA synthesis reaction mix contained ^{3}H-TTP and α^{32}P-TTP, with the specific activity of each isotope equal to 23×10^3 cpm/n mole.

Table I.

Requirements and Inhibitors of Replicative DNA Synthesis in Permeabilized L Cells

Condition	Relative Activity, %	Condition	Relative Activity, %
Complete	100	+ 10mM Hydroxyurea	104
- NaCl	57	- 2 ME	99
- MgCl$_2$	23	- 2 ME + 1mM PHMB	5.2
- ATP	5.6	- 2 ME + 2mM NEM	4.6
- 3 dNTPs	6.8	+ 50μg/ml Act D	12
- dATP	22.2	+ 0.1 mM AraC	100
- dCTP	6.6	+ 1mM AraC	90
- dGTP	8.5	+ 0.1 mM AraCTP	18

Reaction conditions were as described for determining incorporation of ^{3}H-TTP into DNA of permeabilized L cells. Incubations were for 30 minutes at 37°C.

performed with TK⁻ cells to rigorously confirm that the
dNTPs were used for DNA synthesis without intermediate
conversion to nucleosides. These cells were demonstrated
to be unable to incorporate exogenously supplied TdR into
DNA. Thymidine Kinase assays (10) were performed on cell
extracts and confirmed that the TK⁻ cells were devoid of
enzyme activity (7). When the TK⁻ cells were permeabilized, they used the exogenously supplied dNTPs to synthesize DNA. In addition, the rate of incorporation of the
^{3}H-TTP was equal to the rate of incorporation of the α ^{32}P-
TTP (7). Since the TK⁻ cells have no mechanism to phospho-
rylate thymidine, these studies rigorously confirm that the
exogenously supplied deoxynucleotides serve as the substrates for DNA synthesis without intermediate degradation
to nucleosides.

If the permeabilized cells carry out replicative DNA
synthesis, then the newly synthesized DNA should be
located in the nucleus and should be the product of semiconservative replication. Subcellular fractionation studies
(11) demonstrated that 98% of the DNA synthesized by the
permeable cells was located in the nucleus (7). Fig. 2
indicates the results of a neutral CsCl equilibrium density
centrifugation on DNA extracted from cells in which the
bulk DNA was prelabelled by incubating growing cells for
18 hours in the presence of ^{14}C-TdR. The cells were then
permeabilized and incubated with a DNA synthesis mix
containing BrdUTP substituted for TTP and the radioactive
label present as ^{3}H-dCTP. The newly synthesized, ^{3}H
labelled DNA banded at a hybrid density of 1.760g/ml
which is separated from the bulk ^{14}C-DNA at 1.718g/ml,
confirming that semiconservative DNA synthesis occurs in
this permeable cell system.

The requirements and inhibitors of the replicative
DNA polymerase in the permeabilized cells are reported in
Table 1. DNA synthesis is optimal in the presence of
70mM NaCl. Equimolar concentrations of K^+ or NH_4^+ can
substitute for Na^+ suggesting that the requirement for a
monovalent cation is due to an effect of the ionic strength
rather than a requirement for a specific ion. Activity is

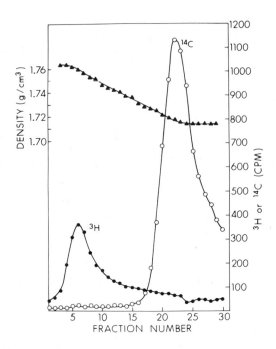

Figure 2. Neutral CsCl equilibrium density gradient centrifugation of newly synthesized, density labeled DNA. Bulk
cell DNA was pre-labeled with ^{14}C-TdR by growing cells
for 18 hours in ^{14}C-TdR. Cells were permeabilized then
allowed to synthesize DNA in a complete reaction mix containing BrdUTP substituted for TTP and ^{3}H-dCTP, 65×10^3
cpm/n mole. Incubation was carried out for 30 minutes at
37^oC. DNA was extracted and banded in CsCl (7).

diminished by 77% in the absence of Mg^{++}. The system has an absolute requirement for ATP for replicative DNA synthesis as has been noted with other permeabilized cell systems (1-6) but not with isolated DNA polymerases.

Replicative DNA polymerases are expected to have a stringent requirement for the presence of all 4dNTPs to maintain fidelity of replication. The DNA polymerase measured in the permeabilized L cells demonstrates these stringent requirements since the reaction is inhibited by more than 90% when either dCTP or dGTP are absent. When dATP is the only dNTP missing the reaction occurrs at 22% of the complete system. This is probably due to conversion of some of the ATP present to the deoxynucleotide.

Hydroxyurea inhibits DNA synthesis in intact cells by inhibition of ribonucleotide reductase (12). Since the dNTPs are supplied directly to the DNA polymerase in the permeabilized cells the action of ribonucleotide reductase is not required and the presence of hydroxyurea has no inhibitory effects.

When the 2ME is deleted from the final reaction system the DNA polymerase still demonstrates full activity. In the absence of 2ME, the sulfhydryl reagents pHMB and NEM inhibit the enzyme activity. Inhibition by sulfhydryl reagents is characteristic of the high molecular weight eukaryotic DNA polymerase presumed to be the replicative form of the enzyme (13,14). Actinomycin D, at 50μg/ml inhibits DNA synthesis presumably by its effect on the DNA template. The nucleoside analog araC inhibits DNA synthesis in intact cells (15) but does not inhibit DNA synthesis in permeabilized cells. In contrast, the nucleotide analog araCTP is a very effective inhibitor of DNA synthesis in the permeabilized cells. These results confirm the observation that araCTP is the active form of the analog that inhibits DNA polymerase (16). A double reciprocal plot of the kinetics of inhibition of DNA synthesis in the permeable cell system indicates that araCTP functions as a competitive inhibitor of the replicative DNA polymerase (Fig. 3).

These studies demonstrate that replicative DNA

725

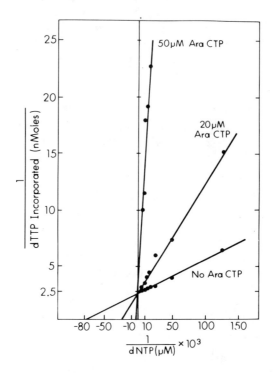

Figure 3. The inhibition of DNA synthesis in permeabilized L cells by araCTP. DNA synthesis reaction mix contained ^{3}H–TTP, 30×10^3 cpm/n mole. The concentrations of all 4 dNTPs were varied while other reaction components and the specific activity of the ^{3}H–TTP were held constant. The reactions were performed in the absence of araCTP, and in the presence of 20 µM and 50 µM araCTP as indicated in the double reciprocal plot. All components were combined in an ice water bath. The reactions were started by transfering tubes to a 37°C water bath and terminated after a 15 minute incubation.

synthesis in permeabilized cells is not inhibited by agents which inhibit DNA synthesis in intact cells by interfering with the synthesis of precursors. DNA synthesis in this system is affected by agents which have a direct effect on the replicative DNA polymerase or on the intrinsic DNA template. The permeabilized cell system is useful to identify and study the mechanism of action of agents which affect the function of the replicative DNA polymerase or the intrinsic DNA template.

ACKNOWLEDGEMENTS

This work was supported by funds from a General Research Support Grant to the Jewish Hospital of St. Louis.

REFERENCES

1. Mordoh, J., Hirota, Y., and Jacob, F., Proc. Natl. Acad. Sci., U.S.A., 67, 773 (1970).
2. Moses, R. E., and Richardson, E. C., Proc. Natl. Acad. Sci., U.S.A., 67, 674 (1970).
3. Vosberg, H.P., and Hoffmann-Berling, H., J. Mol. Biol., 58, 739 (1971).
4. Schaller, H., Otto, B., Nusslein, U., Huf, J., Hermann, R., and Bonhoeffer, F., J. Mol. Biol., 63, 183 (1972).
5. Fox, R. M. and Goulian, M., in press.
6. Seki, S., LeMahieu, M. and Mueller, G.C. Biochem. Biophys. Acta., 378, 333 (1975).
7. Berger, N. A., and Johnson, E. S., in press.
8. Kit, S., Dubbs, D. R., Piekarski, L. J., and Hsu, T. C., Exp. Cell. Res., 31, 297 (1963).
9. Randerath, K. and Randerath, E., J. Chromatog., 16, 111 (1964).
10. Taylor, A. T., Stafford, M. A. and Jones, W. O., J. Biol. Chem. 247, 1930 (1972).

11. Blobel, G., and Potter, V.R., Science, 154, 1662
     (1966).

12. Young, C.W., and Hodas, S., Science, 146, 1172
     (1964).

13. Chiu, J.F., and Sung, S.C., Nature New Biol., 239
     178 (1972).

14. Baril, E.F., Jenkins, M.D., Brown, O.E., Laszio, J.
     and Morris, H.P., Cancer Res., 33, 1187 (1973).

15. Graham, F.L. and Whitmore, G.F., Cancer Res., 30
     2627 (1970).

16. Furth, J.J., and Cohen, S.S., Cancer Res., 28, 2061
     (1968).

Part XI
Viral Nucleic Acid Replication in Eucaryotes

ROLE OF REVERSE TRANSCRIPTASE IN THE LIFE CYCLE OF RNA TUMOR VIRUSES

Inder M. Verma and Wade Gibson
Tumor Virology Laboratory
The Salk Institute
San Diego, California 92112

ABSTRACT. RNA tumor viruses contain a DNA polymerase (reverse transcriptase) which can faithfully transcribe the viral RNA into complementary DNA. In the scheme of events postulated by the pro-virus hypothesis, such an enzyme is required very early after infection. Purified DNA polymerase isolated from two temperature-sensitive mutants of Rous sarcoma virus (RSV), LA337 and LA335, with defects in very early events of the virus growth cycle have been shown to be more thermolabile than the DNA polymerase from the wild-type parent. Furthermore, isolated small subunit, α, from LA337 manifesting both the polymerase and ribonuclease H activities is five to seven times more thermolabile than the isolated α subunit from the wild-type parent. Thus it appears that reverse transcriptase is required to establish infection and at least the α subunit is coded for by the viral RNA.

In order to investigate the nature of the large subunit, β, we purified reverse transcriptase from avian myeloblastosis virus (AMV) and RSV, in vitro radiolabeled it with [^{125}I] and subjected it to SDS-PAGE to separate the two subunits. The tryptic hydrolysates of the α and β subunits were compared by two-dimensional fingerprinting techniques. The results indicate that β and α subunits from both AMV and RSV are structurally related. The possible mechanism of synthesis of α and β and the role of β subunits will be discussed.

730

INTRODUCTION

Transformation of cells by RNA tumor viruses requires the conversion of viral genetic material into complementary DNA (1-3). The virions of RNA tumor viruses contain a DNA polymerase which can transcribe 70S viral RNA into complementary DNA (2,4,5). This enzyme has been called reverse transcriptase or RNA-dependent DNA polymerase or DNA polymerase of RNA tumor viruses. During the life cycle of RNA tumor viruses (Fig. 1), early after infection the viral RNA is transcribed into complementary double-stranded DNA probably through an RNA-DNA hybrid as an intermediate (6). Circular double-stranded DNA (provirus) complementary to viral RNA has been shown to be present in the cytoplasm (7,8). It is then transported to the nucleus where it gets integrated into the host chromosome by some unknown mechanism (2). Transcription of the provirus is probably carried out by cellular DNA-dependent RNA polymerase (10-12). Although neither the nature of the primary transcript of the provirus nor the mechanism of its processing is understood, viral messenger RNA containing poly adenylic acid has been detected in the cytoplasm (13-15). Whether all of the 35S viral messenger RNA present in the cytoplasm acts as messenger RNA is not known. The 35S viral RNA gets encapsidated by virus specific proteins and the maturing virions move to the cell surface from which they bud, acquiring an outer envelope (3). Further steps of maturation may occur in the extra-cellular particles.

Reverse transcriptase is required very early after infection to synthesize provirus (Fig. 1). Purified reverse transcriptase from either avian or mammalian RNA tumor viruses can manifest both synthetic and degradative activities (2, 3). The synthetic activity is characterized by RNA-dependent DNA polymerase activity and DNA-dependent DNA polymerase activity. The degradative activity is characterized by ribonuclease H (RNase H) activity which selectively degrades the RNA moiety of an RNA-DNA hybrid (2, 16-19). Purified reverse transcriptase from avian RNA tumor viruses has an average molecular weight

731

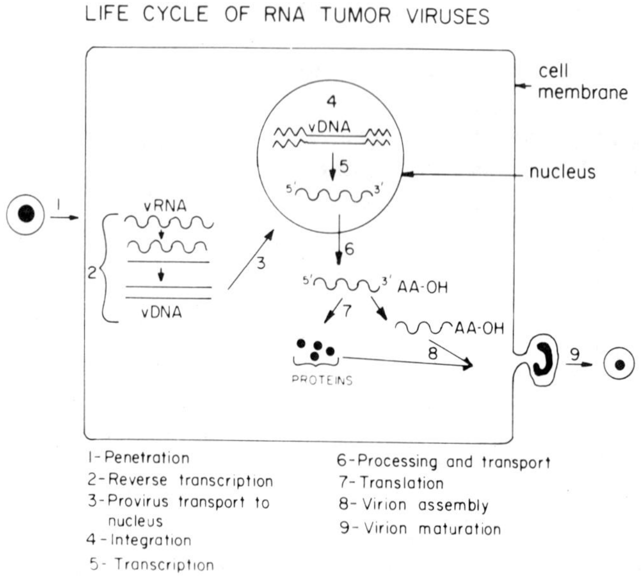

LIFE CYCLE OF RNA TUMOR VIRUSES

Fig. 1

Diagramatic representation of the probable life cycle of RNA tumor viruses.

of 160,000 and consists of two subunits with molecular weights of approximately 95,000 and 65,000 respectively (2, 20-22). The lower molecular weight subunit, α, has been isolated from avian myeloblastosis virus (AMV) and has been shown to contain both the polymerase and nuclease activities (21).

It has not been shown unambiguously whether the virion DNA polymerase is responsible for provirus synthesis or if the enzyme is coded for by the viral genome or is cellular in origin. In order to show that reverse transcriptase is coded for by the viral RNA, we have studied the thermolability of DNA polymerase activity from two temperature-sensitive mutants of Rous sarcoma virus (RSV). Linial and Mason (23) have isolated two such mutants of RSV which have defects in a very early event of the virus growth cycle. These mutants have been designated as ts LA335 PR-C (LA335) and ts LA337 PR-C (LA337)(24). They are unable to transform or replicate at the non-permissive temperature (41 C). The temperature-sensitive function displayed by these mutants is necessary for the initiation but not the maintenance of both cell transformation and viral replication.

In this manuscript, we would like to summarize evidence in support of the following conclusions: a) reverse transcriptase in avian RNA tumor viruses is coded for by the viral RNA, b) both the polymerase and nuclease activities are located on the same polypeptide, c) the polymerase and nuclease activities are probably located at different sites on the same polypeptide, and d) the two subunits are structurally related.

METHODS

Purification of Virion DNA Polymerase: Purified virions of LA335, LA337 and wild-type RSV were obtained as described (23). Chicken plasma containing avian myeloblastosis virus (AMV) was obtained through the Office of Program Resources and Logistics of the Virus Cancer

Program of the National Cancer Institute. The details of purification of virion DNA polymerase have been published elsewhere (25). Briefly, the virions were lysed with non-ionic detergent (Nonidet P-40) and the viral DNA polymerase was then purified from the lysate by sequential column chromatography on DEAE Sephadex and phosphocellulose. A minor peak of activity was eluted from the phosphocellulose column at a salt concentration of 0.1 M, followed by a major peak of activity eluting at a salt concentration of 0.25 M KCl. The details of isolation of small subunit are described in Fig. 2. Unless otherwise indicated, only material from the major peak of enzymatic activity was used.

Assays of Heat Lability: Heat inactivation studies were performed by pre-incubating the DNA polymerase in reaction mixture I, which contained 100 mM Tris-HCl, pH 8.3, 20 mM dithiothreitol, 12 mM magnesium acetate, 120 mM NaCl and about 0.05 to 1.0 unit of purified enzyme per 0.05 ml. (A unit of enzyme activity is defined as the amount of enzyme required to incorporate 100 pmoles of [^{3}H]dGMP at 37 C in 15 min.) The reaction mixture was incubated at the temperature indicated in the figures. At various intervals, aliquots of 0.05 ml were withdrawn and added to 0.05 ml of reaction mixture II which contained template•primer and substrates. This mixture was kept on ice for a maximum of 10 to 15 min, then further incubated at 34 C for 15 min. Acid precipitable radioactivity was measured as described (26). The rate of the enzyme reaction at 34 C was linear for at least 15 min.

Ribonuclease H: Ribonuclease H was assayed by using [^{3}H]poly(A)•poly(dT) as substrate and monitoring the degradation of [^{3}H]poly(A) to acid soluble material (17, 27). Heat inactivation was carried out as described above, except that reaction mixture I contained 40 mM Mg acetate and 60 mM NaCl. Reaction mixture II contained the substrate [^{3}H]poly(A)•poly(dT). The reaction at 34 C was

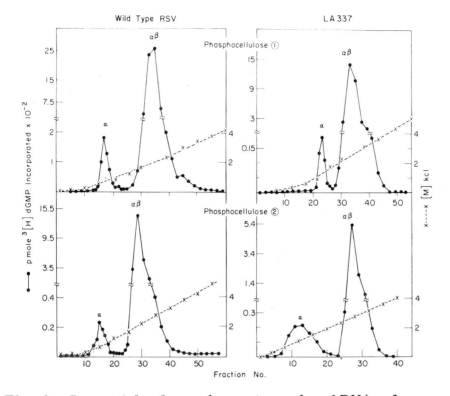

Fig. 2 Sequential column chromatography of DNA polymerase from wild-type RSV and LA337. The major peak of
enzymatic activity from the first phosphocellulose column
was pooled, dialyzed overnight against buffer B, and rechromatographed on a second phosphocellulose column.
Two peaks of enzymatic activities, α and $\alpha\beta$, could be
distinguished. Peak α was stored in 50% glycerol at
-20 C, whereas $\alpha\beta$ was stored at -70 C. The recovery
of the enzymatic activity from the second phosphocellulose column was over 50%. Enzyme assays were performed utilizing poly(C)·oligo(dG) as template·primer
and [^{3}H]dGTP (100 counts per min/pmol) as substrate.
Peaks α and $\alpha\beta$ from LA337 virions were isolated in an
identical fashion, except the dialysis was carried out for
5 to 8 h instead of overnight to minimize thermal inactivation of the enzyme during the purification procedure.
The recovery of total enzyme activity after first phosphocellulose column was over 20%. On the second phosphocellulose column, over 75% of the input enzymatic
activity was recovered. Peak α constituted about 10% of
the total enzyme activity. Both α and $\alpha\beta$ obtained from
LA337 virions were stored at -70 C.

carried out for 60 min under an atmosphere of nitrogen.
The acid-soluble radioactivity was determined by first
adding 10 μl of 10 mg/ml bovine serum albumin and 30 μl
of 50% TCA to the reaction mixture at the termination of
incubation. After 15 min at 4 C, the samples were centrifuged in a bench-top centrifuge for 3 min and 100 μl aliquots were withdrawn from the supernatant and counted
directly in 5.0 ml of dioxane-based scintillation fluid.
The degradation of poly(A) by the unincubated enzyme is
represented as 100%. All other values are normalized to
the value of the unincubated sample.

In vitro Radiolabeling of Reverse Transcriptase: Purified
reverse transcriptase, isolated from either AMV or RSV,
was precipitated from solution using trichloroacetic acid
(TCA) and extracted with acetone as described (22). The
dried protein was then resuspended in 30 μl of 0.05 M
Tris buffer, pH 7.5, containing 2% sodium dodecyl sulfate (SDS) and 4 M urea, and iodinated by the addition of
4 μl (about 2 mCi) of [^{125}I] sodium iodide and 5 μl of
chloramine-T (5 mg/ml in H$_2$O). After 15 min at 23 C,
the reaction was stopped by adding 100 μl of 0.05 M Tris
buffer, pH 7.0, 2% SDS, 20 mM dithiothreitol or 5% β-
mercaptoethanol, 20% glycerol and 0.005% bromophenol
blue (solubilization buffer). The radiolabeled sample was
then subjected to electrophoresis in a preparative polyacrylamide gel.

Polyacrylamide Gel Electrophoresis Techniques: Samples
in solubilization buffer (described above) were subjected
to electrophoresis in polyacrylamide slab gels using a
discontinuous buffer system containing SDS (28-30). Preparative gels were cross-linked with N,N'-diallyltartar-
diamide (31,32); analytical gels were cross-linked with
methylene-bisacrylamide. Following electrophoresis,
the protein bands were located in the gels by staining with
Coomassie brilliant blue (33) and/or by autoradiography
(34). Details of these procedures have been described
previously (32,35).

Peptide Analysis: The procedures used for peptide analysis have been described elsewhere (35). Briefly, radiolabeled proteins were removed from the preparative gel, oxidized with performic acid, and digested with TPCK-treated trypsin. Tryptic hydrolysates were then subjected to two-dimensional separations on thin layer cellulose plates, and the distribution of the $[^{125}I]$-labeled peptides determined by autoradiography.

RESULTS

DNA Polymerase and RNase H Activities of Purified Enzymes from LA335, LA337 and Wild-type RSV. Purified enzymes from the two mutants and from the wild-type parent are indistinguishable as judged by ion exchange chromatography on DEAE-Sephadex and phosphocellulose columns. All three enzymes contain two subunits of approximate molecular weights of 90,000 and 60,000 in equimolar ratios. Table 1 summarizes the average one-half time for inactivation of DNA polymerase and RNase H activities of LA335, LA337 parent RSV. The RNA-directed DNA polymerase activity was measured by using poly(C)·oligo(dG) as template·primer, DNA-directed DNA polymerase activity was measured by using poly(dC)·oligo(dG) as template·primer and for RNase H activity. $[^{3}H]$poly(A)·poly(dT) was used as substrate. The enzymes from the mutants appear to be four- to ten-fold more heat labile than the enzymes from the wild-type virus. The RNase H activity was less sensitive to heat inactivation than the DNA polymerase activity in the enzymes from both the mutant and the wild-type virus, suggesting that the polymerase and nuclease activities may be situated at different sites of the molecule.

Mason et al. (36) have isolated revertants of both LA335 and LA337 which are no longer temperature-sensitive for the initiation of infection at 41 C. To see if the phenotypic reversion was paralleled by a reversion of DNA polymerase to wild-type behavior, the heat inactivation kinetics of the DNA polymerase and RNase H

TABLE 1

$t_{1/2}$ in minutes

Source	DNA Polymerase Activity at 44°		Ribonuclease H Activity at	
	Poly(C)·oligo(dG)	Poly(dC)·oligo(dG)	44°	47°
RSV wild type	11.5	11.5	30	13
LA335	2	3	7.4	--
LA337	1.8	2.8	5	1.5

Average Times for Inactivation of One-half of the Activities of the Purified DNA Polymerase from Mutant and Wild-type RSV.

$t_{1/2}$ is the time required for inactivation of 50% of the relevant activity.

For poly(C)-directed poly(dG) synthesis, 1 μg poly(C) with 0.5 μg oligo(dG) was utilized as template·primer and 20 nmoles [³H]dGTP (100 cpm/pmole) was used as substrate. For poly(dC)-directed poly(dG) synthesis, the same conditions were used except that 1.0 μg of poly(dC) was substituted for poly(C). For RNase H assay, 10 pmole of [³H]-poly(A) (specific activity 1,000 cpm/pmole) and 2.8 nmole poly(dT) were used as substrate.

activities in virions of revertants of LA335 and LA337 were compared to that of mutant LA337. Due to limited quantities of the revertants, we did not purify the DNA polymerase but instead compared the polymerase and nuclease activities from the purified virions. The virion-associated DNA polymerase and RNase H activities of the revertants is three- to four-fold more heat stable than the virion-associated DNA polymerase activity of the mutants (27). Therefore, both the biological and biochemical parameters vary together during selection for reversion of the temperature-sensitive lesion.

Since the purified DNA polymerase studied in these experiments contained both subunits, it could not be conclusively established that the temperature-sensitive lesion resides in the same subunit that contains the enzymatic activities. To demonstrate that the polypeptide chain manifesting the synthetic and degradative activities is thermolabile, the lower molecular weight subunit, α, has been isolated from the temperature-sensitive mutant LA337.

Isolation of α from LA337 and Wild-type RSV: The lower molecular weight subunit α was isolated by sequential chromatography on phosphocellulose columns. Figure 2 shows the chromatographic profiles of DNA polymerases isolated from LA337 and wild-type RSV. The major peak of enzymatic activity obtained from the first phosphocellulose chromatography was dialyzed overnight against buffer B and rechromatographed on a second phosphocellulose column. When assayed with poly(C)·oligo(dG) as template·primer, two peaks of enzymatic activity could be distinguished. The minor peak of enzyme activity, α, eluting at a salt concentration of 0.1 M, represented 3 to 10% of the total enzyme activity. The major peak of enzyme activity elutes at a salt concentration of 0.24 M KCl. Similar separation of two enzyme activities can be obtained from AMV (37, 38). The α subunit used in all subsequent experiments was obtained from the second phosphocellulose column.

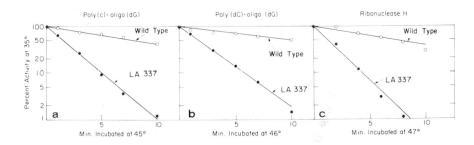

Fig. 3 Thermal inactivation profiles of DNA polymerase
and RNase H activities of isolated α subunit from LA337
and wild-type RSV. Assays of heat lability were performed as described in the Methods. a) 1 µg of poly(C)
and 0.5 µg of oligo(dG) were used as template·primer and
10 nmol of [3H]dGTP (1,000 counts per min/pmol) was
used as substrate; b) the same reaction conditions as described for (a) were used except 1 µg of poly(dC) was
substituted for poly(C) and the pre-incubation was carried
out at 46 C. Approximately 0.005 to 0.05 units of purified α subunit was used per assay. c) For RNase H assays
about one to two units of enzyme activity were used and
pre-incubations were carried out at 47 C.

TABLE 2

| | $t_{1/2}$ (min) | | |
Source	RNA-directed DNA synthesis poly(C)· oligo(dG)	DNA-directed DNA synthesis poly(dC)· oligo(dG)	RNase H [3H] poly(A)· poly(dT)
LA337	1.8	2.1	1.5
Wild-type RSV	8.8	9.5	8.0

Average One-half Time for Inactivation of DNA Polymerase
and RNase H Activity of Isolated α from LA337 and Wild-
type RSV.

$t_{1/2}$ is the time required for inactivation of 50% of activity
at the temperatures indicated in Fig. 3.

The molecular size of α subunit from wild-type RSV and AMV was characterized by chromatography on Sephadex G-100 column and by electrophoresis in poly-acrylamide gels. The results indicate that α from AMV (21) and RSV consists of one polypeptide chain with an average molecular weight of 65,000. The molecular size of α isolated from LA337 was not characterized because of insufficient material. It has, however, been shown previously that the two subunit enzyme, $\alpha\beta$, obtained either from LA337 or wild-type RSV has the same mole-cular size and number of subunits (27).

DNA Polymerase and RNase H Assay of α Subunit of LA337 and Wild-type RSV. The three known enzymatic activities associated with the DNA polymerase of avian RNA tumor viruses were compared using poly(C)·oligo(dG) to measure RNA-directed DNA synthesis, poly(dC)·oligo(dG) to assay DNA-directed DNA synthesis and [^{3}H]poly(A)·poly(dT) to assay for RNase H activity (Fig. 3). The average time of inactivation of one-half of the activity of α from LA337 and wild-type RSV for these three activities has been calculated (Table 2). The α subunit isolated from the virions of LA337 showed a five- to seven-fold greater thermolability than the α subunit isolated from wild-type RSV for all three activities. Thus it can be concluded that i) both the polymerase and nuclease activities reside on the same polypeptide chain, and ii) at least the lower molecular weight subunit α is coded for by the viral RNA.

Structural Relatedness of the Two Subunits: If the small subunit α can manifest both the polymerase and nuclease activities associated with purified avian RNA tumor viruses, the question arises as to what role the large subunit β plays in the process of reverse transcription and what is its origin? The heat inactivation kinetics of enzymatic activity associated with α subunit is unchanged when the pre-incubations are carried out either in the presence or absence of template·primer (38). If,

741

however, the two-subunit complex, $\alpha\beta$, is pre-incubated in the presence of the template·primer, very little loss of enzymatic activity is observed (38). Grandgenett et al. (39) have shown that α subunit associated RNase H acts as a random exonuclease whereas RNase H associated with the native enzyme acts as a processive exonuclease. Thus it appears that the β subunit serves to increase the affinity of native enzyme for template.primer or substrate. However, the source of the genetic information for large subunit is not known. One possibility was that like Qβ-RNA dependent RNA polymerase (40) the large subunit is a host protein. If this were the case, then one would expect the primary structure of the two proteins to be different. A comparison of the tryptic peptides of isolated α and β subunits of AMV and RSV, however, showed that the two subunits shared extensive amino acid homology.

In Vitro Radiolabeling of Reverse Transcriptase of AMV: About 100 µg of purified AMV reverse transcriptase was radiolabeled with $[^{125}I]$, solubilized and subjected to electrophoresis in a 14% preparative polyacrylamide gel, as described above. For purposes of comparison, a non-iodinated portion (10-20 µg) of the same enzyme preparation was subjected to electrophoresis adjacent to the radiolabeled enzyme. As shown in Fig. 4, the iodination procedure had no apparent effect on the electrophoretic mobility of either subunit (compare panels A and B). It did, however, render even the native enzyme functionally inactive. This inactivation was not a consequence of iodination itself, but of the procedure used, since we have recently found that reverse transcriptase iodinated by use of lactoperoxidase (41) retains its enzymatic activity (Verma and Gibson, unpublished results). Estimates based on densitometric measurements indicate that (i) the relative amounts of α and β were also unchanged by the iodination procedure (compare ratios at bottom of panels A and B), and (ii) greater than 95% of the total radioactivity in the gel was contained in the two subunit bands in a ratio of 1.4, β to α (panel C). The apparent

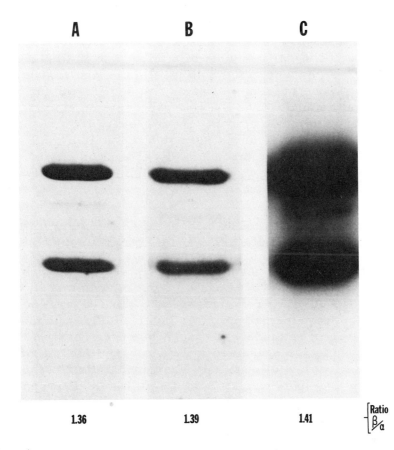

Fig. 4

Polyacrylamide gel analysis of reverse transcriptase of AMV. Photographs are shown here of the Coomassie brilliant blue-stained reverse transcriptase subunits of AMV before (A) and after (B) iodination, and an autoradiogram of the gel containing [^{125}I-labeled sub-units (C). The ratio of the densitometric absorbance (measured with an EC 910 scanning densitometer, E. C. Apparatus Corp., St. Petersburg, Fla.) of the α and β subunits is presented at the bottom of each panel. The direction of the protein's electrophoretic migration was from the top to the bottom in these figures.

molecular weights of the α and β subunits in this polyacrylamide gel system were 62,000 and 95,000, respectively. Small amounts of a third protein (molecular
weight 84,000), detected both by staining and autoradiography, were present between α and β. This minor band
is a contaminant of the enzyme preparation, since it is
present in variable amounts from preparation to preparation, and can be removed almost completely by sedimenting the enzyme through a glycerol gradient or by
chromatography on an oligo(dT)-cellulose column (Verma
and Gibson, unpublished results).

Comparison of the Tryptic Peptides of α and β Subunits of
Reverse Transcriptase Ioslated from AMV: After isolation and separation by electrophoresis, the α and β subunit bands of AMV reverse transcriptase were sectioned
from the gel and processed for tryptic peptide analysis.
The respective hydrolysates were compared by two-
dimensional separation on thin layer cellulose plates, as
described in Methods. Photographs of the resulting autoradiograms are presented in Fig. 5. These data can be
summarized as follows. (1) The general distribution of
radioactive spots was strikingly similar in all preparations. (ii) With the possible exception of several $[^{125}I]$-
labeled peptides in the region of spots n and o, all of the
$[^{125}I]$-labeled peptides of the small subunit of AMV are
also present in the large subunit. It is not yet clear
whether the apparent differences in the region of spots n
and o, observed between the large and small subunits are
reproducible, since some variability has been noted in
this region from preparation to preparation. (iii) Although some quantitative variation was observed between
the two AMV preparations shown here, there were no detectable qualitative differences (compare panels A and B,
Fig. 5). Purified reverse transcriptase from RSV was
subjected to similar analysis and the results show that
there were few differences observed between the reverse
transcriptase preparations of AMV and RSV. As mentioned above, there are apparent differences in the region

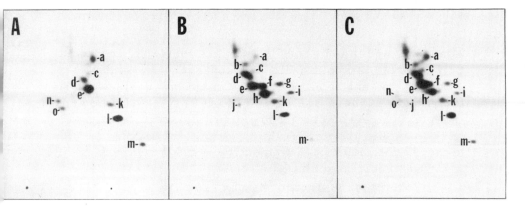

Fig. 5 Autoradiograms prepared from two-dimensional
separations of tryptic hydrolysis of the α and β subunits
of reverse transcriptase of AMV. These photographs
show the autoradiograms of the α subunit (A), the β subunit (B), and a mixture of the two (C). Electrophoretic
and chromatographic separations were from the left to
the right, and from the bottom to the top, respectively.
The origin of sample application appears as a spot in the
lower left-hand corner of these photographs.

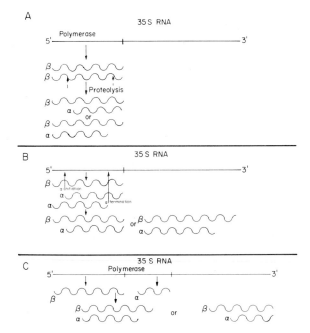

Fig. 6 Diagramatic models for the synthesis of α and β
subunits of reverse transcriptase.

of spots n and o and all AMV preparations examined contain a series of spots to the left and above the spots a and b, which are not observed in RSV preparations. Thus, the results of these analyses indicate that α and β subunits from AMV are structurally related. In addition, the α and β subunits from RSV also appear to be related, both to each other and to the respective α and β subunits of AMV reverse transcriptase (42).

DISCUSSION

RNA tumor viruses contain a DNA polymerase which can utilize polyribonucleotides as templates to synthesize faithful complementary DNA transcript (2). This enzyme has been attributed a central role in the provirus hypothesis (2, 3). In the scheme of events postulated in the provirus hypothesis, such an enzyme needs to be operational just after infection for the synthesis of DNA complementary to viral RNA. However, it has not previously been possible to determine unambiguously if the virion DNA polymerase is responsible for provirus synthesis and if the enzyme is coded for by the viral genome or is cellular in origin. Virus-specificity of the enzyme has been suggested by the finding that it appears only after infection of the cell (43, 44). The finding of non-infectious virions lacking DNA polymerase has been interpreted as evidence for an obligatory role for the enzyme (45-48). The experiments reported here demonstrate co-variation of thermolability of the virion DNA polymerase in vitro and temperature sensitivity of infection and transformation by the mutant virions. Two early temperature-sensitive mutants of Rous sarcoma virus were found to have DNA polymerase which was more thermolabile than that of the wild-type virus. The thermal inactivation profile of the DNA polymerase activity from the wild-type revertants (wt LA335 and wt LA337) is indistinguishable from that of wild-type virus (27). Mason et al. (36) have recently characterized the biological and biochemical properties of these mutants, their revertants, and several recombinants with leukosis viruses and have shown that the lesion in these mutants is the increased thermolability of their DNA polymerase activity. Varmus et al. (Cold Spring Harbor Symp. Quant. Biol., in press) have reported that less viral DNA is produced in LA335 than LA337 at the non-permissive temperature than is produced by the wild-type virus. These results clearly indicate that RSV RNA encodes the information for at least part of the DNA polymerase molecule and the enzyme is obligatory to initiate infection or transformation by RSV.

Since all three known enzyme activities associated with the RNA tumor virus DNA polymerase were thermolabile, it was suggested that all three activities were located on the same polypeptide chain (27). To establish that the polypeptide chain manifesting all three enzyme activities is thermolabile and is encoded by the viral RNA, α subunit from LA337 was isolated and its thermal inactivation properties compared with those of isolated α subunit from wild-type RSV. The α subunit was isolated from purified DNA polymerase containing equimolar amounts of α and β subunits. This approach assures that the isolated lower molecular weight subunit α is part of the viral DNA polymerase. The mechanism of separation of α from $\alpha\beta$ (two subunit complex) is not clear. Both synthetic and degradative activities of the isolated small subunit of LA337 showed heat-inactivation kinetics indistinguishable from those of the native LA337 DNA polymerase (37). It must therefore be that the lower molecular weight subunit is encoded by the viral RNA.

The observed similarities between the large subunit and small subunit [^{125}I]-labeled peptide distributions suggest a close structural relationship between these two proteins. There are at least three possible explanations of their apparent relatedness. (i) The small subunit of reverse transcriptase is derived from the large subunit by proteolytic cleavage (Fig. 6, A). This mode of synthesis of α from β has also been suggested by K. Mölling (Cold Spring Harbor Symp. Quant. Biol., Vol.

XXXIX, in press). The observation that the small subunit may contain two $[^{125}I]$-labeled peptides (e.g., spots n and o) that are absent from the large subunit is not incompatible with this interpretation, since cleavage at the NH_2-terminal or COOH-terminal or both ends could generate one or two new peptides. If the small subunit is derived from the large subunit and the native enzyme is the $\alpha\beta$ complex, then reverse transcriptase from avian RNA tumor viruses may be an example of an enzyme that requires close association between the product and precursor for maximal enzymatic activity. (ii) The two proteins are independently coded for by the same gene (e.g., more than one initiation and/or termination site within the same gene)(Fig. 6, B). And (iii) That each subunit is coded for by separate but related genes (Fig. 6, C). Although neither of these last two possibilities can be ruled out by the data presented here, the last one at least seems unlikely since together these proteins would require nearly 50% of the potential coding capacity of the avian RNA tumor virus genome (49, 50, and D. Baltimore, Summary, Cold Spring Harbor Symp. Quant. Biol., Vol. XXXIX, in press).

A number of questions concerning the synthesis and putative processing of avian reverse transcriptase arise from the suggestion that the small subunit may be derived from the large subunit. For example, (i) At what point during the life cycle of the virus is the small subunit produced? (ii) What is the function of each of the subunits? (iii) Is there a mechanism that regulates the amount of large subunit converted to small subunit? We propose the following possible chain of events leading to the synthesis of enzymatically active reverse transcriptase. (i) The large subunit is the molecular species of reverse transcriptase synthesized in infected cells and packaged into the virions. (ii) After or during maturation of the virions some of the large subunit molecules are cleaved to produce small subunits. (iii) Conformational changes that render the small subunit-large subunit complex more resistant to proteolysis may be involved in limiting the amount of large subunit converted to small subunit, to an equimolar ratio.

The similarities of $[^{125}I]$-labeled peptide distributions of AMV and RSV reverse transcriptases (data not shown) suggest that these enzymes may share extensive amino acid sequence homology. This is consistent with the observation that antisera to reverse transcriptase of AMV inhibit the enzymatic activity of reverse transcriptase of RSV (51). Although the technique used in the experiments reported here allows detection of only those peptides containing tyrosine residues (52), it can be used to make a limited comparative study of reverse transcriptase from other sources to determine whether some portions of its amino acid sequence may have been conserved during the evolution of the RNA tumor viruses.

ACKNOWLEDGMENTS

This work was supported by Research Grant No. CA 16561-01 from the National Cancer Institute and a Research Grant, No. 320, from Jane Coffin Childs Memorial Fund to I.M.V.; a contract from the Virus Cancer Program to Dr. Walter Eckhart and Grant No. CA 14195. W.G. was supported by a postdoctoral fellowship from the Damon Runyon Memorial Fund for Cancer Research.

REFERENCES

1. Temin, H. M. Ann. Rev. Genetics 8, 155 (1974).
2. Temin, H. M. and Baltimore, D. In Advances in Virus Research, V. 17, p. 129. Academic Press, New York (1972).
3. The Molecular Biology of Tumour Viruses, Tooze, J. (ed.) Ch. 11. Cold Spring Harbor Library, Cold Spring Harbor, New York (1973).
4. Baltimore, D. Nature (London) 226, 1209 (1970).
5. Temin, H. and Mizutani, S. Nature (London) 226, (1970).

6. Manly, K., Smoler, D. F., Bromfeld, E. and Baltimore, D. J. Virol. 7, 106 (1971).
7. Guntaka, R. V., Bishop, J. M. and Varmus, H. E. Proc. Nat. Acad. Sci. (USA)(In press)(1975).
8. Gianni, A. M., Smotkin, D. and Weinberg, R. A. Proc. Nat. Acad. Sci. (USA)(In press)(1975).
9. Varmus, H. E., Vogt, P. K. and Bishop, J. M. Proc. Nat. Acad. Sci. (USA) 70, 3067 (1973).
10. Jacquet, M., Groner, Y., Monroy, G. and Hurwitz, J. Proc. Nat. Acad. Sci. (USA) 71, 3045. (1974).
11. Rymo, L., Parsons, J. T., Coffin, J. M. and Weissman, C. Proc. Nat. Acad. Sci. (USA) 71, 2782 (1974).
12. Astrin, S. M. Biochem. Biophys. Res. Commun. 61, 1242 (1974).
13. Leong, J. A., Garapin, A. C., Jackson, N., Fancier, L, Levinson, W. and Bishop, J. M. J. Virol. 9, 891 (1971).
14. Fan, H. and Baltimore, D. J. Mol. Biol. 83, 93 (1973).
15. Gielkens, A. L. J., Salden, M. H. L. and Bloemendal, H. Proc. Nat. Acad. Sci. (USA) 71, 1093 (1974).
16. Mölling, K., Bolognesi, D. P., Bauer, H., Büsen, N., Plassman, H. W. and Hausen, P. Nature N. Biol. 234, 240 (1971).
17. Baltimore, D. and Smoler, D. J. Biol. Chem. 247, 7282 (1972).
18. Leis, J. P., Berkower, I. and Hurwitz, J. Proc. Nat. Acad. Sci. (USA) 70, 466 (1973).
19. Keller, W. and Crouch, R. Proc. Nat. Acad. Sic. (USA) 69, 3360 (1972).
20. Kacian, D. L., Watson, K. F., Burny, A. and Spiegelman, S. Biochim. Biophys. Acta 246, 365 (1971).
21. Grandgenett, D. P., Gerard, G. F. and Green, M. Proc. Nat. Acad. Sci. (USA) 70, 230 (1973).

22. Verma, I.M., Meuth, N. L., Fan, H. and Baltimore, D. J. Virol. 13, 1075 (1974).
23. Linial, M. and Mason, W. S. Virology 53, 258 (1973).
24. Vogt, P. K., Weiss, R. A. and Hanafusa, H. J. Virol. 13, 551 (1974).
25. Verma, I. M. and Baltimore, D. In Methods in Enzymology, Grossman, L. and Moldave, K. (ed.) V. 29, p. 125 (1973).
26. Baltimore, D., Huang, A. S. and Stampfer, M. Proc. Nat. Acad. Sci. (USA) 66, 572 (1970).
27. Verma, I. M., Mason, W. S., Drost, S. D. and Baltimore, D. Nature 251, 27 (1974).
28. Ornstein, L. An. N.Y. Acad. Sci. 121, 321 (1964).
29. Davis, B. J. An. N.Y. Acad. Sci. 121, 404 (1964).
30. Laemmli, U. K. Nature 227, 680 (1970).
31. Anker, H. S. FEBS Letters 7, 293 (1970).
32. Gibson, W. and Roizman, B. J. Virol. 13, 155 (1974).
33. Fairbanks, G., Levinthal, C. and Reedey, R. H. Biochem. Biophys. Res. Commun. 20, 393 (1965).
34. Fairbanks, G., Steck, T. L. and Wallach, D.F.H. Biochem. 10, 2606 (1971).
35. Gibson, W. Virology 62, 319 (1974).
36. Mason, W. S., Friis, R. R., Linial, M. and Vogt, P.K. Virology 61, 559 (1974).
37. Verma, I. M. J. Virol. 15, 121 (1975).
38. Panet, A., Verma, I. M. and Baltimore, D. Cold Spring Harbor Symp. Quant. Biol. XXXIX (In press).
39. Grandgenett, D. P. and Green, M. J. Biol. Chem. 249, 5148 (1974).
40. Blumenthal, T., Landers, T. and Weber, K. Proc. Nat. Acad. Sci. (USA) 69, 1313 (1972).
41. David, G. and Reisfeld, R. A. Biochem. 13, 1014 (1974).
42. Gibson, W. and Verma, I. M. Proc. Nat. Acad. Sci. (USA) 71, 4991 (1974).
43. Ross, J., Scolnick, E. M., Todaro, G. and Aaronson, S. A. Nature N. Biol. 231, 163 (1972).

44. Baltimore, D., McCaffrey, R. and Smoler, D. F. Virus Research, Fox, C. F. and Robinson, W. S. (ed.) p. 51, Academic Press, N. Y. (1973).

45. Hanafusa, H., Baltimore, D., Smoler, D. F., Watson, K. F., Yaniv, A. and Spiegelman, S. Science 177, 1188 (1972).

46. Peebles, P. T., Haapala, D. K. and Gazdar, A. F. J. Gen. Virol. 9, 488 (1972).

47. May, J. T., Somers, K. D. and Kit, S. J. Gen. Virol. 16, 223 (1972).

48. Somers, K. D., May, J. T., Kit, S., McCormick, K. J., Hatch, G. G., Stenback, W. A. and Trentin, J. J. Intervirology 1, 11 (1973).

49. Billeter, M. A., Parsons, J. T. and Coffin, J. M. Proc. Nat. Acad. Sci. (USA) 71, 3560 (1974).

50. Beemon, K., Duesberg, P. and Vogt, P. K. Proc. Nat. Acad. Sci. (USA) 71, 4254 (1974).

51. Nowinski, R. C., Watson, K. F., Yaniv, A. and Spiegelman, S. J. Virol. 10, 959 (1972).

52. Marchalonis, J. J., Cone, R. E. and Santer, V. Biochem. J. 124, 921 (1971).

A THERMOLABILE REVERSE TRANSCRIPTASE FROM A
TEMPERATURE-SENSITIVE MUTANT OF MURINE LEUKEMIA VIRUS

Steven R. Tronick[*], John R. Stephenson[*], Inder M. Verma[+],
and Stuart A. Aaronson[*]
[*]Viral Carcinogenesis Branch
National Cancer Institute
Bethesda, Maryland 20014
[+]Tumor Virology Laboratory
The Salk Institute
La Jolla, Ca.

ABSTRACT. Three temperature-sensitive mutants of the Rauscher strain of murine leukemia virus are defective in early post-penetration functions required both for leukemia
virus infection and for initiation of transformation of
cells by their pseudotypes of murine sarcoma virus. In
the present studies, the reverse transcriptase of one of
these mutants (ts 29) is shown to be thermolabile compared
to the enzymes of the wild-type virus and several other
temperature-sensitive mutants. These findings provide
evidence that the reverse transcriptase is required both
for leukemia virus infection and for initiation of transformation by the replication-defective murine sarcoma
virus genome.

INTRODUCTION

Temperature-sensitive (ts) mutants of Rauscher murine
leukemia virus (R-MuLV) defective in post-penetration functions have been characterized with respect to the stages
at which viral replication is blocked at the non-permissive
temperature (39C) (1-4). Viruses designated as early mutants do not synthesize viral antigens at the non-permissive temperature (39C). In contrast, viruses designated
as late mutants synthesize wild-type (wt) levels of viral
antigens as efficiently as wt at 39C. A MuLV gene product
likely to be required at an early stage of infection is
the RNA-dependent DNA polymerase (reverse transcriptase).

In the present study, we have compared the thermolability of the reverse transcriptase of R-MuLV ts mutants with that of the wt viral enzyme.

RESULTS

R-MuLV Mutants in Early Post Penetration Functions: To investigate the defects associated with 3 early mutants, temperature-shift experiments were performed (3). Exponentially growing BALB/3T3 cultures were infected at 38C with wt or ts R-MuLV pseudotypes of Ki-MSV. Parallel cultures were infected at 31C. After various incubation periods, cultures were exposed to 1.0% trypsin for 5 min, a treatment which inactivated more than 99% of free or absorbed virus. After resuspension in regular medium, cells were transferred at ten-fold dilutions to 50-mm petri dishes containing 10^5 BALB/3T3 cells and incubated for the remainder of the experiment at the permissive temperature. Cultures were scored for Ki-MSV transformed foci on the 12th day following infection.

The results (Fig. 1) show that with Ki-MSV pseudotypes of the wt virus, ts 25, and ts 26 were able to penetrate within 5-6 hr at 39C. The kinetics of infection with Ki-MSV coated by each of three early mutants were very similar during the first 2-3 hr at the nonpermissive temperature indicating they absorbed and penetrated at this temperature. However, a peak of focus formation occurred with incubation for 3 hr at 38C. When infected cells were maintained for longer times at the high temperature prior to downshift, the number of MSV foci that eventually registered markedly decreased. When cells were maintained as long as 24 hrs at 39C only 5-10% of the maximum focus-forming units were achieved. These results indicate that ts 17, ts 19, and ts 29 were each temperature sensitive in MuLV function(s) expressed after virus absorption and penetration yet necessary for fixation of sarcoma virus transformation.

Thermolability of Reverse Transcriptase Activity of Disrupted Virions: Virions of wt and of early and late mutants were disrupted with Triton X-100, incubated at 40C, and then assayed for enzyme activity at 34C using poly(rA)·oligo(dt) as template. The results (Fig. 2A) indicate

754

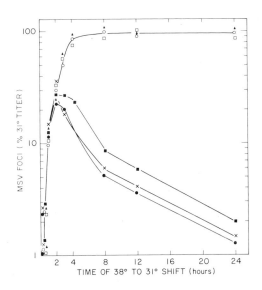

Fig. 1. Transformation of BALB/3T3 cells infected at the nonpermissive temperature with wt and ts R-MuLV pseudotypes of Ki-MSV and shifted to 31C at varying time intervals. Around 10^5 BALB/3T3 cells were infected with each virus at 38C. At subsequent time points, the cells were trypsinized and transferred at tenfold dilutions to empty petri dishes at 31C. Focus formation was scored 12 days later. R-MuLV pseudotypes of Ki-MSV included: wt(O), ts 17 (X), ts 19 (■), ts 25 (▲), ts 26 (□), ts 29 (●).

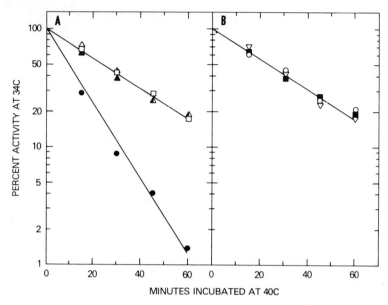

Fig. 2. Thermal inactivation of reverse transcriptase
activity of disrupted virions. Viruses were incubated at
40C in 100mM Tris·HCl, pH 7.8, 180mM KCl, 0.2mM MnCl$_2$.
0.1% Triton, and 4mM dithiothreitol, sampled and assayed
for activity (10) at 34C with 0.1 mCi/ml ^{3}H-TTP, 0.5 µg/ml
poly(rA), and 0.38 µg/µl oligo(dT). A) ts 29,•-•; ts
17,□-□; ts 19,△-△; wt,▲-▲. B) ts 18,▽-▽; ts 25,○-○;
ts 26,▦-▦.

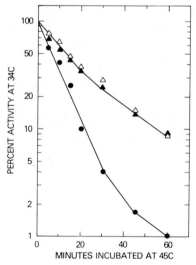

Fig. 3. Thermal inactivation of purified reverse trans-
scriptases. Enzymes from ts 29,●●, and two separate
clones of wt (▲-▲, △-△) were heated at 45C and assayed
at 34C as described in Figure 2.

that the reverse transcriptase of ts 29, was markedly more thermolabile than the enzyme activities of wt virus, other early mutants (ts 17, ts 19) or late mutants (ts 18, ts 25, ts 26) (Fig. 2B). The time required for one-half inactivation ($t_{1/2}$) of the activity of ts 29 was nine minutes in contrast to 23-24 minutes for the enzyme activities of the other virions.

Thermolability of Enzymes Purified from wt and ts 29: In order to show that the thermolability of the ts 29 polymerase activity was due to a defect in the enzyme rather than an indirect effect of using disrupted virions, the enzymes from wild-type and ts 29 virions were purified by the method of Ross, et al. (5). The enzymes were heated at 45C and assayed at 34C using poly(rA)·oligo(dt). The ts 29 enzyme was inactivated considerably faster than the wt enzyme (Fig. 3) ($t_{1/2}$ for ts 29 = 6 min and for wt = 13 min). To provide further evidence that the thermolability of the ts 29 enzyme was not due to an artifact, wt and ts 29 enzymes were mixed, heated at 45C, and assayed at 34C. The activity of the mixture at each time point was the same as the sum of the activities of each enzyme heated individually (data not shown).

Effect of Synthetic Templates on Thermolability of wt and ts 29 Enzymes: Mammalian type-C viral reverse transcriptases utilize a variety of synthetic and natural polyribo- and polydeoxyribonucleotide templates, although with widely varying efficiencies (6,7). Thus, it was of interest to determine whether the thermolability of the ts 29 enzyme would be evident when assayed with other templates. Of several templates tested, only poly(rA)·poly(dT) and poly(rC).oligo(dG) supported sufficient activity with the amounts of purified enzymes available. In each case, the ts 29 polymerase was inactivated two to three times faster than the wt enzyme. The enzymes were also heated at 45C in the presence of templates. Templates stabilized both mutant and wt enzymes (two to four-fold increases in $t_{1/2}$ values), but in each case, the ts 29 polymerase was more thermolabile than the wt enzyme (data not shown).

DISCUSSION
The present report provides the first evidence for a mammalian RNA type C virus with a temperature-sensitive

757

reverse transcriptase. This demonstrates that the viral reverse transcriptase is required for leukemia virus infection and is coded for by the type C viral genome. Three early R-MuLV mutants, including ts 29, were defective in initiating transformation of cells at the non-permissive temperature (3). Thus, the present evidence indicates that one of the helper functions provided the defective sarcoma viral genome by MuLV is the reverse transcriptase.

Recently, two ts mutants of RSV were found to possess thermolabile reverse transcriptases (8,9), which appear to be defective in binding template at the high temperature (9). In the present studies, similar experiments with ts 29 indicated that template protected mutant and wt enzymes to the same extent. This suggests that, in contrast to the avian mutant enzymes, the lesion in the ts 29 enzyme does not involve binding of template.

The thermolabilities of the reverse transcriptase activities in disrupted virions (Fig. 2) and in purified enzyme preparations (unpublished results) of other early mutants (ts 17 and ts 19) were indistinguishable from that of the wt enzyme. These findings suggest that early leukemia viral gene functions in addition to the reverse transcriptase may be required for infection by MuLV. Alternatively, the ts 17 and ts 19 polymerases may be defective in reverse transcriptase activities other than those studied here. For example, the reverse transcriptases of these mutants may fail to copy viral RNA or may not manifest RNase H activity (11,12) at restrictive temperatures. Studies are now in progress to resolve these possibilities.

ACKNOWLEDGEMENT

We thank Marjorie M. Golub and Janet A. Steel for excellent technical assistance. This work was supported in part by Contract No. NCI-E-73-3212 of the Virus Cancer Program of the National Cancer Institute.

REFERENCES

1. Stephenson, J.R., Reynolds, R.K., and Aaronson, S.A. Virology 48, 749 (1972).

2. Stephenson, J.R. and Aaronson, S.A. Virology 54, 53 (1973).
3. Stephenson, J.R. and Aaronson, S.A. Virology 58, 294 (1974).
4. Stephenson, J.R., Tronick, S.R., and Aaronson, S.A. J. Virol. 14, 918 (1974).
5. Ross, J., Scolnick, E.M., Todaro, G.J., and Aaronson, S.A. Nature New Biol. 231, 163 (1971).
6. Baltimore, D. and Smoler, D. Proc. Nat. Acad. Sci. (Wash.) 68, 1507 (1971).
7. Wang, L.H. and Duesberg, P.H. J. Virol. 12, 1512 (1973).
8. Linial, M. and Mason, W.S. Virology 53, 258 (1973).
9. Verma, I.M., Mason, W.S., Drost, S.D., and Baltimore, D. Nature (London) 251, 27 (1974).
10. Tronick, S.R., Scolnick, E.M., and Parks, W.P. J. Virol. 10, 885 (1972).
11. Moelling, K. Virology 62, 46 (1974).
12. Verma, I.M. J. Virol., in press.

CORRELATION OF THE BINDING PROPERTIES OF α AND αβ DNA POLYMERASE OF AVIAN MYELOBLASTOSIS VIRUS WITH THEIR DIFFERENT MODE OF RIBONUCLEASE H ACTIVITY

Duane Grandgenett
St. Louis University School of Medicine
Institute for Molecular Virology
3681 Park Avenue
St. Louis, Missouri 63110

ABSTRACT. αβ DNA polymerase from avian myeloblastosis virus forms more stable polymerase-nucleic acid complexes than α DNA polymerase isolated from the same virus. The stability of these polymerase-nucleic acid complexes were measured by competition experiments using first labeled and subsequently an excess of homologous unlabeled nucleic acid and by measuring the stability of these complexes under various ionic conditions. The polymerase-nucleic acid complexes were measured by utilizing the nitrocellulose membrane filter technique. These results provided direct evidence which support the conclusion that the random and processive mode of exoribonuclease H activity associated with α and αβ, respectively, is the result of the polymerase ability to bind to nucleic acids.

INTRODUCTION

Two DNA polymerases which have associated exoribonuclease H (RNase H) activity have been isolated from avian myeloblastosis virus (AMV) (1). The single subunit enzyme, α, has a random exoribonuclease H activity (2) while αβ, a two subunit enzyme, has a processive mode of nuclease activity (2,3). These different modes of nuclease activity suggested that the αβ enzyme bound tighter to nucleic acids than α DNA polymerase. This assumption was further supported by affinity chromatography studies (4). In this report, we present data showing that the DNA binding site on each respective enzyme behaved the same as its binding site for a DNA-RNA hybrid, the substrate for RNase H activity.

METHODS

Purification of viral DNA polymerases: The α and αβ DNA polymerases from AMV were purified through DEAE-cellulose,

760

phosphocellulose, and glycerol gradient centrifugation as described except that the detergent lysed-virus was precipitated by ammonium sulfate prior to the DEAE-cellulose chromatography (1).

Nitrocellulose filter assay: Binding of α and $\alpha\beta$ DNA polymerase to various nucleic acids was performed at 37C in the following mixture (0.1 ml) with 20 mM Tris-HCl (pH 8.0), 10 mM MgCl$_2$, 5 mM dithiothreitol, and varying amounts of KCl as indicated. Usually, both enzymes were allowed to bind for 4 min. The samples were diluted with 2 ml of the above binding buffer (40 mM KCl) and immediately filtered through Schleicher and Schuell B6 or Millipore HA nitrocellulose filters. The filters were then washed with an additional 2 ml of buffer. Filtration was carried out at room temperature at a flow rate of 4 ml per min. The filters were dried and the radioactivity determined in a toluene-based solvent in a liquid scintillation counter. The maximum amount of [^{3}H]poly(dA-dT) and [^{3}H]labeled adenovirus DNA that was bound by the enzymes were different. With several commercial preparations of [^{3}H]poly(dA-dT), the maximum amount bound by either enzyme was 80% while with isolated [^{3}H]-adenovirus DNA it was 100%.

RESULTS

Properties of DNA polymerase-nucleic acid complexes: Binding of enzymes to nucleic acids has been extensively studied using the nitrocellulose filter assay technique (5). Various parameters of the assay technique were investigated. The nucleic acids utilized in this study did not bind significantly (<5%) to the filters by themselves. Both α and $\alpha\beta$ DNA polymerase bound to the filters in the absence of any nucleic acid at all the KCl concentrations utilized for this study (0.05 M to 0.3 M). No polymerase activity could be detected in the filtrates and the enzyme retained on the filter was enzymatically active. The time required for formation of a stable non-filterable polymerase-nucleic acid complex in the reaction mixture was less than 1 min and the enzymes were enzymatically stable at 37C in 40 mM KCl binding buffer with [^{3}H]poly(dA-dT) for at least 5 min. Both enzymes were stable at least 3 min upon dilution and once on the filter, the complexes were stable and washing with 0.5 or up to 12 ml of 40 mM KCl binding buffer did not dissociate them.

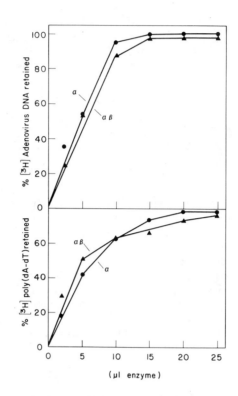

Figure 1.   Retention of labeled nucleic acids on
nitrocellulose filters by α and αβ DNA polymerase.

A typical binding curve is shown in Figure 1. Glycerol gradient purified α and $\alpha\beta$ DNA polymerase retained approximately equivalent amounts of both [^{3}H]poly(dA-dT) (400 pmoles, 12 cpm/pmole) and ^{3}H-labeled adenovirus DNA (0.024 μg, 1.9×10^5 cpm/μg) at 40 mM KCl. α and $\alpha\beta$ DNA polymerase (5 μl) incorporated 640 and 955 pmoles, respectively, of [^{3}H]TTP in 15 min at 37C using poly(A)·oligo(dT) as template-primer; and the protein concentration was 221 and 42 μg/ml, respectively.

Effect of ionic strength on formation of non-filterable DNA polymerase-nucleic acid complexes: The formation of non-filterable polymerase-nucleic acid complexes under varying ionic conditions with α or $\alpha\beta$ were different. The formation of non-filterable complexes with α DNA polymerase and [^{3}H]poly(dA-dT) was more easily prevented with increasing ionic strength in the reaction mixture when compared to $\alpha\beta$ DNA polymerase (Figure 2). When the ionic strength was raised above 100 mM KCl, the $\alpha\beta$ DNA polymerase-[^{3}H]poly(dA-dT) complex was slightly affected while the binding efficiency of α was reduced 60%. Similar results were obtained when adenovirus DNA was used (Figure 2, lower panel).

To provide direct evidence that there was a stronger interaction between $\alpha\beta$ and nucleic acids in comparison to α, the following study was initiated. First, labeled nucleic acid was bound to a limiting amount of enzyme and subsequently an excess of homologous but unlabeled polymer was added. If the labeled nucleic acid was firmly bound to the polymerase, it would remain so, even in the presence of the competitor. In contrast, if the labeled nucleic acid dissociates easily from the polymerase, complexes between the cold competitor and polymerase would be formed and the amount of radioactivity in the original complex would decrease. Upon binding of a limited concentration of α and $\alpha\beta$ DNA polymerase to [^{3}H]poly(dA-dT) in 100 mM KCl binding buffer, a 25-fold excess of cold polymer was added and samples were taken at various intervals. As shown in Figure 3, the amount of [^{3}H]labeled polymer which remained complex with α decreased rapidly while $\alpha\beta$ retained the labeled polymer. Measuring the rate of dissociation under conditions similar to that above, the time required to dissociate 50% of the [^{3}H]poly(dA-dT) molecules from α at 100 mM KCl was approximately 3 min while for $\alpha\beta$ it

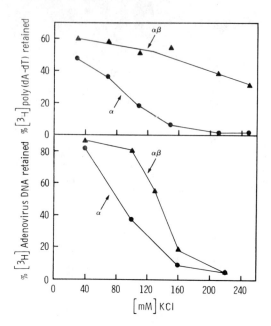

Figure 2. Formation of polymerase-nucleic acid complexes at different ionic concentrations.

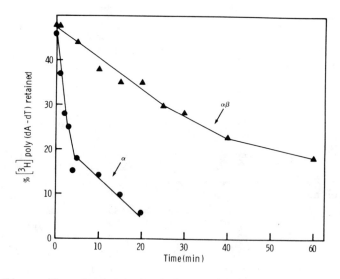

Figure 3. Competition of labeled nucleic acid-DNA polymerase complexes with cold polymer.

was approximately 40 min. Similar data was obtained if the enzymes were allowed to initially bind to [^{3}H]poly(dA-dT) at 40 mM KCl, cold polymer added and then the salt concentration immediately adjusted to 130 mM KCl (data not shown). In addition, similar competition data was obtained by measuring the rate of solubilization of [^{3}H]poly(A)· poly(dT), a substrate for RNase H activity, with α and αβ DNA polymerase in the presence or absence of an excess of unlabeled polymer (2).

DISCUSSION

Both α and αβ DNA polymerase from AMV bind to nitro-cellulose filters in the absence of nucleic acids which is similar to that found with Escherichia coli RNA polymerase (6). Utilizing this property, we have demonstrated that the α DNA polymerase is more readily dissociated from various nucleic acids by either increasing the salt concentration or by competition of bound labeled polymer with additional cold polymer, while αβ forms relatively resistant nucleic acid-enzyme complexes.

ACKNOWLEDGMENT

This work was supported by Grant 1-RO1-CA16312-01 from the National Cancer Institute, Contract NO1 CP 43359 within the Virus Cancer Program, and NIH General Research Support Grant 324.

REFERENCES

1. Grandgenett, D.P., Gerard, G.F., and Green, M. Proc. Nat. Acad. Sci. 70, 230 (1973).
2. Grandgenett, D.P., and Green, M. J. Biol. Chem. 249, 5148 (1974).
3. Leis, J.P., Berkower, I., and Hurwitz, J. Proc. Nat. Acad. Sci. 70, 466 (1973).
4. Grandgenett, D.P., and Rho, H.M. J. Virol. 15, in press.
5. Jones, D.W., and Berg, P. J. Mol. Biol. 22, 199 (1966).
6. Hinkle, D.C., and Chamberlin, M.J. J. Mol. Biol. 70, 157 (1972).

ROLLING CIRCULAR DNA ASSOCIATED WITH A HUMAN HEPATITIS B CANDIDATE VIRUS (DANE PARTICLES)

P.P. Hung, J.C. Mao, C.M. Ling, L.R. Overby and T. Kakefuda

Division of Experimental Biology
Abbott Laboratories
North Chicago, Illinois and
Chemistry Branch
National Cancer Institute
Bethesda, Maryland

ABSTRACT

DNA isolated from Dane particles appeared in electron micrographs as double-stranded open circles and linear structures. Short double-stranded tails were found on a few (less than 2%) of the open circles. However, DNA isolated from Dane particles after incubation for DNA synthesis by the endogenous DNA polymerase showed that the number of the tailed circles increased to about 15 to 20% of the circles and the tail lengthened greatly, sometimes longer than a unit circle. These observations suggest that the circular genome of Dane particles is replicated by the rolling circle model.

INTRODUCTION

In the plasma of hepatitis B antigen (HB_SAg) carriers, there are predominant spherical particles of 20 nm, and a much smaller number of filamentous structures with the same diameter. In addition, larger spherical particles, 42 nm in diameter with a 28 nm core, first described by Dane et al. (1), are routinely found in crude plasma or purified preparations of HB_SAg. Dane particles also contain a primed DNA polymerase activity (2) capable of synthesizing new DNA. Robinson et al. (3) examined DNA from Dane particles by electron microscopy and found open, double-stranded circular configurations approximately 0.78 microns in length. We have examined DNA isolated from Dane particles directly and also after formation of radioactive products by the endogeneous DNA polymerase reaction. The DNA structures appeared in electron micrographs as double-stranded linear strands and circles with examples of: (A) completely open circles; (B) partially open circles; and (C) circles with attached linear segments in different lengths. The open circles with attached linear segments are suggestive of a replicative form, and this is consistant with a rolling circle model (4) for replication. This model for replication was strengthened by the comparison of the types of DNA found in particles before and after reaction of the polymerase. The number of rolling circles and the length of the tails were vastly increased after DNA synthesis. This type of replication has not been reported heretofore for animal viruses.

MATERIALS AND METHODS

Plasmas from commercial blood donors, containing HBsAg of ad or ay subtype, were used to isolate and purify Dane particles by density gradient centrifugation and isopycnic banding techniques (2). DNA polymerase activity and electron microscopy were used to monitor fractions during the isolation and purification. Three Dane particle preparations, shown to be essentially free of 20 nm spherical and filamentous particles by electron microscopy were purified from approximately 6 liters each of HB_sAg-positive plasmas of subtypes ay and ad. One ad subtype and one ay subtype particle preparation, containing DNA polymerase activity, were then radiolabeled with ^{3}H-thymidine triphosphate for 6 hours at 37°. The reaction mixture, in a final volume of 12.8 ml, contained 1.6 ml of Dane particle suspension, plus the following reagents at the indicated concentrations: NP-40 detergent, 1%; mercaptoethanol, 0.3%; $MgCl_2$, 40 mM; NH_4Cl, 120 mM; Tris-HCl, pH 7.5, 160 mM; dATP, dCTP and dGTP, 0.5 mM each; and ^{3}H-TTP (17 Ci/mMole), 300 μCi.

Radioactive DNA was isolated according to the method of Kaplan, et al. (2). There was insufficient nucleic acid to characterize by absorbancy; however, the total radiolabeled DNA recovered in the two preparations was 2.2×10^6 dpm for the ad plasma and 3.6×10^6 dpm for the ay plasma.

The third preparation (ay subtype) was divided into two portions. One was incubated for 16 hours and extracted as above to yield radio-labeled DNA. The other portion was extracted directly to obtain native particle DNA. The four DNA preparations were submitted to one of us (TK) for electron microscopy by the protein monolayer with formamide technique (5).

RESULTS

DNA of Dane Particles. Double-stranded open circles and linear strands were seen by electron microscopy in DNA extracted directly from purified Dane particles. The circular structures (Figure 1) were approximately 0.78 microns in contour length, similar to the findings of Robinson et al. (3). The linear structures (not shown) varied in length from 0.5 to 4 microns. There were many simple open circles. Three structures suggestive of tails were observed (Figure 1, bottom row) in the Dane particle DNA. One of these (bottom right) appeared to be an unequivocal rolling circle with a short tail.

DNA of Dane Particles After DNA Synthesis. The DNA configurations seen by electron microscopy were similar for the three preparations obtained after the polymerase reaction. The linear structures became shorter after incubation, suggesting possible degradation by endogeneous nuclease. The most frequently observed structures were the double-stranded open circles. Most interestingly, about 15 to 20% of these circular DNA molecules were found as readily identified rolling circles. These structures (Figure 2) possessed linear portions with different lengths with an average of 0.55 microns suggesting various stages of DNA synthesis. Several circles were found with tails longer than unit length. At the junction of the circle and the linear filament, a small segment of single strandedness was observed (arrows, Fig. 2). The presence of tailed circles in rudimentary form in native particles, and an abundance of them after a DNA polymerase reaction, suggested that the polymerase-primed DNA in some of the Dane particles replicated by the rolling circle model (4). This structure is based on one closed circle representing the negative strand to serve as a template for elongation of the positive strand, which grows by the addition of nucleotides to its 3'-hydroxyl end. The model for a double

768

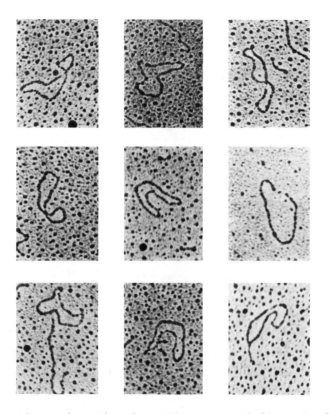

Figure 1. Circular DNA extracted directly from purified Dane particles. Magnification: 45,000 X.

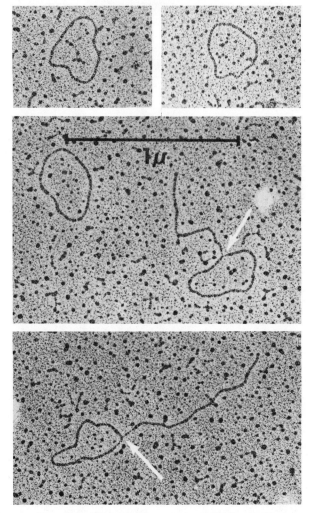

Figure 2.   Circular DNA extracted from Dane particles
after incorporation of ^{3}H–thymidine monophosphate with
endogeneous DNA polymerase. Magnification: 45,000 X.

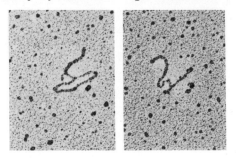

Figure 3.   Structures resembling superhelicies in
the extracted Dane particle DNA. Magnification: 45,000 X.

stranded tail requires that complementary fragments are synthesized on the growing positive strand. The small segments of single strand at the junction of the circle and the linear filament would indicate the site of new synthesis of complementary fragments.

Structures suggestive of superhelicies were observed (Figure 3) but could not be demonstrated reproducibly in these studies. It is not clear whether this is due to the artificial replication conditions in the test tube where no sites are available for the tail of the rolling circle to attach, thus producing a superhelical progeny (4).

DISCUSSION

Although the infectious form of hepatitis B virus has not been identified, the Dane particle is a leading candidate for the infectious agent. The findings that a unique closed circular DNA and a DNA polymerase are associated with purified Dane particles are consistant with the hypothesis that a nucleic acid-containing Dane particle is an infectious virion. The molecular weight of the circular DNA calculated from the mean contour length of 0.78 microns was about 1.6×10^6. This is similar to the findings of Robinson et al. (3) and indicates that the DNA is smaller than the double-stranded DNA of any known virus. The genetic information carried by DNA of this size therefore is very limited. If this DNA represents the viral genome, some unique additional system, such as complementation or helper virus, may be required to program the life cycle of the virus. The presence of an active, primed DNA polymerase in the native particles from plasma suggests the possibility of initiation of DNA synthesis in some Dane particles in blood or plasma. Although they were infrequent, short tails were observed in circular structures obtained directly from the particles prior to the in vitro incubation for DNA synthesis, indicating initiation of replication by rolling circles had already taken place in a few particles. The quantity of rolling circles and the length of the tails increased significantly after incorporation of nucleotides. Thus, the in vitro polymerase reaction could result in elongation of pre-existing tailed circles as well as in new initiation of replication by rolling circles. The in-vitro reaction is a highly artificial incubation condition, and therefore the rolling circles reported here may or may not represent the actual mechanism in vivo. Confirmation of the rolling circle mechanism of replication should be of interest because it would represent the first example, known to the authors, of this mechanism of replication of DNA obtained from a mammalian virus system.

REFERENCES

1. Dane, D.S., Cameron, C.H. and Briggs, M. Lancet 1, 695 (1970).

2. Kaplan, P., Greenman, R.L, Gerin, J.L., Purcell, R.H., and Robinson, W.S. J. Virol. 12, 995 (1973).

3. Robinson, W.S., Clayton, D.A., and Greenman, R.L. J. Virol. 14, 383 (1974).

4. Gilbert, W., and Dressler, D. Cold Spring Harbor Symp. Quant. Biol. 33, 473 (1968).

5. Davis, R.W., Simon, M., and Davidson, N. Methods in Enzymol. 21, 413 (1971), Academic Press, N.Y. and London.

Part XII
DNA Repair Synthesis

THE REPAIR MODE OF DNA REPLICATION IN *ESCHERICHIA COLI*

Philip Hanawalt, Ann Burrell, Priscilla Cooper
and Warren Masker[*]

Department of Biological Sciences
Stanford University
Stanford, California 94305

ABSTRACT. Current models for the dark repair of ultraviolet light induced pyrimidine dimers and other structural defects in DNA include the excision-repair mode and a recombinational mode. Both of these general schemes have been well documented in bacteria and have been shown to include several alternative pathways. One of the pathways for excision-repair utilizes DNA polymerase I for the gap filling step while another invokes DNA polymerase II or III. This latter pathway is characterized by a longer average patch size than that mediated by DNA polymerase I. The 5'-3' exonuclease of DNA polymerase I does not appear to be essential for repair replication as demonstrated in mutants deficient in this enzyme activity. Such mutants are more sensitive to ultraviolet light than the parent strain but less sensitive than a mutant deficient in the polymerase activity.

The Pol I independent repair replication, like semiconservative synthesis, proceeds in the presence of ATP, as shown in toluene permeabilized cells. Unlike semiconservative synthesis, the repair mode can proceed if other nucleoside triphosphates are substituted for ATP. The nucleoside triphosphate requirement is not due to an involvement of the *recBC* nuclease in excision-repair although that nuclease could, in principle, operate in the excision step. The *recA* function is also not required for repair replication in toluene-treated polymerase I deficient cells although it may modulate such synthesis *in vivo*.

---

[*] Present address: Biology Division, Oak Ridge National Laboratory, Oak Ridge, Tennessee 37830

774

INTRODUCTION

It is now well established that the process of DNA synthesis does not always result in a net increase in the amount of DNA. A nonconservative mode of replication has been identified in a number of biological systems and it has been shown to occur in response to damage in the DNA. This synthesis involves the filling of gaps in parental DNA strands and it is associated with the removal of damaged nucleotides. Accordingly, it has been termed "repair replication" (1,2).

DNA repair processes are best understood in bacterial systems, from which many of the relevant enzymes have now been isolated and characterized. The repair response of the bacterium *Escherichia coli* has been examined for many types of DNA damage although the most commonly studied lesion is the pyrimidine dimer, induced in the DNA by ultraviolet light (UV[1]). The photoproduct involves the covalent linking of adjacent pyrimidines in a DNA strand through their respective 5 and 6 carbons to form a cyclo-butane ring. The thymine-thymine dimer is the most readily formed of the various possible combinations and this lesion will be used as a model in the present discussion. The dimerized thymines distort the secondary helical structure of the DNA, resulting in a region of local denaturation since they are unable to form hydrogen bonds with the adenines in the otherwise complementary strand. The dimerization of 30% of the thymines results in a 10° drop in the melting temperature but little change in the renaturation rate of bacteriophage T7 DNA (3). The thymine dimer poses an obstacle to the normal processes of DNA replication and transcription. In the absence of repair the presence of a single dimer in the genome may be lethal to an *E. coli* cell (4).

In this paper we will discuss the principal repair schemes by which *E. coli* is able to circumvent or deal with thymine dimers in its DNA (not including photore-activation). We will review the results of experiments on repair replication carried out in toluene permeabilized cells. Then we will focus upon the role of DNA polymerase

---

[1]Abbreviations used are: UV, ultraviolet light; 5-BU, bromouracil; Ara CTP, 1-β-D arabinofuranosylcytosine triphosphate.

I and its associated 5' exonuclease activity in the excision-repair process in *E. coli*.

EFFECT OF DIMERS ON REPLICATION

There is little understanding of how a dimer inhibits replication and whether, in fact, all DNA polymerases are totally blocked at the site of a dimer. One would predict that parental strand unwinding should proceed normally through a dimer-containing region. What happens subsequently might depend upon *which* strand contains the dimer, and perhaps whether chain elongation is continuous or discontinuous along that strand. The effect of a dimer may depend upon whether it is in an initiation region (i.e., RNA primer region) or in the DNA chain elongation region of an "Okazaki fragment". In the one case a dimer in the template strand would be first encountered by an RNA polymerase and then by a DNA polymerase (e.g., Pol I) as the RNA primer is displaced. In the other case the dimer would presumbably be encountered by DNA polymerase III. If chain elongation by DNA polymerase III is terminated at a dimer, then it would seem logical that synthesis would restart at the next RNA primer, to initiate the next "Okazaki fragment" along that template strand as parental strands continue to unwind. It is improbable that DNA polymerase III could hurdle the dimer and then return to the template strand downstream to resume replication without the prerequisite 3'-OH primer site.

It is an experimental fact that gaps appear in the newly-replicated daughter strands during replication of dimer-containing DNA. This phenomenon has led to a model for the repair of such gaps by recombination, in which a segment of the parental strand in one sister chromatid is utilized to fill the corresponding gap in the daughter strand of the other (5). One possible representation of this model is illustrated in Figure 1. The heavy lines in the figure represent parental DNA strands containing a dimer (inverted "v") and the lighter lines represent daughter strands synthesized after the UV irradiation. Presumably the resulting gap in the parental strand is filled by repair replication, as shown by the wavy line. Also it is likely that some "tidying up" synthesis must occur in the fitting of the exchanged segment of DNA into

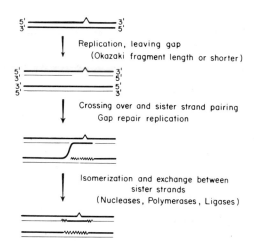

Fig. 1. Post-replication repair in *E. coli* by recombination.

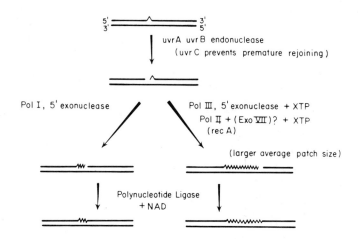

Fig. 2. Multiple pathways for excision-repair in *E. coli*.

the gap opposite the dimer. The model has been supported
by the finding that the gaps in daughter strands eventually disappear and also by density labeling and transfer
experiments in which the physical exchange of DNA is
demonstrated (6,7). In principle this repair mode could
result in dimer-free template DNA within one generation
for the subsequent round of normal replication. Definitive proof of the post-replication DNA exchange model as
well as a complication has been raised by the studies of
Ann Ganesan in our laboratory. Dr. Ganesan has developed
a sensitive assay for pyrimidine dimers utilizing the
dimer-specific endonuclease from T4 bacteriophage (8).
She has shown that many dimers are in fact transferred
from parental strands into the daughter strands synthesized after irradiation (9). Such transfer is probably
fortuitous but it postpones the generation of dimer-free
templates. Nevertheless, the repetition of the DNA
exchange process should result in a gradual diluting out
of the dimers in progeny cells. In addition the delay
afforded by the retarded rate of replication on dimer-
containing templates may allow time for the cell to deal
with the dimers by the other principal dark-repair process,
excision-repair. The post-replication recombination mode
of repair appears to be absolutely dependent upon the
recA gene product (6).

EXCISION-REPAIR OF DIMERS

The second general repair scheme, excision-repair,
requires enzymatic recognition of the dimer or other
structural distortion in the DNA as illustrated in Figure
2. The process is initiated by specific endonucleolytic
attack on the 5' side of the dimer: the *uvrA* and *uvrB* loci
specify the essential nuclease in *E. coli* (10). A third
locus, *uvrC*, appears to function in a regulatory sense to
prevent the premature closure of the incision by ligase
until the excision-resynthesis sequence has been completed
(11). At least two pathways are possible for the next
steps. One of these involves DNA polymerase I and its
associated 5' exonuclease activity to simultaneously
remove the dimer and fill the resulting gap, a reaction
that has been demonstrated *in vitro* (12). The other pathway utilizes DNA polymerase II and/or III for the resynthesis step (13-16) and as yet unassigned nucleases to

perform the excision. Possible candidates for dimer
excision include the 5' exonuclease activity associated
with DNA polymerase III (17), exonuclease VII (18), and
exonuclease V (the *recBC* enzyme) (19).

The final step in the excision-repair scheme must be
the ligase-catalyzed joining of the 3' OH end of the repair patch to the 5' end of the contiguous parental DNA
strand. The amount of parental strand degradation and
the correlated distribution of repair patch lengths can
in principle be regulated by the level of ligase activity.
Until the gap is sealed it may serve as substrate for
continued nuclease and polymerase activity to result in
larger patches.

A heterogeneity in patch size has been observed in
UV irradiated bacteria in which much of the repair
synthesis is found in a few very large patches, several
thousand nucleotides in extent (20). Our earlier studies
on repair replication in a *polA1* mutant showed that it
performed an *increased* amount of UV stimulated repair
synthesis relative to that in the *pol⁺* parent strain, a
result that we interpreted as indicating a longer average
patch size in some polymerase I independent pathway (21).
In contrast we found that a *recArecB* mutant exhibited very
little repair synthesis and that synthesis was predominantly in very short patches, suggesting that *recA* and
possibly the *recBC* nuclease might function in a polymerase
I independent pathway (21). The studies reported in the
following sections were designed to examine some of the
characteristics of the alternative pathways in excision-
repair.

REPAIR SYNTHESIS IN TOLUENE-TREATED *polA* STRAINS

One of the experimental problems in repair replication studies is the quantitation of the process in the
presence of extensive DNA degradation, since the degradation products compete with exogenously supplied radioactive precursors such as 5-BU. The intracellular pools of
low molecular weight precursors and cofactors can be
controlled through the use of toluene permeabilized cells
(22). In this *quasi in vitro* system we have characterized
a mode of DNA replication which resembles *in vivo* repair
replication in that it is non-conservative, stimulated by
UV irradiation, and persists under conditions of inhibited

semiconservative replication (23). This mode of synthesis requires the four deoxyribonucleoside triphosphates and millimolar concentrations of ATP or another nucleoside triphsophate (23,24).

These studies were all carried out in *polA* mutants to examine specifically the polymerase I independent pathway of excision-repair. Additionally, this eliminated the high backgrduond of ATP-independent nonconservative synthesis seen in unirradiated *pol*+ cells (22). *In vivo* experiments had shown the dependence of UV-stimulated repair replication upon *uvrA* (25) and *uvrB* (26) gene products. Similarly, the repair mode in the toluenized cells was shown to depend upon the *uvrA* gene product (24). Repair replication in a *dnaB* mutant at the nonpermissive temperature (preventing semiconservative synthesis) had been shown *in vivo* (27) and this result was confirmed in the toluenized cells (23). Semiconservative synthesis in this system was also found to be selectively suppressed by nalidixic acid (28) or by araCTP (29). In fact we found it generally advantageous to selectively inhibit semi-conservative synthesis (e.g., in *dnaB* mutant placed at the nonpermissive temperature) in order to examine repair synthesis.

UV stimulated repair synthesis was demonstrated in a *polA polB* double-mutant, indicating that DNA polymerase III can participate in this mode (16). It also was shown to occur in a *polA dnaE (polC)* mutant at the non-permissive temperature, but it was totally eliminated in a triple mutant additionally deficient in DNA polymerase II (14). This result ruled out the possible contribution of residual DNA polymerase I in these *polA* mutants (30) to the ATP-dependent repair synthesis. It also has provided the first demonstrable phenotype for DNA polymerase II. Thus, either polymerase II or polymerase III is capable of performing UV-stimulated repair synthesis in this system. Of course these results do not provide information on the relative importance of the respective polymerases in biologically significant repair events.

Moses and Richardson (22) originally demonstrated the ATP dependence of semiconservative replication in toluenized bacteria. This is a stringent requirement for semiconservative replication (31) but not for UV-stimulated repair replication, for which the other nucleoside triphosphates can partially substitute (24). The use

of GTP in place of ATP in the assay mixture reduces the
extent of semiconservative synthesis by about 95% while
reducing repair synthesis by only 30%. This substitution,
then, provides yet another means for selectively inhibiting
the semiconservative mode (16,24). In view of our earlier
suggestion from *in vivo* experiments that the ATP-dependent
recBC nuclease might participate in excision-repair (21)
and the observed nucleoside triphosphate dependence of
repair synthesis in the toluenized cells, it was of
interest to examine the effect of *recB* in this system.
A toluene treated preparation of a *recB21 polA12* mutant
was incubated with araCTP to selectively block semiconservative replication and the reaction was carried out at
44°C to inactivate the thermally labile DNA polymerase I.
Repair synthesis was observed after UV irradiation with
500 ergs/mm^2. Thus, there is no evident connection
between the *recB* genotype and the nucleoside triphosphate
dependence of repair replication (24).

Figure 3 illustrates a typical repair replication
experiment with toluene permeabilized cells. Strain
MM 454 (*polA12 recA56 recB21*) (32) was used to determine
whether the polymerase I independent repair synthesis
requires the *recA* gene product. The general protocol for
determining repair replication by density labeling has
been described by Hanawalt and Cooper (2) and the tolueni-
zation procedure has been described by Masker and Hanawalt
(23,24). The experiment was carried out at 44°C to
inactivate DNA polymerase I and araCTP was used as a
selective inhibitor of semiconservative synthesis as shown
in the unirradiated control, frame B. It is clear from
frames C and D that a dose dependent UV stimulation of
repair synthesis does occur in this *recA recB* mutant and
from frame E that this synthesis is ATP dependent.
Therefore, the *recA* gene product is not essential for the
polymerase I independent pathway of excision-repair. Note,
however, a characteristic general feature of DNA synthesis
in toluenized cells; that after 60 minutes incubation only
a very tiny fraction of the parental DNA ^{14}C-label has
appeared in hybrid DNA (frame A) indicating that synthesis
has not proceeded very far. The reason for this early
termination of synthesis is not known and it is likely
that the observed repair synthesis is also limited in the
toluenized cells. We hesitate, therefore, to make
quantitative statements about the extent (or patch size

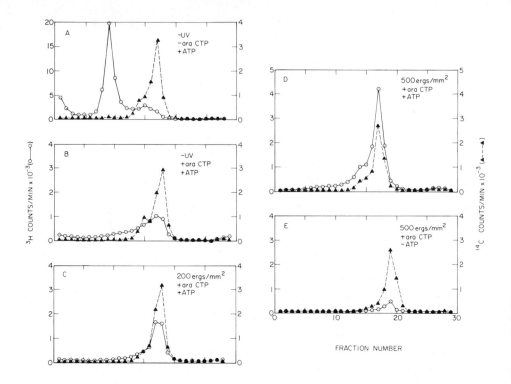

Fig. 3. Repair replication in toluenized *E. coli*;
isopycnic analysis.

An exponentially growing culture of strain MM 454
(*polA12 recA56 recB21*) was uniformly labeled in its DNA
by three generations of growth in the presence of 6n
Ci/20 μg [^{14}C] thymine/ml and 100 μg deoxyadenosine/ml.
The culture was divided into separate equal portions for
UV irradiations and unirradiated controls. The irradiated
cells and the unirradiated controls were treated with
toluene and incubated at 44°C with BrdUTP in place of dTTP
and 17.5 μ Ci/20 nmol [^{3}H] dATP/ml. ATP, when present,
was at 1.2 mM. Ara CTP, when present, was at 200 μM.
After 60 min the reaction was stopped, the cells were
lysed, and their DNA was analyzed in isopycnic CsCl
gradients. Density increases from right to left. Note
the factor of 5 difference in [^{3}H] scale between the
ordinates of (A) and those of the other frames. Prelabel
[^{14}C] thymine is indicated by the solid triangles and the
[^{3}H] dATP by the open circles. (A) No UV, No araCTP, +
ATP; (B) No UV, + araCTP, + ATP; (C) 200 ergs/mm^2 UV, +
araCTP, + ATP; (D) 500 ergs/mm^2 UV, + araCTP, + ATP; (E)
500 ergs/mm^2 UV, + araCTP, No ATP.

distribution) of repair synthesis in this system. It is
possible that the *recA* gene product may operate in the
polymerase I independent pathway in a regulatory sense,
perhaps to modulate the extent of repair synthesis.

ROLE OF *rec* GENES IN EXCISION-REPAIR

It is evident that the absence of the *recBC* nuclease
and the *recA* gene product does not eliminate repair synthesis in the *polI* independent pathway in toluene-treated
cells. Our previous *in vivo* studies had indicated a
greatly reduced level of repair in a *recArecB* mutant and
we had hypothesized that the remaining synthesis was due
to the "short patch" repair mediated by DNA polymerase I
(21). In these experiments the actual quantitation of
repair is made difficult by the background of labelled DNA
in the intermediate density region of the CsCl gradient
(between parental density and hybrid) for the unirradiated
control. This problem is particularly serious in the *rec*⁻
strains where the background level in the control is unusually high. Some of this synthesis, which is suppressed
by UV, may, in the *rec*⁻ cells, be an aberrant mode occurring in the non-viable segment of the population since a
high degree of lethal sectoring is known to exist in these
strains (33). We found that this synthesis could be
completely abolished at the restrictive temperature in
dnaB strains. We have now examined UV stimulated repair
replication *in vivo* in a pair of nearly isogenic *dnaB*
mutants one of which was *rec*⁺ and the other, *recA56 recB21*
(34). UV irradiation stimulated nonconservative synthesis
in both strains at the restrictive temperature but the
quantitative extent was variable. In some experiments a
similar amount of repair synthesis was seen in the *recA
recB* strain and in the *rec*⁺ strain whereas in other
experiments the *recArecB* strain showed a reduced level of
repair. For both strains there was a correlation between
increasing repair synthesis and increasing UV dose. The
rec⁺ strain showed a slight density shift in the repair
peak in the CsCl gradients, indicative of some "long
patches", while the *recArecB* strain did not exhibit such
a shift after any of the doses examined (34). This result
is consistent with the view that *recA* affects the patch
size distribution in excision-repair. The studies of
Youngs *et al.*, (35) have implicated the *recA*, *recB* and

exrA genes in a growth medium dependent branch of excision-repair and *polI* in a growth medium independent pathway.

ROLE OF DNA POLYMERASE I IN EXCISION-REPAIR

We have reported that a *polA1* mutant (36) performs more extensive repair replication than its parent strain following the same UV dose and have ascribed this to the production of more "long patches" in the alternative pathway shown in Figure 2 (21). The *polA1* mutant is UV sensitive but not as sensitive as are *uvrA* mutants. Dimers are excised but the final rejoining step is delayed in the *polA* mutant (37,38). More extensive DNA degradation is seen in the irradiated *polA1* mutant (37,21) as consistent with our hypothesis of longer repair patches. Although the DNA polymerase I activity is only about 1% of the normal level in the *polA1* mutant, there are nearly normal amounts of the 5' exonuclease activity that is part of the DNA polymerase I enzyme (30) and one might wonder whether this nuclease is essential for dimer excision *in vivo*.

Friedberg and Lehman (39) have shown that the 5' exonuclease "small fragment" of DNA polymerase I is capable of excising dimers from DNA *in vitro* albeit at only one-tenth the rate of the intact enzyme. The rate was increased by the addition of the "large fragment" and dramatically increased by the further addition of the four deoxynucleoside triphosphates to permit simultaneous DNA synthesis. A similar result was obtained with the isolated *amber* peptide from the *polA1* mutant. Thus, it appears that the dimer excision activity of DNA polymerase I may be driven by the polymerase activity as repair replication proceeds. It is not known whether that 5' exonuclease activity can be driven by one of the other two DNA polymerases in the absence of DNA polymerase I.

Survival, DNA degradation, and repair replication following UV irradiation have now been examined in several mutants deficient in the 5' exonuclease activity of DNA polymerase I, obtained through the courtesy of I.R. Lehman. One of these strains, RS5064 (*polAex1*), exhibits about 3% of the wild type 5' exonuclease activity at 30° and 1% of the wild type activity at 44° (40). Under certain plating conditions, for example tryptone-yeast extract agar, the strain is conditionally lethal at 43° (40). However, it will form colonies on nutrient agar plates

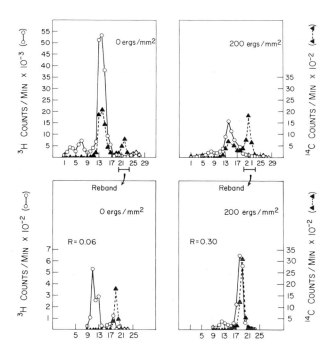

Fig. 4.   Repair replication in *E. coli in vivo*;
isopycnic analysis.
      An exponentially growing culture of strain RS 5064
thy⁻ [W3110 F⁻ trp A₃₃ polAex1, made thymine-requiring
by trimethoprim selection] was uniformly labeled in its
DNA by more than three generations of growth at 33° in
the presence of 0.025 μ Ci [¹⁴C] thymine/ml and 2 μgm/ml
thymine.   The culture was divided into equal portions for
UV irradiation and unirradiated control.   Following irradiation the cells were incubated for 45 minutes at 44°
in medium containing 5 μ Ci [³H]-5BU/ml and 10 μ gm/ml
5BU in place of thymine.   Culture lysates were analyzed
in isopycnic CsCl gradients.   Fraction numbers are indicated on the abcissa with density increasing from right
to left.   Upper frames show initial CsCl bandings for
control and irradiated cultures.   Lower frames show re-
bandings of the respective parental density regions (as
indicated) in second CsCl gradients.   Prelabel [¹⁴C] thymine is indicated by the solid triangles and the [³H]-5BU
repair label by the open circles.

at 44° (41). The strain is UV sensitive at either 33° or 44° on nutrient agar but not as sensitive as a *polA1* mutant (41). Figure 4 shows the result of an *in vivo* repair replication experiment with a thymine auxotroph of strain RS5064, following the usual protocol (2) except that the experiment was carried out at 44°. It is of interest that a substantial amount of semiconservative DNA synthesis has occurred in this strain in 45 minutes in spite of the low level of 5' exonuclease; most of the parental [^{14}C]-thymine prelabel has been transferred to the hybrid density position and some of the [^{3}H] 5-BU label appears at the heavy density position (indicating that some of the hybrid DNA has replicated again). As expected, the 200 ergs/mm^2 UV dose markedly suppresses the semiconservative mode of replication. A hybrid density band still appears at the expected position in the gradient indicating little dilution of the 5-BU repair label with endogenous thymidine. The rebanding of the parental density DNA in a second CsCl isopycnic density gradient permits a comparison between the control and irradiated cultures of the amount of [^{3}H] 5-BU label at this position. It is clear that the UV irradiation has stimulated the nonconservative repair mode of DNA synthesis (note the ratio of ^{3}H to ^{14}C of 0.06 in the control parental peak vs. the ratio of 0.30 in the parental peak from the irradiated culture). Therefore, the 5' exonuclease of DNA polymerase I does not appear to be essential for excision-repair. A quantitative comparison of the extent of repair synthesis after the same UV dose to the parent strain, a *polA1* strain, RS5064, and another 5' exonuclease deficient strain RS5069 (*polAex2*) has been carried out. The 5' exonuclease deficient strains and the *polA1* strain all exhibited an enhanced amount of repair synthesis relative to that in the parent strain whether incubated at 33° or at 44° (41). During incubation at 44° following UV irradiation, the *polex2* strain undergoes an extent of DNA degradation intermediate between that of the parent strain and the *polA1* strain (41). Similar results were obtained with a 5' exonuclease deficient mutant, *polA107*, obtained from Barry Glickman (42). We were unable to confirm Glickman's report of decreased repair replication in this strain (43).

In summarizing conclusion, our results suggest that DNA polymerase I and its associated 5' exonuclease participate in one of the excision-repair pathways following the initial incision by the *uvrAB* endonuclease but that neither activity is essential for an alternative pathway. The alternative excision-resynthesis pathways appear to involve more extensive DNA degradation and a longer average patch size. An alternative pathway requiring ribonucleoside triphospates as cofactors (of unknown function) may be mediated by either DNA polymerase II or III.

ACKNOWLEDGMENTS

This research was supported by a grant GM 09901 from the U.S. Public Health Service and a contract AT(04-3)326-7 with the U.S. Atomic Energy Commission. Tom Simon participated in some of the experiments on toluenized cells, Rolando Arrabal maintained glassware and media stocks, and Chris Spraker typed the manuscript. We are indebted to Ann Ganesan, Mike Kahn, Pat Seawell, Allen Smith and others in the lab encounter group for their constructive criticism.

REFERENCES

1. Pettijohn, D. and Hanawalt, P. *J. Mol. Biol. 9*, 395 (1964).

2. Hanawalt, P. and Cooper, P. in *Methods in Enzymology* (L. Grossman and K. Moldave, eds.) XXI Part D, p. 221 Academic Press, N.Y. (1971).

3. Kahn, M. *Biopolymers 13*, 669 (1974).

4. Howard-Flanders, P. and Boyce, R. P. *Radiation Research Suppl. 6*, 156 (1966).

5. Rupp, W. D. and Howard-Flanders, P. *J. Mol. Biol. 31*, 291 (1968).

6. Smith, K. C. and Meun, D. H. C. *J. Mol. Biol. 51*, 459 (1970).

7. Rupp, W. D., Wilde, C. E., III, Reno, D. L. and Howard-Flanders, P. *J. Mol. Biol. 61*, 25 (1971).

8. Ganesan, A. K. *Proc. Nat. Acad. Sci. U.S.A. 70*, 2753 (1973).

9. Ganesan, A. K. *J. Mol. Biol. 87*, 103 (1974).

10. Braun, A. and Grossman, L. *Proc. Nat. Acad. Sci. 71*, 1838 (1974).

11. Seeberg, E. and Rupp, W. D. in *Molecular Mechanisms for Repair of DNA* (P. Hanawalt and R. Setlow, eds.) Part B, Plenum Press N. Y. (1975).

12. Kelly, R. B., Atkinson, M. R., Huberman, J. A. and Kornberg, A. *Nature 224*, 495 (1969).

13. Youngs, D. A. and Smith, K. C. *Nature New Biol. 244*, 240 (1973).

14. Masker, W., Hanawalt, P., and Shizuya, H. *Nature New Biol. 244*, 242 (1973).

15. Tait, R. C., Harris, A. L., and Smith, D. W. *Proc. Nat. Acad. Sci. U.S.A. 71*, (1974).

16. Masker, W. E., Simon, T. J. and Hanawalt, P. C. in *Molecular Mechanisms for the Repair of DNA* (P. Hanawalt and R. Setlow, eds.) Plenum Press, N. Y. p. 245 (1975).

17. Livingston, D. M. and Richardson, C. C. *J. Biol. Chem. 250*, 470 (1975).

18. Chase, J. and Richardson, C. C. *J. Biol. Chem. 249*, 4553 (1974).

19. Karu, A. E., MacKay, V., Goldmark, P. J., and Linn, S. *J. Biol. Chem. 248*, 4874 (1973).

20. Cooper, P. K. and Hanawalt, P. C. *J. Mol. Biol. 67*, 1 (1972).

21. Cooper, P. K. and Hanawalt, P. C. *Proc. Nat. Acad. Sci. U.S.A. 69*, 1156 (1972).

22. Moses, R. E. and Richardson, C. C. *Proc. Nat. Acad. Sci. 67*, (1970).

23. Masker, W. E. and Hanawalt, P. C. *Proc. Nat. Acad. Sci. U.S.A. 70,* 129 (1973).

24. Masker, W. E. and Hanawalt, P. C. *J.Mol.Biol. 88,* 13 (1974).

25. Cooper, P. K. and Hanawalt, P. C. *Photochem. Photobiol. 13,* 83 (1971).

26. Nakayama, H. C., Pratt, A. and Hanawalt, P. C. *J. Mol. Biol. 70,* 281 (1972).

27. Couch, J. and Hanawalt, P. *Biochem. Biophys. Res. Comm. 29,* 779 (1967).

28. Simon, T. J., Masker, W. E. and Hanawalt, P. C. *Biochim. Biophys. Acta 349,* 271 (1974).

29. Masker, W. E. and Hanawalt, P. C. *Biochim. Biophys. Acta 340,* 229 (1974).

30. Lehman, I. R. and Chien, J. R. *J. Biol. Chem. 248,* 7717 (1973).

31. Pisetsky, D., Berkower, I., Wickner, R., and Hurwitz, J. *J. Mol. Biol. 71,* 557 (1972).

32. Monk, M., Kinross, J. and Town, C. *J. Bacteriol. 114,* 1014 (1973).

33. Capaldo, F. N. and Barbour, S. D. in *Molecular Mechanisms for the Repair of DNA* (P. Hanawalt and R. Setlow, eds.) p. 403, Plenum Press, N.Y. (1975).

34. Burrell, A. D. and Hanawalt, P. C. (to be published).

35. Youngs, D. A., van der Schueren, E. and Smith, K. C., *J. Bacteriol. 117,* 717 (1974).

36. De Lucia, P., and Cairns, J. *Nature 224,* 1164 (1969).

37. Boyle, J. M., Patterson, M. C., and Setlow, R. B. *Nature 226,* 708 (1970).

38. Kanner, L. C. and Hanawalt, P. C. *Biochem. Biophys. Res. Comm. 39,* 149 (1970).

39. Friedberg, E. C. and Lehman, I. R. *Biochem. Biophys. Res. Comm. 58,* 132 (1974).

40. Konrad, E. B., and Lehman, I. R. *Proc. Nat. Acad. Sci. U.S.A. 71,* 2048 (1974).

41. Cooper, P. K. (manuscript in preparation).

42. Glickman, B. W., van Sluis, C. A., Heijneker, H. L., and Rörsch, A. *Mol. Gen. Genet. 124,* 69 (1973).

43. Glickman, B. W. in *Molecular Mechanisms for the Repair of DNA* (P. Hanawalt and R. Setlow, eds.) p. 213, Plenum Press, N. Y. (1975).

EXCISION REPAIR OF DNA
Lawrence Grossman
Brandeis University

INTRODUCTION

The ability of cells to survive in an environment
specifically damaging to its DNA can be attributed to a
variety of inherent repair mechanisms. Reversibility is
best exemplified by the photoreactivation of ultraviolet
(UV)-induced pyrimidine dimers by photolyases (photore-
activating enzymes)(1-4). Dilution of damage can be effected through a series of sister chromatid exchanges,
controlled by recombinational mechanisms as a postreplication event (5). The specific removal of damage from modified DNA involves the sequential excision of damaged
nucleotides by a variety of nucleases.

Much of the current interest and knowledge of excision repair mechanisms has been confined to UV-induced
photoproducts, their effects, and subsequent removal. Any
progress made is a reflection of a good understanding of
the fundamentals of UV photochemistry which allows for
control of photoproduct formation in DNA and ease of analysis. Therefore in the ensuing discussion the removal of
the primary UV photoproducts, pyrimidine-pyrimidine dimers
(intrastrand diadducts), is emphasized.

EXCISION REPAIR - GENE PRODUCTS

The general mechanism shown in Figure 1 represents a
current view of the excision repair cycle present in Escherichia coli and Micrococcus luteus in which intrastrand
diadduct damaged DNA is sequentially removed; the individual steps are detailed in each section of this paper.

The excision event represents a series of enzymatic
steps including incision by rather specific correndonucleases (endonucleases acting exclusively on damaged DNA in
which correctional pathways are eventually followed) under

DNA Intrastrand Diadduct Repair Cycle

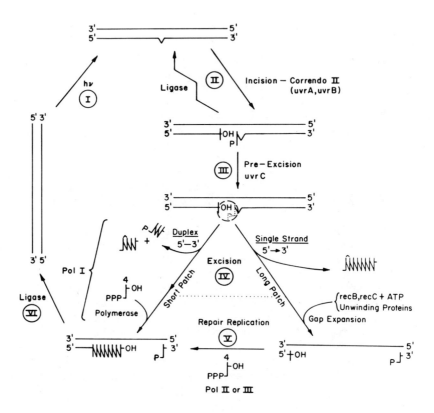

Figure 1

potential control by a subsequent post-incision step in
which abortive repair is prevented. The concluding step
in the excision of damage can occur via two interdependent
pathways, which may involve either polymerase-associated
exonucleases or unassociated exonucleases involved in the
removal of photoproducts. The reinsertion mechanisms may
occur either concomitantly in polymerase I-associated excision pathways (short patch repair) or it can proceed as
a result of a stepwise mechanism (long patch repair) when
single stranded $5' \to 3'$ exonucleases catalyze the removal of
photoproduct regions of DNA. Long patch pathways of exci-
sion-reinsertion may be controlled by polymerases II and/
or III. The restoration of the strand continuity is under
the control of a single polynucleotide ligase step.

INCISION STEP

General

Although many correndonucleases (endonucleases whose
activity is dependent on damaged DNA as a substrate eventually leading to correctional mechanisms) have not been
fully characterized, a similarity is apparent among most
of those activities studied. Correndonucleases generally
are small proteins, having molecular weights of less than
about 30,000. All act in the absence of divalent cation,
although some enzymes are stimulated by Mg^{2+} or NaCl. All
appear to incise close to the damage on the DNA strand.

Correndonuclease II (Specific in diadduct-type damage, e.g., dimers, crosslinks etc.)

All known correndonuclease II-type enzymes incise UV-
irradiated DNA. In general, some of these enzymes also
can incise DNA containing damage involving only a single
nitrogenous base. The enzymes described below are classified as correndonuclease II-type enzymes, since it is believed that they are involved in the repair of pyrimidine
dimer damage (intrastrand diadduct), as well as crosslink
damage (interstrand diadduct).

Escherichia coli Correndonuclease II (uvrA, uvrB endonu-clease) - The endonuclease involved in the first step of
excision repair has recently been isolated and partially
characterized (6). The involvement of this enzyme in UV
repair is indicated by its absence in uvrA and uvrB excision defective mutants. The enzyme is small, appearing to

793

have a molecular weight of less than 14,000. It acts in the presence of 10^{-3} M EDTA and requires an ionic strength of about 50 mM. In an irradiated:unirradiated DNA heteroduplex, the enzyme acts only on the damaged strand (7). The incision event occurs 5' to a pyrimidine dimer and leaves a 3'-hydroxyl terminus, which renders the break sealable by polynucleotide ligase (8,9).

Binding of the enzyme to its substrate can be measured by using the membrane filter technique of Riggs et al. (10). The specific binding of the enzyme to UV irradiated DNA can be prevented by prior treatment of the irradiated DNA with yeast photolyase and light (6). The purified enzyme, which has a K_m of 1.5 x 10^{-8} M (pyrimidine dimers) (11), is competitively inhibited by caffeine with a K_i of about 10^{-2} M; this is comparable to the inhibitory concentration of this drug in vivo (12). Caffeine inhibits the binding stage of the enzyme reaction. Preliminary data indicate that inhibition of the endonuclease by acriflavin also occurs at concentrations comparable to those found to be inhibitory in vivo (13).

Since uvrA and uvrB mutants of E. coli are more sensitive than wild-type to crosslinking agents such as mitomycin C (14) and psoralen plus light (15), there is reason to believe that the E. coli correndonuclease II acts on many diadduct forms of damage. However, there is in vivo evidence that the enzyme also acts on monoadduct damage. Otsiji and Murayama reported that uvrA and uvrB mutants of E. coli are sensitive to monofunctional analogs of mitomycin (16).

Another endonuclease acting on UV-irradiated DNA from E. coli has recently been described by Radman (17). This enzyme, referred to as endonuclease III by the author, does not appear to be involved in pyrimidine dimer repair, since it was found to be present in excision defective cells. Many of the properties of the enzyme were similar to the uvrA uvrB gene product.

The recent observation by Waldstein et al. (18) that UV-specific incision rates in toluene treated E. coli cells can be significantly increased by the addition of ATP suggests that another step in the excision program may exist. Since ATP does not appear to enhance the reaction rate of the E. coli UV correndonuclease in vitro (19), the ATP enhanced incision rate may be due to either a preparative incision step or to some unknown phosphorylated activator of the system in vivo.

Micrococcus luteus Correndonuclease II (UV endonuclease)
At least two UV-specific endonuclease activities exist in
M. luteus. A general radiation endonuclease, which acts
on both UV- and X-irradiated DNA (a UV-X-correndonuclease)
has been described (20). There is also an activity that
is similar in its properties to the E. coli uvrA, uvrB
correndonuclease and does not incise X-irradiated DNA (21).
This correndonuclease II incises 5' to the pyrimidine dimer, leaves a 3'-hydroxyl terminus (22), requires 50 mM
salt, and acts in the presence of 10^{-3} M EDTA (6). The
general radiation correndonuclease also incises UV irradiated DNA 5' to a pyrimidine dimer; however, it incises
by a b mechanism, leaving a 3'-phosphoryl terminus. This
is not a trivial distinction, since a 3'-hydroxyl terminus provides a priming site for the DNA polymerase and a
substrate site for polynucleotide ligase, whereas a 3'-
phosphoryl terminus is refractory to the action of both
enzymes.

"UV endonucleases," active on UV-irradiated DNA,
have been isolated from extracts of M. luteus by several
laboratories (23-25). Since it has recently become clear
that at least two such activities are present in extracts
of M. luteus, it is difficult to ascribe substrate specificities to enzymes described by a given laboratory. For
example, Patrick and Harm (26) have shown that the M. luteus UV endonuclease and yeast photolyase act on all pyrimidine dimers (T̂T̂, ĈT̂, ĈĈ). Unfortunately, it is not
possible to determine with certainty which of the two
correndonucleases was used in these experiments.

Bacteriophage T4 Correndonuclease II (V^{+} gene UV endonuclease) - Correndonuclease II, coded for by the phage T4
gene, is present in large amounts in T4-infected E. coli
(27-31). This enzyme is similar in many of its properties to the E. coli uvrA, uvrB correndonuclease II. It
is a small protein of molecular weight about 15,000,
which acts in the presence of 10^{-3} M EDTA. Incision is
5' to the pyrimidine dimer and a 3'-hydroxyl terminus is
generated (32). The enzyme acts both on single and double stranded UV irradiated DNA (32).

An interesting property of the T4 correndonuclease II
may be an ability to incise gross distortions of double
stranded DNA. Berger and Benz reported that heteroduplex

T4 DNA with one strand containing a deletion is repaired
by extracts of T4-infected E. coli (33,34). Extracts of
T4 V⁻ infected E. coli did not repair these heteroduplexes,
implying that the V gene product, the T4 correndonuclease
II, incises at the large noncomplementary loops formed in
these heteroduplex DNA structures.

Kozinski presented biochemical evidence indicating
that the T4 virion contains the V gene product, correndonuclease II (35,36). In vivo evidence indicates that the
T4 virion injects the V gene product into the infected
cell (35).

PRE-EXCISION STEP

Incision Mechanism [a] (5'-phosphoryl group)

Although the incision produced by the E. coli correndonuclease II provides an initiation site for excision and
nucleotide incorporation by DNA polymerase I (22), indirect evidence suggests that an intermediate step between
incision and excision exists. This evidence is based on
the properties of the excision defective uvrC mutant.
Such mutants are unable to remove pyrimidine dimers from
DNA in vivo, although there are indications that single
strand incision events are operative in only a transient
manner. Since exonuclease VII, correndonuclease II, and
polymerase levels appear to be normal in such mutants, it
can be assumed that the capacity for incision and excision
is unaffected. The progression of single strand molecular
weight changes in uvrC mutants is peculiar in that the extent of such breakage is low, whereas the initial rate of
single strand break formation is indistinguishable from
wild-type cells (37). What is seen is a rapid, partial
decrease in single strand molecular weight, which is
quickly restored to that of control chain lengths. DNA
isolated from wild-type cells under similar conditions,
however, exhibits a rapid and more extensive loss in its
single stranded molecular weight, which is fully restored
during extended postirradiation time periods.

Seeberg and Rupp (37) prepared double uvrC and temperature-sensitive polynucleotide ligase mutants. At

temperatures restrictive for ligase, the early rate and extent of postirradiation molecular weight losses are similar to those of wild-type cells, followed, however, by the expected accumulation of low molecular weight DNA. At permissive temperatures, however, the molecular weight is rapidly and fully restored. These data, in conjunction with the observations that polynucleotide ligase is capable of resealing the phosphodiester bonds of incised DNA, places polynucleotide ligase in a controlling position during early repair steps. The juxtaposition of a 3'-hydroxyl group and the 5'-phosphoryl group, in association with a pyrimidine dimer containing nucleotide, appears to be sensitive to E. coli polynucleotide ligase.

From these two observations, it can be inferred that the uvrC gene product prevents resealing of the correndonuclease II (incision) before excision of the pyrimidine dimer has occurred. A number of mechanisms, such as a 5'-polynucleotidase, nuclease, or perhaps a binding protein affecting the conformation of the damaged strand, can be suggested for its molecular mechanism in restricting ligase activity at this step of repair.

Incision Mechanism[b] (3'-phosphoryl group)

The incisions produced by the M. luteus general radiation correndonuclease II enzyme result in a 3'-phosphoryl terminus, 5' to a pyrimidine dimer (38). This terminus is not suitable as a nucleophilic site necessary for priming by polymerases. Exonuclease III, because it can act as a phosphomonoesterase at such a 3'-phosphoryl double stranded terminus prior to phosphodiester bond hydrolysis, has the potential for repair capacity (39).

The 3'-polynucleotidase reported by Becker and Hurwitz (40) could conceivably participate at this step in the repair cycle in providing such a nucleophilic site. Mutants defective in exonuclease III, isolated by Milcarek and Weiss (41), show normal UV sensitivity, implying that the correndonuclease II enzymes repairing UV-irradiation damage by endonucleolytic activity either do not act by a b-type mechanism or, if 3'-phosphoryl groups were formed, other exonucleases may participate in this step in the cycle.

EXCISION MECHANISMS

It was implicit in the early observations of Nakayama et al. (42) that excision was at least a two-step process in M. luteus. They found that the purification of a UV-dependent nuclease activity by TEAE-cellulose chromatography resulted in the loss of activity, which could be recovered by the reconstitution of two different fractions eluted from such a column. The first fraction, fraction A, did not exhibit any nuclease activity; however, acid-soluble nucleotide release specifically from irradiated DNA required the presence of a second TEAE fraction which contained a nonspecific nuclease activity, fraction B. Fraction B was active on unirradiated or irradiated denatured DNA, and it was correctly assumed by these investigators that fraction A was an endonuclease.

An analogous multistep excision repair system was demonstrated in phage T4-infected E. coli (31). Correndonuclease II, isolated and purified to homogeneity, is the V gene product of phage T4. To demonstrate the role of such an endonuclease in excision, crude extracts of cells infected with phage T4 V⁻ were used as the source of exonuclease. The exonucleolytic properties associated with such crude extracts were later identified by Onshima and Sekiguchi (43). Similar types of excision systems seem to exist in M. luteus, E. coli, and phage T4-infected E. coli.

Two types of exonucleases can execute excision: unassociated exonucleases may act on incised DNA, or DNA polymerase-associated exonucleases may react with such incised irradiated DNA intermediates.

DNA Polymerase-Associated Exonucleases

3'→5' Exonucleolytic Activity - DNA polymerase I of E. coli is a multifunctional enzyme contained on a single polypeptide chain of molecular weight 109,000 (44). The M. luteus DNA polymerase for all intents and purposes is identical to the E. coli enzyme, having a similar molecular weight, nuclease properties, and N-ethyl maleimide insensitivity. Preliminary immunological experiments have demonstrated crossreactivity with antibodies specifically directed against E. coli DNA polymerase (45).

In addition to their polymerizing properties, these

enzymes have two associated exonucleolytic activities, one of which is a 3'→5' nuclease, which at 37°C prefers single stranded DNA possessing a 3'-hydroxyl terminus in which 5' nucleotides are the exclusive product of digestion. An important role for this nuclease activity, in conjunction with the enzyme's polymerizing properties, is its "editing" functions, in which the exonuclease activity is capable of digesting in a direction opposite to that of polymerization up to the point of hydrogen bond stability. Brutlag and Kornberg (46) demonstrated that synthetic duplex model polymers containing noncomplementary nucleotides at the 3' end of the primer template are removed by the polymerase's 3'→5' exonucleolytic activity.

A series of mutator and antimutator mutations of bacteriophage T4 has been isolated in gene 43 (47), the structural gene of the T4 DNA polymerase. Muzyczka et al. (48) have shown a close correlation between the mutator properties of the polymerase and reduced 3'→5' exonuclease activity. Conversely, antimutator polymerase has an increased 3'→5' activity. Hall and Lehman (49) have shown in vitro that the rate of misincorporation of nucleotides by mutator T4 DNA polymerase is statistically greater than that by wild-type polymerases.

5'→3' Exonucleolytic Activity - The 5'→3' nuclease activity associated with polymerase I of E. coli and M. luteus is specific for DNA duplexes. This exonuclease activity is stimulated by neighboring 3'-hydroxyl groups provided during polymerization of the four deoxynucleoside triphosphates (22,50,51) included in reaction mixtures in which nicked DNA provides priming and potential templating sites. When synthetic and hydrolytic activities proceed at comparable rates, synthesis leads to an extension of the 3'-hydroxyl terminus of a nick and simultaneous removal of nucleotides at the 5' terminus, which is referred to as nick translation (51). The translation continues in a 5'→3' direction until the nick is located at the extreme 3' end of the template strand, at which time both activities cease. This process is probably the dominant reaction in vitro at temperatures below 22°C (52). At higher temperatures, however, another process dominates which is expressed by the synthesis of branched, nondenaturable structures. Such aberrant synthetic mechanisms are reflected in an extent of synthesis in excess of one template equivalent (53).

The 5'→3' nuclease initiates hydrolysis of DNA duplexes at internal rather than terminal phosphodiester
bonds, regardless of whether such termini are 5' esteri-
fied (54). About 20-25% of the products of 5'-terminated
DNA are dinucleotides and longer oligonucleotides with the
majority represented by 5'-nucleotides. Low pH favors the
production of dinucleotides; at pH 6, approximately 50%
of the acid-soluble products are dinucleotides (55).

Exposure of the polymerase to primer templates with a
short 5' noncomplementary terminus results in release of
such noncomplementary nucleotides as a fragment (56).

Irradiated poly(dA:dT) incised nonspecifically with
pancreatic DNase was hydrolyzed by polymerase I, libera-
ting dimer-containing fragments. The products of exonucleolytic hydrolysis by the polymerase were oligo-
thymidylates of chain lengths ranging from dinucleotides
to heptanucleotides containing photoproducts. The spectrum of products released with the labeled irradiated
poly(dA:dT) was somewhat different than that with the unirradiated control in which all the radioactive label was
isolated as mononucleotides. Comparative results were obtained with irradiated DNA in which dimer-containing fragments larger than trinucleotides were located chromato-
graphically.

Any assignment of cellular involvement in polymerase-
mediated repair is dependent on experiments obtained with
those mutants lacking the respective nuclease activities.
Mutant strains of E. coli with negligible levels of DNA
polymerase I (pol A), although increasingly sensitive to
UV, are considerably more resistant than uvr or rec mutants (57,58). polA Strains of E. coli, such as P3478,
excise thymine-containing dimers to the same rate and extent as do wild-type cells (59). The inference derived
from such results is that the UV sensitivity of these mutants may be related to the reduced polymerizing capabilities during reinsertion, and that other mechanisms of excision must be operative in these strains of E. coli.
Polymerase I has been purified from a variety of polA mutants. Polymerase I activity in these E. coli mutants
was between 0.5-3% of wild-type activity, whereas the
associated 5'→3' exonuclease activities remained unaffected (60). These findings account for the normal levels
of pyrimidine dimer excision in such mutants.

The size of reinserted regions of DNA from experi-

ments in vivo ranges from 10 nucleotides (short patch repair) to about 3,000 nucleotides in length (long patch
repair). PolAl mutations result in a preponderance of
long patch repair synthesis (61). These data may be ex-
plicable in terms of the elevated ratio of 5'→3' hydrolysis to polymerizing activity in such mutants. However,
the resolved 5'→3' exonuclease fragment (3.3S) excises
slowly in the absence of concomitant polymerization. The
addition of the large (7.7S) polymerizing fragment markedly stimulates pyrimidine dimer removal (62). These
findings are in accord with the known properties of the
enzyme in nick translation in vitro (51).

It is difficult to conclusively assign an in vivo
excision role to the 5'→3' exonuclease of polymerase I,
since temperature-sensitive mutants in this exonucleolytic activity have been found to be conditionally lethal
by a number of laboratories (63,64). However, a mutant
of E. coli with a substantial defect in the 5'→3' exonuclease, but with normal polymerizing properties, has been
described by Glickman et al. (65,66). This mutant is
moderately UV sensitive and shows a somewhat reduced
pyrimidine dimer excision capability.

Polymerase III Excision Capabilities

E. coli polymerase III has associated 3'→5' and 5'→
3' hydrolytic activities (67). The 3'→5' activity prefers
single stranded DNA, releases 5' mononucleotides, and its
activity is inhibited by a 3'-phosphoryl group. The 5'→
3' activity can initiate hydrolysis on single strands and
proceed into a duplex region once initiation has occurred.
The 5'→3' activity can also initiate hydrolysis of double
strand DNA possessing a single stranded 5' terminus. The
limit products of hydrolysis of the 5'→3' activity are
dinucleotides and mononucleotides which may arise from
the combined 3'→5' and 5'→3' activities of polymerase III.
The structure of UV-irradiated DNA incised by the M. luteus correndonuclease II is ideally suited for 5'→3' exonucleolytic removal of pyrimidine dimers.

Unassociated Exonucleases in Excision (Correxonucleases)

The hydrolysis of the phosphodiester bond 5' to the
photoproduct by the correndonuclease II requires that
exonucleases involved in excision initiate hydrolysis in
a 5'→3' directional manner. Moreover, an additional re-

quirement for such exonucleases is that they must be able
to hydrolyze internal phosphodiester bonds in order to
catalyze the removal of pyrimidine dimers. It would appear that those exonucleases involved in the excision of
pyrimidine dimers do not hydrolyze the phosphodiester bond
linking the pyrimidine nucleotide dimer residues (68).

The catalytic properties of two generally distinguishable but functionally related exonucleases capable of excising pyrimidine dimers have been identified in M. luteus
(38), E. coli (69), T4-infected E. coli (43,70), rabbit
liver nuclei (71,72), human placenta (73), and mammalian
cell lines (74). Such exonucleases have specificities
limited to denatured DNA and are capable of hydrolyzing
UV damaged and undamaged substrates at comparable rates
and to the same extent (75). This characteristic property
can be used to distinguish those single stranded exonucleases with potential pyrimidine dimer excision capabilities
in vitro.

The correxonuclease from M. luteus can be distinguished from functionally unrelated exonucleases, such as
E. coli exonuclease I, venom phosphodiesterase, and bovine
spleen phosphodiesterase, whose hydrolytic activities are
restrained by photoproducts containing DNA (75). The prokaryotic correxonucleases already characterized have the
unusual capacity of initiating hydrolysis from either 3'
or 5' termini. In all cases, the products, whether mononucleotides or oligonucleotides, contain 5' phosphorylated
termini. The M. luteus correxonuclease is unique in that
it requires divalent cation and has no limit hydrolysis
product (38). However, E. coli exonuclease VII (69) and
a similar type of enzyme purified from M. luteus (76) do
not require magnesium, thereby allowing for their ability
to function optimally in the presence of 10^{-3} M EDTA, permitting its easy detection in crude extracts.

Exonuclease VII can act on denatured DNA, single
stranded regions extending from duplex DNA treated with
exonuclease III (69) or λDNA. An exonuclease VII type of
enzyme from M. luteus cleaves part of the 12 nucleotide
long single stranded cohesive ends of λDNA, leaving
two nucleotides on each end (77). None of these exonucleases are able to use RNA as a substrate, nor are poly-
ribonucleotide:polydeoxyribonucleotide hybrid polymers
hydrolyzed by exonuclease VII.

The eukaryotic exonuclease activity that presumably

constitutes the second step in the excision of pyrimidine
dimers has been purified from rabbit tissue by Lindahl
(71,72). The enzyme, termed DNase IV, hydrolyzes DNA in
a 5'→3' direction, releasing oligonucleotides containing
five to eight residues. A similar activity has been detected in crude extracts of human cell cultures (74). The
importance of such exonucleolytic activities in mammalian
cells is emphasized by the fact that, unlike the bacterial
polymerases, none of the mammalian DNA polymerases purified to homogeneity have been found to contain any nuclease activity.

The M. luteus correxonuclease (38) and DNase IV (71)
of rabbit liver nuclei produce 5' mononucleotides during
digestion with no oligonucleotide intermediates. The
limit products of exonuclease VII action are oligonucleotides bearing 5'-phosphoryl and 3'-hydroxyl termini in
which approximately one third are in the range of dinucleotides to trinucleotides and the majority are in the range
of tetranucleotides to dodecamers, in which no mononucleotides have been observed as intermediates during hydrolysis.

The exonuclease VII-type enzymes and polymerase-
associated 5'→3' exonucleases yield oligonucleotides as an
ultimate product, and the UV exonuclease catalyzes the
release of dimer-containing oligonucleotides from correndonuclease II incised irradiated DNA (78). Since the M.
luteus UV exonuclease does not hydrolyze the phosphodiester bond located between thymidylate dimer residues, the
initial hydrolytic event must have occurred approximately
six nucleotides 3' to the photochemical damage (68). The
ratio of nucleotides released per phosphodiester bond
originally hydrolyzed by the correndonuclease II provides
an estimation of the approximate size of the region distorted by the formation of photoproducts. Although such
ratios are somewhat dependent on the source of irradiated
DNA substrates, they do correlate moderately well with
the distortion sizes predicted spectroscopically (79) and
from model building (56). From data obtained with the
correndonuclease II of M. luteus, the ratio of the number
of nucleotides released to the number of phosphodiester
bonds broken indicates that the size of the excised region is approximately 6-10 nucleotides in length (78).

The conformational specificities of the polymerase
associated and unassociated 5'→3' exonucleases are dif-

ferent, which may influence the ability of these enzymes
to excise pyrimidine dimers in regions of differing hydrogen bond stabilities. For example, the combined reaction
of correndonuclease II and the single stranded specific
M. luteus correxonuclease seems to function preferentially
in AT-rich regions, judging from the distribution of nucleotides 3' to dimers in excised fragments(78). It would
be of interest, therefore, to determine not only the nucleotide compositions of fragments excised by polymerase-
associated and -unassociated correxonucleases, but environmental effects as well on the course of action of such
nucleases. Parenthetically, neither exonuclease VII mutants nor polymerase mutants affected in their 5'→3' nuclease activity show marked UV sensitivity or impaired
dimer excision capabilities (65,80). The construction of
double mutants will, therefore, be of considerable value
in assessing the involvement in vivo of these various
enzymatic activities. That significant increases in UV
sensitivity are not observed in single mutants may reflect
either the interdependency or overlapping specificities
of the polymerase-associated exonucleases and the unassociated exonucleases.

REINSERTION MECHANISMS

Following excision of the damaged region, reinsertion
of nucleotides is catalyzed by DNA polymerases using the
complementary strand as a template. The reinsertion of
nucleotides into UV-irradiated E. coli was first studied
by Pettijohn and Hanawalt (81) and shown to be distinct
from semiconservative DNA replication.

Reinsertion events in their experiments are distinguishable from semiconservative replication through the
uptake of $[^3H]$-BrdU into DNA having light or intermediate
buoyant densities. Since 5-BrdU is more dense than its
homolog thymidine, its incorporation into DNA in a semiconservative manner (i.e. normal replication) results in
the newly synthesized strand of DNA having a greater density than the template strand. Measurement after shearing
to small fragments (0.5 x 10^6 daltons) of the density of
the double stranded DNA by isopycnic centrifugation in
CsCl or NaI reveals radioactivity banding as a "hybrid"
area of the gradients. After UV irradiation, incorporation of nucleotides takes place in relatively short gaps

exposed during excision of the damaged bases. Radioactive 5-BrdU incorporated into small gaps does not significantly increase the density of the DNA and thus appears in the normal density regions of the gradient (81). This operational definition of short patch repair is restricted to regions of approximately 10-30 nucleotides in length (82, 81). Under conditions in which short gap repair is inhibited, as in polA mutants, the labeled 5-BrdU is associated with intermediate regions of the gradient which are less dense than the hybrid semiconservatively replicated DNA and more dense than the normal density regions of the gradient. Such regions are approximately 1000-3000 nucleotides long and represent long patch repair (61). Both short and long patch repair occur in wild-type cells, but are absent in excision defective cells. Therefore, two branches of reinsertion are available to E. coli cells.

One route of reinsertion is considerably more complicated, more extensive in extent and perhaps, as a consequence, less efficient than the other. This long patch repair (61), or growth medium-dependent (83) pathway of repair, requires uvrA (84) and rec genes (61,83), polymerase II (85), and/or polymerase III (86), is ATP dependent in toluene treated cells (87), and may require the further support of specific unwinding proteins (88). An additional pathway involving recF genes (89,90), which is an exonuclease I-sensitive locus (91), can supplement both the polA and recA, recBC branch of repair, and is perhaps similar to recA, recBC postreplicative repair mechanisms (92,93).

It is difficult to assess the specific molecular determinants that influence the direction of excision and reinsertion into any one of the enzymatic branches of repair. Such determinants may be architectural; for example, enzymatic juxtapositions to damaged DNA might govern repair directions. They might also be influenced by the structural conformations of incised DNA's.

Short Patch Pathway of Repair (uvr, polA, lig)

There is sufficient evidence that the sequence of events controlled by the uvr genes is necessary for excision and reinsertion of nucleotides via this pathway of repair (94,84). The size of the patches observed in vivo from 5-BrdU density labeling is of the same order of magnitude expected from studies in vitro (95). Although it

is difficult to determine under normal conditions what
propprtion of the DNA is repaired by polymerase I, it can
be deduced from the data of Cooper and Hanawalt (84) that
the majority of repaired DNA is in short patches synthesized via the polymerase I pathway of reinsertion.

Short patch repair has been mimicked under controlled
in vitro conditions. Heijneker et al. (96) partially
restored the transforming activity of UV-irradiated B.
subtilis DNA by incising the DNA with the M. luteus correndonuclease II and then incubating the incised DNA with
a mixture of E. coli polymerase I and polynucleotide
ligase. Roughly 20 nucleotides per pyrimidine dimer were
excised at 34°C.

Hamilton et al. (22) examined the control in vitro of
both excision of pyrimidine dimers and the reinsertion of
nucleotides catalyzed by DNA polymerase I of M. luteus.
Polymerase I-type enzymes, unlike polymerases II and III
and phage-infected DNA polymerases, specifically bind at
nicks, satisfying an important and specific requirement
for the short patch excision process (97). It would appear, therefore, that the nick binding and translating
properties of these enzymes provide for the necessary
concerted mechanisms of reinsertion and excision. Nick
translational conditions result in a stoichiometry of
equivalence between removal of photoproducts with associated nucleotides and the reinsertion of nucleotides.
Furthermore, polymerase is capable of repairing gaps (98).
Incubation of this polymerase with excised DNA, for example, at 10°C in optimal Mg^{2+} leads to a stoichiometric
reinsertion-excision reaction which may be controlled by
polynucleotide ligase. Under conditions of strand displacement at 37°C, the presence of ligase appears to restrict the stoichiometric ratio to one nucleotide excised
per nucleotide polymerized. Therefore, in vitro, the
polymerase:correndonuclease II combination in conjunction
with ligase is sufficient for the complete repair of
single strand breaks associated with incision and restoration of biological activities of UV-irradiated transforming DNA.

When transforming DNA is 50% or less inactivated by
UV, there is a quantitative restoration of biological
activity by these three enzymes. At higher doses, the
maximum repair capabilities by this multi-enzyme system
decline considerably. This is attributed to the formation

of double strand breaks, arising not from irradiation, but as a consequence of the repair process. The unidirectionality (5'→3') of repair on strands of opposite polarity leads to double strand break formation during the excision process. Hamilton et al. (22) have suggested that polyamines, such as spermine, may stabilize the DNA repair intermediates and limit their degradation to prevent double strand breaks and thus increase the reactivation of B. subtilis DNA at high UV doses.

Long Patch Pathway of Repair

uvr, recBC, unwinding protein, polB, dnaE, lig Delineation of repair pathways is operational and defined according to the density position of [^{3}H]-BrdU in gradients where long patches are at the heavy end and short patches are located with the light end of such a gradient among nonreplicated DNAs. There are clearly no well-separated peaks of label, but rather a spectrum of densities which must be visualized, in molecular terms, as path average rather than path specific. Cooper and Hanawalt (61) found, for example, that the UV-induced nonconservative DNA synthesis, rather than becoming limited in polA mutants, was in fact considerably stimulated. This stimulated synthesis was dependent on the uvrA gene product (87), indicating a divergent, rather than two separate paths of reinsertion. The size of the reinserted patch was at least 100 times greater than that observed in a polA-dependent short patch pathway of reinsertion and, furthermore, was dependent on the presence of the recA, recBC gene products.

Experiments in toluene treated E. coli suggest that both polymerases II and III are involved in UV-induced repair. Since polymerase II lacks a 5'→3' exonuclease function (99), its role in repair must be confined to reinsertion reactions. Repair in toluene treated cells seems to be of the long patch type, which is missing if both polymerases II and III are inactivated (84).

A controlling feature of long patch repair may be attributable to the size of the fragment excised by the unassociated exonuclease VII, UV exonucleases or polymerase III initiation of hydrolysis of the single stranded regions of incised DNA. The fragments released by the unassociated exonucleases vary between 6-8 nucleotides in length (78,67,69). Once the single strand fragment is

removed, the duplex nature of the excised DNA limits
further hydrolysis by the exonucleases. The binding
properties, however, of polymerases II and III are limited
to much larger regions, such that in order for these
polymerases to carry out reinsertion, a gap expansion
step is necessary (97). This gap expansion is possibly
catalyzable by the ATP-stimulated double stranded exonuclease activity associated with the multifunctional recBC
enzyme (exonuclease V) (100,101-103). It is the ATP-dependent double stranded exonuclease activity that cannot
hydrolyze nicked DNAs, but is able to degrade duplex circles containing gaps as short as five nucleotides in
length(103). This activity is bidirectional such that
hydrolysis occurs in a 5' and 3' direction (104). This
gap expansion activity is thermolabile in recB temperature sensitive mutants. Furthermore, at restrictive
temperatures, it is this recB function that confers both
UV sensitivity and reduced viability to such mutant E.
coli cells (105).

The E. coli unwinding protein required for the processive polymerization by polymerase II on single stranded
DNA (106) is potentially able to assume a protective role
in the gap expansion process. The complementary strand of
the excised DNA should be susceptible to endonucleolytic
attack by the ATP-dependent single stranded endonuclease
of the recBC enzyme. The E. coli unwinding protein, however, is able to inhibit this activity specifically in
vitro without affecting the ATP-dependent duplex exonuclease activity, thereby preserving the integrity of the
strand opposite to that which is being repaired (107). It
may be suggested that the recA gene product serves similarly in protecting the complementary strand exposed at
this locus in repair or during postreplication repair in
an analogous manner to the T4 gene 32 protein, which
functions in recombination and replication. Parenthetically, temperature-sensitive gene 32 proteins are extremely UV sensitive under restrictive conditions (108).

ATP is required in addition to the four deoxynucleoside triphosphates to stimulate long patch repair in toluene treated polA mutants (85). Such cofactor requirements may partially represent the need for a number of
different processes in toluene treated cells. For example
ATP is needed for the enlargement of gaps, required in
toluene treated cells for preincision steps, and may also

be needed for recombinational mechanisms yet to be elucidated.

COMMENTS ON EUKARYOTIC REPAIR MECHANISMS

Progress in elucidating prokaryotic repair mechanisms has been enhanced by the available mutants affecting various loci of the excision or reinsertion pathways of repair. In the absence of equivalent isogenic strains of cloned mammalian cell lines, those interested in eukaryotic repair pathways are exploiting related mutant skin fibroblasts lines obtained from patients with the photosensitive disease xeroderma pigmentosum (XP). At least five complementation groups of XP and related cell lines have been reported (109,110), indicating that the numbers of genes controlling repair in mammalian systems must be at least as numerous as the genes controlling UV sensitivity in E. coli. In spite of the fact that cell lines from the various complementation groups show some physiological defect in a UV repair system, the relationship between clinical symptoms and biochemical abnormalities in repair have not been clearly established. This lack of correlation is the result of a paucity of information dealing with the enzymatic repair of DNA in normal mammalian cells. The photobiology of mammalian cells, however, has been effectively reviewed recently and for those readers interested in mammalian repair, the review of Cleaver is recommended (111).

ACKNOWLEDGMENTS

The work of this laboratory has been supported by grants from the American Cancer Society (NP-8D), the Atomic Energy Commission [AT(11-1)3232-2], the National Institutes of Health (GM-15881-14), and the National Science Foundation (GB-29172X). This is contribution No. 1006 of the Graduate Department of Biochemistry, Brandeis University.

REFERENCES

1. Kelner, A.,Proc. Nat. Acad. Sci. USA 35, 73 (1949).
.2. Dulbecco, R., J. Bacteriol. 59, 329 (1950).
3. Sutherland, B., Nature 248, 109 (1974).

4. Regan, J. D., <u>Ann</u>. <u>Ist</u>. <u>Super</u>. <u>Sanita</u> <u>5</u>, 355 (1969).
5. Rupp, W.D., Wilde, C.E., Reno, D.L., Howard-Flanders, P., <u>J</u>. <u>Mol</u>. <u>Biol</u>. <u>61</u>, 25 (1971).
6. Braun, A., Grossman, L., <u>Proc</u>. <u>Nat</u>. <u>Acad</u>. <u>Sci</u>. <u>USA</u> <u>71</u>, 1838 (1974).
7. Radman, M., Braun, A., unpublished data.
8. Braun, A., Grossman, L., <u>Fed</u>. <u>Proc</u>. <u>33</u>, 1599 (1974).
9. Rupp, D., Seeburg, E., Braun, A., in preparation.
10. Riggs, A. D., Suzuki, H., Bourgeois, S., <u>J</u>. <u>Mol</u>. <u>Biol</u>. <u>48</u>, 67 (1970).
11. Braun, A., Hopper, P., Grossman, L., <u>Molecular</u> <u>Mechanisms</u> <u>for</u> <u>the</u> <u>Repair</u> <u>of</u> <u>DNA</u>, ed. P. C. Hanawalt, R. B. Setlow. Basic Life Science Series, Vol. 5. General Ed. A. Hollander, in press.
12. Sideropoulous, A. S., Shankel, D. M., <u>J</u>. <u>Bacteriol</u>. <u>96</u>, 198 (1968).
13. Smith, B., Braun, A., Grossman, L., unpublished data.
14. Boyce, R. P., Howard-Flanders, P., <u>Z</u>. <u>Verebungslehre</u> <u>94</u>, 345 (1964).
15. Kohn, K., Steigbigel, N., Spears, C., <u>Proc</u>. <u>Nat</u>. <u>Acad</u>. <u>Sci</u>. <u>USA</u> <u>53</u>, 1154 (1965).
16. Otsiji, N., Murayama, I., <u>J</u>. <u>Bacteriol</u>. <u>109</u>, 475 (1972).
17. Radman, M., <u>Molecular</u> <u>Mechanisms</u> <u>for</u> <u>the</u> <u>Repair</u> <u>of</u> <u>DNA</u>, ed. P. C. Hanawalt, R. B. Setlow. Basic Life Science Series, Vol. 5. General Ed. A. Hollander, in press.
18. Waldstein, E. A., Sharon, R., Ben-Ishai, R., <u>Proc</u>. <u>Nat</u>. <u>Acad</u>. <u>Sci</u>. <u>USA</u> <u>71</u>, 2651 (1974).
19. Braun, A., unpublished data.
20. Kushner, S. R., <u>UV-Endonuclease: purification, properties and specificity of the enzyme from Micrococcus luteus. The photoproduct excision system of Micrococcus luteus.</u> Ph.D. thesis, Brandeis University, (1970).
21. Riazuddin, S., Grossman, L., unpublished data.
22. Hamilton, L. D. G., Mahler, I., Grossman, L., <u>Biochemistry</u> <u>13</u>, 1886 (1974).
23. Carrier, W. L., Setlow, R. B., <u>J</u>. <u>Bacteriol</u>. <u>102</u>, 178 (1970).
24. Nakayama, H., Okubo, S., Takagi, Y., <u>Biochim</u>. <u>Biophys</u>. <u>Acta</u> <u>228</u>, 67 (1971).
25. Kaplan, J. C., Kushner, S. R., Grossman, L., <u>Proc</u>. <u>Nat</u>. <u>Acad</u>. <u>Sci</u>. <u>USA</u> <u>63</u>, 144 (1969).

26. Patrick, M. H., Harm, H., Photochem. Photobiol. 18, 371 (1973).
27. Friedberg, E. C., King, J. J. Biochem. Biophys. Res. Commun. 37, 646 (1969).
28. Sekiguchi, M., Yasuda, S., Okubo, S., Nakayama, H., Shimada, K., Takagi, Y., J. Mol. Biol. 47, 231 (1970).
29. Yasuda, S., Sekiguchi, Y., J. Mol. Biol. 47, 243 (1970).
30. Yasuda, S., Sekiguchi, Y. Proc. Nat. Acad. Sci. USA 67, 1839 (1970).
31. Friedberg, E. C., King, J. J., J. Bacteriol. 106, 500 (1971).
32. Friedberg, E. C., Minton, K., Durphy, M., Clayton, D. A., Molecular Mechanisms for the Repair of DNA, ed. P. C. Hanawalt, R. B. Setlow. Basic Life Science Series, Vol. 5. General Ed. A. Hollander, in press.
33. Berger, H., Benz, W., Molecular Mechanisms for the Repair of DNA, ed. P. C. Hanawalt, R. B. Setlow. Basic Life Science Series, Vol. 5. General Ed. A. Hollander, in press.
34. Benz, W., Berger, H., Genetics 73, 1 (1973).
35. Shames, R., Lorkiewicz, K., Kozinski, A., J. Virol. 12, 1 (1973).
36. Shames, R., Kozinski, A., Fed. Proc. 33, 1493 (1974).
37. Seeberg, E., Rupp, W. D., Molecular Mechanisms for the Repair of DNA, ed. P. C. Hanawalt, R. B. Setlow. Basic Life Science Series, Vol. 5, General Ed. A. Hollander, in press.
38. Kaplan, J. C., Kushner, S. R., Grossman, L., Biochemistry 10, 3315 (1971).
39. Richardson, C. C., Kornberg, A., J. Biol. Chem. 239, 242 (1964).
40. Becker, A., Hurwitz, J., J. Biol. Chem. 242, 936 (1967).
41. Milcarek, C., Weiss, B., J. Mol. Biol. 68, 303 (1972).
42. Nakayama, H., Okubo, S., Takagi, Y., Biochim. Biophys. Acta 228, 67 (1971).
43. Onshima, S., Sekiguchi, M., Biochem. Biophys. Res. Commun. 47, 1126 (1972).
44. Englund, P. T. et al., Cold Spring Harbor Symp. Quant. Biol. 33, 1 (1968).
45. Hamilton, L, unpublished data (1974).
46. Brutlag, D., Kornberg, A., J. Biol. Chem. 247, 241 (1972).

47. Drake, J. W., Abbey, E. F., Cold Spring Harbor Symp. Quant. Biol. 33,339 (1968).
48. Muzyczka, N., Poland, R. I., Bessman, M. J., J. Biol. Chem. 247, 7116 (1972).
49. Hall, Z. W., Lehman, I. R., J. Mol. Biol. 36, 321 (1968).
50. Zimmerman, B. K., J. Biol. Chem. 241, 2035 (1966).
51. Kelly, R. B., Cozzarelli, N. R., Deutscher, M. P., Lehman, I. R., Kornberg, A., J. Biol. Chem. 245, 39 (1970).
52. Dumas, L. B., Darby, G., Sinsheimer, R. L., Biochim. Biophys. Acta 228, 407 (1971).
53. Wu, R., J. Mol. Biol. 51, 501 (1970).
54. Cozzarelli, N. R., Kelly, R. B., Kornberg, A., J. Mol. Biol. 45, 513 (1969).
55. Deutscher, M. P., Kornberg, A., J. Biol. Chem. 244, 3029 (1969).
56. Kelly, R. B., Atkinson, M. R., Huberman, J. A., Kornberg, A.,Nature 224, 495 (1969).
57. deLucia, P., Cairns, J., Nature 224, 1164 (1969).
58. Gross, J. D., Gross, M., Nature 224, 1166 (1969).
59. Boyle, J. M., Patterson, M. C., Setlow, R. B., Nature 226, 708 (1970).
60. Lehman, I. R., Chien, J. R., J. Biol. Chem., 248, 7717 (1973).
61. Cooper, P. K., Hanawalt, P. C., Proc. Nat. Acad. Sci. USA 69, 1156 (1972).
62. Friedberg, E., Lehman, I. R., Molecular Mechanisms for the Repair of DNA, ed. P. C. Hanawalt, R. B. Setlow. Basic Life Science Series, Vol. 5. General Ed. A. Hollander, in press.
63. Konrad, E. B., Lehman, I. R., Proc. Nat. Acad. Sci. USA 71, 2048 (1974).
64. Olivera, B., Bonhoffer, F., Nature 250, 513 (1974).
65. Glickman, B. W., van Sluis, C. A., Heijneker, H. C., Rorsch, A., Mol. Gen. Genet. 124, 69 (1973).
66. Glickman, B. W., Molecular Mechanisms for the Repair of DNA, ed. P. C. Hanawalt, R. B. Setlow. Basic Life Science Series, Vol. 5. General Ed. A. Hollander, in press.
67. Livingston, D. M., Richardson, C. C., J. Biol. Chem. (1975).
68. Kaplan, J. C., UV-Exonuclease: purification, properties and specificity of the enzyme from M. luteus,

Ph.D. thesis, Brandeis University (1971).
69. Chase, J., Richardson, C. C., J. Biol. Chem. 249, 4553 (1974).
70. Sekiguchi, M., Shimizu, K., Sato, K., Yasuda, S., Onshima, S., Molecular Mechanisms for the Repair of DNA, ed. P. C. Hanawalt, R. B. Setlow. Basic Life Science Series, Vol. 5. General Ed. A. Hollander, in press.
71. Lindahl, T., Eur. J. Biochem. 18, 407 (1971).
72. Lindahl, T., Eur. J. Biochem. 18, 415 (1971).
73. Doniger, J., Grossman, L., Biophys. J. 15, 297a (1975).
74. Duncan, J., Slor, H. H., Cook, K., Friedberg, E. C., Molecular Mechanisms for the Repair of DNA, ed. P. C. Hanawalt, R. B. Setlow. Basic Life Science Series, Vol. 5. General Ed. A. Hollander, in press.
75. Grossman, L., Kaplan, J. C., Kushner, J. C., Mahler, I., Cold Spring Harbor Symp. Quant. Biol. 33, 229 (1968).
76. Garvik, B., Grossman, L., unpublished data (1974).
77. Ghangas, G. S., Wu, R., J. Biol. Chem. (1975).
78. Kushner, S. R., Kaplan, J. C., Ono, H., Grossman, L., Biochemistry 10, 3325 (1971).
79. Pearson, M., Johns, H. E., J. Mol. Biol. 20, 215 (1966).
80. Chase, J., Richardson, C. C., personal communication (1974).
81. Pettijohn, D. F., Hanawalt, P. C., J. Mol. Biol. 9, 395 (1964).
82. Setlow, R., Carrier, W., Proc. Nat. Acad. Sci. USA 51, 293 (1964).
83. Youngs, D. A., van der Schuren, E., Smith, K. C., J. Bacteriol. 117, 717 (1974).
84. Cooper, P. K., Hanawalt, P. C., J. Mol. Biol. 67, 1 (1972).
85. Masker, W. E., Hanawalt, P. C., Shizuya, H., Nature New Biol. 244, 242 (1973).
86. Youngs, D. A., Smith, K. C., Nature New Biol. 244, 240 (1973).
87. Masker, W. E., Hanawalt, P. C., Proc. Nat. Acad. Sci. USA 70, 129 (1973).
88. Gefter, M. L., Ann. Rev. Biochem. 44 (1975).
89. Horii, A., Clark, A. J., J. Mol. Biol. 80, 327 (1973).
90. Clark, A. J., Genet. Suppl. XIII Int. Congr. Genet., in press.

91. Kushner, S. R., Nagaishi, H., Templin, A., Clark, A. J., Proc. Nat. Acad. Sci. USA 68, 824 (1971).
92. Hertman, I., Genet. Res. 14, 291 (1969).
93. Clark, A. J., Margulies, A. D., Proc. Nat. Acad. Sci. USA 53, 451 (1965).
94. Howard-Flanders, P., Boyce, R. P., Theriot, L., Genetics 53, 1119 (1966).
95. Cerutti, P. A., Photochemistry and Photobiology of Nucleic Acids, ed. S. Y. Wang and M. Patrick (1974).
96. Heijneker, H. L., Pannekock, H., Oosterbaan, R. A., Pouwels, P. H., Bron, S., Arwert, F., Venema, G., Proc. Nat. Acad. Sci. USA 68, 2967 (1971).
97. Gefter, M. L., Prog. Nucl. Acid Res. 14, 101 (1974).
98. Wu, R., Kaiser, A. D., J. Mol. Biol. 35, 523 (1968).
99. Wickner, R. B., Ginsberg, B., Berkower, I., Hurwitz, J., J. Biol. Chem. 247, 489 (1972).
100. Goldmark, P. J., Linn, S., Proc. Nat. Acad. Sci. USA 67, 434 (1970).
101. Goldmark, P. J., Linn, S., J. Biol. Chem. 247, 1849 (1972).
102. Karu, A. E., Mackay, V., Goldmark, P. J., Linn, S., J. Biol. Chem. 248, 4874 (1973).
103. Linn, S., Mackay, V., Molecular Mechanisms for the Repair of DNA, ed. P. C. Hanawalt, R. B. Setlow. Basic Life Science Series, Vol. 5. General Ed. A. Hollander, in press.
104. Mackay, V., Linn, S., J. Biol. Chem. 249, 4286 (1974).
105. Kushner, S. R., J. Bacteriol. 120, 1219 (1974).
106. Mackay, V., Linn, S., personal communication (1974).
107. Molineaux, I. J., Gefter, M. C., Proc. Nat. Acad. Sci. USA 71, 3858 (1974).
108. Baldy, M. W., Virology 40, 272 (1970).
109. Robbins, J. H., Kraemer, K. H., Lutzner, M. A., Festoff, B. W., Coon, H. G., Annals Intern. Med. 80, 221 (1974).
110. Kraemer, K. H., Coon, H. G., Robbins, J. H., J. Cell Biol. 59, 176a (1973).
111. Cleaver, J. E., Advances in Radiation Biology, ed. J. T. Lett, H. Adler, M. Zelle, pp. 1-75 (1973).

DNA POLYMERASE I INVOLVEMENT IN REPAIR IN TOLUENE-TREATED ESCHERICHIA COLI

John W. Dorson and Robb E. Moses
Marrs McLean Department of Biochemistry
Baylor College of Medicine
Houston, Texas 77025

ABSTRACT. Toluene-treated cells have been used in the study of DNA repair after ultraviolet irradiation. Repair synthesis is lower in DNA polymerase I-containing strains than in deficient strains, possibly indicating more efficient repair. Alkaline sucrose gradient analysis reveals a faster rejoining of the parental DNA strands during repair in polA+ strains suggesting that DNA polymerase I-mediated repair is more readily sealed by ligase. The ligation can be blocked by the addition of nicotinamide mononucleotide resulting in a pronounced uv-specific repair synthesis which is catalyzed by DNA polymerase I. The 5'→3' exonuclease function of DNA polymerase I is also required for this synthesis.

INTRODUCTION

There are three DNA polymerases in Escherichia coli. DNA polymerase III is required for replication and also can participate in repair (1,2). No function is clearly evident for DNA polymerase II. This report will be concerned with DNA polymerase I. This enzyme has been implicated in DNA replication, as deficient strains, although viable (3), show a reduced ability to process Okazaki fragments (4). However, the major role for DNA polymerase I appears to be in the repair of various types of damage to the bacterial genome (2). Mutants in DNA polymerase I are sensitive to uv irradiation but still retain their ability to excise pyrimidine dimers from their DNA (5). The purified enzyme utilizes an irradiated DNA template that has been incised and will both excise pyrimidine dimers and polymerize new nucleotides in their place (6).

We have studied DNA repair of uv damage in cells rendered permeable by exposure to toluene (7). We show that DNA polymerase I participates in a repair that is both rapid and efficient in that relatively few nucleotides are incorporated. Also, we present a method by which the amount of incorporation catalyzed by DNA polymerase I can be amplified for better characterization of the reaction.

METHODS

Assay: Cells were grown and toluene-treated as described (7). After toluene-treatment cells were exposed for various times to uv irradiation at a constant dose of 6 joules meter^{-2} minute^{-1}. Reaction conditions were as described with [^{3}H]-dTTP (10 cpm pmole^{-1}). Thirty minute incubations were terminated with trichloracetic acid and acid insoluble material collected by filtration. Incorporation was determined by liquid scintillation counting.

Alkaline Sucrose Gradients: For sucrose gradients cells were prelabeled by growth in the presence of [^{3}H]-thymidine. Gradients of 5→20% alkaline sucrose were overlayed with an alkaline lysis solution (7). At the end of the prescribed incubation time the cells were lysed directly on the gradients. The gradients were centrifuged for two hours at 25×10^3 rpm at $20°$, collected from the bottom onto continuous paper strips (8) and radioactivity determined by liquid scintillation counting.

RESULTS

DNA Polymerase I Reduces Incorporation: In toluene-treated cells, uv irradiation inhibits replicative DNA synthesis. This is demonstrated (Fig. 1) by the rapid loss of incorporation after even short exposures to uv. In P3478 cells which lack DNA polymerase I (3), this synthesis falls until it reaches a constant level. Actually, the level of synthesis is a composite of two reactions: replication, which is high in the absence of uv but falls exponentially with increasing exposure and repair, which is very low initially but increases with uv exposure. This is best seen in the situation where replication is blocked and repair induced with uv exposure (1).

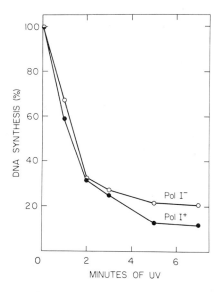

Fig. 1. DNA synthesis following ultraviolet irradiation in P3478 (polA⁻) and a revertant P3478R (polA⁺). At the indicated times samples were withdrawn and incubated for thirty minutes in a complete reaction mixture (1).

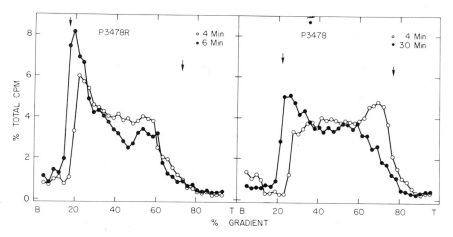

Fig. 2. Alkaline sucrose gradient analysis of pre-labeled DNA following uv exposure of 30 J/M². Cells were incubated for indicated times in complete reaction mixtures containing unlabeled dNTPs. The left arrow indicates the position of DNA from unirradiated cells and the right arrow indicates the position of incised DNA.

However, in P3478R, a revertant containing DNA polymerase I (9), this behavior is altered. Replication again
falls with uv exposure. But the resultant level of repair
is consistently lower and in some cases at background.
This "inhibitory" effect of DNA polymerase I on repair synthesis has also been observed by others (10).
This is just the opposite of the expected result
following uv exposure. There should be many nicks in the
DNA and DNA polymerase I should utilize these for extensive
synthesis. This is indeed the case when toluene-treated
cells are exposed to DNase which initiates a burst of synthesis (7).

Sedimentation Analysis: Molecular weight changes in the
parental DNA are a very sensitive measure of both the
initial incision as well as the final rejoining of the
repaired strands during the repair process. It is possible
to observe both of these functions in this in vitro system.
The experiments (Fig. 2) demonstrate that the increase to
high molecular weight following incision is much faster in
cells containing DNA polymerase I. In this in vitro system
polA$^-$ strains do not achieve as high a molecular weight
during reformation as polA$^+$ strains.

NMN Stimulation: It is evident from Fig. 1 and earlier
discussion that there is insufficient incorporation for
adequate analysis of repair in polA$^+$ strains. The sucrose
gradient analysis indicates a positive role for DNA polymerase I and ligase. If ligase action were inhibited then
the resulting synthesis might be sufficient for analysis.
This may be achieved by the addition of nicotinamide
mononucleotide (NMN) to prevent formation of an active
ligase by NAD. The addition of NMN to reaction mixtures of
polA$^+$ cells results in a large stimulation of DNA synthesis (Fig. 3). NMN stimulation is not present in the
isogenic strain which lacks DNA polymerase I. The lack of
any NMN stimulation in polA$^-$ cells indicates that ligase
does not exert a strong modulating role in this type of
repair.
Because the elevated synthesis in NMN following uv
might be nick translational, it seems likely that it would
be dependent on the 5'→3' exonucleolytic activity of DNA
polymerase I. A mutant conditionally defective in the
5'→3' exonuclease function of DNA polymerase I (11) allows

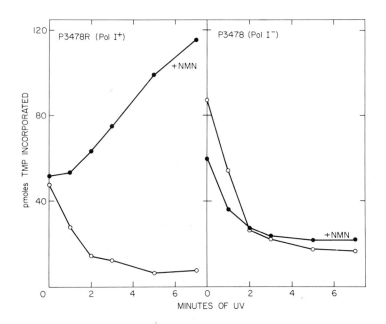

Fig. 3. Effect of NMN on synthesis following irradiation in P3478R and P3478.

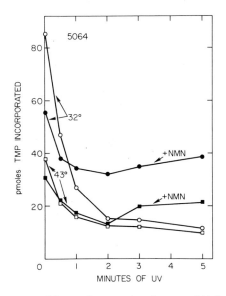

Fig. 4. NMN effect in a strain conditionally defective in the 5'→3' exonuclease function of DNA polymerase I.

its evaluation in this process. At the restrictive temperature (Fig. 4) there is almost no NMN stimulation of repair
synthesis. This indicates a requirement for both the polymerase and 5'→3' exonuclease functions of DNA polymerase I
in the NMN-stimulated, uv-dependent repair synthesis.

DISCUSSION

There are several lines of evidence indicating a role
for DNA polymerase I in DNA repair: deficient strains are
somewhat uv-sensitive, the enzyme has properties in vitro
which make it very suitable for a repair role, and finally,
it is present in abundance which might be a requirement
for simultaneous repair at many sites. With this information it is somewhat surprising to observe that the presence
of DNA polymerase I actually reduces the amount of synthesis after uv exposure in toluene-treated cells. However,
this is consistent with a repair synthesis that is very
efficient, replacing a minimum number of bases in the
repaired region in vivo. Experiments to determine the size
of the repair patch agree with this hypothesis (12). The
patch size in DNA polymerase I-containing cells is small
compared to polA⁻ cells whose patch size is about one hundred times larger.

The above observations prompted us to investigate the
activity of DNA polymerase I during repair in toluene-
treated cells. Our experiments indicate that DNA polymerase I is available to act at sites of repair and that it
can play a role in normal repair as manifested by increased
efficiency of reformation. The conclusion from these
data is that strains containing DNA polymerase I synthesize
less DNA during repair than deficient strains.

It is known from studies in vitro that DNA polymerase
I will nick translate (13) and the observation that wild-
type patch size is small indicates a strong influence of
ligase on terminating the reaction. Inhibition of ligase
with NMN removes this inhibition and allows extensive
synthesis by DNA polymerase I, probably by a nick translation mechanism. The 5'→3' exonuclease function is
required for this NMN-stimulated uv-specific DNA polymerase
I-mediated synthesis.

ACKNOWLEDGEMENTS

This work was supported by grants from the U.S. Public Health Service (GM-19122), the American Cancer Society (NP-153A) and the Robert A. Welch Foundation (Q-543). J. W. D. is a postdoctoral trainee (GM-00081-01). R. E. M. is a recipient of U.S.P.H.S. Research Career Program Award (GM-70314).

REFERENCES

1. Bowersock, D. and Moses, R. E., J. Biol. Chem. 248, 7449 (1973).
2. Tait, R. C., Harris, A. L. and Smith, D. W., Proc. Nat. Acad. Sci. 71, 675 (1974).
3. DeLucia, P. and Cairns, J., Nature 224, 1164 (1969).
4. Okazaki, R., Arisawa, M. and Sugino, A., Proc. Nat. Acad. Sci. 68, 2954 (1971).
5. Boyle, J. M., Paterson, M. C. and Setlow, R. B., Nature 226, 708 (1970).
6. Kelly, R. B., Atkinson, M. R., Huberman, J. A. and Kornberg, A., Nature 224, 495 (1969).
7. Moses, R. E. and Richardson, C. C., Proc. Nat. Acad. Sci. 67, 674 (1970).
8. Carrier, W. L. and Setlow, R. B., Anal. Biochem. 43, 427 (1971).
9. Coukell, M. B. and Yanofsky, C., Nature 228, 633 (1970).
10. Masker, W. E. and Hanawalt, P. C., J. Mol. Biol. 88, 13 (1974).
11. Konrad, E. B. and Lehman, I. R., Proc. Nat. Acad. Sci. 71, 2048 (1974).
12. Cooper, P. K. and Hanawalt, P. C., Proc. Nat. Acad. Sci. 69, 1156 (1972).
13. Masamune, Y. and Richardson, C. C., J. Biol. Chem. 246, 2692 (1971).

DNA SYNTHESIS IN PHLEOMYCIN-TREATED LYSATES OF
Go LYMPHOCYTE NUCLEI

Harvard Reiter and Pei-Ling Hsu

Department of Microbiology
University of Illinois at the Medical Center
Chicago, Illinois 60680

ABSTRACT. A DNA-synthesizing, DNA polymerase β-template
complex has been prepared from lysed nuclei of Go, resting
lymphocytes. ATP or GTP are required for synthesis by the
untreated complex. Pancreatic endonuclease I increases
only ATP or GTP-dependent synthesis, whereas phleomycin
induces some synthesis in the absence of ATP, and even
more synthesis in the presence of ATP or GTP. Phleomycin
causes only repair synthesis in the absence of ATP but
both repair and semiconservative synthesis in the presence
of ATP.

INTRODUCTION

Phleomycin arrests DNA synthesis in normal bacteria,
and in mammalian cells (1, 2, 3, 4), but it stimulates
both long-strand repair DNA synthesis by polymerase II,
and semiconservative DNA synthesis by polymerase III in
toluenized bacterial cells (5). In this paper we present
evidence that phleomycin also induces semiconservative as
well as repair DNA synthesis by polymerase β in lysates of
nuclei from resting, Go rabbit lymphocytes. The data
indicate that polymerase β is attached to its template in
Go cells, and suggest that polymerase activation depends
primarily on the availability of a suitable template
initiation site.

METHODS

Rabbit lysenteric lymph nodes were teased apart in
Eagle's minimal essential medium (MEM) and strained

through a 200 mesh/inch stainless steel screen. This preparation was then filtered through cotton to remove clumped cells, and the filtered single cells were layered, and centrifuged at 400Xg for 20 minutes at 4° C, on 2 ml Hypaque-Ficoll (H-F: 10 parts of 34% Hypaque plus 24 parts of 9% Ficoll) to remove red blood cells, granulocytes, and dead cells. The lymphocytes on top of the H-F were collected, washed 3 times with cold MEM, and used immediately or after storage overnight in a refrigerator.

Approximately 5×10^8 of these cells were washed with 10 ml volumes of cold buffer (0.05 M KH_2PO_4 pH 8.0, 0.013 M $MgSO_4$, 0.002 M dithiothreitol) and then resuspended in 1 ml of buffer. Nonidet NP40 was added to a concentration of 0.2%, and the cells were passed through a 25 gauge needle, two times, under maximum thumb pressure. The volume was brought to 15 ml with cold buffer, and the nuclei were collected by centrifugation at 170Xg for 8 minutes. The nuclei were washed once more with 10 ml of buffer, resuspended in 1 ml of buffer, and broken by 10 seconds of sonication. This material was either used directly as the enzyme-template component of the DNA synthesis reaction, or it was brought to 15 ml with cold buffer, centrifuged for 5 minutes at 2500 Xg, resuspended in 1 ml of fresh buffer, and then used as enzyme-template. No significant functional difference was found between the two preparations.

The reaction mix was 0.2 ml of enzyme-template plus 0.2 to 0.25 ml of 30 µM dCTP, 30 µM dGTP, 30 µM dATP, 2 µM dTTP, and 16 µc ^{3}H-dTTP (21 c/mM). When present, ATP was 1.5 mM, GTP was 1 mM, and phleomycin was 20 µg/ml. DNA synthesis was measured by determining the amounts of ^{3}H activity in 30 µl samples that precipitated in 5% cold trichloroacetic acid.

RESULTS

DNA synthesis. Figure 1 shows that a very small amount of DNA synthesis occurred in the control mix in the absence of ATP, and that ATP or GTP increased synthesis to 8-fold over background in forty minutes. Phleomycin added to the culture at zero time caused a 10-fold increase in DNA synthesis in the absence of ATP, and it caused a 50-

823

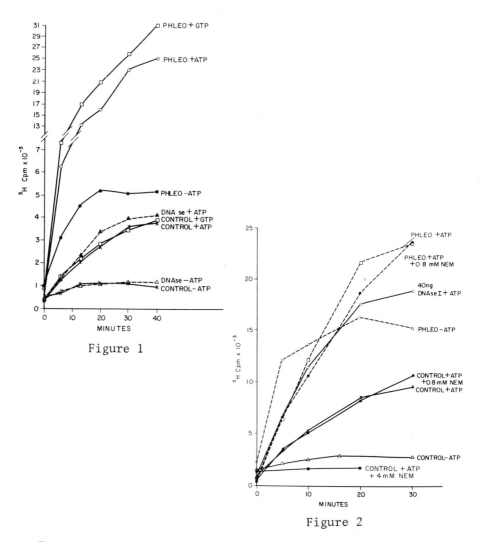

Figure 1

Figure 2

Fig. 1.   DNA synthesis in a reaction mix containing 0.2 ml of nuclear lysate plus 0.2 ml of dXTP, ^{3}H-TTP, and ATP. GTP, phleomycin, or 20 ng/ml pancreatic DNAse I, as indicated.

Fig. 2.   DNA synthesis in a reaction mix containing 0.2 ml of nuclear lysate plus 0.2 ml of dXTP, ^{3}H-TTP, and ATP, GTP, phleomycin, DNAse I, or N-ethylmaleimide (NEM), as indicated.

fold increase in the presence of ATP. In this experiment, as in others, GTP was somewhat more efficient than ATP in enhancing synthesis. Unlike phleomycin, pancreatic DNase I, added to cultures at zero time, never enhanced synthesis in the absence of ATP, but it did increase synthesis, in proportion to its concentration, in the presence of ATP (Figure 2).

Polymerase β. Since both polymerase α, and polymerase β could have been present in the reaction mixes, we tested the sensitivity of the nuclear enzyme-template complex to N-ethylmaleimide (NEM) to determine which of the two polymerases was functioning in the system. We first established that 0.6mM NEM inhibited 80% of the DNA synthesis mediated by the polymerase in the cytoplasmic fraction of an NP40 cell lysate. Two-tenths ml of cytosol was added to a normal phleomycin plus ATP reaction mix containing 50 ug salmon sperm DNA. The cytoplasmic polymerase is presumably polymerase α, which is relatively more sensitive than polymerase β to NEM inhibition. The sensitivity of DNA synthesis in the nuclear enzyme-template system is shown in Figure 2. Eight-tenths mM NEM did not reduce DNA synthesis, but 2 to 8 mM generally inhibited synthesis almost completely.

The DNA. In order to determine whether phleomycin stimulated repair or semiconservative synthesis we ran an experiment with a reaction mix in which ^{3}H-dATP was substituted for dATP, and BrdUTP for TTP. We then analyzed the DNA in shallow density gradients of CsCl made by centrifugation in a fixed-angle rotor. Figure 3 shows that in control cultures most of the DNA banded in fractions 26 to 28, the position of normal, 1.700 gm/cm^3 density, rabbit DNA. In control plus ATP cultures the DNA was distributed in two wide bands, one (fractions 21 to 28) with a density of 1.701 to 1.703 gm/cm^3, and the other (fractions 6 to 12) with a density of 1.711 to 1.717 gm/cm^3. The phleomycin culture had a single wide band of DNA (fractions 11 to 20) with a density range of 1.706 to 1.714 gm/cm^3. The phleomycin plus ATP culture was noticeably different from all of the others. Most of the DNA in this culture banded in fractions 1 to 9, and thus had a density greater than 1.715 gm/cm^3. The true density of this DNA could not be determined from these gradients, so an

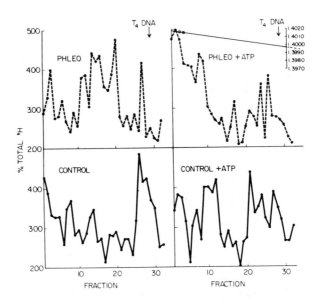

Fig. 3. Distribution of DNA in shallow gradients of CsCl. Two-tenths ml of a nuclear enzyme-template preparation was incubated at 37° C, for 28 minutes, with 0.25 ml dGTP, dCTP, BrdUTP, ^{3}H-dATP, and phleomycin, and ATP as indicated. Two-tenths ml of the culture was then added to 0.2 ml of 0.01 M EDTA-0.01 M Tris, pH 8.0 + 0.2 mg/ml pronase, at 37° C, for 1 minute, and then 0.1% SDS was added. The entire lysate, containing 40 to 50,000 acid precipitable ^{3}H counts/min, was diluted to 5 ml with EDTA-Tris plus 10 mM NaCl, and sufficient CsCl to have a refractive index of exactly 1.4020. The samples were centrifuged in a Spinco fixed-angle, type 50 rotor, at 40,000 rpm, 18° C, for 48 hours. Forty-six to forty-eight, 6-drop fractions were collected. Refractive indices were read on portions of the first, and last three fractions. All fractions were assayed for ^{3}H activity.

additional experiment was performed in which the DNA was analyzed on steeper CsCl gradients made by centrifugation in a swinging-bucket rotor. In this experiment (Figure 4) most of the DNA banded at the position of normal density DNA, but a significant fraction of the DNA (fraction 18) had the 1.756 gm/cm^3 density of hybrid DNA.

DISCUSSION

The data show that DNA polymerase β is present in Go cells. The data also suggest that enzyme activity depends on the presence of an adequate template substrate. The definition of an adequate template, however, is not clear. DNA polymerase activity can be stimulated in Go cells by UV or X-irradiation, or by methyl-methanesulfonate (MMS) (6). But, these agents induce only repair synthesis which occurs in whole cells after a considerable time lag. We have found that pretreating whole cells or nuclei with 400 to 800 $ergs/mm^2$ UV light, or with 3 mM MMS did not change ^3H-incorporation in nuclear lysate preparations.

The resting, Go stage in lymphocytes is a natural state which may be significantly different from the arrested or resting Go stage of HeLa cells, mouse L cells, or rat liver cells. This difference may be reflected in the fact that calf thymus DNA or salmon sperm DNA can be used as templates to demonstrate polymerase β activity throughout the cell cycles of the cultured cells (7, 8, 9) and polymerase β activity in transformed or mitogen stimulated lymphocytes, but they show little or no polymerase activity in Go lymphocytes (10) which we now know have active polymerase.

It is possible that ATP which is required for DNA synthesis in a number of nuclear systems (11, 12), serves simply to protect the deoxytriphosphates from destruction by phosphatases. Our observation that GTP also enhances DNA synthesis is consistent with this mechanism. However, ATP and GTP may play a role in the preparation of the "adequate template". They may be components of an RNA segment needed to prime DNA synthesis. This mechanism fits nicely with some preliminary data indicating that the DNA made in our system is 4S size, i.e., Okazaki size, segments. If ATP and GTP are required as primers for DNA

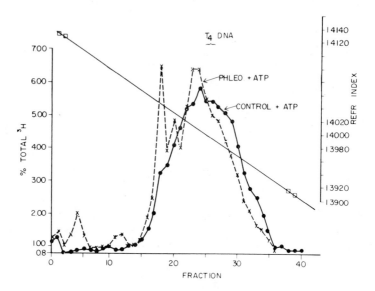

Fig. 4. Distribution of DNA in steep gradients of CsCl. The procedure outlined in Fig. 3 was followed, except that the 5 ml CsCl samples were centrifuged in a Spinco SW 50.1 rotor.

synthesis, a simple and intriguing hypothesis to explain the stimulation of DNA synthesis by phleomycin in the absence of ATP or GTP, is that phleomycin itself acts both to make single strand breaks and to prime DNA synthesis at the breaks in place of RNA. This hypothesis is testable and it is now being investigated.

ACKNOWLEDGEMENTS

These studies were supported by Public Health Service Research Grant CA 17123-03A2 from the National Cancer Institute.

REFERENCES

1. Maeda, K., H. Kosada, K. Yagishita, and H. Umezowa, J. Antibiot. (Tokyo) Ser. A. 9, 82 (1956).
2. Shuve, S. J., and A. M. Rauth. Cancer Res. 31, 1422 (1971).
3. Tanaka, N., H. Yamaguchi, and H. Umezowa, J. Antibiot. (Tokyo) Ser. A. 16, 86 (1963).
4. Reiter, H., M. Milewskiy, and P. Kelley, J. Bacteriol. 111, 586 (1972).
5. Reiter, H. Biochem. Biophys. Res. Commun. 60, 1371 (1974).
6. Clarkson, J. M., and H. J. Evans, Mutation Res. 14, 413 (1972).
7. Spadari, S. and A. Weissbach, J. Mol. Biol. 86, 11 (1974).
8. Chang, L.M.S., McK. Brown, and F. J. Bollum, J. Mol. Biol. 74, 1 (1973).
9. Baril, E. F., M. D. Jenkins, O. E. Brown, J. Laszlo, and H. P. Morris, Cancer Res. 33, 1187 (1973).
10. Lock, L. A. and S. S. Agarwal. Exp. Cell Res. 66, 299 (1971).
11. Friedman, D. L. and G. C. Mueller, Biochem. Biophys. Acta 161, 455 (1968)
12. Hallick, L. M., and M. Namba, Biochem, 13, 3152 (1974).

Part XIII
Replication Intermediates

DISCONTINUOUS REPLICATION IN PROKARYOTIC SYSTEMS

Reiji Okazaki, Tuneko Okazaki, Susumu Hirose, Akio Sugino,
Tohru Ogawa, Yoshikazu Kurosawa, Kazuo Shinozaki,
Fuyuhiko Tamanoi, Tetsunori Seki, Yasunori Machida,
Asao Fujiyama and Yuji Kohara

Institute of Molecular Biology
Faculty of Science, Nagoya University
Nagoya, Japan 464

ABSTRACT.   A new method for the detection and assay of
RNA-linked nascent DNA pieces has been developed. The
method relies on selective degradation by spleen exonuclease of radioactive 5'-hydroxyl-terminated DNA that is
produced from the pulse-labeled nascent pieces upon alkaline hydrolysis. Analysis with this method in wild-type E.
coli shows relatively high proportions of the RNA-linked
molecules after shorter pulses and in the smaller pieces,
supporting the transient nature of the RNA attachment to
the nascent pieces. The RNA-linked nascent pieces are
accumulated by both E. coli polAex1 (defective in 5' → 3'
exonuclease of DNA polymerase I) and E. coli polA12 (defective in polymerase of DNA polymerase I), suggesting the
requirement of the concerted action of both 5' → 3' exonuclease and polymerase of DNA polymerase I for the removal
of the RNA attached to the nascent pieces. Most of the
nascent DNA pieces accumulated by E. coli ligts7 (defective
in DNA ligase) are not linked to RNA, as expected from the
direct role of DNA ligase in joining of the pieces.
     The analysis with spleen exonuclease also shows that
a large portion of the nascent DNA pieces present in the
cell under the normal steady-state conditions are not
linked to RNA. The level of these RNA-free DNA pieces, as
well as the level of the RNA-linked DNA pieces, is
increased in polA mutants. These findings suggest that
the removal of RNA from the nascent pieces is a relatively
rapid process and the joining of the DNA pieces is the
rate-limiting step that requires the concurrent action of
DNA polymerase and DNA ligase.
     The 5'-hydroxyl end of DNA produced from the RNA-
linked nascent DNA pieces on alkaline hydrolysis can also
be assayed using T4 polynucleotide kinase. Interference
of the assay by the exchange between the γ-phosphate of
ATP and the 5'-phosphate of DNA catalyzed by the kinase is

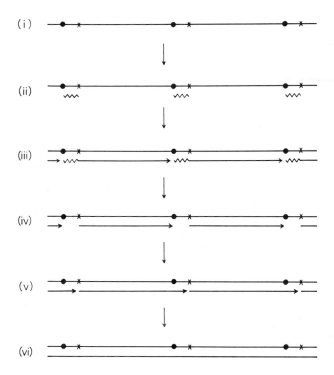

Fig. 1.   A hypothetical mechanism for discontinuous replication.   The description of each step is given in the text.   The temporal sequence of the reactions may not necessarily be as depicted.   For example, the removal of RNA from each piece may occur before the completion of its DNA part (iii - iv), and the RNA removal and gap filling may take place concurrently by a nick translation mechanism (iii - v).

avoided by performing the kinase reaction at 0°C, where little exchange reaction occurs. Using the kinase reaction under these conditions, the accumulation of RNA-linked nascent DNA pieces in E. coli polAex1 is also demonstrated.

Analysis of the deoxyribonucleotide at the 5' end of the DNA segment of E. coli RNA-linked nascent DNA pieces with polynucleotide kinase shows the occurrence of all four deoxyribonucleotides. Analysis of the nucleotides sequence by ^{32}P transfer from [α-^{32}P]deoxyribonucleoside triphosphates to ribonucleotides upon alkaline hydrolysis of the nascent DNA synthesized by toluene-treated cells also shows the presence of all 16 possible nucleotide sequences at the RNA-DNA junction. These results do not support the previous conclusion that the RNA-DNA linkage of E. coli nascent pieces has specific sequences.

Analyses of P2 phage DNA replicating in E. coli wild-type, polA12 and polAex1 reveal that both H- and L-strands of phage DNA are formed discontinuously, but the rate of joining of the pieces is much faster in the L-strand elongating in the overall 5' → 3' direction than in the H-strand elongating in the overall 3' → 5' direction. RNA is shown to be linked to the nascent pieces of both strands of P2 DNA by both spleen exonuclease and polynucleotide kinase methods.

The double-stranded replicative form of φX174 phage DNA is also shown to be synthesized discontinuously via short piece intermediates to which RNA is attached. Evidence is also presented for the RNA attachment to the nascent DNA pieces of T4 phage.

INTRODUCTION

We have presented the following evidence for the discontinuous mechanism of DNA replication involving synthesis and joining of short DNA chains. (a) Nascent DNA is isolated as short chains (1-3). (b) These short chains accumulate upon inhibition of DNA ligase activity (3,4). (c) The direction of synthesis of DNA in vivo is 5' → 3' (5,6).

The important question raised by the finding of the discontinuous mode of DNA replication is how is the discontinuity of chain growth produced; that is, what is the mechanism of initiation of the short DNA chains (nascent

834

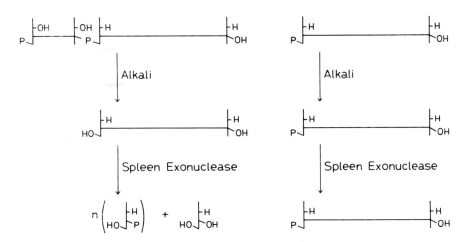

RNA-linked DNA   RNA-free DNA

Fig. 2. Principle of the spleen exonuclease assay of RNA-linked nascent DNA pieces.

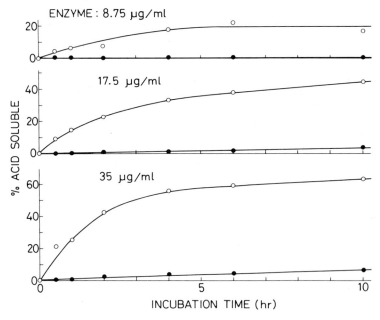

Fig. 3. Degradation of 5'-OH and 5'-P DNA by spleen exonuclease (41). ^{3}H-labeled 5'-OH DNA (o; $\underline{n}$ = 800–1500) and ^{14}C-labeled 5'-P DNA (•; $\underline{n}$ = 800–1500) were heated at 90°C for 2 min and then incubated at 37°C with spleen exonuclease in a 20-μl reaction mixture containing 0.1 M potassium phosphate (pH 6.0), 50 mM Na_2SO_4 and 10 mM EDTA. Similar results were obtained with 5'-OH and 5'-P DNA of various sizes from 100 to 2500 nucleotides.

DNA pieces). All known DNA polymerases lack the ability to initiate a new chain and only extend pre-existing poly-nucleotide chains (7-9). Unlike DNA polymerases, RNA poly-merases have the ability to start new chains (10). Furthermore, RNA as well as DNA chains can be extended by DNA polymerases (11-13). Therefore, DNA chain initiation might involve synthesis of an RNA stretch by RNA polymerase and its extension by DNA polymerase. In support of this possibility, Kornberg and his collaborators (14) found that the conversion of M13 single-stranded circular DNA to the double-stranded replicative form is blocked by rifampicin, a specific inhibitor of host RNA polymerase.

Although replication of E. coli chromosome, once initiated at the chromosome origin, is insensitive to rifampicin, we obtained evidence during 1972 that RNA is covalently attached to the 5' end of the nascent short DNA chains of E. coli (15-18). This suggests that RNA synthe-sis is involved in the initiation of synthesis of short DNA pieces during discontinuous replication, and that the reaction at the replication fork consists of several steps as shown in Fig. 1. (i) Unwinding of the parental strands. (ii) Synthesis of short RNA chains along the DNA template by an RNA polymerase. The enzyme may first bind speci-fically at an initiation signal (•) and stop polymerizing at a termination signal (x). (iii) Elongation of the chains by a DNA polymerase using the RNA as primer. (iv) Removal of the RNA segments by an RNase activity. (v) The filling of the gaps between DNA pieces by a DNA polymerase. (vi) Covalent joining of the DNA pieces by DNA ligase.

The major findings which indicated the RNA attachment to the nascent DNA were the following: (a) The buoyant density of nascent DNA pieces labeled by a very brief pulse of [^{3}H]thymidine is greater than that of ordinary single-stranded DNA and some [^{3}H]uridine-labeled RNA bands in the DNA region in the Cs_2SO_4 gradient (15-17). (b) Alkaline hydrolysis of the nascent piece results in the creation of a 5'-OH end of DNA that can be labeled with [γ-^{32}P]ATP by the polynucleotide kinase reaction (17). (c) ^{32}P is trans-ferred from [α-^{32}P]dNTP's to ribonucleotides upon alkaline hydrolysis of the nascent DNA synthesized by toluene-treated cells (18).

However, subsequent studies revealed various technical difficulties involved in these experiments. The density analysis of nucleic acids is complicated by noncovalent

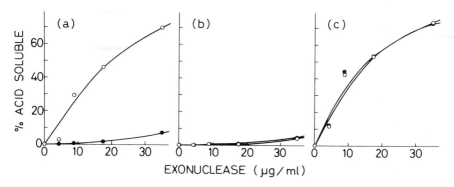

Fig. 4. Effect of polynucleotide kinase and alkaline phosphatase treatment on the susceptibility of DNA to spleen exonuclease (41). Mixtures of ^{3}H-labeled 5'-OH DNA (o; $\underline{n}$ = 400-800) and ^{14}C-labeled 5'-P DNA (•; $\underline{n}$ = 400-800) were subjected to spleen exonuclease digestion directly (a), after polynucleotide kinase treatment (b), and after alkaline phosphatase treatment (c).

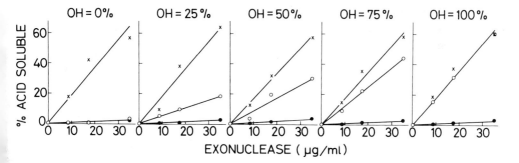

Fig. 5. Spleen exonuclease digestion of samples with a known proportion of 5'-OH and 5'-P DNA (41). Mixtures of ^{14}C-labeled 5'-OH and 5'-P DNA ($\underline{n}$ = 400-800) in the indicated proportions were digested with spleen exonuclease directly (o), after polynucleotide kinase treatment (•), and after alkaline phosphatase treatment (x).

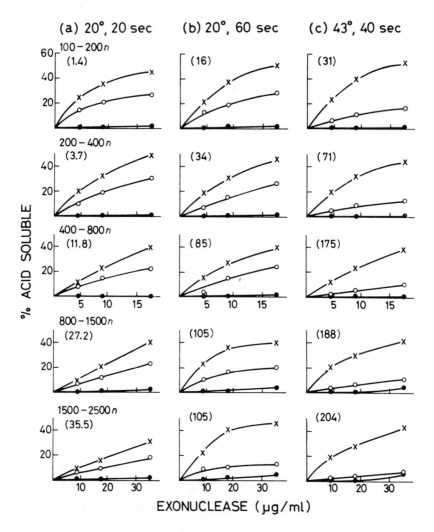

Figure 6.

Fig. 6.   Assay for the 5'-OH ends in alkali-treated nascent pieces from E. coli W3110 (wild type)(41).   (a) A culture grown at 37°C to 4 x 10^8 cells/ml and then at 20°C to 8 x 10^8 cells/ml was pulse labeled with [^{3}H]thymidine for 20 sec at 20°C.   (b) A culture grown as in (a) was pulse labeled for 60 sec at 20°C.   (c) A culture grown at 30°C to 8 x 10^8 cells/ml was transferred to 43°C and 3 min later pulse labeled for 40 sec.   Nucleic acid was extracted by the modified Thomas procedure (43) and heat denatured. The nascent pieces were isolated by centrifugation in neutral sucrose and Cs_2SO_4 gradients, treated with poly-nucleotide kinase and ATP, and then hydrolyzed with 0.15 M NaOH at 37°C for 20 hr.   The alkali-treated nascent DNA pieces were fractionated by Sepharose 4B gel filtration into five size classes as indicated and digested with spleen exonuclease directly (o), after polynucleotide kinase treatment (•) and after alkaline phosphatase treat-ment (x).   Numbers in parentheses indicate the amount of radioactive DNA (cpm x 10^{-2}) of each size class per milli-liter of culture.

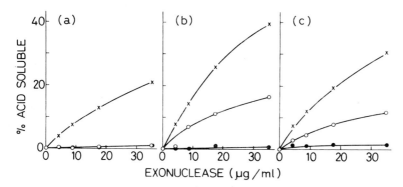

Fig. 7.   Assay for the 5'-OH ends in alkali- or RNase-treated nascent pieces from E. coli W3110 (41).   A portion of the sample of Fig. 6a was withdrawn before alkali treatment and divided into three portions (a, b and c); (a) was left untreated, (b) was treated with 0.15 M NaOH at 37°C for 20 hr, and (c) was treated with 180 µg/ml of RNase IA in 10 mM Tris·HCl (pH 8.0)-0.1 mM EDTA at 37°C for 2 hr.   Samples from each portion were digested with spleen exonuclease directly (o), after polynucleotide kinase treatment (•) and after alkaline phosphatase treat-ment (x).

RNA-DNA interactions. Such RNA-DNA interactions can also confound the purification of the nascent DNA pieces. Furthermore, the exchange between the γ-phosphate of ATP and 5'-phosphate of DNA catalyzed by polynucleotide kinase (19) necessitates special precautions for the analysis of RNA-linked DNA pieces with the kinase method.

Therefore, we have worked for improvement of the methods for the isolation and analysis of the RNA-linked nascent DNA pieces and investigated this problem further with the improved methods in E. coli and some bacteriophages.

RESULTS

Analysis of E. coli RNA-linked nascent DNA pieces by the spleen exonuclease method: Fig. 2 shows the principle of a new method developed for the detection and assay of RNA-linked nascent DNA pieces. The pulse-labeled pieces are isolated and treated with polynucleotide kinase and ATP to mask any pre-existing 5'-OH ends of DNA. The sample is then subjected to alkaline hydrolysis, which results in the production of radioactive 5'-OH DNA[1] from the RNA-linked nascent pieces. The radioactive 5'-OH DNA thus produced is degraded selectively by spleen exonuclease.

Under the appropriate conditions, 5'-OH DNA, but little 5'-P DNA, of the size of the nascent pieces (100 to 2500 nucleotides), was in fact degraded by spleen exonuclease (Fig. 3). The exonucleolytic degradation of 5'-OH DNA did not reach completion and 5'-P DNA was also degraded to some extent (Fig. 4a). However, if 5'-OH and 5'-P DNA were subjected to exonuclease digestion after treatment with polynucleotide kinase and ATP, the degradation of both was at the same low level (Fig. 4b), whereas after alkaline phosphatase treatment both were degraded at the same rate (Fig. 4c). Therefore, the proportion of 5'-OH and 5'-P DNA in a DNA sample can be determined by digesting it by spleen exonuclease directly, after alkaline phosphatase treatment, and after polynucleotide kinase treatment. As shown in Fig. 5, this was substantiated with samples containing 5'-OH and 5'-P DNA in known proportions. With each sample, the relative extent of degradation without pre-treatment

---

[1]Abbreviations used: 5'-OH DNA, 5'-hydroxyl-terminated DNA; 5'-P DNA, 5'-phosphoryl-terminated DNA

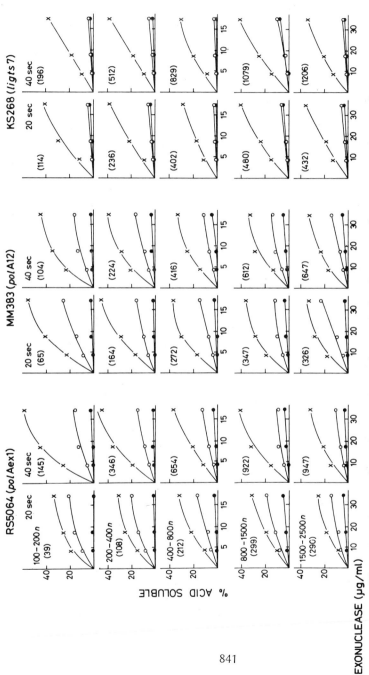

Fig. 8. Assay for the 5'-OH ends in alkali-treated nascent pieces from temperature-sensitive E. coli strains (41). Cultures were grown at 30°C to 8 x 10^8 cells/ml, transferred to 43°C and 3 min later pulse labeled with [^{3}H]thymidine for 20 sec or 40 sec. Alkali-treated nascent DNA pieces were prepared as in Fig. 6 and digested with spleen exonuclease directly (o), after polynucleotide kinase treatment (●) and after alkaline phosphatase treatment (x). Numbers in parentheses indicate the amount of radioactive DNA (cpm x 10^{-2}) of each size class per milliliter of culture; since the rate of incorporation of [^{3}H]thymidine in the thy$^-$ strains was 2/3 of that in the thy$^+$ strains, corrections for this factor were made with MM383 and KS268.

was in proportion to the amount of 5'-OH DNA.

Fig. 6 shows the results of the assay of RNA-linked nascent DNA pieces in E. coli W3110, a wild-type strain. The nascent pieces were isolated from three cultures pulse labeled (a) for 20 sec at 20°C, (b) for 60 sec at 20°C, and (c) for 40 sec at 43°C. After polynucleotide kinase treatment, the sample was hydrolyzed with alkali and fractionated into five size classes by Sepharose 4B gel filtration, and radioactive 5'-OH DNA in each size class was assayed with spleen exonuclease. In a, b and c, about 57%, 51% and 23% of the labeled DNA pieces, respectively, were found to be linked to RNA, indicating relatively high proportions of the RNA-linked molecules after shorter pulses. With b and c, the smaller pieces generally had a relatively high proportion of the RNA-linked molecules. These findings support the transient attachment of RNA to the nascent DNA pieces.

As shown in Fig. 7, similar results were obtained with RNase IA as with alkali. Furthermore, no 5'-OH DNA was detected without alkali or RNase treatment. These facts indicate that the labeled 5'-OH DNA measured after alkaline hydrolysis by this procedure in fact represents the RNA-linked nascent DNA pieces.

E. coli mutants defective in either the polymerase or the 5' → 3' exonuclease activity of polymerase I show retarded joining of the nascent DNA pieces (20–22), as do the mutants defective in DNA ligase (23,24). To explore RNA attachment to the nascent pieces accumulated in these mutants, the ts mutants of each type, E. coli RS5064 (polAexl), MM383 (polA12) and KS268 (ligts7), were pulse labeled at 43°C for 20 or 40 sec. During the 20- and 40-sec pulse, one-half and all of the accumulated pieces were labeled respectively. Assay of the 5'-OH ends (Fig. 8) after alkaline hydrolysis revealed that RNA is linked to approximately 52% and 24% of the DNA pieces from E. coli polAexl labeled during the 20- and 40-sec pulses. With E. coli polA12, RNA was shown to be linked to 30% and 28% of the pieces labeled by the 20- and 40-sec pulses, respectively. In contrast, only about 5% of the DNA pieces from E. coli ligts7, labeled by either type of pulse, contained RNA.

From the data with the 40-sec pulse, the polAexl and polA12 mutants were estimated to contain 5.2 and 4 times as many RNA-linked pieces as the wild type, whereas the

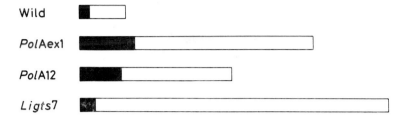

Wild

PolAex1

PolA12

Ligts7

Fig. 9.   RNA-linked and RNA-free nascent DNA pieces
in various strains (41).   The relative number of the molecules was calculated from the radioactivity and the ratio
of 5'-OH and 5'-P DNA in each size class in Figs. 6c and 8,
assuming uniform labeling of all the nascent pieces during
the 40-sec pulse at 43°C. ▨ , RNA-linked; ☐ , RNA-free.

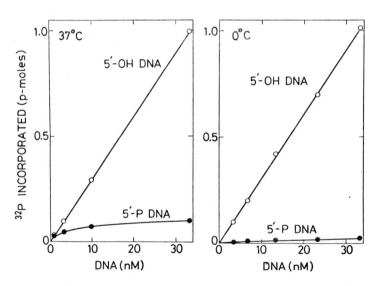

Fig. 10.   Labeling of 5'-OH and 5'-P DNA with [γ-
^{32}P]ATP by the polynucleotide kinase reaction at 37°C and
0°C (28).   Single-stranded 5'-OH or 5'-P DNA (200-2800
nucleotides) at the indicated concentrations was incubated
for 30 min with 33 units/ml of enzyme.

843

TABLE 1

AMOUNTS OF RNA-LINKED DNA PIECES IN VARIOUS E. COLI
STRAINS DETERMINED BY THE POLYNUCLEOTIDE KINASE METHOD

Expt. No	Strain	Number per cell
1	RS5064 (polAex1)	48
2	"	43
3	"	50
4	"	49
5	W3110 (wild type)	8
6	"	8
7	KS268 (ligts7)	10

The nascent pieces were isolated from cells which had
been exposed to 43°C for 3 min and pulse labeled with
[^{3}H]thymidine for 20 to 40 sec. In experiments 1-3, and
5-7, the pieces were isolated as described previously (28).
The sample of experiment 4 was purified by passing through
a column of nitrocellulose (39) instead of Cs_2SO_4 density
gradient centrifugation. After incubation with unlabeled
ATP and polynucleotide kinase and alkaline hydrolysis,
each sample was subjected to the polynucleotide kinase
reaction with [γ-^{32}P]ATP. The amount of 5'-OH DNA per
milliliter of culture was obtained as described previously
(28). The number of RNA-linked DNA pieces per cell was
calculated from the amount of 5'-OH DNA and the number of
cells per milliliter of culture.

level of the RNA-linked pieces in the ligts7 mutant is similar to that in the wild type (Fig. 9).

The accumulation of RNA-linked nascent DNA pieces by both E. coli polAex1 (defective in 5' → 3' exonuclease of polymerase I) and E. coli polA12 (defective in the polymerase activity) suggests that the removal of the primer RNA of the nascent DNA pieces (Fig. 1 step iv) and gap filling (step v) are achieved by the concerted action of exonuclease and polymerase of DNA polymerase I (25-27). The finding that most of the nascent DNA pieces accumulated by E. coli ligts7 (defective in DNA ligase) are not linked to RNA was expected from the direct role of DNA ligase in joining of the pieces.

These analyses also showed that a large portion of the nascent DNA pieces present in the cell under the normal steady-state conditions are not linked to RNA. This suggests that the removal of RNA from the nascent pieces (step iv) is a relatively rapid process and that the joining reaction (step vi) is rate limiting. The number of the RNA-free DNA pieces was further increased in the polA mutants, suggesting that the joining reaction requires the function of DNA polymerase I concurrent with that of ligase. The retardation of the joining reaction in the polA mutants may be due to inadequate substrates for ligase action produced by the unbalanced exonuclease and polymerase activities. In the polAex1 mutant, the 5' terminus of the DNA piece would be displaced because of the deficiency in exonuclease activity, whereas in the polA12 mutant single-stranded gaps would persist between the pieces because of the deficiency in polymerase activity.

Analysis of E. coli RNA-linked nascent DNA pieces by the polynucleotide kinase method: As noted in Introduction, the exchange reaction between the γ-phosphate of ATP and the 5'-phosphate of DNA complicates the polynucleotide kinase assay of the 5'-OH group of DNA produced from the RNA-linked DNA pieces upon alkaline hydrolysis. However, we found that the labeling of 5'-P DNA by the exchange reaction is sufficiently suppressed at 0°C as shown in Fig. 10. Using these conditions for the kinase reaction, the accumulation of RNA-linked nascent DNA pieces by the polAex1 mutant was also demonstrated.

The data in Table 1 show that in the polAex1 mutant the number of RNA-linked DNA pieces per cell was about 47

TABLE 2

DEOXYRIBONUCLEOTIDES AT THE JUNCTION OF RNA-LINKED DNA
PIECES IN E. COLI RS5064 (polAex1) ANALYZED BY THE
POLYNUCLEOTIDE KINASE METHOD

Expt. No	dGMP	dTMP	dAMP	dCMP
1	15%	28%	16%	41%
2	17	34	16	33
3	17	33	16	33
4	13	33	15	39
mean	15.5	32.0	15.7	36.5

The 5' termini of DNA of the RNA-linked DNA pieces
were labeled with ^{32}P by the polynucleotide kinase reaction as described in Table 1. The DNA was hydrolyzed
successively with pancreatic DNase and snake venom phosphodiesterase. The resulting radioactive deoxyribonucleoside
monophosphates were analyzed by PEI-cellulose thin-layer
chromatography (40).

Notes to Table 3.

(a) The previous experiment with E. coli P3478 (18).
(b) A new experiment with E. coli H560. Cells treated
with 1% toluene for 10 min at 37°C were incubated in a
reaction mixture (18) with one of the four dNTP's labeled
with ^{32}P at the α position as indicated. The labeled DNA
was extracted, banded in Cs_2SO_4 and hydrolyzed with alkali
as described previously (18). The acid-soluble hydrolysate
was fractionated by DEAE-cellulose-7 M urea chromatography
and the mononucleoside monophosphate fraction was subjected
to Dowex AG1 chromatography. The radioactive material
eluted with each of the four 2'(3')-rNMP markers was pooled,
desalted and analyzed by PEI-cellulose thin-layer chromatography (40).

TABLE 3

TRANSFER OF ^{32}P FROM [α-^{32}P]dNTP'S TO RIBONUCLEOTIDES
UPON ALKALINE HYDROLYSIS OF DNA SYNTHESIZED
BY TOLUENE-TREATED CELLS

(a) E. coli P3478 (polA1) (Sugino & Okazaki, 1973)

[α-^{32}P]dNTP substrate	^{32}P-labeled ribonucleotides (number per cell)				
	CMP	AMP	GMP	UMP	Total
dCTP	6.0	2.5	4.0	25.7	38.2 (58.2%)
dGTP	——	0.8	1.3	23.6	25.8 (39.3%)
dTTP	0.4	0.2	0.3	0.6	1.4 (2.1%)
dATP	——	——	——	0.3	0.3 (0.4%)
Total	6.4	3.5	5.6	50.2	65.7

(b) E. coli H560 (polA1, endI)

[α-^{32}P]dNTP substrate	^{32}P-labeled ribonucleotides (number per cell)				
	CMP	AMP	GMP	UMP	Total
dCTP	8.3	15.5	24.3	15.0	63.0 (45.5%)
dGTP	7.0	10.4	14.1	8.4	39.9 (28.8%)
dTTP	3.7	4.8	8.1	9.4	26.1 (18.8%)
dATP	1.8	0.9	2.2	4.7	9.6 (6.9%)
Total	20.8	31.6	48.7	37.5	138.6

—— : not detected

(Notes on facing page.)

847

compared with 8 in the wild type and 10 in the ligts7
mutant. These data are in good agreement with the previous
results obtained with the spleen exonuclease assay.

The labeling of the 5' end of the DNA after alkaline
hydrolysis allows the identification of the 5'-terminal
deoxyribonucleotide of the DNA segment that is involved in
the RNA-DNA linkage. Our previous experiment with E. coli
Q13 (17) indicated that the deoxyribonucleotide at the
junction is exclusively dCMP. However, as shown in Table
2, recent analysis with RNA-linked DNA pieces from E. coli
polAex1 showed that other deoxyribonucleotides are also in
the RNA-DNA link, although dCMP is the most abundant
residue.

Analysis of E. coli RNA-linked nascent DNA pieces by the
^{32}P transfer method: The covalent attachment of RNA to the
5' end of the nascent DNA pieces can also be demonstrated
by transfer of ^{32}P from [α-^{32}P]dNTP's to ribonucleotides
upon alkaline hydrolysis of the nascent DNA synthesized in
toluene-treated cells. The ^{32}P transfer also provides
information about the nucleotide sequence at the RNA-DNA
link. In our previous experiments of this type (18), shown
in Table 3a, the predominant ^{32}P transfer was observed from
dCTP to UMP and dGTP to UMP and little or no other transfer
was found. We have recently repeated these experiments
many times with various strains. These experiments confirmed the ^{32}P transfer and hence the synthesis of the RNA-
linked DNA pieces by toluene-treated cells, but gave different transfer patterns. As shown in Table 3b, although
the ^{32}P transfer from dCTP was most frequent as before, all
four ribonucleotides received significant amounts of the
label, and all 16 possible combinations were observed,
though at unequal frequencies. In repetitions of these
experiments, the frequency of the transfer from dCTP to the
four ribonucleotides, particularly that to UMP, varied.
The reason for this variation is unclear at this time.
These results, as well as the results of the kinase experiment presented above, do not support the previous conclusion that the RNA-DNA linkage of the E. coli nascent pieces
has specific sequences.

Discontinuous replication of P2 phage DNA: In many systems
replication proceeds bidirectionally from the origin (29).
Replication of P2 phage DNA, however, is initiated near

Genetic map

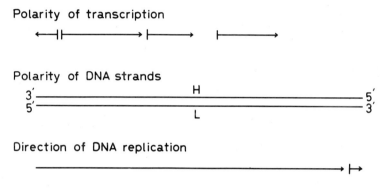

Fig. 11.   Genetic and biochemical map of P2 phage genome.

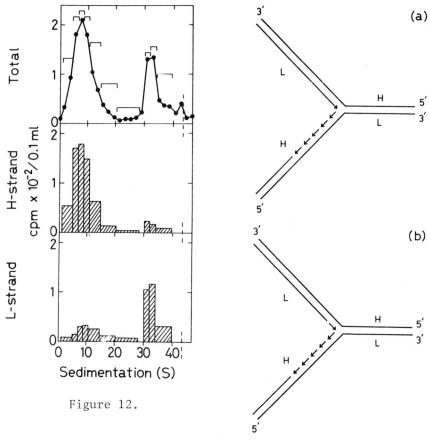

Figure 12.

Figure 13.

Fig. 12.   Size distribution of the H- and L-strand components of nascent DNA from P2-infected E. coli (42). P2 vir_1-infected E. coli C-1a was pulse labeled with [^{3}H]thymidine for 7 sec at 100 min after infection at 20°C. The pulse-labeling was terminated with an ethanol-phenol mixture (43). DNA was extracted and sedimented through alkaline sucrose gradients and fractions were collected from the bottom. Acid-insoluble radioactivity in each fraction was measured. The bracketed fractions were pooled and tested for the ability to hybridize with the separated L- and H-strands of P2 DNA to obtain the amounts of the radioactive H- and L-strand components.

Fig. 13.   Possible structures of the replication fork of P2 DNA. (a) Only the H-strand is formed discontinuously. (b) Both strands are formed discontinuously, but the rate of joining of short chains is much faster in the L-strand than in the H-strand.

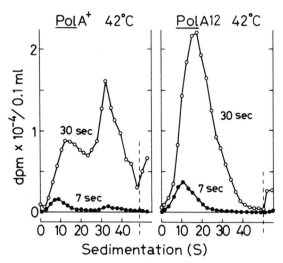

Fig. 14. Alkaline sucrose gradient sedimentation of
nascent DNA from P2-infected polA[+] and polA12 cells (42).
E. coli C-1200 (polA[+]) and C-2107 (polA12) were infected
with P2 vir[1] at 20°C. After incubation for 120 min at
20°C, cells were transferred to 42°C and 2 min later pulse
labeled with [³H]thymidine for the times indicated. The
pulse was terminated with the ethanol-phenol mixture
and DNA was extracted and analyzed by alkaline sucrose
gradient sedimentation.

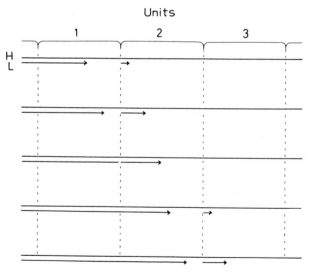

Fig. 15. A model for discontinuous replication of
the L-strand elongating in the overall 5' → 3' direction
(42). Each unit is initiated as a short chain but linked
to the older part of the strand in the middle of its
synthesis.

cistron A and proceeds unidirectionally from left to right through cistron Q (30)(Fig. 11). The direction of transcription of the P2 chromosome is also mainly from left to right on the genetic map (31,32) and the H-strand of the phage DNA is predominantly transcribed (33). Since mRNA synthesis proceeds in the 5' → 3' direction, the 3' end of the H-strand must be at the left end of the P2 genetic map. Accordingly, the overall direction of elongation of the H- and L-strand during replication is 3' → 5' and 5' → 3', respectively.

When P2-infected cells were labeled by a very brief [^{3}H]thymidine pulse, the label was incorporated mainly into two size classes of DNA chains: short chains of about 10 S, and long chains of one genome length with a sedimentation coefficient of about 32 S. The pulse-labeled short chains hybridized predominantly with the L-strand of the phage DNA, whereas the labeled genome-length DNA hybridized preferentially with the H-strand (Fig. 12). This suggests the two possibilities shown in Fig. 13. (a) Only the H-strand, elongating in the 3' → 5' direction, is synthesized discontinuously. (b) Both H- and L-strands are formed discontinuously, but the rate of joining of the pieces is much faster for the L-strand, elongating in the 5' → 3' direction, than for the H-strand, elongating in the 3' → 5' direction.

The nascent short chains of L-strand do exist under the normal steady-state conditions, though in a small amount. Furthermore, if P2-infected E. coli polA12 or polAex1 was pulse labeled at the restrictive temperature, virtually all the label incorporated was found in short DNA chains (Fig. 14). These short chains accumulated during inhibition of either polymerase or 5' → 3' exonuclease activity of the host DNA polymerase I contained equal amounts of H- and L-strand components. Therefore, both strands of P2 DNA are replicated discontinuously, but the rate of joining of the L-strand short chains is very much faster than that of the H-strand short chains.

As can be seen from Fig. 12, even after a very brief pulse, two-thirds of the pulse label incorporated into the L-strand was in long DNA chains in the wild-type host. This fact can be explained by the possibility illustrated in Fig. 15. Since the overall direction of elongation of the L-strand is 5' → 3' as is the direction of synthesis of the short chains, each unit of discontinuous synthesis of

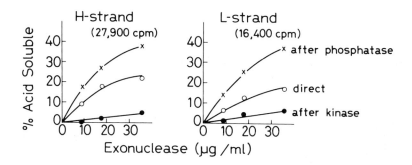

Fig. 16.   RNA attachment to P2 nascent pieces of both strands demonstrated by the spleen exonuclease method. E. coli C-N3 (polAex1) was infected with P2 vir$_1$ at 20°C. After incubation for 100 min, cells were transferred to 43°C and 2.5 min later pulse labeled with [^{3}H]thymidine for 30 sec. Nucleic acid was extracted and heat denatured. Nascent pieces (<15 S) were isolated by centrifugation in neutral sucrose and Cs$_2$SO$_4$ gradients, treated with poly-nucleotide kinase and ATP, and hydrolyzed with 0.15 M NaOH at 37°C for 20 hr. The sample was divided into three portions (a, b and c); (a) was left untreated, (b) was treated with polynucleotide kinase and ATP, and (c) was treated with alkaline phosphatase. Each portion was allowed to anneal to two membrane filters, each of which was loaded with L- or H-strand of P2 DNA. The material annealed to each strand was eluted with 0.1 M NaOH–10 mM EDTA, and subjected to digestion with spleen exonuclease.

this strand is initiated as a short chain but is linked to
the older part of the strand in the middle of its synthesis, namely before its completion. Consequently, there
exist both short and long growing chains of the L-strand at
a given moment and both are labeled by a brief pulse.

As shown in Fig. 16 and Table 4, a large portion of P2
nascent DNA pieces of both H- and L-strands accumulated in
the polAex1 cell were demonstrated to be linked to RNA by
both spleen exonuclease and polynucleotide kinase methods.
Thus, the RNA priming mechanism appears to be involved in
discontinuous synthesis of both the 3' → 5' and 5' → 3'
strands.

Discontinuous replication of the replicative form of φX174
phage DNA: In the 1968 Cold Spring Harbor Symposium, we
suggested that the discontinuous mechanism might be a
device evolved for replication of genomes which exceed a
certain size, so that a small circular DNA such as φX174
phage DNA might be formed by a continuous mechanism (3).
To test this idea, we have performed experiments on replication of the φX174 replicative form (RF).

In the experiment shown in Fig. 17, φX174-infected
wild-type cells were pulse labeled with [^{3}H]thymidine
during RF replication. After cell lysis with lysozyme and
sarkosyl, the RF molecules were isolated by neutral sucrose
gradient sedimentation. Analysis of these molecules by
alkaline sucrose gradient sedimentation indicated the
presence of nascent short DNA pieces of about 8 S. The
presence of these short chain intermediates was demonstrated more clearly by a similar experiment with E. coli
polAex1 as the host, in which pulse labeling was performed
at 30°C or 43°C during RF replication (Fig. 18). Using the
spleen exonuclease method, RNA was shown to be attached to
these nascent DNA pieces of the φX174 RF molecules (Fig.
19). These results strongly suggest that one or both
strands of the φX174 RF molecules are replicated by the
discontinuous mechanism involving the synthesis of DNA
pieces with RNA primer. Discontinuous synthesis of the
minus strand of the φX174 RF molecules has been suggested
by Eisenberg and Denhardt (34) based on a different type
of experiment.

RNA attachment to T4 nascent DNA pieces: Earlier studies
(2-6) showed that both strands of T4 phage DNA are repli-

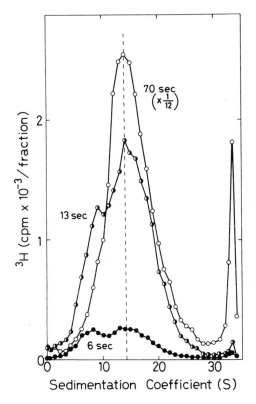

Fig. 17.   Alkaline sucrose gradient sedimentation of
pulse-labeled DNA from φX174-infected E. coli C-1200
(polA⁺).   Cells grown in TPG-CAA medium (44) at 30°C to 4
x 10⁸ cells/ml were suspended, after washing, in starvation
buffer (45) at 8 x 10⁸ cells/ml and incubated at 30°C for
100 min. The cells were then infected with φX174 am3 at a
multiplicity of 10. After 5 min at 30°C, an equal volume
of 2 x TPG-CAA was added and the culture was incubated for
20 min at 30°C. The cells were then exposed to [³H]thymi-
dine for the times indicated and the pulse labeling was
terminated with the ethanol-phenol mixture. The cells
were lysed by lysozyme and sarkosyl treatment, and the
replicative forms were isolated by neutral sucrose gradient
sedimentation (46) and subjected to alkaline sucrose gradient sedimentation. The dashed line indicates the position
of the genome-length linear DNA.

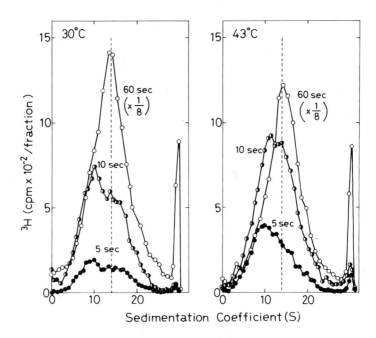

Fig. 18.   Alkaline sucrose gradient sedimentation of pulse-labeled DNA from φX174-infected E. coli C-N3 (polAex1).   The experiment was performed as in Fig. 17. For pulse labeling at 43°C, temperature was shifted from 30°C to 43°C at 17.5 min after infection, and 2.5 min later cells were exposed to [³H]thymidine for the times indicated.   The dashed lines indicate the position of the genome-length linear DNA.

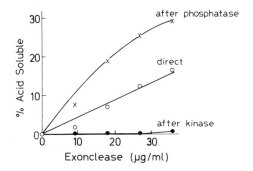

Fig. 19.   Demonstration of RNA attachment to φX174
nascent pieces by the spleen exonuclease method.   φX174
am3-infected E. coli C-N3 (polAex1) was pulse-labeled for
5 sec at 43°C and the RF molecules were isolated as in Fig.
18.   After treatment with pronase and phenol and heat
denaturation, 5-10 S nascent pieces were isolated by
centrifugation in neutral sucrose and Cs$_2$SO$_4$ gradients.
The nascent pieces were treated with polynucleotide kinase
and ATP and then subjected to alkaline hydrolysis.   The
alkali-treated pieces were digested with spleen exonuclease
directly, after polynucleotide kinase treatment, and after
alkaline phosphatase treatment.

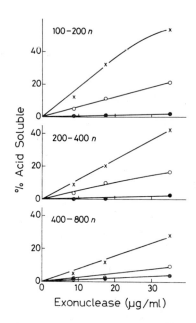

Fig. 20. Demonstration of RNA attachment to T4
nascent pieces by the spleen exonuclease method. E. coli
W4597 (galU) was infected with T4D$^+$ at a multiplicity of 10
at 37°C. At 15 min after infection, the cells were transferred to 20°C and 15 min later pulse labeled with [^{3}H]thy-
midine for 5 sec. Alkali-treated nascent DNA pieces were
prepared as in Fig. 6 and degraded with spleen exonuclease
directly (o), after polynucleotide kinase treatment (•) and
after alkaline phosphatase treatment (x).

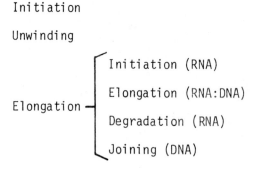

Initiation

Unwinding

Elongation — Initiation (RNA)

Elongation (RNA:DNA)

Degradation (RNA)

Joining (DNA)

Fig. 21. Reactions involved in DNA replication.

cated discontinuously. Analysis with spleen exonuclease demonstrated that RNA is attached to nascent short DNA pieces of T4 labeled by a very brief pulse (Fig. 20).

CONCLUSION

Although much remains to be elucidated, studies during the past 10 years have demonstrated with little doubt that DNA chain elongation, initially thought to be mere polymerization of deoxyribonucleotides by DNA polymerase, is a multistep process consisting of: initiation by RNA, elongation of RNA, two-step elongation of DNA, degradation of RNA, and joining of DNA (Fig. 21).

Several replication enzymes and proteins have been isolated from E. coli and other sources (this volume). In replication of DNA of E. coli and certain coli phages dependent on host functions, it seems likely that the dnaG protein functions for RNA initiation and synthesis (35,36), DNA polymerase III (37) or polymerase III holoenzyme (36, 38) for the synthesis of the main body of the nascent piece, DNA polymerase I for the RNA degradation and gap filling, and DNA ligase for the joining of DNA pieces.

ACKNOWLEDGMENTS

This work was supported by research grants from the Ministry of Education of Japan, Toray Science Foundation, Yamaji Science Foundation and Matsunaga Science Foundation. We thank Drs. I. R. Lehman, E. B. Konrad, R. Calendar and M. Sunshine for providing us with bacterial and phage strains and Dr. N. R. Cozzarelli for critical reading of the manuscript.

REFERENCES

1. Sakabe, K. and Okazaki, R., Biochim. Biophys. Acta 129, 651 (1966).
2. Okazaki, R., Okazaki, T., Sakabe, K., Sugimoto, K. and Sugino, A., Proc. Nat. Acad. Sci. (USA) 59, 598 (1968).
3. Okazaki, R., Okazaki, T., Sakabe, K., Sugimoto, K., Kainuma, R., Sugino, A. and Iwatsuki, N., Cold Spring Harbor Symp. Quant. Biol. 33, 129 (1968).
4. Sugimoto, K., Okazaki, T. and Okazaki, R., Proc. Nat.

TABLE 4

RNA ATTACHMENT TO P2 NASCENT PIECES OF BOTH STRANDS
DEMONSTRATED BY THE POLYNUCLEOTIDE KINASE METHOD

5'-OH DNA pieces after alkaline hydrolysis (number/cell)	
H–strand	L–strand
80	30

Alkali-treated nascent DNA pieces were prepared from P2 vir$_1$-infected E. coli C-N3 (polAex1) as in Fig. 16. The 5'-OH end of the DNA was phosphorylated by the polynucleotide kinase reaction with [γ-^{32}P]ATP at 0°C (28). After passing through a Sephadex G100 column, the sample was allowed to anneal to two membrane filters, each of which was loaded with L- or H-strand of P2 DNA. The number of 5'-OH DNA pieces of each strand per cell was calculated as follows:

Amount of ^{32}P annealed to the complementary strand (mol)

$$\times \frac{\text{total } ^3\text{H label incorporated into the pieces (cpm)}}{^3\text{H label annealed to the two strands (cpm)}}$$

$$\times \frac{\text{Avogadro number}}{\text{total cell number}}$$

Acad. Sci. (USA) 60, 1356 (1968).

5. Okazaki, T. and Okazaki, R., Proc. Nat. Acad. Sci. (USA) 64, 1242 (1969).
6. Sugino, A. and Okazaki, R., J. Mol. Biol. 64, 61 (1972).
7. Goulian, M., Proc. Nat. Acad. Sci. (USA) 61, 284 (1968).
8. Wickner, R. B., Ginsberg, B. and Hurwitz, J., J. Biol. Chem. 247, 498 (1972).
9. Kornberg, T. and Gefter, M. L., J. Biol. Chem. 247, 5369 (1972).
10. Maitra, U. and Hurwitz, J., Proc. Nat. Acad. Sci. (USA) 54, 815 (1965).
11. Wells, R. D., Flügel, R. M., Larson, J. E., Schendel, P. F. and Sweet, R. W., Biochemistry 11, 621 (1972).
12. Chang, L. M. S. and Bollum, F. J., Biochem. Biophys. Res. Commun. 41, 1354 (1972).
13. Keller, W., Proc. Nat. Acad. Sci. (USA) 69, 1560 (1972).
14. Brutlag, D., Schekman, R. and Kornberg, A., Proc. Nat. Acad. Sci. (USA) 68, 2826 (1971),
15. Sugino, A., Hirose, S. and Okazaki, R., Proc. Nat. Acad. Sci. (USA) 69, 1863 (1972).
16. Okazaki, R., Sugino, A., Hirose, S., Okazaki, T., Imae, Y., Kainuma-Kuroda, R., Ogawa, T., Arisawa, M. and Kurosawa, Y. In DNA Synthesis In Vitro (Wells, R. D. & Inman, R. B., eds.), pp. 83-104, University Park Press, Baltimore (1973).
17. Hirose, S., Okazaki, R. and Tamanoi, F., J. Mol. Biol. 77, 501 (1973).
18. Sugino, A. and Okazaki, R., Proc. Nat. Acad. Sci. (USA) 70, 88 (1973).
19. van de Sande, J. H., Kleppe, K. and Khorana, H. G., Biochemistry 12, 5050 (1973).
20. Kuempel, P. L. and Veomett, G. E., Biochem. Biophys. Res. Commun. 41, 973 (1970).
21. Okazaki, R., Arisawa, M. and Sugino, A., Proc. Nat. Acad. Sci. (USA) 68, 2954 (1971).
22. Konrad, E. B. and Lehman, I. R., Proc. Nat. Acad. Sci. (USA) 71, 2048 (1974).
23. Gottesman, M. M., Hicks, M. L. and Gellert, M., J. Mol. Biol. 77, 531 (1973).
24. Konrad, E. B., Modrich, P. and Lehman, I. R., J. Mol. Biol. 77, 519 (1973).

25. Kelly, R. B., Cozzarelli, N. R., Deutscher, M. P., Lehman, I. R. and Kornberg, A., J. Biol. Chem. <u>245</u>, 39 (1970).
26. Westergaard, O., Brutlag, D. and Kornberg, A., J. Biol. Chem. <u>248</u> (1973).
27. Roychoudhury, R., J. Biol. Chem. <u>248</u>, 8465 (1973).
28. Okazaki, R., Hirose, S., Okazaki, T., Ogawa, T. and Kurosawa, Y., Biochem. Biophys. Res. Commun. <u>62</u>, 1018 (1975).
29. Klein, A. and Bonhoeffer, F., Ann. Rev. Biochem. <u>41</u>, 301 (1972).
30. Schnös, M. and Inman, R. B., J. Mol. Biol. <u>55</u>, 31 (1971).
31. Sunshine, M. G., Thorn, M., Gibbs, W. and Calendar, R., Virology <u>46</u>, 691 (1971).
32. Lindahl, G., Virology <u>46</u>, 620 (1971).
33. Lindqvist, B. H. and Bøvre, K., Virology <u>49</u>, 690 (1971).
34. Eisenberg, S. and Denhardt, D. T., Proc. Nat. Acad. Sci. (USA) <u>71</u>, 984 (1974).
35. Lark, K. G., Nature New Biol. <u>240</u>, 237 (1972).
36. Schekman, R., Weiner, A. and Kornberg, A., Science <u>186</u>, 987 (1974).
37. Gefter, M. L., Hirota, Y., Kornberg, T., Wechsler, J. A. and Barnoux, C., Proc. Nat. Acad. Sci. (USA) <u>68</u>, 3150 (1971).
38. Wickner, W. and Kornberg, A., J. Biol. Chem. <u>249</u>, 6244 (1974).
39. Boezi, J. A. and Armstrong, R. L., In Methods in Enzymology (Grossman, L. & Moldave, K., eds.), vol. 12A, pp. 684-686, Academic Press, New York and London (1967).
40. Mirzabekov, A. D. and Griffin, B. E., J. Mol. Biol. <u>72</u>, 633 (1972).
41. Kurosawa, Y., Ogawa, T., Hirose, S., Okazaki, T. and Okazaki, R., J. Mol. Biol. in press.
42. Kurosawa, Y. and Okazaki, R., J. Mol. Biol. <u>94</u>, 229 (1975).
43. Okazaki, R., In DNA Replication (Methods in Molecular Biology vol. 7), (Wickner, R. B. ed.), pp. 1-32, Marcel Dekker, Inc., New York (1974).
44. Calendar, R., Lindqvist, B., Sironi, G. and Clark, A. J., Virology <u>40</u>, 722 (1970).
45. Denhardt, D. T. and Sinsheimer, R. L., J. Mol. Biol. <u>12</u>, 641 (1965).
46. Schekman, R. W. and Ray, D. S., Nature New Biol. <u>231</u>, 170 (1971).

LIST OF CONTRIBUTORS

Aaronson, Stuart A, Viral Carcinogenesis Branch, National Cancer Institute, Building 37 — Room 2D24, Bethesda, Maryland 20014

Alberts, Bruce (Senior Author), Department of Biochemical Sciences, Princeton University, Frick Chemical Laboratory, Princeton, New Jersey 08540

Appelbaum, Edward R., Laboratory of Molecular Biology, 1525 Linden Drive, University of Wisconsin, Madison, Wisconsin 53706

Arnberg, Annika A., Biochemical Laboratory, The University, Zernikelaan, Groningen, The Netherlands

Berk, Arnold J., Department of Pathology, Stanford University, School of Medicine, Stanford, California 94305

Berger, Nathan A. (Senior Author), Washington University Medical Center and The Jewish Hospital of St. Louis, St. Louis, Missouri 63110

Bittner, Michael, Department of Biochemical Sciences, Princeton University, Frick Chemical Laboratory, Princeton, New Jersey 08540

Blair, D.G., Department of Biology, University of California, San Diego, P.O. Box 109. La Jolla, California 92037

Borst, Piet (Senior Author), Section for Medical Enzymology and Molecular Biology, Laboratory of Biochemistry, Jan Swammerdam Institute, 1e Constantijn Huygensstraat 20, Amsterdam, The Netherlands

Bouche, Jean-Pierre, Department of Biochemistry, Stanford University, Stanford, California 94305

Bowes, J. Michael (Senior Author), Department of Bacteriology, University of California, Davis, California 95616

Brewer, Eugene N., Division of Radiation Biology, Wearn Research Building, Case Western Reserve University, Cleveland, Ohio 44106

Brown, Neal, Department of Pharmacology, University of Massachusetts Medical School, Worcester, Massachusetts

Burrell, Ann, Department of Biological Sciences, Stanford University, Stanford, California 94305

Carl, Philip L. (Senior Author), Department of Microbiology, 131 Burrill Hall, University of Illinois, Urbana, Illinois 61801

Caro, L., Département de Biologie Moléculaire, 30, Quai Ernest-Ansermet, Ch — 121 — Geneve 4, Switzerland

Carroll, Lynn E., Department of Molecular, Cellular and Developmental Biology, University of Colorado, Boulder, Colorado 80302

Champoux, James J. (Senior Author), Department of Microbiology SC-42, School of Medicine, University of Washington, Seattle, Washington 98195

Chandler, M. (Senior Author), Département de Biologie Moléculaire 30, quai Ernest-Ansermet, CH — 1211 — Genève 4, Switzerland

Clayton, David A. (Senior Author), Department of Pathology, Stanford University School of Medicine, Stanford, California 94305

Collins, J., Department of Biology, University of California, San Diego, P.O. Box 109, La Jolla, California 92037

Cooper, Priscilla, Department of Biological Sciences, Stanford University, Stanford, California 94305

Cozzarelli, Nicholas R. (Senior Author), Department of Biochemistry, The University of Chicago, 920 East 58th Street, Chicago, Illinois 60637

Dankwarth, Ludolf (Co-Author), Max-Planck-Institut Für Molekulare Genetik, Berlin 33 Dahlem, Ihnestr. 63-73, Germany

Das, Jyotirmoy, Department of Microbiology, University of Rochester Medical Center, Rochester, New York 14642

Denhardt, David T. (Senior Author), Department of Biochemistry, McGill University, McIntyre Medical Sciences Building, 3655 Drummond Street, Montreal, Quebec, Canada H3G 1Y6

Donegan, James J., Department of Biochemistry, State University of New York, Stony Brook, New York 11794

Dorson, John W., Baylor College of Medicine, Marrs McLean Department of Biochemistry, Texas Medical Center, Houston, Texas 77025

Drlica, Karl, Department of Biochemical Sciences, Princeton University, Princeton, New Jersey 08540

Dueber, Jeanene, Department of Biology, University of California, Los Angeles, California 90024

Durnford, Joyce M., Department of Microbiology, School of Medicine, University of Washington, Seattle, Washington 98195

Eisenberg, Shlomo, Department of Biochemistry, McGill University, McIntyre Medical Sciences Building, 3655 Drummond Street, Montreal, Quebec, Canada H3G 1Y6

Evans, Helen H. (Senior Author), Division of Radiation Biology, Wearn Research Building, Case Western Reserve University, Cleveland, Ohio 44106

Evans, Thomas E., Division of Radiation Biology, Wearn Research Building, Case Western Reserve University, Cleveland, Ohio 44106

Fong Liu, Leroy, Department of Chemistry, University of California, Berkeley, California 94720

Fox, Richard M., Department of Medicine, Monash University Medical School, Prince Henry's Hospital, St. Kilda Road, Melbourne, Victoria 3004, Australia

Fujiyama, Asao, Institute of Molecular Biology, Faculty of Science, Nagoya University, Chikusa-ku, Nagoya, Japan 464

Funderburgh, Martha, Département de Biologie Moléculaire, 30, quai Ernest-Ansermet, CH − 1211 − Genève 4, Switzerland

Gefter, Malcolm L. (Senior Author), Department of Biology, Massachusetts Institute of Technology, Room 56-705, Cambridge, Massachusetts 02139

Gibson, Wade, Tumor Virology Laboratory, P.O. Box 1809, The Salk Institute, San Diego, California 92112

Godson, G. Nigel, Department of Radiobiology, Yale University School of Medicine, New Haven, Connecticut 06510

Goldbach, Rob W., Section for Medical Enzymology and Molecular Biology, Laboratory of Biochemistry, Jan Swammerdam Institute, 1e Constantijn Huygensstraat 20, Amsterdam, The Netherlands

Goto, Nobuichi, Department of Microbiology, School of Medicine, Tokyo Medical and Dental University, Tokyo 113, Japan

Goulian, Mehran, Room 5062, Basic Science Building, Division of Hematology, Department of Medicine, University of California, San Diego, La Jolla, California 92037

Grandgenett, Duane P. (Senior Author), St. Louis University School of Medicine, Institute for Molecular Virology, 3681 Park Avenue, St. Louis, Missouri 63110

Greenstein, M., Roche Institute of Molecular Biology, Nutley, New Jersey 07110

Griffith, Jack D., Department of Biochemistry, Stanford University School of Medicine, Stanford, California 94305

Grossman, Lawrence (Senior Author), Department of Biochemical and Biophysical Sciences, School of Hygiene and Public Health, The Johns Hopkins University, Baltimore, Maryland 21205

Guiney, D.G., Department of Biology, University of California, San Diego, P.O. Box 109, La Jolla, California 92037

Hanawalt, Philip (Senior Author), Department of Biological Sciences, Stanford University, Stanford, California 94305

Hanna, Michael H., Department of Biology, Guyot Hall, Princeton University, Princeton, New Jersey 08540

Haralson, Michael A., Department of Biochemistry, Baylor College of Medicine, 1200 Moursund Avenue, Houston, Texas 77025

Harbers, Barbara, Department of Biochemistry, McGill University, McIntyre Medical Sciences Building, 3655 Drummond Street, Montreal, Quebec, Canada H3G 1Y6

Hayes, Sidney (Senior Author), Department of Molecular Biology and Biochemistry, University of California, Irvine, California 92664

Hecht, Ralph, Department of Biophysics and Genetics, University of Colorado Medical Center, 4200 East Ninth Avenue, Denver, Colorado 80220

Helinski, Donald R. (Senior Author), Department of Biology, University of California, San Diego, P.O. Box 109, La Jolla, California 92037

Hershfield, V., Department of Biology, University of California, San Diego, P.O. Box 109, La Jolla, California 92037

Hinkle, David C., Department of Biology, University of Rochester, Rochester, New York 14627

Hirose, Susumu, Institute of Molecular Biology, Faculty of Science, Nagoya University, Chikusa-ku, Nagoya, Japan 464

Hsu, Pei-Ling, Department of Microbiology, University of Illinois Medical Center, Chicago, Illinois 60680

Huberman, Joel A. (Senior Author), Department of Biology, Room 56-639, Massachusetts Institute of Technology, Cambridge, Massachusetts 02139

Hung, P.P. (Senior Author), Experimental Biology, Abbott Laboratories, North Chicago, Illinois 60064

Hurwitz, Jerard, Developmental Biology and Cancer, Albert Einstein College of Medicine, 1300 Morris Park Avenue, Bronx, New York 10461

Imada, Sumi, Department of Molecular, Cellular and Developmental Biology, University of Colorado, Boulder, Colorado 80302

Inman, Ross B. (Senior Author), Biophysics Laboratory, 1525 Linden Drive, University of Wisconsin, Madison, Wisconsin 53706

Johnson, Elizabeth S., Washington University Medical Center and The Jewish Hospital of St. Louis, St. Louis, Missouri 63110

Kakefuda, T., Chemistry Branch, National Cancer Institute, Bethesda, Maryland 20014

Katz, L., Department of Biology, University of California, San Diego, P.O. Box 109, La Jolla, California 92037

Kavenoff, Ruth, Department of Chemistry, University of California, San Diego, La Jolla, California 92037

Kemper, Catherine, Department of Biology, University of California, San Diego, La Jolla, California 92037

Khachatourians, George, Department of Microbiology, University of Saskatchewan, Saskatoon, Saskatchewan, Canada

Knopf, Karl-Werner, Roche Institute of Molecular Biology, Nutley, New Jersey 07110

Kobayashi, Midori, Cancer Institute (Japanese Foundation for Cancer Research), Kami-Ikebukuro, Toshimaku, Tokyo 170, Japan

Kohara, Yuji, Institute of Molecular Biology, Faculty of Science, Nagoya University, Chikusa-ku, Nagoya, Japan 464

Koike, Katsuro (Senior Author), Cancer Institute (Japanese Foundation for Cancer Research), Kami-Ikebukuro, Toshimaku, Tokyo 170, Japan

Konrad, E. Bruce (Postdoctoral Fellow of the Jane Coffin Childs Memorial Fund for Medical Research), Building 4, Room 105, National Institutes of Health, Bethesda, Maryland 20014

Kornberg, Arthur (Senior Author), Department of Biochemistry, Stanford University, Stanford California 94305

Kornberg, Thomas (Senior Author), Department of Biochemical Sciences, Princeton University, Princeton, New Jersey 08540

Kupersztoch-Portnoy, Y., Department of Biology, University of California, San Diego, P.O. Box 109, La Jolla, California 92037

Kurosawa, Yoshikazu, Institute of Molecular Biology, Faculty of Science, Nagoya University, Chikusa-ku, Nagoya, Japan 464

Laipis, Philip J. (Senior Author), Box 724, J. Hillis Miller Health Center, University of Florida, Gainesville, Florida 32610

Lane, David, Department of Biochemistry, McGill University, McIntyre Medical Sciences Building, 3655 Drummond Street, Montreal, Quebec, Canada H3G 1Y6

Leavitt, R.W., Department of Biology, University of California, San Diego, P.O. Box 109, La Jolla, California 92037

Lehman, I.R. (Senior Author), Chairman, Department of Biochemistry, Stanford University School of Medicine, Stanford, California 94305

Leighton, Terrance (Senior Author), Department of Bacteriology and Immunology, 3573 Life Science Building, University of California, Berkeley, California 94720

Levine, Arnold J., Department of Biochemistry, Moffett Labs, Princeton University, Princeton, New Jersey 08540

Ling, C.M., Experimental Biology, Abbott Laboratories, North Chicago, Illinois 60064

Lovett, M.A., Department of Biology, University of California, San Diego, P.O. Box 109, La Jolla, California 92037

Low, Robert L., Department of Biochemistry, The University of Chicago, 920 East 58th Street, Chicago, Illinois 60637

Mace, David, Department of Biochemical Sciences, Princeton University, Frick Chemical Laboratory, Princeton, New Jersey 08540

Machida, Yasunori, Institute of Molecular Biology, Faculty of Science, Nagoya University, Chikusa-ku, Nagoya, Japan 464

Maniloff, Jack (Senior Author), Department of Microbiology, University of Rochester Medical Center, Rochester, New York 14642

Mao, J.C., Experimental Biology, Abbott Laboratories, North Chicago, Illinois 60064

Mark, David, Department of Biological Chemistry, Harvard Medical School, 25 Shattuck Street, Boston, Massachusetts 02115

Masker, Warren, Biology Division, Oak Ridge National Laboratory, Oak Ridge, Tennessee 37830

McFadden, Grant, Department of Biochemistry, McGill University, McIntyre Medical Sciences Building, 3655 Drummond Street, Montreal, Quebec, Canada H3G 1Y6

Messer, Walter (Senior Author and Co-Author), Max-Planck-Institut Für Molekulare Genetik, Berlin 33 Dahlem, Ihnestr. 68-73, Germany

Minkoff, Robert, University of California Dental School, Los Angeles, California 90024

Modrich, Paul, Department of Chemistry, University of California, Berkeley, California 94720

Molineux, Ian J., Imperial Cancer Research Fund, Burtonhole Road, London NW7 1AD, England

Moran, Laurence, c/o Dr. A. Tissières, Université de Genève, Institut de Biologie Moléculaire, 24 Quai de L'Ecole de Médecine, 1211 Genève 4, Switzerland

Morris, C. Fred, Department of Biochemical Sciences, Princeton University, Frick Chemical Laboratory, Princeton, New Jersey 08540

Moses, Robb E. (Senior Author), Baylor College of Medicine, Marrs McLean Department of Biochemistry, Texas Medical Center, Houston, Texas 77025

Mulder, Carel, Cold Spring Harbor Laboratories, P.O. Box 100, Cold Spring Harbor, New York 11724

Oertel, Wolfgang, Institut für Mikrobiologie der Universität Würzburg, 8700 Würzburg, Röntgenring 11, West Germany

Ogawa, Tohru, Institute of Molecular Biology, Faculty of Science, Nagoya University, Chikusa-ku, Nagoya, Japan 464

Okazaki, Reiji (Senior Author), Institute of Molecular Biology, Faculty of Science, Nagoya University, Chikusa-ku, Nagoya, Japan 464

Okazaki, Tuneko, Institute of Molecular Biology, Faculty of Science, Nagoya University, Chikusa-ku, Nagoya, Japan 464

Overby, L.R., Experimental Biology, Abbott Laboratories, North Chicago, Illinois 60064

Pauli, Andrew, Department of Biology, Massachusetts Institute of Technology, Room 56-708, Cambridge, Massachusetts 02139

Peebles, Craig L., Department of Biochemistry, The University of Chicago, 920 East 58th Street, Chicago, Illinois 60637

Perlman, Daniel, Department of Biology, Massachusetts Institute of Technology, Cambridge, Massachusetts 02138

Pettijohn, David (Senior Author), Department of Biophysics and Genetics, University of Colorado Medical Center, 4200 East Ninth Avenue, Denver, Colorado 80220

Rashbaum, Stephan A., Department of Biochemistry, The University of Chicago, 920 East 58th Street, Chicago, Illinois 60637

Ray, Dan S. (Senior Author), Department of Biology and Molecular Biology Institute, University of California, Los Angeles, California 90024

Reichard, Peter, Medical Nobel Institute, Department of Biochemistry, Karolinsha Institute, Stockholm, Sweden

Reiter, Harvard (Senior Author), Department of Microbiology Medical Center, University of Illinois, Chicago, Illinois 60680

Reuben, R., Roche Institute of Molecular Biology, Nutley, New Jersey 07110

Richardson, Charles C. (Senior Author), Department of Biological Chemistry, Harvard Medical School, 25 Shattuck Street, Boston, Massachusetts 02115

Roufa, Donald J. (Senior Author), Assistant Professor of Biochemistry and Medicine, Department of Biochemistry, Baylor College of Medicine, Moursund Avenue, Houston, Texas 77025

Rownd, Robert H. (Senior Author), Laboratory of Molecular Biology, 1525 Linden Drive, University of Wisconsin, Madison, Wisconsin 53706

Ryder, Oliver A. (Senior Author), Department of Biology, University of California, San Diego, La Jolla, California 92037

Sato, S., Department of Biology, University of California, San Diego, P.O. Box 109, La Jolla California 92037

Schekman, Randy, Department of Biology, University of California, San Diego, La Jolla, California 92037

Schindler-Tippe, Ruth (Co-Author), Max-Planck-Institut Für Molekulare Genetik, Berlin 33 Dahlem, Ihnestr. 63-73, Germany

Seki, Tetsunori, Institute of Molecular Biology, Faculty of Science, Nagoya University, Chikusa-ku, Nagoya, Japan 464

Sen, Arup, Meloy Laboratories, Inc., Springfield, Virginia 22151

Sherman, Linda, Department of Biology, Massachusetts Institute of Technology, Room 56-708, Cambridge, Massachusetts 02139

Shinozaki, Kazuo, Institute of Molecular Biology, Faculty of Science, Nagoya University, Chikusa-ku, Nagoya, Japan 464

Sinha, Navin, Department of Biochemical Sciences, Princeton University, Frick Chemical Laboratory, Princeton, New Jersey 08540

Skalka, A. (Senior Author), Roche Institute of Molecular Biology, Nutley, New Jersey 07110

Smith, Douglas W. (Senior Author), Department of Biology, University of California, San Diego, La Jolla, California 92037

Soucek, L. Stephanie, Department of Microbiology, 131 Burrill Hall, University of Illinois, Urbana, Illinois 61801

Spadari, Silvio, Laboratorio Di Genetica Biochimica Ed Evoluzionistica, Via S. Epifanio, 14, 2710 Pavia, Italy

Stephenson, John R., Viral Carcinogenesis Branch, National Cancer Institute, Building 37 — Room 2D24, Bethesda, Maryland 20014

Sternglanz, Rolf (Senior Author), Department of Biochemistry, State University of New York, Stony Brook, New York 11794

Sueoka, Noboru (Senior Author), Department of Molecular, Cellular and Developmental Biology, University of Colorado, Boulder, Colorado 80302

Sugino, Akio, Institute of Molecular Biology, Faculty of Science, Nagoya University, Chikusa-ku, Nagoya, Japan 464

Szybalski, Waclaw (Senior Author), McArdle Laboratory for Cancer Research, University of Wisconsin, Madison, Wisconsin 53706

Tamanoi, Fuyuhiko, Institute of Molecular Biology, Faculty of Science, Nagoya University, Chikusa-ku, Nagoya, Japan 464

Tanaka, Shigeaki, Cancer Institute (Japanese Foundation for Cancer Research), Kami-Ikebukuro, Toshimaku, Tokyo 170, Japan

Tronick, Steven R. (Senior Author), Viral Carcinogenesis Branch, National Cancer Institute, Building 37, Room 2D24, Bethesda, Maryland 20014

Tsai, Alice, Department of Biology, Room 56-639, Massachusetts Institute of Technology, Cambridge, Massachusetts 02139

Tseng, Ben Y. (Senior Author), Room 5062, Basic Science Building, Division of Hematology, Department of Medicine, University of California, San Diego, La Jolla, California 92037

Valenzuela, Manuel S., Graduate Department of Biochemistry, Brandeis University, Waltham, Massachusetts 02154

Van Beveren, Charles P. (Senior Author), Department of Medicine, M-013, University of California, San Diego, La Jolla, California 92037

Van Bruggen, Ernst F.J., Biochemical Laboratory, The University, Zernikelaan, Groningen, The Netherlands

Verma, Inder M. (Senior Author), Tumor Virology Laboratory, The Salk Institute, P.O. Box 1809, San Diego, California 92112

Vinograd, Jerome (Senior Author), Division of Chemistry and Chemical Engineering, California Institute of Technology, 1201 East California Boulevard, Pasadena, California 91125

Vosberg, Hans-Peter, Division of Biology, California Institute of Technology, 1201 East California Boulevard, Pasadena, California 91125

Wake, R.G. (Senior Author), Biochemistry Department, University of Sydney, N.S.W., 2006, Australia

Wang, Helen F., Department of Biochemistry, State University of New York, Stony Brook, New York 11794

Wang, James C. (Senior Author), Department of Chemistry, University of California, Berkeley, California 94720

Waqar, M. Anwar, Department of Biology, Room 56-639, Massachusetts Institute of Technology, Cambridge, Massachusetts 02139

Wechsler, James A. (Senior Author), Department of Biological Sciences, Columbia University, New York, New York 10027

Weiner, Joel, Department of Biochemistry, Stanford University, Stanford, California 94305

Weissbach, Arthur (Senior Author), Roche Institute of Molecular Biology, Department of Cell Biology, Nutley, New Jersey 07110

Wickner, Sue (Senior Author), Building 37, Room 2C09, National Cancer Institute, National Institutes of Health, Bethesda, Maryland 20014

Williams, P.H., Department of Biology, University of California, San Diego, La Jolla, California 92037

Womack, John E. (Senior Author and Co-Author), Max-Planck-Institut Für Molekulare Genetik, Berlin 33 Dahlem, Ihnestr. 63-73, Germany

Worcel, A. (Senior Author), Department of Biochemical Sciences, Princeton University, Princeton, New Jersey 08540

Wright, Andrew, Department of Molecular Biology and Microbiology, Tufts University School of Medicine, 136 Harrison Avenue, Boston, Massachusetts 02111

Zahn, Gabriele (Co-Author), Max-Planck-Institut Fur Molekulare Genetik, Berlin 33 Dahlem, Ihnestr. 63-73, Germany

Zdzienicka, Malgorzata, Department of Biological Sciences, Columbia University, New York, New York 10027

Zechel, Kasper, Max-Planck-Institut Für Biochysikalische Chemie, D-3400 Göttingen, Postfach 968, Germany

Zyskind, Judith W., Department of Biology, University of California, San Diego, La Jolla, California 92037

SUBJECT INDEX

Italicized numbers refer to illustrative matter, figures, tables, plates.

A conformation, 420
A-DNA, 419-420
abortive repair, 793
acriflavin, 794
actinomycin D, 725
alkaline phosphatase, 273-274
amino acid starvation, 162, 165-169
antipodal position, 663
araCTP, 725, 780
arylazopyrimidines, 688, 690
arylhydrazinopyrimidines, 14
    inhibition of, 24, 27-35
Aspergillus orizae, 49
assembly and maturation, 449
ATP, 774, 780-*782*
    dependent repair synthesis, 780,
      822-825
ATP, γ-phosphate, 831-832, 842
ATPase, DNA dependent, 227, 233,
    235-237, 241, 244, 266
autoradiographic studies, 658-664, 667
avian myeloblastosis virus (AMV), 742

B-DNA, 419-420
Bacillus subtilis
    bidirectional replication in, 650-674
    DNA membrane complex, 187-198
    pulse labeling studies, 309-320
    sporulation, 677-686
"back transition," 547
bacteria, *see Bacillus subtilis, Escherichia*
    coli, Micrococcus luteus
bacterial cells
    DNA packaging in, 138
    semiconservative DNA replication,
      677-686
    sporulation, 678-686
bacterial chromosomes, replication, 652
bacterial nucleoids (*E. coli*), 122

bacterial replicons, 547
bacteriophage, *see* phages
beads, 201, *203*
βgalactosidase, 273-*274*, 292
βprotein, 475
bidirectional replication, 509
    in *Bacillus subtilis*, 650-674
    complete, 651
    consequences of, 670-674
    extent of, 668-670
    partial, 651, 665, 667-669
    recent studies, 664-670
binding ability, 760
binding protein, *228*-231, 236
    complex with pol II, 5-10
    effect on eukaryotic DNA poly-
      merases, 75-81
    interaction with DNA, 3-13
    role in DNA replication, 2-13
binding site, DNA-RNA hybrid, 760
branch migration, *456-458*

C(3')-endo conformation, 419
C(3')-exo conformation, 419
caffeine, 794
catenated dimers, 561, 567-571, 574-575
catenated obligometric forms, 561, 563,
    572
cdd genes, 619-623
cell division
    conventional, 677
    normal developmental sequence, 678
    post exponential, 677, 679-681,
      683-685
        and semiconservative repli-
          cation, 677, 684
cell extracts, 270-271, 273, 288
cell-free protein synthesizing extracts, 702
cell growth rates, 695-700

871

Chinese hamster lung (CHL) cell mutants,
702-711
see also TS-14 mutant
chloramphenicol, 275-276, 447, 507,
518, 603, 609-610, 640, 643, 645
CHO cells, 341-347, *350, 351-353*
chromatography, 739
chromosomes
condensed,
domains of DNA coding in,
123
stabilization of, 122
eukaryotic, 201
folded,
aggregation of, 147-148
DNA in, 123
effects of treatment, 138-153
membrane associated, 159-184
membrane free, 139, 162
supercoils in, 129
-membrane complex, 187-188
with multiple origins of replication, 556-557
mutant, growing forks, 414-418
reinitiated (dichotomous) 657
replication rounds, 159
initiation of, 160-161, 175-181
segregation of, 671
shear stress responses, 141-143
terminalized, 178
unfolded, 138-157
circular duplex DNA, 514
see also plasmids
circularization, 593, *597*
cI-rex transcript, 487
closed circular molecules, 94, 560, 563,
567, 575
covalent, 467
replicating, 578, 580
co-DNA polymerase III*, 297-298
cold sensitivity, 629, 635
see also temperature sensitivity
ColEI replication, 517
cloning and amplification of DNA
segments, 520
competition, 410
annealing, 410
experiments, 760
-hybridization, 411
complementary strand, 370-384, 404
concatemers, 462
copolymerase III, 210, 212, 214, 221
correndonuclease, 791, 793
E. coli UV, 794
general radiation, 795

correndonuclease II
in *E. coli*, 793, 794, 796
and T4 phage, 795-796
correxonucleases, 801
crosslinking, 793
agents, 794
cross-strand exchange, 457
cryolethal suppressors, 626, 630, 635
reversion of, *631*
CsCl gradients, 341, 343, *348-349*
CsCl-sucrose double gradient, 187-189,
191
Cs_2SO_4 gradients, 336-343, *348-349*, 498
cycloheximide, 702, 713-718
cytidine (deoxycytidine) deaminase, 619
see also cdd genes
cytidine-triphosphate synthetase
see pyrG genes
Cytosine Arabinoside Triphosphate, 719
cytosol fraction, 325

Dane particles, 776-772
DEAE-Dextran treated cells, 323-324, 329,
331
DEAE-Sephadex, 737
degredation
of concatemers, 473
of parental strand, post UV, 784, 786
rate of, 469
denaturation mapping, 452, 537
of composite *NR1* DNA, 544
de novo DNA chain initiation, 241, 256-
258, 265
deoxynucleoside triphosphate level, 660
deoxyribonucleoside triphosphate, 780,
784, 787
dichotomous (reinitiated) chromosomes,
657
dimers
catenated, 561, 567-571, 574-575
thymine containing, 800
discontinuous nascent DNA, 324-*326*
discontinuous replication, 578
in both strands, 848-852
of P2 phage DNA, 845
of ϕX174 phage DNA, 852
discontinuous synthesis, 357, 359, 408,
417
displacement loops (D-loops), 505, 587
role of, 588-590
DNA
chain growth, 325, 327-331
de novo initiation, 241, 256-
258, 265
chromosomal, 138
closed circular, 94, 467, 560, 563,
567, 575, 578, 580

damaged, 793
double strand replicative intermediates, 446, 449
double strand templates, 241, *249*-250, 255-265
elongation factors I and II, *228*-231, 236-237
endonuclease, 273-274, 278
extrachromosomal, 538
fold, 123
heteroduplex, irradiated:unirradiated, 794
initiation, RNA primed, 357
ligase, 296, 298-*299*, 304-306
        mutants, 305-306
maturation, 483
membrane complex, 187-198
mitochondrial, 561-575, 578-584
molecules, branched, 586
nascent chain, 402
negative superhelical, 38-39
        and loop enlargement, 60
        preparation of, 42-48
packaged, 129
R plasmid, 537
        transition of, 540-543
regulation mechanisms, 702-711
renaturation of, 775
repair and UV radiation, 815-818
repetitive, 401
replication
        correctly scheduled, 678
        factors x, y and z, *229*-231, 233, *235*-237
        mammalian, 702-711
        Okazaki fragments in, 304-307
        regulatory coordinating mechanisms, 702-703, 706-709
        semiconservative, 643, 645, 677-686
        swivel for, 83
        see also replication
replicative form (RF), 425
        see also replication, replicative form
strand elongation and cycloheximide, 713-718
transforming, 806
x-irradiated, 795
        see also A-DNA, B-DNA, circular duplex DNA, discontinuous nascent DNA, DNA synthesis, excision-repair of DNA, linear DNA, mammalian DNA,

mtDNA, polyoma DNA, R plasmid DNA
DNA-DNA hybridization, 404, *415*, 578, 579, 582, 695
        see also hybridization
DNA-RNA complexes, 133
DNA-RNA copolymer, 603, 608
DNA synthesis
        asymmetric, 384
        bidirectional, 399, *416*
        continuous, 408, 418
        discontinuous, 241, 250, 260, 265, 309-320, 324, 357, 359, 408, 417
        double stranded origins, 386, 394
        eukaryotic polymerases in, 79-81
        initiation, 72, 713
        in permeabilized Mouse L cells, 719-727
        in permeabilized TK⁻ cells, 723
        in vitro, human lymphocytes, 322-332
        kinetics of inhibition, 725-726
        rate of, 469, *472*
        replicative, ATP requirement, 725
        RNA primed, 72-75, 327-332, 335
        semiconservative, in permeable cell systems, 719-727, 822
        single stranded, origins, 386, 390-394
        unidirectional, 399, *400*, 418
*dna*A mutants, 412, 431, 517, 624-637
        regulation of o-RNA, 602-615
        suppressor, 625, 629-630, 636
*dna*B mutant (product), 210-211, *215*, *223*-224, 227-237, 306, 412, 423-442, 780, 783
*dna*C(D) mutant (product), 159, 173, 175, 184, 210-212, *215*-216, 223-224, 227-237, 412, 414, 431, 517, 602, 640-643
*dna*E mutant (product), 210-211, 214, 227, 230, *412*, 414, 431
        *see also pol*C mutant
*dna*F mutant (product), 517
*dna*G mutant (product), 210-212, *215*-224, 227-231, 236, *412*, 414, 431, 517
*dna*S mutant, 304-307
*dna*S₁ mutant, 305-306
DNase IV, 803
double branched circles (θ structures), 539, 544
double strand breaks, 807
drug resistance gene amplification, 538
duplication inversion, 586, *592*

Dye-Agarose Gell electrophoresis, 579, 582

EcoRI restriction
    endonuclease, 553, 579, 582, 584
    enzymes, 478
EDTA lysozyme spheroplasts
    see lysozyme spheroplasts
EGTA lysozyme spheroplasts
    see lysozyme spheroplasts
electron microscopy, 248, 260, *262-263*, 579, 582-*583*
    and denaturation mapping, 452
    of denatured mtDNA, 586, *591*
    of interphase chromosomes, 202-205
    of membrane associated folded chromosomes, 159, 164, 169-173
    of replicating forms, 586
endonucleases
    and damaged DNA, 791
    digestion, 579, 582, 584
    general radiation, 795
    polymerase associated, 798-801
Escherichia coli,
    cell envelope, 270
    chromosome, 202, 304, 414
        in vitro replication, 296-301, 695
    correndonuclease II, 793
    DNA binding protein, 2-13
    dna mutant, 270, 285, 291
    DNA replication, 414
    *dna*C mutant, 640
    exonuclease III, 478
    isolated nucleoide of, 122
    *ligts*7, 840
    periplasmic factors, 8
        in vitro synthesis by, 270-293
    plasmid DNA replication in, 514-534
    *pol*A mutants, 291, 800-801
        *pol*AexI, 840
        *pol*A12, 840
    *pol*B mutants, 291
    pulse labeling studies, 309-320
    rep⁻ cells, 370, 372, 380-384
    repair response of, 775, *777*
    replication fork velocity, 695-700
    T4 V⁻ infected, 796
    tsnC mutants, 239
ethidium bromide
    chromosome relaxation, 143-145, *154-155*
    differential binding of, 87
    intercalation of, 122
    unwinding angle of, 45, 46, 84

ethylmethane sulfonates, 305
eukaryotic cells
    chromosome analogs, 556
    DNA polymerases of, 64, 75-79
    nomenclature, 65-67
    possible role in DNA replication, 64, 67, 323
    repair mechanisms, 809
    replication units, 419
    see also HeLa cells
excision-repair of DNA, 774, *777-778*, 791-809
    cycle, 791-*792*
    pol I associated pathways, 793
    pol III capabilities, 801
    role of *rec* genes in, 783-784
        see also repair
exonuclease
    DNA polymerase associated, 793, 798-801
    eukaryotic, 802
    intrinsic, 24-27
    processive, 742
    random, 742
    unassociated, 793, 801
exonuclease I, 2, 5, 7
exonuclease III, 478, 797
exonuclease V, 779
    see also recBC mutant
exonuclease VII, 779, 796, 802-804
exonuclease 3' →5', 798-799
    activity, 799-801
    pol associated, 475
    single stranded, 793
    unidirectionality of repair, 807
    UV survival, 784
exoribonuclease H activity, 760
exponential cells, 645-*646*
*exr*A genes, 784

fd phage, *8*, 10-*11*
feb⁻ conditions, 473
fec⁻ conditions, 462
    relaxed circles in, 478
"forced transition," 550
forespore chromosomes, 682
    asymmetric septation of, 677, 684

G4 phage DNA
    cleavage map, 386, 388, *395*
    double stranded origins, 386, 394
    physical map and replication, 386-396
    replication of, 210-224
    resolution of multienzymatic systems, 216-218
    single stranded origins, 386, 390-394
    structure and replication, 370-384

*gam*ts protein, 464
gap, *400*
    base composition of, 404-*405*
    -circle molecules, 364-368
    expansion, 808
    ϕXRFII, 401
    repair, 476
gene A protein, 419
gene amplification, 540
    mechanisms of, 546-547
gene 32 protein, 241-244, 251, *253-256*
gene 41 protein, 241-243, *253-256*
gene 43 protein, 241-246, *253-256*
gene 44 protein, 241-251, *253-256*
gene 45 protein, 241-246, *253-256*
gene 62 protein, 241-251, *253-256*
general recombination, 306
Go rabbit lymphocytes, 822
Go resting lymphocytes, 822
growing forks in mutant chromosomes,
    414, 418
GTP dependent synthesis, 781, 822-825

Harford-Lepesant map, 665-667
HeLa cells, 337, 341, 351, *353*
    see also eukaryotic cells
hepatitis B and rolling circular DNA, 766-
    772
"Hershey" circles, 478
Hfr strains, 306
HinII and III restriction enzymes, 478
histone F, 202
histones, and eukaryotic polymerases, 75-79
host modification and restriction, 447-449
HPUra, 688, 690, 693
human lymphocyte cells, 351, *353*
    in vitro study systems, 322-332
human placenta exonucleases, 802
hy42 DNA template, 505
hybridization, 309-310, 316-320, 411
    see also DNA-DNA hybridization
hydrolysis, 14, 24-28
hydroxylapatite, *26*
6-(p-hydroxyphenylazo)-uracil, *26*, 27
6-(p-hydroxyphenylhydrazino)-isocystocine
    sensitivity of *B.subtilis* pol III to, 15
    in ternary complex, 28-31, *33*
6-(p-hydroxyphenylhydrazino)-uracil
    sensitivity of *B.subtilis* pol III to, 15
    structure of, *26*, 27-28
    in ternary complex, 28-35
hydroxyurea (HU), 361, 364, *366-367*,
    725
"hyper rec" mutants, 304-306

I-RNA
    see o-RNA

incision, 791, 793
    UV-specific rates, 794
initiating primer RNA, 527
    see also primer RNA
initiation of DNA replication, 408, 537,
    645
    *dna*C mutant, 640
    multiple, *400*
    negative control of, 636
    possible membrane involvement, 637
    region, 776
    site, 578-584
"initiator," 487
initiator RNA (iRNA) 357-359
    see also primer RNA
intrastrand diadducts,
    see pyrimidine-pyrimidine dimers
intrastrand diadduct damaged DNA, 791-
    792
in vitro DNA repair system, 362-*363*
in vitro synthesis, 270-293
    ATP dependent, semiconservative,
        275-278, 280-283
    of closed circular molecules, 578,
        579
    effects of added cell extracts, 288-290
    of *E. coli* chromosomes, 296-301
    in human lymphocyte cells, 322-332
    lysozyme agar systems, 273-275
    periplasmic factor stimulation by
        toluenized cells, 280
    procedures, 271-272
    RNA, 643-*644*
    stimulation by osmotic shock fluid,
        278-280
ionophoresis/homochromatography, 404,
    406-407
irradiated poly (dA:dT), 799
isolated nuclei, 357

"kinetic proof reading" scheme, 266

L-cell mitochondrial DNA replication, 560-
    575
l strand λDNA template, 505
LA, 337
    small subunit α, 731, 739-742
    *tsLA*337PR-C(*LA337*) (24), 733
labeled species and activated BC nuclease,
    467
λb221cl26, 454
λb221cl26red270a42, 454-459
λDNA, cohesive ends of, 802
    replication, 452-459
        early stage, 462-*463*
        molecular mechanisms, 460-483
        turnoff of, 462

synthesis and *oop*RNA primer, 486-509
λexonuclease, 475
λgene *O*, λgene *P*, 487-491
λphage, 205, 306, 696-699
λred⁻, 306
large subunit *β*, 730, 741-742, 744
leukemia virus infection, 758
ligase, 399, *409*, 778-779, 797, 815
    defectives, 304-307
ligation of DNA fragments, 713, 716
ligts 7, 408, *409*
linear DNA, 588
    mature ends of, 481
*lit*RNA transcription, 486
long patch
    pathways, 793, 805, 807-809
    repair, 793
loop enlargement, 39, *44*, 46-48, *50*, 60
lysed nuclei, 822
lysozyme spheroplasts as *in vitro* synthesis
    systems, 273-*279*, *284*, *286-287*,
    289-290

M13 phage replication, 423-442
mammalian cells, 335
    density shift experiments, 341-346
    exonuclease in, 802
    nearest neighbor experiments, 347-
    353
mammalian DNA replication, 702-711
mammalian RNA type C virus, 757
marker frequencies, 414-*415*
membrane expansion, 671
methyl-methanesulfonate (MMS), 827
Micrococcus luteus, 791
    correndonuclease II, 795-797,
    801-804, 806
    UV endonuclease, 795
microinjection, 339
mitochondrial DNA replication, 578-584
    L-cell, 560-575
mitomycin C monofunctional analogue,
    794
monomeric circular molecules, 462
mother cell, asymmetric septation of, 677,
    684
Mouse L cells, 719-727
    permeabilized, DNA synthesis in,
    719-727
mtDNA
    evolution of sequences, 594
    in *Tetrahymena pyriformis*, 586-599
    nascent, 579
    replication, 578
Mu-1 phage, 696-699

multiforked forms, 657
murine sarcoma virus, 753-758
mycoplasma virus replication, 445-449

N. crassa nuclease, 411
    single stranded specific, *410*
nalidixic acid, 275-*276*, 296-298, 780
nascent chains, DNA bound, 123
nascent DNA, 325, *328*-331
    fragments, 304
nearest neighbor analysis, 336-*342*,
    347-353
nearest neighbor transfer, 327
negative control of initiation, 602, 610-
    615
"nick," *400*, 417, 419
    binding, 806
    translation, 39, 50, 61, 799
        conditions, 806
        mechanism, 820
nicking-closing protein, 94-96
    and replicating DNA, 95
        see also relaxation protein
nicotinimide mononucleotide (NMN), 815-
    820
nitrocellulose membrane filter techniques,
    761-763
nitrosoguanidine, 305
nonconservative replication, 775, 779
    see also repair, replication
nontransitioned state, 547
novel junction, 454
nucleases
    activated BC, 467
    polymerase associated, 5
nucleoids
    membrane associated, 162
    stabilization of, 134-135
nucleotide triphosphate, 774, 780-781
120 nucleotide fragments, 327
nucleotides, reinsertion of, 806

3'OH primer, 417
o-RNA, 602-615
    specific labeling in *E.coli*, 618-623
Okazaki pieces, 298, 300-301, 304-307,
    309-320, 323, 357, 359, 362-364,
    509, 528, 776
    in vitro synthesized, 296
ω untwisting enzyme, 83
    filter binding assay, 84-87
    fluorometric assay, 87
        see also relaxation protein, un-
        winding protein
*oop*RNA, 486-509
    OSA (*oop* specifying activity), 487
    transcription, 486

ori site, 486
    leftward transcription of, 487, *492*
origin
    region, 671-672
    of replication, 527, 537
    -RNA, *see* o-RNA
osmotic shock fluid, 270-293

P. mirabilis transitioned cells, 547
P2 phage DNA, 845
packing ratio, 202
palindrome, 586
pancreatic endonuclease I, 822
panhandle, 594
parental RF, 417
patches
    large, 779
    long, 783, 793, 807-809
    short, 783, 793, 805-807
    size, 779, 781, 787
    small, 779
peptide analysis, 737, 744-746
periplasmic factors, 270-271, 280,
    285, 288
periplasmic space, 271
phage *A* protein, 483
phage-host function interactions, 460-
    483
phages,
    see fd phage, G4 phage, λphage,
    M13 phage, Mu-1 phage, φX174
    DNA, P2 phage DNA, PM2 DNA,
    T4 phage, T7 phage
φX A gene, 417
φX cistron, *412*
φX-like G phages, 386-387
φX174 DNA, 227-237
    binding of protein n to, 214-216
    conversion to RF II, 210-224
    RF-replication, 399-*400*, *412*
φXRFII gaps, 401
    composition of, *405*
φXtB DNA, *228*, 230
phleomycin, 822-828
5'-phosphate of DNA, 831-832, 842
phosphocellulose columns, 737, 739
phospholipase A, 193
photlyases, 791, 794
    yeast, 795
photoreacting enzymes,
    see photolyases
Physarum polycephalum, 334-354
    cycloheximide and DNA strand
        elongation, 713-718
    RNA-DNA links in, 335
physical restriction enzyme cleavage map
    see restriction enzyme map

π protein (hypothetical), 401
    see also RNA polymerase
plasmids, 514-534
    colicinogenic, 517
    R6K, 517
PM2 DNA
    superhelical
        DNA primed, 41, *44*-60
        RNA primed, 41-42, *44*-60
pol I, 296, *299*, 305-307, *400*, 517,
    775-776, 784
    antibodies specific for, 38
    defectives, 304-307
    DNA synthesis by, 40, *44*-48, 52
    and endonucleolytic scission, 45,
        55-56, 61
    enzymatic activity, 2-6
    5' exonuclease of, 778, 786
    5' →3' exonuclease function, 487,
        774, 815, 820
    independent pathway, 779-783
    ligase repair system, 473
    *pol*A mutants, 800
    repair involvement, 815-820
pol II, 774, 778, 780, 787
    DNA binding protein interactions,
        2-13
    excision-reinsertion, 793, 807
    exonuclease activities, 2
pol III, 210, 212, 214, 221, *228*-231,
    236, *400*, 517, 774, 776, 778-780,
    787
    arylhydrazinopyrimidine inhibition,
        14, 27-35
    base specificity, 14, 28, 31, 33
    dissociation rate, 14, 31, 33
    enzymatic activity, 2, *4*, 5, *8*, 12
    excision-reinsertion, 793, 801, 807
    holoenzyme form, 212, 214-216,
        218-*219*, 223
    hydrodynamic properties, 14
    molecular weight, 14, 16, 21
    purification and properties, 16-21
    reactions catalyzed by, 21-27
    sedimentation coefficient, 14, *22*
    stokes radius, 14, 16, 21, *22*
pol A mutants, 401, *412*
    *pol*A1, 779
    *pol*A12, 781
    *pol*107, 786
    *pol*A*dna*E (*pol*C), 780
    *pol*Aex1, 784
    *pol*A*pol*B double mutant, 780
    repair synthesis in, 779
    T7 infected hosts, 239
pol α
    correlation with αβ polymerase, 760-765

possible role in DNA replication, 71-72

sedimentation behavior of, 71

pol αβ (AMV), 760-765

*pol*B gene, *412*

pol β, 822, 825, 827

sedimentation behavior of, 71

*pol*C mutant, 15, 412, 517

and arylhydrazinopyrimidine inhibitors, 27

cotransduction with *ura* and cysCl, 16, *18*

mapping by transduction, *18*

polarity, 411

poly-r-determinant, 537

polyacrylamide gel electrophoresis, 736

polyamines, 205, 807

polymerases, 368, 399, *400*, 420, 486

dissimilarity of, 68

DNA dependent, 731

immunological relationships, 68-71

replicative, competitive inhibition, 725-726

ribonucleotide incorporation, 21-24

RNA dependent, 731, 753

polynucleotide chain elongation, 361

polynucleotide kinase, 404, *405*

polynucleotide ligase, 399

temperature sensitive mutants, 796

3'-polynucleotidase, 797

polyoma DNA, 357-359

post-incision step, 793

primer RNA, 210, 251-252, 357, 402, 420, 596, 852

*dna*G protein synthesis of, 218-224

primer-template, 420

processiveness, 244, 248

prokaryotic system discontinuous replication, 831-859

promotors, 386

see also RNA polymerase, binding sites

protease, effect on folded chromosomes, 151-153

protein i, 210, 212-215, 224

protein n, 210, 212-216, 218-221, *223*

provirus, 731-*732*

hypothesis, 730, 746

psoralen plus light and crosslinking, 794

puromycin, 702

*pyr*E locus, 306-307

*pyr*G genes, 619-623

pyrimidine, 619

pyrimidine dimers, 784, 791, 795, 796

assay for, 778

repair of, 793

UV light induced, 774-776

pyrimidine-pyrimidine dimers, 791

pyrimidine tract, 404, *406-407*

complexity of, 418

pyrophosphate exchange, 14, *23*

pyrophosphorolysis, 14, 21, *23*

r-determinants

amplified segments of, 553

autonomous replication of, 540

multimeric forms of, 548

origin, 546

tandem copies of, 540

R plasmid DNA, 537

composite, 540

dissociation of, 540, 547

density profile, 542, 548

NR1, *550*

NR84, *550*

percentage, 550

replicating, 537, 543

initiation, 556

rabbit liver exonuclease, 802

randomly nicked circles, 478

rat liver mitochondria, 579

*rec*A function, 774, 778, 781, 783

*rec*A⁻ induction, 478

*rec*A*rec*B mutant, 779, 783

*rec*B, 781

*rec*BC mutant, 528, 779, 807-808

*rec*BC nuclease, 464, 774, 779, 781, 783

*rec*Bts, 808

recombinational intermediates, 457

mechanisms, 791

mode, 774

red-promoted recombination, 473

frequency of, 476

possible intermediate in, 476

reinserted regions, 800

reinsertion-excision, 806

reinsertion mechanisms, 793, 804-809

relaxation complex, 528

relaxation protein, 83, 94

enzymatic mode of action, 111-112

mechanisms, 113-118

properties of, 107-111

reaction with positive superhelical DNA, 112-115

see also nicking-closing protein, ω untwisting enzyme

relaxed circles, 467

released single strands, 325

rep mutant, *412*-414, *416*, 417

repair

nucleases, 420

patch lengths, 779

replication, 774-776

pol I independent, 774

synthesis, 822

replicating fork, 414, *416*, 464, 487

movement at terminus, 673
rate of movement, 660, *662*
velocity and cell growth rate, 695-700
replicating molecules, 544
replication,
    asymmetrical, 578, 580-582
    autonomous, 540
    bacterial chromosomes, 650
    bidirectional, 509, 650-674
    direction of, 538
    discontinuous, 578, 584, 831-859
    factors, 291-293
    initiation of, 537, 543, 546-550
    model, 586
    mutants, 517
    nickases, 419
    nonconservative, 775, 779
    of mitochondrial DNA, 560-575,
      578-584
    order map, *652-653*
    ordered, 651
    origins of, 187, 537-539, 556-557
    plasmid, control of initiation, 546-550
    reinitiation of, 657, 670-671
    replicative form, 370, 375, 410, 417
      425
    RF →RF, 372, 375, 378, 381, 384
    terminus, 187, 546
    time, 695, *697*-700
    unidirectional, 509, 654-655
    see also repair, replicating fork
replicative intermediates, 306, 446, 449,
    517, 587
    SV40, 361-362, 368
"replicator," 487
replicon
    bacterial, 547
    cis-dominant target site on, 487
resistance transfer factor (RTF), 537
    TC origin, 546
restriction endonuclease EcoR1, 364, 518
restriction enzyme, 404, 419
    map, 386, 388, *395*
reverse transcriptase
    in vitro radiolabeling of, 736, 742-
    744
    of AMV, 742
    role in RNA tumor virus life cycle,
      730-749
    thermolabile, 753-758
RF replication, *410*, 417
RF→RF replication, 272, 275, 378, 381,
    384
ribonuclease chromosome unfolding,
    138, 145-147, *152*, 155-156
ribonucleaseU (RNase H), 730-731
ribonuclease I, 275, 278, *287*

ribonucleotide incorporation, 21-24
rifampicin, *130*, 131, 133, 212, *217*,
    223, 275-*276*, 447, 493, 603, *609*,
    640, 643, 645, 834
RNA-DNA
    covalent linkage, 210, 211, 221,
      327-332, 845
    duplex, 419
    junction, 329-332, 339-341
    lability of, 329-*330*
RNA linked nascent pieces, 831
    in ϕX174 RF molecules, 852, *857*
    polynucleotide kinase analysis,
      842-845
    spleen exonuclease analysis, 836-842
    32P transfer analysis, 845
RNA molecules
    DNA bound, 131-134
      binding sites, 135
      as nascent chains, 123
      rifampicin and, *130*, 131, 133
    nucleoid bound, 123, 125-131
    and nucleoid stabilization, 135
RNA nucleotides, 329-332
RNA polymerase, 210-212, 216, *223*,
    402, 419
    βsubunit, 486
    binding sites, 386, 388-390, 395
    enzymatic activity, 2, 12
RNA synthesis *in vitro*, 643-*644*
RNA tumor viruses
    DNA polymerases of, 731
    role of reverse transcriptase, 730-749
    transformation of cells by, 731
rolling circle mechanism, 241, 256-261,
    414, 442-*463*
    hepatitis B associated, 766-772
Rous sarcoma virus (RSV), 730
    *LA*335, 730, 733
    *LA*337, 730, 733
    mutants, 733
      revertants and recombinants of,
      747
rU_6 sequence, 505

^{35}S-labeled protein
Salmonella typhimurium, 688-693
satellite band, 551
SDS-polyacrylamide gel electrophoresis,
    189, 193-*194*
sensitive transitional intermediates, 473
sheared fragments, 481
short patch repair, 793
    pathways, 805-807
single strand (SS-) circular DNA, 425, 445,
    449
single strand specific nuclease, 411

sister chromatid exchange, 791
small subunit α, 730, 739-742, 744
sonication and relaxation protein, 103
spermine, 807
spleen exonuclease, 831
    analysis of RNA linked nascent
        pieces, 836-842
ST-1 DNA^2, *228*, 230
stationary phase, 641, 645-*646*
sucrose step gradients, 187, 189, 191
sulfhydryl reagents, 725
supercoiling, 201
superhelix formation
    density of, 202
    free energy of, 38-39, 43
    relaxation protein, 83
surface protein iodination, 195
SV40 virus, 202, 361-368

T4 phage
    correndonuclease II, 795-796
    DNA polymerase, 402, 478
    pyrimidine dimer assay, 778
    replication aparatus, 241-267
    RNA attached nascent DNA pieces,
        852
    UV sensitivity of gene 32 protein,
        808
T7 phage, 775
    dimer specific endonuclease from,
        778
    DNA polymerase, 239-240
    in vitro synthesis of, 239-240
temperature and chromosome unfolding,
    147, *152*
temperature sensitivity, 645
    of CHL cell mutants, 702-711
    and initiation, 663
    in polynucleotide ligase mutants, 796
    of mutants, 753, 757-758
temperature shift experiments, 464
template switching, 38
    and secondary initiation, 48-50
termination position, 452
termination site, 663, 667
ternary complex, 14, 24, 28-35
(θ) type structures, 539, 544
thioredoxin (*E.coli*), 239
thymidine kinase deficient L cells, 720
    DNA synthesis in permeabilized-, 723
toluene permeabilized cells, 270, 272,
    280-*281*, 290, 640-*642*, *644*, 688,
    690-693, 719-727, 774, 775, 779-
    782, 783, 815-818, 845
toxic amino acid analogs, 702
trans-acting proteins, 487
transfection system, 445

transformation of cells, 753
transformation recipient, 188
transitioned state, 550
    level of drug resistance in, 551
Triton X-100, *642-643*
trypitc hydrolysates, 730
TS-14 mutant
    analysis by sucrose gradient centri-
        fugation, 709-711
    protein biosynthetic aparatus, 709

ultraviolet light, 774-775
    irradiation with, 827
    photoproducts induced by, 791
    and repair of DNA, 815-818
unique discontinuity, 370, 378, 382,
    384
unique gene sequence, 586
unique terminus, 667
unit fibers, universal nature of, 201-208
unwinding protein, 210-224, 242, *249*,
    251, *297*-298, 417, 807-808
    see also relaxation protein,
        ω untwisting enzyme
*uvr*A
    endonuclease, 793, 794, 807
    excision defective mutants
    locus, 778, 780, 784
    mutant, 528
*uvr*AB endonuclease, 787
*uvr*B
    endonuclease, 794
    excision defective mutants, 793
    locus, 778, 780
*uvr*C
    locus, 778
    mutant, 796-797

V gene product, 796
viral genes, 418
viral protein gamma (gam), 462
viral strand, 404
    origin, 370-384
virion proteins, 446-449
viruses,
    see avian myeloblastosis virus, eukaryotic
        cells, murine sarcoma virus, mycoplasma
        virus, phages, polyoma DNA, RNA tumor
        virus, Rous sarcoma virus, SV40 virus,
        viral genes
viscometry, 138, 140
    aggregation of folded chromosomes
        during, 147-*149*, 153-157

whole cell lysate, 324-329

x-irradiation, 827
xeroderma pigmentosum (XP), 809